STP 1198

Nondestructive Testing of Pavements and Backcalculation of Moduli: Second Volume

Harold L. Von Quintus, Albert J. Bush, III, and Gilbert Y. Baladi, Editors

ASTM Publication Code Number (PCN)
04-011980-08

ASTM
1916 Race Street
Philadelphia, PA 19103
Printed in the U.S.A.

Library of Congress Cataloging-in-Publication Data

Nondestructive testing of pavements of backcalculation of moduli,
    Second volume/Harold L. Von Quintus, Albert J. Bush, III, and
    Gilbert Y. Baladi, editors.
        p.   cm. — (STP 1198)
    Contains papers presented at the symposium held in Atlanta, GA on
    23–24 June 1993, sponsored by ASTM Committee D-18 on Soil and Rock and
    its Subcommittee D4 on Road Paving Materials.
    "ASTM publication code number (PCN) 04-011980-08."
    Includes bibliographical references and index.
    ISBN 0-8031-1865-1
    1. Pavements—Testing—Congresses.   2. Nondestructive testing—Congresses.
    I. Von Quintus, H. L. (Harold L.)   II. Bush, A. J. (Albert Jasper)   III.
    Baladi, Gilbert Y., 1943–   .   IV. ASTM Committee D18 on Soil and Rock.
    Subcommittee D-4 on Road and Paving Materials.   V. Series:
    ASTM special technical publication: 1198
    TE250.N572   1994
    625.8′028′7—dc20   94-24308
                                                                          CIP

Photocopy Rights

Peer Review Policy

Each paper published in this volume was evaluated by three peer reviewers. The authors addressed all of the reviewers' comments to the satisfaction of both the technical editor(s) and the ASTM Committee on Publications.

To make technical information available as quickly as possible, the peer-reviewed papers in this publication were printed "camera-ready" as submitted by the authors.

The quality of the papers in this publication reflects not only the obvious efforts of the authors and the technical editor(s), but also the work of these peer reviewers. The ASTM Committee on Publications acknowledges with appreciation their dedication and contribution to time and effort on behalf of ASTM.

Printed in Ann Arbor, MI
December 1994

Foreword

This publication, *Nondestructive Testing of Pavements and Backcalculation of Moduli (Second Volume),* contains papers presented at the symposium of the same name held in Atlanta, GA on 23–24 June 1993. The symposium was sponsored by ASTM Committee D18 on Soil and Rock and its Subcommittee D4 on Road and Paving Materials. Albert J. Bush, III, of U.S. Army Corps of Engineers in Vicksburg, MS, Harold L. Von Quintus of Brent Rauhut Engineering in Austin, TX, and Gilbert Y. Baladi of Michigan State University in East Lansing, MI presided as symposium chairmen and are the editors of the resulting publication.

Contents

PROBLEMS/ERRORS ASSOCIATED WITH BACKCALCULATION METHODS AND DESIGN PARAMETERS

NDT FOR OTHER PAVEMENT USES

PROPOSED STANDARD GUIDE

Overview

In June 1988, the first International Symposium on Nondestructive Testing (NDT) of Pavements and Backcalculation of layer moduli was held. Since then, another symposium on NDT and backcalculation of layer moduli was held in August of 1991 and was sponsored by the Transportation Research Board. Both of these symposia were well attended, and showed that there was a strong interest within the transportation community in the area of NDT and the use of deflection data for evaluating and designing pavement structures. Unfortunately, these two symposia also showed that the industry was divided regarding the adequacy and use of state-of-the-art evaluation procedures for determining structural capacity of pavement structures.

As a result of the first symposium in 1988, ASTM Subcommittees D18.10 and D04.39 have been extensively involved in the preparation of standardized procedures for NDT and the evaluation of deflection data. Standardized procedures have been prepared and approved for collecting deflection data with different devices. These are listed below for reference purposes:

D 4602   Standard Guide for Nondestructive Testing of Pavements Using Cyclic Loading Dynamic Deflection Equipment

D 4694   Standard Test Method for Deflections with a Falling-Weight-Type Impulse Load Device

D 4695   Standard Guide for General Pavement Deflection Measurements

The task of standardizing backcalculation procedures, however, has been more difficult, because of the diversity of opinions and procedures currently in use by the transportation industry. The first draft of a standard guide for backcalculation of layer moduli from deflection measurements was balloted in 1986. The latest draft balloted in 1992 received numerous negative ballots that were found to be persuasive. More recently, there have been numerous research projects completed by individual transportation agencies and as part of the Strategic Highway Research Program (SHRP).

With these recent advancements and the need to develop concurrence within the transportation industry to develop a standardized evaluation procedure, Subcommittees D18.10 and D04.39 suggested to the Executive Committees that ASTM sponsor the second International Symposium on Nondestructive Testing of Pavements and Backcalculation of Moduli. This Second International Symposium was held in Atlanta, Georgia in June, 1993. The attendance at this symposium exceeded 80, representing 12 different countries and 25 states in the United States. An attendance list is included at the end of this publication.

The symposium was divided into four sessions (two sessions per day) and one panel workshop or discussion on issues related to standardization of backcalculation procedures. The papers presented at this Second International Symposium focused in the area of backcalculation of layer moduli techniques and comparisons of material moduli as measured in the laboratory to values calculated from field deflection measurements. Information from these papers and discussion were used to establish whether a backcalculation procedure could be standardized based upon the current state-of-the-art technology. The format of the presentations was divided into four sessions followed by a panel discussion. Each of the sessions were subdivided into two parts as follows:

SESSION 1—Analytical Models and Techniques for Backcalculation of Layer Moduli (5 Papers).

Chairman—Dr. Albert J. Bush III, U.S. Army Corps of Engineers, Waterways Experiment Station, Vicksburg, MS.

Part 1 of Session 1: Recent Developments and Tools to be Used in the Future for Evaluating Pavements Based on Backcalculation Techniques (2 Papers).

Keynote Speaker—Dr. Jacob Uzan, Professor, Israel Institute of Technology (Technion), Israel, ''Advanced Backcalculation Techniques.''

Part 2 of Session 1: Methods and Procedures Used for Backcalculation of Material and Pavement Properties (4 Papers).

SESSION 2—Measurement and Calculation Techniques in the Field and Laboratory

Chairman—Mr. Harold L. Von Quintus, President, Brent Rauhut Engineering Inc., Austin, TX.

Part 1 of Session 2: Verification of backcalculation techniques and comparisons of laboratory measured values with those calculated from field measurements or deflections (4 papers).

Part 2 of Session 2: Characterization of Pavement Materials and the Effects of Non Linearity on Backcalculation of Layer Moduli (4 papers).

SESSION 3—NDT for Pavement Structural Evaluation, Design and Rehabilitation.

Chairman—Dr. Albert J. Bush III, U.S. Army Corps of Engineers, Waterways Experiment Station, Vicksburg, MS.

Part 1 of Session 3: Problems/errors associated with backcalculation methods in terms of pavement evaluation, and backcalculation of design parameters for concrete pavements (4 papers).

Part 2 of Session 3: Analysis of deflection measurements and effects of load distributions on pavement response (4 papers).

SESSION 4—NDT for Other Pavement Uses: Use of the Results From NDT to Determine Layer Thickness, Joint Efficiency, and Void Detection (5 Papers).

Chairman—Dr. Gilbert Y. Baladi, Professor, Michigan State University, East Lansing, MI.

SESSION 5—Panel Discussion on Backcalculation of Layer Moduli

Chairman—Dr. Gilbert Y. Baladi, Professor, Michigan State University, East Lansing, MI.

Discussion paper presented by Richard May, Asphalt Institute, Lexington, KY and Harold L. Von Quintus, Brent Rauhut Engineering, Austin, TS entitled ''The Quest for a Standard Guide to NDT Backcalculation''.

Panel participants: Dr. Albert J. Bush III., U.S. Army of Engineers, Waterways Experiment Station, Vicksburg, MS. Dr. Jacob Uzan, Israel Institute of Technology (Technion), Israel; Richter, Federal Highway Administration, Turner Fairbanks, Washington, DC; Dr. Ullditz, Technical University of Denmark, Denmark, and Luckanen, Braun Intertec, Minneapolis, MN.

Papers in this STP are presented on those topics in the four sessions listed previously. These papers include examples of different backcalculation of layer moduli procedures, comparisons between laboratory measured and field calculated values, as well as, the more common examples on the use of deflection testing to evaluate pavement structures. The papers published represent eight different countries, eleven different states, and thirteen different educational agencies. It is the hope of the organizers of this symposium that the papers presented will provide the readers with much of the latest information in the areas of pavement evaluation using NDT techniques, and application of that data for use in pavement design.

One of the goals and objectives of this symposium was to determine if the industry could find a common ground to standardize a backcalculation procedure. In specific, this was the focus of the panel discussion at the end of the symposium. This panel discussion was preceded by a paper entitled ''The Quest for a Standard Guide to NDT Backcalculation'' (presented by Mr. Richard May) and a presentation by Dr. Albert Bush (Symposium Cochairman and D4.39 Subcommittee Chairman) entitled ''Where We Go From Here.''

From the question and answers during the panel discussion, it was the general consensus that backcalculation of layer moduli from deflection measurements will definitely be used in the future for the rehabilitation design and evaluation of pavement structures. The question however, is still: what is the reliability of these values? Specifically, it was the general consensus of the panel and attendees that the accuracy of backcalculated moduli is model dependent and unknown, as well as those values measured in the laboratory because there is a diversity of opinion on the simulation of field conditions in the laboratory. For example, there is controversy within the industry on whether backcalculation procedures should be based on a dynamic or static analysis, and what values actually represent the ''truth,'' both in the laboratory or from field measurements.

In summary, most participants, concurred that there needs to be a standard ''baseline'' of values from which to compare on a project, material, or pavement bases, and that one should not become paralyzed by the imperfection of the procedures. More importantly, research must be merged into practice on a consistent basis and one way to accomplish this is through the standardization process. As such, a procedure needs to be standardized and that procedure should concentrate on user oriented issues. Thus, the editors, panel, as well as most symposium participants involved in these discussions, believe that some standardized procedure should be pursued to ensure that a common set of values can be compared.

The editors wish to thank all those who participated in this symposium and who contributed to this STP. Special thanks are given to the authors, the reviewers of the papers, ASTM Committees D18 and D4 for sponsoring the symposium, and to the members of Subcommittees D18.10 and D04.39 for their valuable input and efforts. Last but not least, the editors would like to express their deep appreciation to the ASTM staff for their assistance in preparing for this symposium and in its preparation. The high professional quality of ASTM publications would not be possible without their dedicated and professional efforts.

Dr. Albert J. Bush III

U.S. Army Corps of Engineers, Waterways
   Experiment Station, Vicksburg, MS; symposium
   cochairman and coeditor.

Mr. Harold L. Von Quintus

President, Brent Rauhut Engineering, Austin, Texas,
   symposium cochairman and editor

Dr. Gilbert Y. Baladi

Professor of Civil Engineering, Michigan State
   University, East Lansing, Michigan, symposium
   cochairman and coeditor

Analytical Models and Techniques

Jacob Uzan[1]

ADVANCED BACKCALCULATION TECHNIQUES

REFERENCE: Uzan, J., **"Advanced Backcalculation Techniques,"** _Nondestructive Testing of Pavements and Backcalculation of Moduli (Second Volume), ASTM STP 1198_, Harold L. Von Quintus, Albert J. Bush, III, and Gilbert Y. Baladi, Eds., American Society for Testing and Materials, Philadelphia, 1994.

ABSTRACT: The backcalculation procedures are separated into five categories; (a) static linear elastic, (b) static nonlinear elastic, (c) dynamic linear using frequency domain fitting, (d) dynamic linear using time domain fitting and (e) dynamic nonlinear analysis. In this paper each category is described and case studies are presented comparing their results. Advanced techniques require more complete material characterization models. In the nonlinear elastic procedure a universal $k_1 - k_5$ model (an extension of the bulk modulus model) is used. In the dynamic analysis technique a generalized power law relationship is used for the asphaltic layer. Two case studies are presented. In the first the nonlinear elastic scheme was found to give excellent results at matching deflection bowls at four different load levels for each of two test sites analyzed. In these analyses only the k_1 of the $k_1 - k_5$ model was backcalculated. It was found that the backcalculated k_1 in all layers including the asphalt concrete are larger than those measured in the laboratory. In the second case study dynamic analysis techniques are applied to full wave shape data obtained from the SHRP data base. Both the frequency and time domain procedures are shown to yield reasonable results.

KEYWORDS: linear, nonlinear, dynamic backcalculation, elastic and viscoelastic material, flexible pavements

---

INTRODUCTION

Backcalculation of moduli of pavement material, is nowadays widely used for structural evaluation and rehabilitation. The number of existing procedures and computer programs for this purpose are relatively large (Rada et al. 1992). Moreover different modulus values may be obtained from these different programs (Lytton 1989; Chou and Lytton 1991). This indicates that backcalculation is very sensitive to

---

[1] Associate Professor of Civil Engineering, TECHNION, Israel Institute of Technology, Haifa, Israel 32000

the kind of analysis and assumptions underlying the analysis. It is the intent of the paper to present a synthesis of the backcalculation procedures and a discussion of their limitations.

All backcalculation procedures use error minimization techniques to minimize either the absolute or the squared error, with or without weighing factors. The most common backcalculation procedure is based on static loading type of analysis and linear elastic material response. However, most of the loading devices apply either a vibratory load or an impact load. Also, the pavement materials are in most cases far from being linear elastic. In recognition of these conditions, the need for applying more advanced techniques using nonlinear and dynamic analysis is becoming stronger everyday. Researchers and engineers are more aware of the badness (not goodness) of fit, and of assumptions leading to large errors in the fit. This paper presents the author's evaluation of these advanced techniques and recommendations for future work. However, it should be kept in mind that these analysis procedures are not currently ready to be implemented due to the lack of knowledge on the behavior of the pavement materials and to the lack of confidence in the accuracy of the time history deflection bowls. A section of the paper will be devoted to discuss the material characterization most appropriate for backcalculation.

Examples are presented to illustrate the procedures developed. These are based on data collected using Falling Weight Deflectometer (FWD) equipment. The procedures work with relatively low frequency components (such as those induced under both FWD and truck loading). They do not include high frequency loading procedures such as spectral analysis of surface waves (SASW).

BACKCALCULATION CATEGORIES

The backcalculation procedures can be separated into several categories, depending on the type of load representation - static versus dynamic and on the type of material characterization - linear versus nonlinear for elastic, viscoelastic and/or plastic materials. A discussion of each is presented below. All backcalculation procedures use error minimization techniques to minimize either the absolute or the squared error, with or without weighing factors.

Static Linear Backcalculation

In the simplest case which is widely used today, the load is assumed to be static and the material is assumed to be linear elastic. In this case only the peak load and peak surface response deflections are used in the backcalculations (Figure 1a). The problem reduces to finding the unit response of the pavement that will correspond to the measured response. The unit response of the pavement is computed using appropriate computer programs for linear elastic multi-layer systems. The unit response is defined by the set of moduli of the pavement layers. Therefore, the problem is reduced again to finding the set of moduli that produces a unit pavement response close to the measured one.

When several load levels are applied in the test, each load level is analyzed separately and separate sets of moduli are obtained. The procedures for backcalculation using the above scheme are numerous. However, most have two essential differences. These differences are: (1) the forward computation of the unit response which are based on numerical integration (such as Peutz et al. 1968; BISAR User's Manual 1972; WESLEA - Van Cauwelaert et al. 1989) or some approximation (such as the Method of Equivalent Thickness MET - Ullidtz 1977; Lytton 1989), and (2) the error minimization scheme. These differences can lead to different backcalculated moduli. However, when both the forward computation programs and the minimization schemes are correct, the

Linear Elastic - Static

Given:

Peak applied load $\qquad$ $I = \max P$

Peak surface deflections $\quad R = \max d_k$

(k= number of sensors)

Find:

$H_k(E)$ such as

$\max d_k \approx H_k(E) * \max P$

Nonlinear Elastic - Static

Given:

Peak applied loads $\qquad$ $I = \max P_j$

Peak surface deflections $\quad R = \max d_{kj}$

(k= number of sensors)

(j= number of load levels)

Find:

$H_k(E)$ such as

$\max d_{kj} \approx H_k(E) * \max P_j$

Linear - Dynamic
Steady state vibratory load

Given:

Load functions $\qquad$ $I = P \exp(i\omega t)$

Deflection functions $\quad R = d_k \exp(i\omega t + \Phi_k)$

(k= number of sensors)

Find:

$H_k(E^*)$ such as

$R \approx H_k(E^*) * I$

(a) $\qquad$ **(b)** $\qquad$ **(c)**

FIG. 1 a-c-- Backcalculation schemes (a) linear elastic (b) nonlinear elastic and (c) dynamic for steady state vibratory load backcalculated moduli are in general similar (Lytton 1989). It is worth mentioning that the emphasis is on the correctness of the computation, not on the ease or speed of computation. For example, the use of MET may in some cases (for example, for varying, decreasing and increasing moduli with depth) may lead to an unacceptable error in the forward computation of the response of the pavement, and thus in the backcalculation.

Static, Nonlinear Backcalculation

In the static, non-linear elastic backcalculation, only the peaks of the loads and surface deflections are used (Figure 1b). In contrast with the linear scheme, all load levels are used simultaneously. In other words, the problem is to find the response function of the pavement that will correspond to the measured response at all load levels. The unit response is defined by the set of material parameters and is usually computed using Finite Element (FE) computer programs (Uzan and Scullion 1990). The material characterization is of prime importance in this category of backcalculation and will be discussed in a separate section. There is only one such computer program for non-linear elastic backcalculation developed by the writer. Results of analysis with this program will be presented later. However, other computer programs using linear elastic multi-layer systems and different kinds of approximations exist and are rather widely used (ELMOD - Ullidtz 1977; PADAL - Brown et al. 1987; MODCOMP - Irwin and Szebenyi 1991; FWDCHECK - PCS/Law Engineering 1990). These programs incorporate an inter-relationship between surface deflection at a particular radial location on a deflection bowl and the elastic stiffness of a particular pavement layer. It is difficult to evaluate the correctness of these programs, because they are not truly non-linear analyses. Moreover, they do not represent true pavement materials because they do not account for dilation and lack of tensile strength effects whenever these conditions prevail.

The non-linear elastic backcalculation still assumes that the permanent deformation is small compared to the resilient one. In other words, the state of stress within the pavement structure is low relative to the ultimate strength. This situation applies to pavements with a moderate to thick AC layer, not with thin surfacing of less than 50 mm (2 inches). In these cases of thin pavements and/or relatively heavy applied loads, the non-linear elastic theory should be replaced by the non-linear elasto-plastic theory. This will account for any permanent deformation and stress redistribution caused during loading as compared to the elastic behavior. It must be noted that the response function of the pavement depends on both the non-linear elastic and the plastic parameters. The number of parameters may become too large to be resolved by the backcalculation procedure alone, and additional information concerning the material properties must be supplied before the analysis is initiated.

Dynamic Linear Backcalculation

The dynamic backcalculation applies to the NDT equipment that apply either a steady state vibratory load or an impact load. In the case of steady state vibratory load with a finite number of frequencies, the problem is reduced to finding the unit response of the pavement that will correspond to the measured response (Figure 1c). The unit response is defined by the set of complex moduli of the pavement layers. Therefore, the problem is reduced again to finding the set of complex moduli (or viscoelastic parameters) that produces a unit response close to the measured one.

When several responses at different frequencies are measured, each

set of data may be analyzed separately to obtain separate sets of moduli
for each frequency and load level. Alternatively, it is possible to
normalize the response with respect to load level and analyze all
frequencies simultaneously to get one set of moduli (independent of
frequency) for all load level. The unit response function at a given
frequency can be computed using computer programs (such as UTFWIBM –
Roesset 1987; SCALPOT – Magnuson 1988).
   In the case of an impact load, two approaches may be used:

a.   <u>Using frequency domain fitting</u> (Figure 1d). In this case, the
      applied load and deflection response time histories are
      transformed into the frequency domain by using fourier transform.
      Dividing the complex deflection by the complex load function gives
      the measured complex unit response of the pavement at several
      particulate frequencies. The backcalculation problem is then
      similar to the case of steady state vibratory load at several
      frequencies, i.e. to find the set of complex moduli of the
      pavement layers that will generate a unit response close to the
      measured one, at several particulate frequencies (Magnuson et al.
      1991; Torpunuri 1990).

b.   <u>Using time domain fitting</u> (Figure 1e). In this case the measured
      time history deflection responses are directly compared by fitting
      with the computed ones. It is possible to obtain a forward
      solution (given the complex moduli of the layers) by direct
      integration of the load history or by using the frequency domain
      unit response and inverse transform techniques. The direct
      integration method may be time consuming compared to the frequency
      and inverse transform method. In the later procedure, the load
      history is transformed in the frequency domain. Then using the
      forward computer program, unit response deflection are computed at
      several particulate frequencies. The complex multiplication of
      the unit response with the load gives the response of the pavement
      in the frequency domain for the particular load applied. Then the
      inverse transform of this result gives the pavement response in
      the time domain. It is seen that the fitting in the time domain
      is more time and computer consuming than the fitting in the
      frequency domain. However, it has several advantages which make
      it more attractive than the fitting in the frequency domain.
      Detailed description of the above dynamic backcalculation
      procedures can be found in Uzan (1993).

Dynamic Non-linear Backcalculation

   The dynamic non-linear backcalculation formulation follows the
previous static non-linear and dynamic linear backcalculations. As the
material is non-linear, the frequency domain transformation is not
applicable. Therefore, the forward computer program must be of the type
of direct integration, with a separate run for every load time history.
The set of non-linear complex moduli of the pavement materials should
predict time history deflection responses similar to the measured one
for all load levels simultaneously. Because of its complexity, the
dynamic non-linear backcalculation does not seem to be practical in the
near future, and therefore will not be dealt with in this paper.

PROBLEM AREAS IN PAVEMENT MODULI BACKCALCULATION

   Backcalculation of moduli is very sensitive to the type of
analysis and the assumptions underlying the analysis. This section
presents a brief discussion of some limitations of the existing
procedures.

Linear - Dynamic

Frequency domain for impact load

Given:

Load time history $I = P(t)$

Surface deflection histories $R = d_k(t)$

(k = *number of sensors*)

Apply:

$P(t) \rightarrow |\,FFT\,| \rightarrow P(\omega_j)$

$d_k(t) \rightarrow |\,FFT\,| \rightarrow d_k(\omega_j)$

$d_k^{\,o}(\omega_j) = d_k(\omega_j) / P(\omega_j)$

Find:

$H_k(E^*)$ such as

$d_k^{\,o}(\omega_j) \approx H_k(E^*) * I^o(\omega_j)$

(d)

Linear - Dynamic

Time domain for impact load

Given:

Load time history $I = P(t)$

Surface deflection histories $R = d_k(t)$

(k = *number of sensors*)

Apply:

$P(t) \rightarrow |\,FFT\,| \rightarrow P(\omega_j) = I(\omega)$

Find:

$H_k(E^*)$ such as

$H_k(E^*) * I(\omega) \rightarrow |\,IFFT\,| \rightarrow R^c(t)$

$R^c(t) \approx R$

(e)

FIG. 1 d-e-- Dynamic backcalculation schemes for impact load (d) frequency domain (e) time domain

Cracking

The current procedures use forward programs which assume that the layers extend to infinity. Therefore, a cracked pavement with longitudinal and/or transversal cracks cannot be reliably analyzed. A systematic error with unpredictable results would be introduced, unless such cracking is taken into account in the forward deflection computations. Such cases exist in rigid pavements where deflections along a free edge are analyzed (Uzan 1992; Uzan et al. 1993). This is achieved by developing a special program for integrating Westergaard equations.

A similar situation to the cracking condition is usually encountered when deflection bowls are measured near the edge of a flexible pavement, for example in the outer wheelpath of sections without paved shoulders. The edge effect induces a systematic error.

Sensitivity to Variable

In order to backcalculate reliably a parameter, the data used in the backcalculation should be sensitive enough to variations of the parameter. For example, when a thin layer of less than 80 mm (3 inches) exists, it would be difficult to backcalculate its modulus. In some cases, it would be wrong to backcalculate the modulus of such thin layer, because the result strongly reflects the errors (random or systematic errors) of the measurement, modeling and computing. In order to avoid situations where unacceptable results are obtained, the following improvements are suggested:

a. Reduce to minimum the random error of the data acquisition by repeating the test several times and taking the average.

b. Reduce the systematic error of the modeling by choosing a more accurate model.

c. Use an analysis procedure capable of identifying and eventually correcting the backcalculation results. Such a procedure exists and is known as singular value decomposition (SVD - see Press et al. 1989). Usually the backcalculation scheme arrives at an optimization problem which requires solving an overdetermined set of equations by least squares techniques. When the data is not sensitive to the parameter to be backcalculated, the solution of the set of equations is problematic, because it involves an ill-conditioned matrix. The SVD method can be used to diagnose the problem and to solve it in the sense that it gives a meaningful numerical answer. In the case of the thin layer, the method will reduce the number of variable by dropping the modulus of the thin layer (for which the data is not sensitive). It is noted that the "backcalculated" modulus remains close to the seed value.

The SVD method can also be used to identify co-linearity between variables. In other words, when two variables have similar effects on the data, the system will then be singular or at worse ill-conditioned. For example, if both the layer thickness and the modulus are backcalculated, one may encounter the ill-conditioned case. The SVD algorithm will then again drop one variable. It is worth mentioning that the process requires special attention in deciding when and which variable should be dropped from the list of variables.

Input Variables

The backcalculation of moduli is sensitive to the input variables of the pavement, such as layer thicknesses and the depth of bedrock. Any user would be unwise to proceed with the backcalculation process, if any of these variables are unknown. Using engineering judgement, one may combine layers of similar quality into one layer in order to avoid the problem of ill-conditioned solution. The depth of bedrock must be known, especially if the bedrock is within 6 meters (20 ft) from the surface. The modulus of the subgrade is very sensitive to the depth of the bedrock for relatively shallow bedrock.

It is worth mentioning that an estimation of the bedrock depth is included in versions 3 and 4 of the program MODULUS (Rohde 1990). This prediction is based on several runs of a linear elastic multilayered computer program with different pavement and subgrade thicknesses. A correlation between combinations of the surface deflections, parameters of the pavement structure and depth of the bedrock was derived. The correlation works well as long as a true rigid layer exists underneath the subgrade, and the linear elastic behavior is predominant. However, the correlation will predict a fictitious depth of bedrock (such a bedrock does not really exist) in most cases of a sandy subgrade. This is because the modulus of the sandy subgrade increases with depth (as the overburden increases). The resulting effect of the increasing modulus on the surface deflection is similar to the effect of a linear elastic subgrade of a finite extent. When the non-linear elastic backcalculation is used, the increasing modulus with depth is included in the pavement response and no fictitious bedrock is needed for fitting the surface deflection bowl.

An alternative way to account for an increasing modulus of the subgrade within the framework of linear material behavior is proposed and used in the dynamic linear backcalculation. It is suggested to subdivide the subgrade layer known to extend beyond 6 meters (20 feet) into two sublayers with the upper sublayer being of finite thickness and the lower one being semi-infinite. The thickness of the upper sublayer may be assumed to be 1.2 to 1.8 meters (4-6 feet).

Forward Computation Programs

Backcalculation of pavement moduli is an inverse problem solution. It is defined as to find the set of moduli which produce a pavement response similar to the measured one. The direct problem solution is the forward computer program which is used to compute the pavement response for the given set of moduli. Therefore, the correctness of the backcalculation results is in direct relation with the correctness of the forward solution. For example, the method of equivalent thickness used in several backcalculation procedures may produce an unacceptable error when the moduli vary in a non-monotonously decreasing way, in the case of buried stiff layer.

Other programs, such as CHEVRON (Michelow 1963) may not converge to the correct solution for a rigid pavement with a weak subgrade or for a deep bedrock. The program MODULUS version 1 (the program was developed by the author, Uzan et al. 1988; Uzan et al. 1989) was released with CHEVRON for the forward solution (Lytton et al. 1990). Because situations with incorrect solutions may be encountered, its use should be avoided. When developing the dynamic backcalculation algorithm, the author checked the program SCALPOT for dynamic analysis and found it incorrect when used with the power law. It was corrected by the developer and seems to produce now correct solutions. However, any results published before this correction should be disregarded (Magnuson et al. 1991; Torpunuri 1990).

It is the author's opinion that any program should be verified before use. Even then, there are no guarantees that it is correct for all conditions for which the program was not verified.

MATERIAL CHARACTERIZATION

An overview of the material characterization associated with each backcalculation procedure is given below.

1. <u>Linear Elastic Material</u>
In general the material is assumed to be homogeneous and isotropic. Therefore only two parameters - the modulus of elasticity and Poisson's ratio are needed to describe the stress-strain relation (constitutive equation). Usually the Poisson's ratios are assumed and only the moduli of elasticity of the pavement materials are backcalculated.

2. <u>Non-linear Elastic Material</u>
The behavior of non-linear elastic materials can be described by a variety of constitutive equations (Eringen 1962; Desai and Siriwardane 1984; Chen and Mizuno 1990). A simple extension of the linear elastic stress-strain relation will be to replace the elastic constants by scalar functions of the stress and/or strain invariants. The constitutive models formulated on this basis are of the Cauchy elastic type which does not imply that the strain energy density functions calculated from these relationships are path independent. The Cauchy type of elastic models may generate energy for certain loading-unloading cycles, thus violating the laws of thermodynamics. Another kind of elastic model which will not generate energy over any loading-unloading cycles has been formulated on the basis of the first law of thermodynamics and the existence of the strain energy density functions. The formulation leads to the hyperelastic model also called Green elastic model.

A shortcoming of the above formulations is the implied stress or strain path independent behavior which is not true for soils. A further improved description of soil behavior is provided by the hypoelastic formulation in incremental terms of stress and strain.

Examples of second order stress strain relationships are given below for illustration purposes.

<u>Cauchy Elastic Type</u>

$$\sigma_{ij} = (c_1 I_1 + c_2 I_1^2 + c_3 I_2)\, \delta_{ij} + (c_4 + c_5 I_1)\, \epsilon_{ij} + c_6 \epsilon_{ik}\epsilon_{kj} \qquad (1)$$

<u>Green (hyperelastic) Elastic Type</u>

$$\sigma_{ij} = (2 c_1 I_1 + 3 c_2 I_1^2 + c_3 I_2)\, \delta_{ij} + (c_4 + c_3 I_1)\, \epsilon_{ij} + c_5 \epsilon_{ik}\epsilon_{kj} \qquad (2)$$

<u>Hypoelastic Type</u>

$$\dot{\sigma}_{ij} = c_0 \dot{\epsilon}_{kk}\delta_{ij} + c_1 \dot{\epsilon}_{ij} + c_2 \sigma_{nn}\dot{\epsilon}_{kk}\delta_{ij} + c_3 \sigma_{nn}\dot{\epsilon}_{ij} + c_4 \sigma_{ij}\dot{\epsilon}_{kk}$$

$$+ c_5 (\sigma_{im}\dot{\epsilon}_{mj} + \dot{\epsilon}_{im}\sigma_{mj}) + c_6 \sigma_{mn}\dot{\epsilon}_{nm}\delta_{ij} \qquad (3)$$

where:

$\sigma_{ij}, \epsilon_{ij}$ =   components of stress and strain tensor

$\dot{\sigma}_{ij}, \dot{\epsilon}_{ij}$ =   components of stress and strain rate (increment) tensor

I_1, I_2 =   stress invariants

δ_{ij} =   Kroniker delta

c_0 to c_6 =   material parameters

The above relations reduce to the case of linear elasticity with only two parameters. The above formulations have been known to the geotechnical researchers, for more than two decades (Chang et al. 1967). Despite the strong mechanistic basis of these formulations, they are not widely used mainly because the material constants (c's in Equations 1 to 3) have no direct physical interpretation in most cases. Moreover, the hyperelastic and hypoelastic models requires complicated testing programs with controlled strain conditions which are very difficult to perform.

The above formulations do not directly account for the observed behavior of geotechnical materials (granular base and subbase and subgrade materials). For example, it is well established that (a) the secant modulus of pavement granular materials is proportional to the first stress invariant to the power k_2 and (b) the secant modulus of clayey materials is proportional to the second stress invariant to the power k_3. In general, the secant modulus of pavement materials would be (Uzan 1985; Witczak and Uzan 1988):

$$MR = k_1 p_a \left(\frac{\theta}{p_a} \right)^{k_2} \left(\frac{\tau_{oct}}{p_a} \right)^{k_3} \tag{4}$$

where:

MR =   resilient modulus or secant modulus

θ =   bulk stress = sum of principal stresses = first stress invariant

τ_{oct} =   octahedral shear stress = second deviatoric stress invariant

p_a =   atmospheric pressure

k_1, k_2, k_3 =   material parameters

The above model is a simplified version of the non-linear Cauchy type model where the strain is a function of the state of stress. The model is mathematically and conceptually simple and attractive to the geotechnical and pavement researchers. In fact, a data bank of the k_1, k_2, and k_3 parameters exists and continues to grow. Equation 4 is usually used with a constant Poisson's ratio. In this case, the formulation may violate the laws of thermodynamics. Therefore, Equation 4 was complemented by an equation for the Poisson's ratio (Uzan 1992; Uzan et al. 1992). This equation which adds two materials constants k_4 and k_5 was derived by imposing the path independence of the strain energy density function.

The full stress-strain relationship will be called the k_1-k_5 model and will be implemented in the static non-linear backcalculation described later in this paper. The model may not be as powerful as the previously mentioned models (especially when higher orders are used), however it does account for non-linearity, stiffening under increasing hydrostatic stresses, softening under increasing deviatoric stresses, and more importantly, it does account for dilation under shear stresses.

The material parameters can be easily determined from conventional triaxial test results as well as from more sophisticated tests, such as the true triaxial and shear tests. It is worth mentioning here that the material parameters cannot be properly derived if the material response is not fully measured. More specifically, the k_1 to k_5 parameters cannot be determined from a "resilient modulus test" if the lateral deformation is not measured. In order to insure path independence of the strain energy density function, all five parameters must be determined simultaneously from the test results, because the Poisson's ratio depends on k_2 to k_5 and the modulus depends on k_1 to k_3. Therefore, it is strongly recommended that lateral deformation be measured in all triaxial tests.

In the non-linear backcalculation, the number of parameters is too large to be effectively and reliably backcalculated. It is recommended to only backcalculate the k_1-constant and get the k_2 to k_5-constants from lab tests or data bank results.

Linear Viscoelasticity

Dynamic analysis includes three types of material properties: the mass, the damping and the stiffness. The stiffness and damping define the viscoelastic behavior which is in general expressed in terms of the complex modulus as follows (Wolf 1985):

$$E^*(\omega) = E'(\omega) + i\, E''(\omega) = E'(\omega)[1 + i2\zeta] \qquad (5a)$$

$$2\zeta(\omega) = \tan\phi(\omega) = E''(\omega)/E'(\omega) \qquad (5b)$$

where:

$$E^*,\ E',\ E'' \quad = \quad \text{complex, real and imaginary modulus}$$

$$\begin{aligned}
\omega \quad &= \quad \text{frequency} \\
\zeta \quad &= \quad \text{damping ratio} \\
\phi \quad &= \quad \text{damping lag angle}
\end{aligned}$$

Usually, the granular base-subbase and the subgrade materials are assumed to have a constant (independent of the frequency) damping ratio. The asphalt concrete material may be modelled using series of the Kelvin, Maxwell models or the power law (used in the worked examples) where the creep compliance is given by (Shapery 1987; Lytton 1989):

$$D(t) = D_o + D_1\, t^m \qquad (6)$$

where

$$\begin{aligned}
D(t) \quad &= \quad \text{creep compliance} \\
t \quad &= \quad \text{time} \\
D_o,\ D_1,\ m \quad &= \quad \text{material properties}
\end{aligned}$$

The behavior of these models can be expressed in terms of complex moduli to be used by the computer program.

In the backcalculation, one may derive the real modulus and the damping in the case of given complex modulus, or the components of the models for the cases of Kelvin or Maxwell models, or the D_o, D_1 and m parameters in the power law equation. However, it should be remembered that the number of backcalculated parameters should be kept to minimum using only parameters that will not produce instability of the set of equations to be solved.

The non-linear elasto-viscoplasticity is not addressed in the paper because of its complexity and its current unavailability.

CHOICE OF THE BACKCALCULATION PROCEDURE

Pavement material behavior is known to be non-linear for both
granular (stiffening) and clayey (softening) materials. For example,
the modulus of a sandy subgrade will increase with depth or increasing
overburden pressure. The modulus of a clayey subgrade (and for some
sandy clayey materials) will increase as the distance from the load
increases or the deviatoric stress decreases. The composite modulus of
the pavement is often used as indication of non-linear behavior of the
pavement. The composite modulus is the equivalent modulus of the multi-
layer system represented by homogeneous half space.

For linear elastic materials, the composite modulus should
decrease with increasing distance from the load and tend to the value of
the subgrade modulus. For non-linear subgrade materials, the composite
modulus will decrease first and then increase with increasing distance
from the load. Therefore, when such increasing composite modulus
conditions are encountered, non-linear backcalculation should be
preferred to the linear one. The MODULUS program which is based on
linear elastic backcalculation uses this indication of non-linear
behavior to modify the backcalculation procedure and obtain a subgrade
modulus representative of the material in the vicinity of the load. In
this procedure, it is assumed that by dropping the outer sensors where
the composite modulus increases with increasing distance from the load,
the backcalculation uses a deflection bowl representative of the near
load region.

The most commonly used NDT equipment is the FWD which applies an
impact load. Several papers have shown that when a rigid layer exists
at some finite depth in the subgrade, the dynamic effect cannot be
ignored (Davies and Mamlouk 1985; Chang et al. 1991). Different
amplification factors apply to the different sensors, causing a
distortion of the deflection bowl as compared to the "static" or non-
rigid layer condition. In these cases, there is no doubt that the
dynamic analysis should be used from the correctness of analysis point
of view. However, the time history or at least the load must be
available. (In the worst case, its shape could be assumed on the basis
of previous measurements). Then the procedure of the dynamic linear
backcalculation can be used, and only peak deflections (both measured
and computed) can be fitted, instead of using the whole time deflection
histories.

The use of time domain fitting in the dynamic linear
backcalculation is recommended over that of the frequency domain fitting
for the following reasons: (a) The data acquisition time window is
rather short - 60 msec in the SHRP NDT results. Usually, the deflection
does not decay to zero (see deflection bowl histories presented in the
section illustration of the dynamic backcalculation). Therefore, a tail
correction is used to bring the deflection at 60 msec to zero (and stack
zeros afterwards for the Fast Fourier Transform). This correction would
be correct if all the deflection at 60 msec was due to electronic drift
(integration error). However, in the case of flexible pavements with
viscoelastic materials, a certain amount of residual deformation
(viscous and delayed elasticity) may not be recovered within the time
window of 60 msec. In other words, under certain circumstances, the
measured deflection at 60 msec may be real and should not be corrected
in order to obtain a more realistic evaluation of the parameters; (b)
The drift error increases as elapsed time increases (probably in a
parabolic shape). Therefore, one can reduce the drift error effect by
using only the first part of the record which corresponds to the loading
sequence. However, unless there is an indication of a large drift
error, none should use all or most of the information in the record (all
60 msec or only the first 40 msec).

The above discussion and the drift correction apply to the

deflection time histories obtained from velocity data integration. The
backcalculation could have been made on the velocity (instead of
deflection) time history of the sensors, thus eliminating the
integration drift. Moreover, since the velocity will decay to zero
within a reasonable time window, the frequency domain analysis would
then be applicable and preferred. However, velocity data is not
typically available.

The time histories of the load and of the deflections contain
information which is not used in the static backcalculation and is
valuable in the dynamic backcalculation. A simple case is presented to
illustrate the difference in the two approaches. The time history of
the load is given in Figure 2a. The load is applied uniformly over a
circular area with a radius of 150 mm. The structure is an homogeneous
half space with constant damping. The half space response was computed
using UTFWIBM.

The calculated deflection time histories of the surface at r=0 are
shown in Figure 2b for 3 cases: real moduli of 50, 100 and 50 ksi and
damping ratios of 1, 1 and 20 percent for case 1, 2, and 3 respectively.
It is seen that the amplitude of the peak deflection is halved where the
modulus is doubled. Increasing the damping from 1 to 20 percent has a
mild effect in reducing the peak deflection. Figure 2c shows the
deflection results normalized, each curve with its maximum value. It is
seen that the first two cases (with the same damping ratio) overlap and
that the shape of the response in the third case of large damping
differs from the first two cases. The large damping tends to smooth the
response (reducing the relative size of the two peaks) and to produce a
larger time lag.

Figures 2d and 2e present the calculated response of the sensor
located 36 inches from the load. The smoothing effect of increasing the
damping is more pronounced than for the sensor under the load. It is
also seen that the time of arrival of the wave is shorter with the
higher modulus cause a shifting to the right of the response history.
Therefore, the magnitude of the modulus affects both the magnitude and
the time positioning of the response. This is an additional piece of
information that cannot be used in the static backcalculation.

Chang et al. (1991) proposed that the depth to bedrock may be
obtained from the time histories of the deflection. Where such a
bedrock exists below the subgrade, the deflection time history will show
a main pulse followed by several oscillations with decaying amplitude.
The depth of the bedrock can be estimated from the period of these
oscillations and the compressional wave velocity of the subgrade
material. However, a longer data acquisition window than 60 msec may be
required to make use of this additional information in the time history
deflections.

At this stage, the most reliable backcalculation procedure, i.e.
the dynamic non-linear backcalculation is not readily available. When a
rigid layer-bedrock is known to exist, it is recommended to use the
dynamic analysis which takes care of the different amplification effects
at different radial distances. One may improve the backcalculation
results by subdividing the subgrade into two layers (with a top layer of
about 1.50 meters). When no rigid layer is expected within the top 6
meters in the subgrade and the composite modulus indicator shows strong
non-linear pavement response, it is recommended to use the static non-
linear backcalculation procedure.

It is worth mentioning that the static linear analysis can be used
when the granular base and the subgrade materials are not subjected to
high load levels, (i.e. when the top asphalt layer is stiff and thick
enough to distribute the load) and no bedrock exists within about the
top 6 meters.

The above recommendations on the choice on when and where to use
the different backcalculation scheme are based on the author's

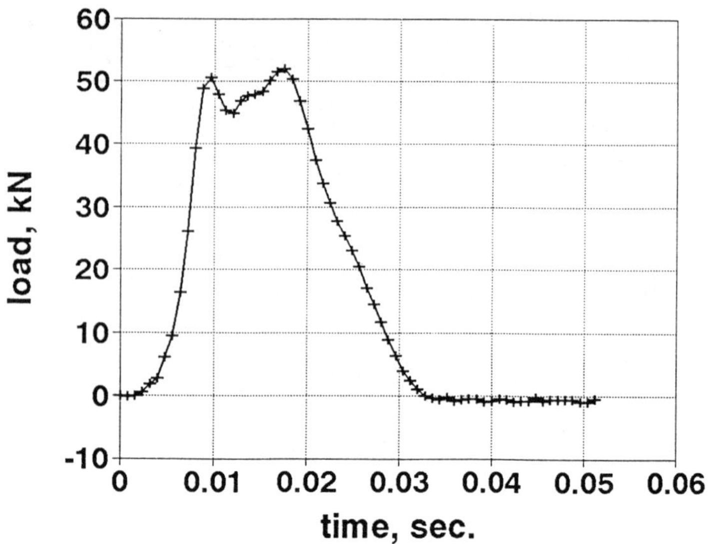

FIG. 2a--Load History

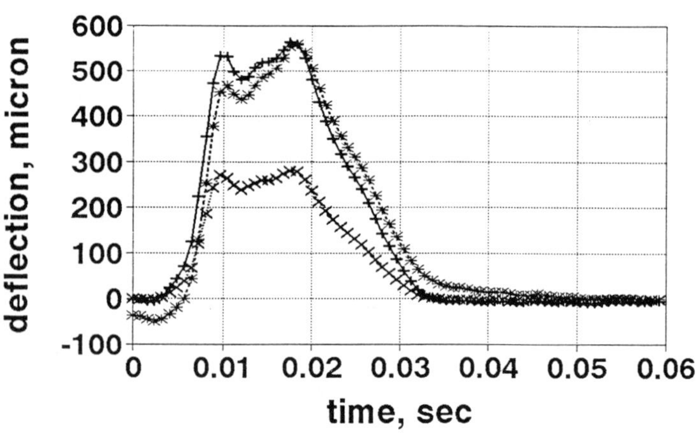

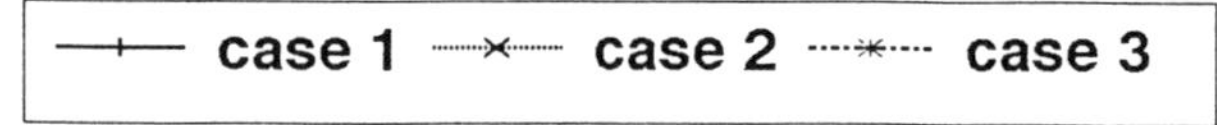

FIG. 2b--Calculated Response Deflection Histories for Sensor at r=0

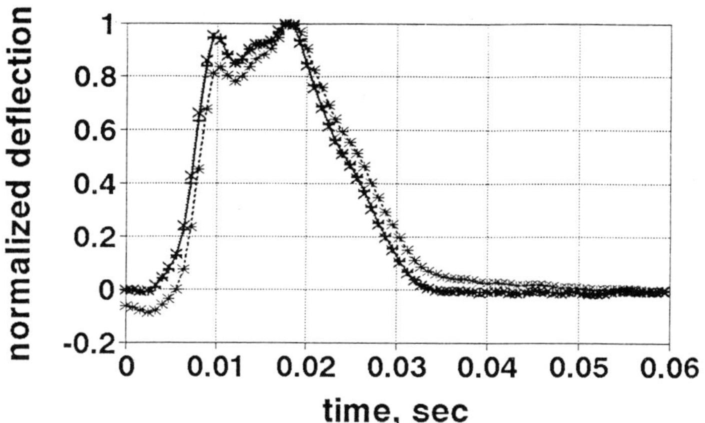

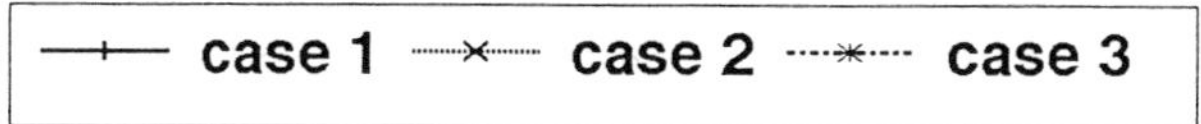

FIG. 2c--Normalized Deflection Histories for Sensor at r=0

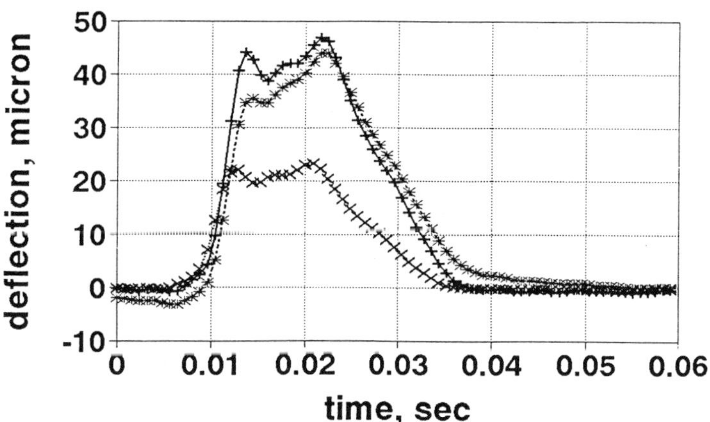

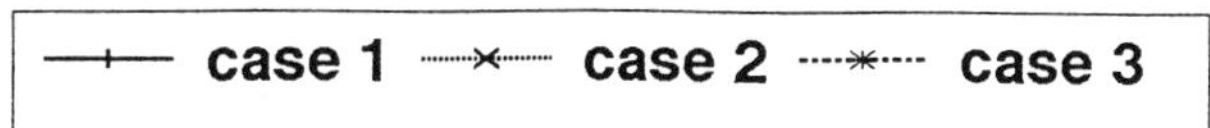

FIG. 2d--Calculated Response Deflection Histories for Sensor at r=0.9 m

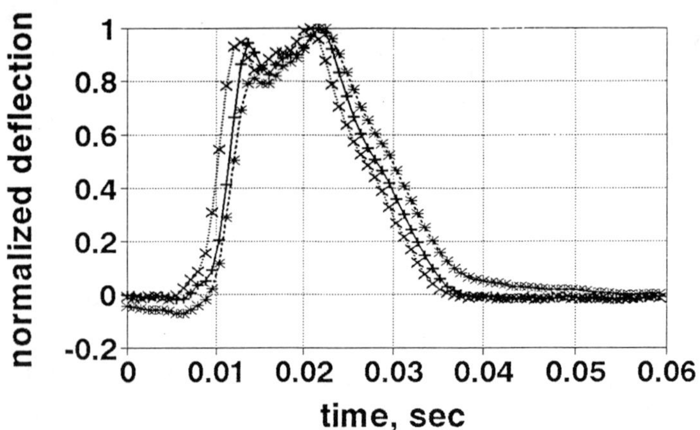
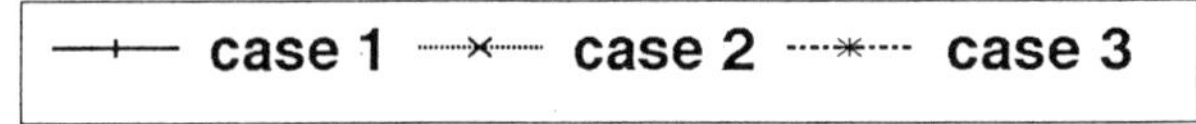

FIG. 2e--Normalized Deflection Histories for Sensor at r=0.9m

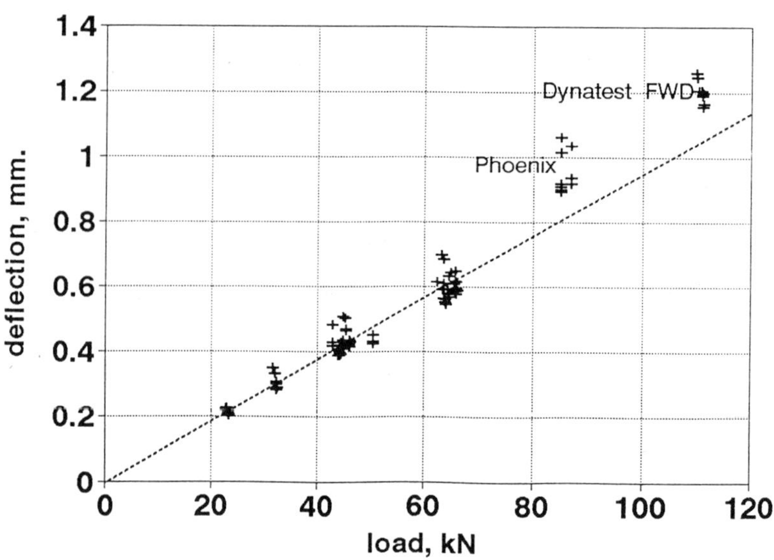

FIG. 3--Measured Deflections versus load for different FWD at Site 4

experience. Additional validation work is needed to support or reorder
the above recommendations. One such validation was presented by Uzan
and Scullion 1990, who showed the clear superiority of the nonlinear
over the linear elastic backcalculation procedures in the case of
flexible pavements with thin pavements. With the introduction of the
advanced backcalculation techniques, the use of instrumented pavements
is strongly recommended for validating the results of each technique.

ILLUSTRATION OF NON-LINEAR ELASTIC BACKCALCULATION

The laboratory and field test results were made available by Dr.
Al J. Bush III from WES. Two sections site 4 and site 12 were selected
from the study "Evaluation of NDT equipment for airfield pavements"
(Bentsen et al. 1989). Site 4 from Pensacola NAS is made of 140 mm (5.5
in) AC, 343 mm (13.5 in) of gravelly silty sand base on top of a silty
sand subgrade. Site 12 from Sheppard AFB is made of 178 mm (7 in) AC,
508 mm (20 in) of sandy silty gravel base on top of a sandy clay
subgrade. Resilient modulus tests were conducted on base and subgrade
materials. Dynamic Cone Penetration (DCP) test were made within the
0.94 mm (37 in) base and subgrade top. There is no indication of
bedrock near the surface. The sections were tested with different NDT
equipments: KUAB FWD, Dynatest HWD, Dynaflect, Dynatest FWD, Road
Rater, WES 16-kip, Phoenix FWD. The following analysis deals only with
FWD equipments. Figure 3 shows the center deflection at site 4 versus
load. It is seen that when the load is heavier than 80 kN (18 kip), the
deflection per unit load increases indicating that damage was being
caused to the pavement. For the non-linear elastic backcalculation,
without damage, the results at loads larger than 80 kN should not be
used. Therefore it was decided to use the results obtained with the
KUAB, at four load levels, at about 20, 30, 45, and 65 kN.

The composite modulus (computed as the load divided by the
distance and divided by the deflection) versus distance is shown in
Figures 4a and 4b for both site 4 and site 12. It is seen that the
composite modulus decreases first and increases afterwards. The
distance at which the composite modulus begins increasing is 450 mm (18
inch) only, indicating a strong non-linearity in the subgrade.
Moreover, the subgrade materials are silty sand and sandy clay for sites
4 and 12 respectively. Their modulus will increase with increasing
depth. When the program MODULUS is used to backcalculate moduli for
each load level separately, the two indicators of non-linearity showed
up very clearly by (a) dropping the sensors in the zone where the
composite modulus increased and (b) generating a fictitious bedrock at
1.50 m (about 60 inches) below the subgrade.

The laboratory test results included resilient modulus testing at
different confining pressures and deviatoric stresses. The vertical
deformation of the central third of the specimen was recorded. The
lateral deformation was not measured. In order to derive the k_1 to k_5
material parameters, the Poisson's ratio (ratio of lateral to vertical
deformation) was assumed to vary with the stress ratio (ratio of
principal stresses) as shown in Figure 5 for site 4, and constant (equal
to 0.3) for site 12. The use of the varying Poisson's ratio for the
material at site 4 as compared to constant Poisson's ratio at site 12
was suggested by the Dynamic Cone Penetration (DCP) results. They show
that the base and subgrade are strong at site 4 (with a corresponding
CBR of the base of 58-100) and weak at site 12 (with a corresponding CBR
of the base of 7-18). It must be emphasized that these assumptions were
made because of the lack of lateral deformation measurements in the
resilient modulus test. The k_1 to k_5 parameters were derived
simultaneously from the test results (with the additional assumption of
the lateral deformation). They are presented in Table 1.

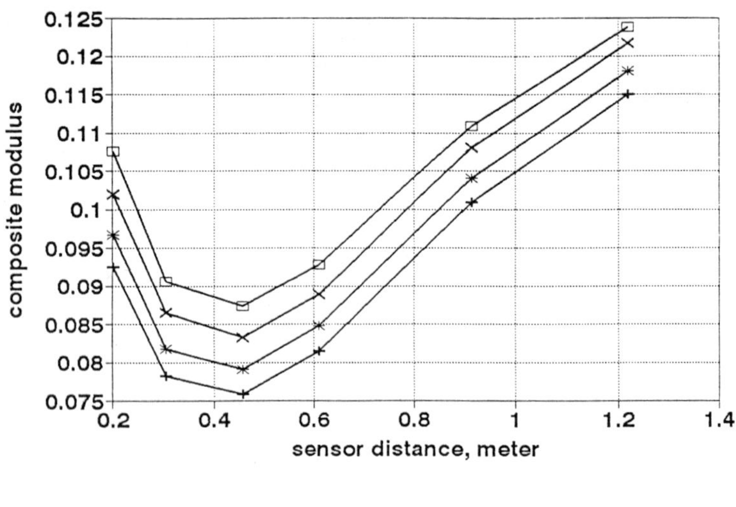
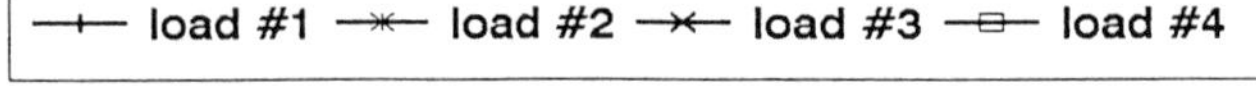

FIG. 4a--Composite Modulus Versus Sensor Distance at Different Load Levels, Site 4

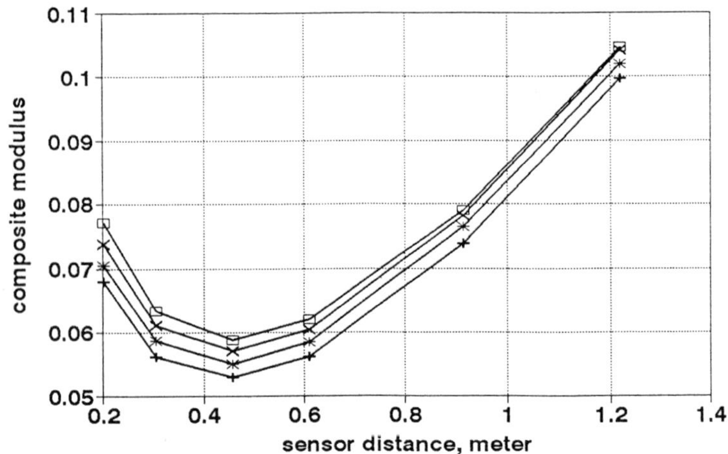
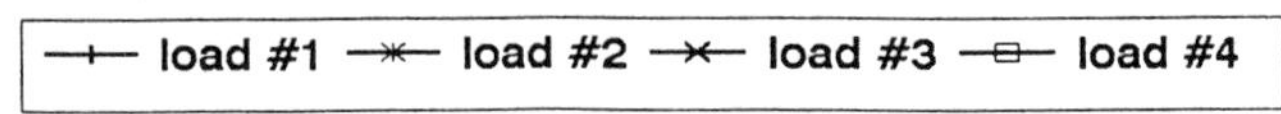

FIG. 4b--Composite Modulus Versus Sensor Distance at Different Load Levels, Site 12

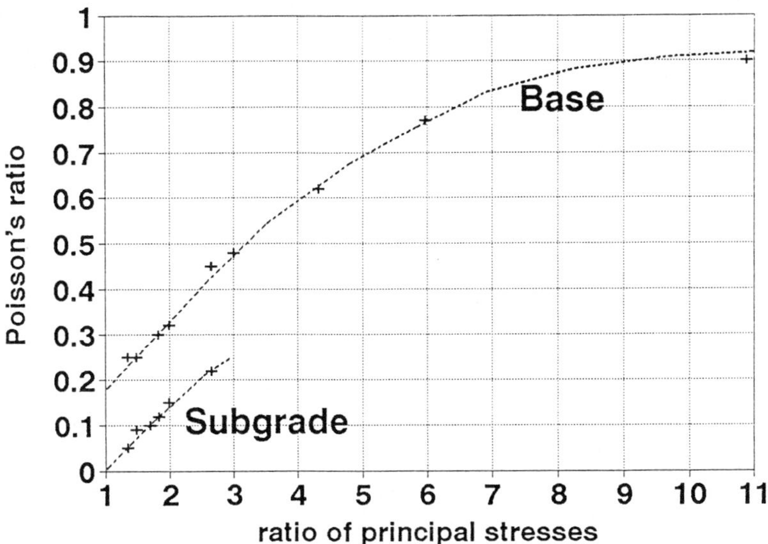

FIG. 5--Assumed Poisson's Ratio Versus Principal Stress Ratio for the
Base and Subgrade Materials at Site 4

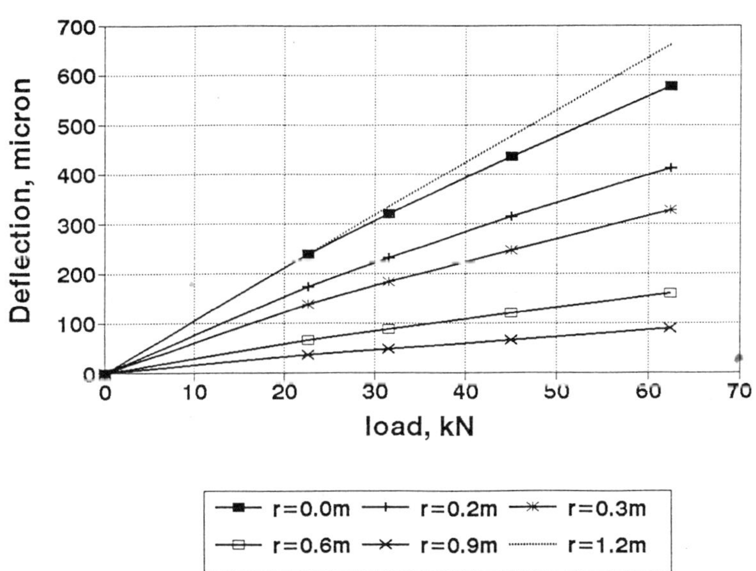

FIG. 6a--Measured and Predicted Deflections Versus Loads at Different
Radial Distances at Site 4

Non-linear backcalculation was conducted for the two sites using a finite element computer program as the forward deflection computation tool. No bedrock was assumed in the subgrade which is of infinite depth. However, due to limitations of the FE discretization, the subgrade was considered non-linear in the upper 5.08 meters (200 inches) and linear elastic material below the 5.08 meters from the surface.

TABLE 1-- Results of laboratory tests in terms of $k_1 - k_5$ model.

Site	Layer	k_1	k_2	k_3	k_4	k_5
4	base	719	0.3827	−0.0001	1.300	−0.2415
	subgrade	170	0.8647	−0.1094	1.540	−0.2830
12	base	1885	0.6650	−0.0587	1.413	−0.1944
	subgrade	878	0.9040	−0.3752	1.830	−0.2358

The results of backcalculation using linear (with MODULUS computer program) and non-linear elastic materials are presented below. In the runs with MODULUS, a fictitious bedrock was introduced at 1.3 m (51.40 inch) and 0.94 m (37.10 inch) below the subgrade interface for site 4 and site 12 respectively. Also, due to the increasing composite modulus with distance (see Figures 4a and 4b), the last sensor was dropped from the analysis by assigning a zero weighing factor. It should be noted that the KUAB sensor positions were 0, 200, 300, 450, 600, 900, 1200 mm (0, 8, 12, 24, 36, 48 inches) from the load. The backcalculated moduli are shown in Table 2. It is seen that the moduli are constant or increase slightly with the load level. The fitting is very good, as indicated by the small error per sensor. It is noted that the error of the last sensor which was dropped from the analysis is not included in the averaging.

TABLE 2--Results of linear elastic backcalculation (Using MODULUS computer program).

Site	Load in kN (kip)	Modulus in MPa (Ksi) of			Error per Sensor %
		AC	Base	Subgrade	
4	22.7 (5.089)	1.84 (267)	0.181 (26.2)	0.083 (12.1)	1.23
	31.6 (7.089)	1.90 (275)	0.192 (27.8)	0.087 (12.6)	1.30
	45.1(10.119)	1.88 (272)	0.211 (30.6)	0.090 (13.0)	1.26
	62.6(14.029)	1.90 (276)	0.230 (33.3)	0.092 (13.4)	1.33
12	22.6 (5.074)	1.03 (149)	0.097 (14.0)	0.051 (7.4)	0.28
	31.4 (7.046)	1.05 (152)	0.101 (14.7)	0.052 (7.6)	0.19
	45.0(10.100)	1.08 (156)	0.109 (15.8)	0.052 (7.6)	0.32
	63.8(14.306)	1.12 (162)	0.117 (16.9)	0.052 (7.6)	0.45

The results of the non-linear backcalculation are shown in Table 3 and Figures 6a-b. The Table 3 presents the backcalculated moduli, for all loads using all seven sensors. The backcalculated k_1 are different from those computed using the lab results, especially for the subgrade at site 4. This may be attributed to different anisotropic consolidation ratio (the ratio of the principal maximum stresses) in the

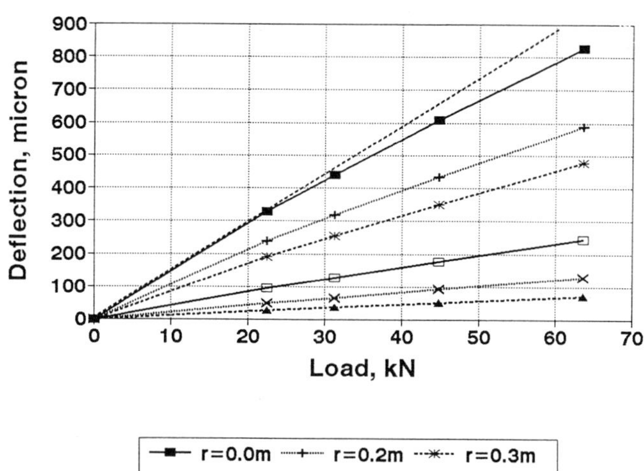

FIG. 6b--Measured and Predicted Deflections Versus Loads at different
Radial Distances at Site 12

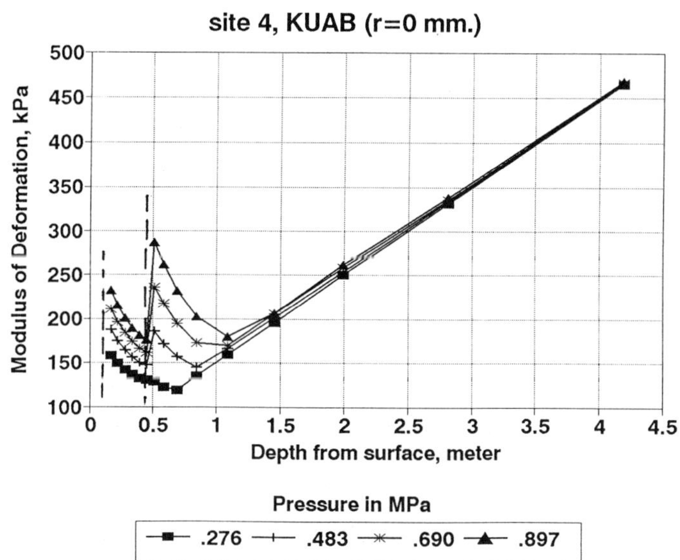

FIG. 7a--Variation of Computed Modulus of Deformation with Depth and
Load Level, at r=0 (under the center of loaded area), at site 4

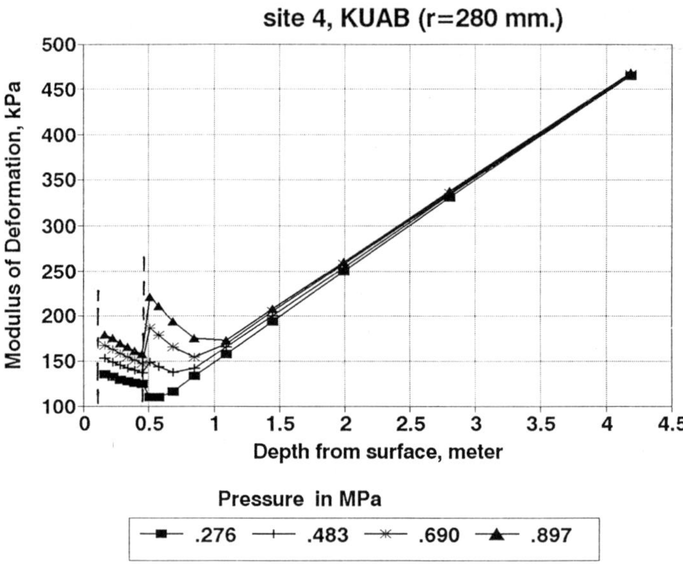

FIG. 7b--Variation of Computed Modulus of Deformation with Depth and
Load Level, r=0.28 m from the Load, at Site 4

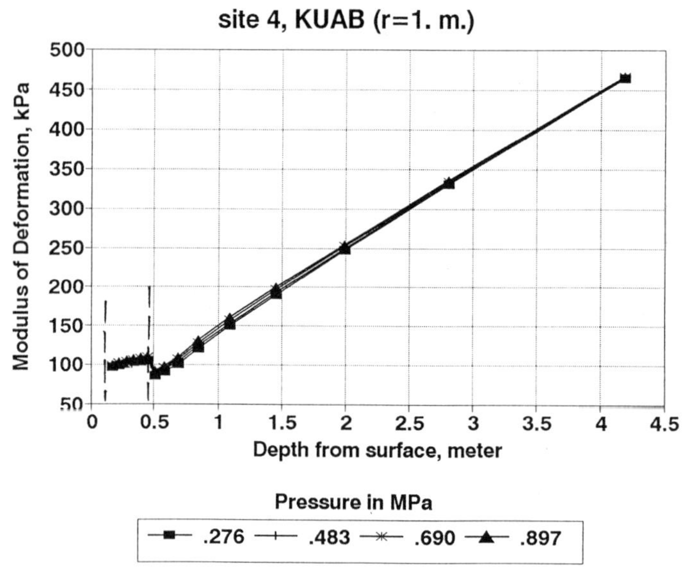

FIG. 7c--Variation of Computed Modulus of Deformation with Depth and
Load Level, 1.0 m from the Load, at Site 4

field and in the laboratory. The goodness of fit is excellent, keeping in mind that all sensors were used in the analysis and that no fictitious bedrock is introduced. Figures 6a-b show the deflections versus load of all sensors. The symbols represent the measured values and the lines are the lines drawn through the computed ones (the symbols of the computed values could not be shown; they fall on the symbols of the measured ones because of scale effects). It is seen that the deflection per unit load decreases with increasing load, thus supporting all indicators of the non-linear behavior of the structure.

It is interesting to look at the distribution of the modulus within the structure. Figures 7a-c for site 4 and 8a-c for site 12 show the calculated variation of the modulus of deformation with depth for four load levels, and at the center under the load, 0.280 and 1.02 meters (11 and 40 inches) from the load. The stress dependence of the base and subgrade materials is very well illustrated from the following viewpoints: (a) The modulus increases with increasing load level, at all locations where the load induced stresses are important, i.e. within a radius of about 0.40 m (18 in) and to a depth of about 1.0 m (40 in); (b) the modulus of the subgrade increases with increasing depth where the overburden pressure is predominant and the load induced stresses are negligible; (c) the modulus varies both vertically and horizontally; (d) the modulus in the top subgrade at site 4 increases strongly with load level. This increase is induced by the dilation of the material. Under the load, large principal stress ratios develop, leading to Poisson's ratio in excess of O.5. It should be mentioned that such state of stress was not reproduced in the laboratory test results.

TABLE 3-- <u>Results of non-linear backcalculation.</u>

Site	AC* Modulus MPa (ksi)	k_1 for Base	k_1 for Subgrade	Error per Sensor %
4	2.18 (316)	1453	2063	0.50
12	1.08 (157)	892	507	0.18

* The AC material was assumed linear elastic.

The moduli of the asphalt concrete (assumed to be linear elastic) are similar in both cases of backcalculation (see Tables 2 and 3). The moduli of the base layer backcalculated with MODULUS are representative of the moduli in the base under the load. The moduli of the subgrade from MODULUS are lower by about 20 to 40 percent than those obtained from the FE analysis, at the top of the subgrade (the lowest values in the entire subgrade). The discrepancies between the two backcalculation schemes seem to be important as they would give different pavement performance prediction. It is interesting to note that the modulus in the base (and top subgrade at site 4) is smaller by 50 percent and more, far from the load than under the load.

During the NDT testing, the measured average pavement temperature of the AC layer was 37°C (98°F) and 32°C (89°F) at sites 4 and 12 respectively. Resilient modulus tests conducted at 25°C (77°F) and 40°C (104°F) on cores from these sites show relatively high moduli. By linear interpolation of the laboratory results, the expected moduli would be 4.0 and 6.35 MPa (580 and 920 ksi) for sites 4 and 12 respectively. These results are 3 times larger than the backcalculated ones. It is worth mentioning that the lab results are obtained from small samples (cores) which may not be representative of the entire AC layer.

In summary, two test sections are analyzed using both the linear and non-linear static backcalculation schemes. The results obtained

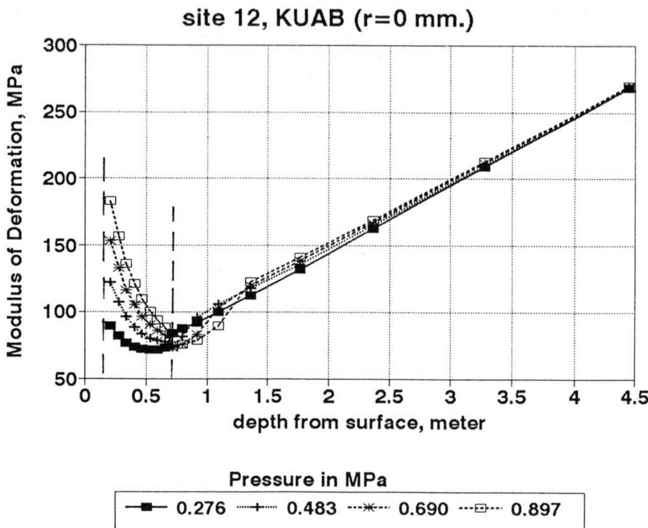

FIG. 8a--Variation of Computed Modulus of Deformation with Depth and Load Level, at r=0 (under the center of loaded area), at Site 12

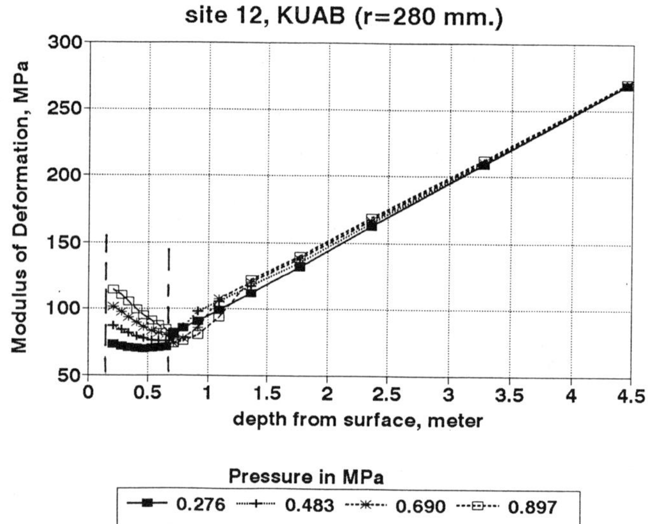

FIG. 8b--Variation of Computed Modulus of Deformation with Depth and Load Level, 0.28 m from the Load, at Site 12

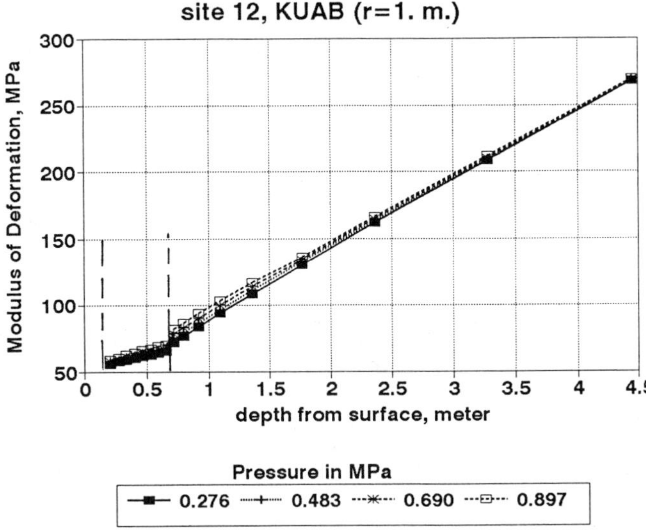

FIG. 8c--Variation of Computed Modulus of Deformation with Depth and
Load Level, 1.0 m from the Load, at Site 12

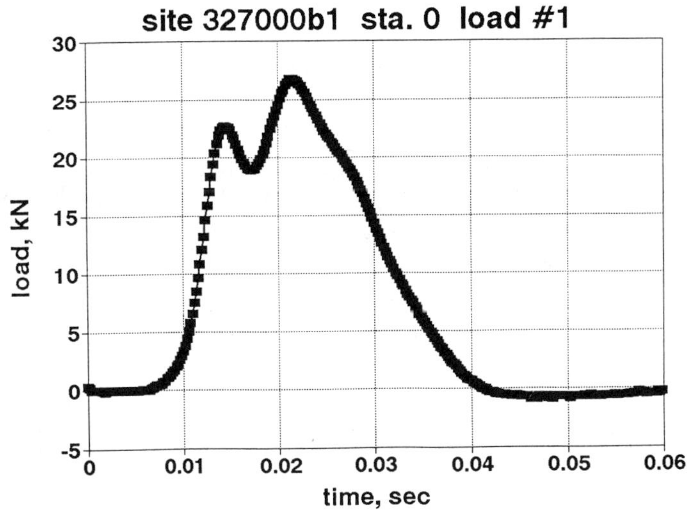

FIG. 9a--Load History at Site 327000b1

with the computer program MODULUS seem to underpredict the subgrade
modulus by a factor of 50 percent and more. The modulus of the base
layer seem to represent the layer underneath the load, at its most
favorable condition of confinement.

ILLUSTRATION OF THE DYNAMIC LINEAR BACKCALCULATION PROCEDURE

     Two sections from the SHRP LTPP data base were selected for
analysis, one of which was analyzed using both the frequency and time
domain analysis. The second one analyzed in the time domain only was
chosen because of the shallow depth of bedrock. The measured load time
histories are presented in Figure 9a and 9b. The pavement at site
327000b1 is made of 122 mm (4.8 in) AC layer, 122 mm (4.8 in) asphalt
treated base and 142 mm (5.6 in) of gravel on top of a poorly graded
gravel with silt and sand subgrade. In the analyses, the two top layers
were combined into one layer. The pavement at site 341030A3 is made of
150 mm (6 in) AC layer, 198 mm (7.8 in) crushed stone base, 594 mm (23.4
in) silty sand with gravel subbase on top of a poorly graded sand with a
silt and gravel subgrade. A hard layer which could not be penetrated by
the auger drill was detected at about 1.37 m (4.5 ft) below the surface.
     The results of the linear elastic backcalculation (using only peak
load and deflections) are given in Table 4. It is noted that the depth
of bedrock was inputted as infinite for site 327000b1 and 1.37 m (4.5
ft) for site 341030A3.

TABLE 4--Results of backcalculation using MODULUS program.

Site	Modulus in MPa (Ksi) of			
	AC	Base	Subbase	Subgrade
327000b1 (infinite depth)	4.16 (604)	0.057 (8.3)	--- ---	0.437 (63.4)
341030A3	1.248 (181)	0.168 (24.3)	0.063 (9.2)	0.057 (8.2)

     In the dynamic backcalculation analyses, the asphalt layers was
assumed to behave according to the power law with three parameters (see
Equation 4) and the granular base-subbase and subgrade materials were
assigned a constant damping ratio of 3 and 5 percent at site 327000b1
and 34103043 respectively.
     The results of the dynamic linear backcalculation are presented
below, for each site.

<u>Site 32700b1</u>

     The measured deflection bowls are shown in Figures 10a-c by the
symbols. It is seen that the deflection does not decay to zero for all
sensors. This requires a tail correction in order to get a zero
deflection at the 0.06 sec time and be able to use the fast fourier
transform. A linear tail correction was used, i.e. a deflection
proportional to the value accumulated at 0.06 sec was subtracted from
the deflection history of each sensor. No tail correction was made for
the analysis in the time domain.
     The results of the dynamic backcalculation are summarized in Table
5. Comparing the results using the frequency and time domains, it is
seen that (a) the parameters of the AC layer are similar, (b) the

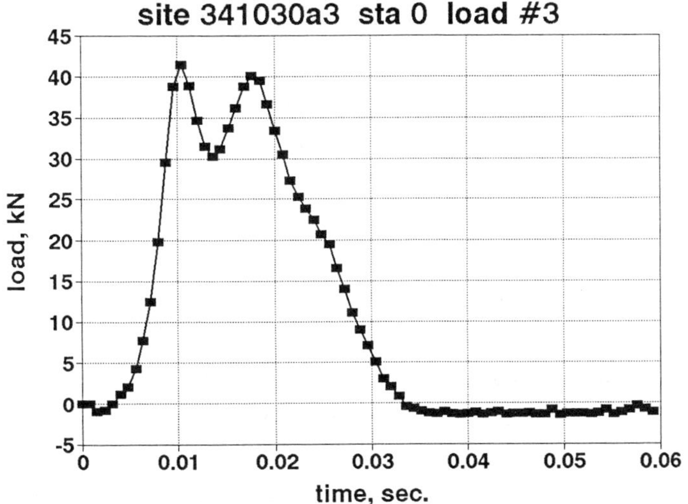

FIG. 9b--Load History at Site 34103093

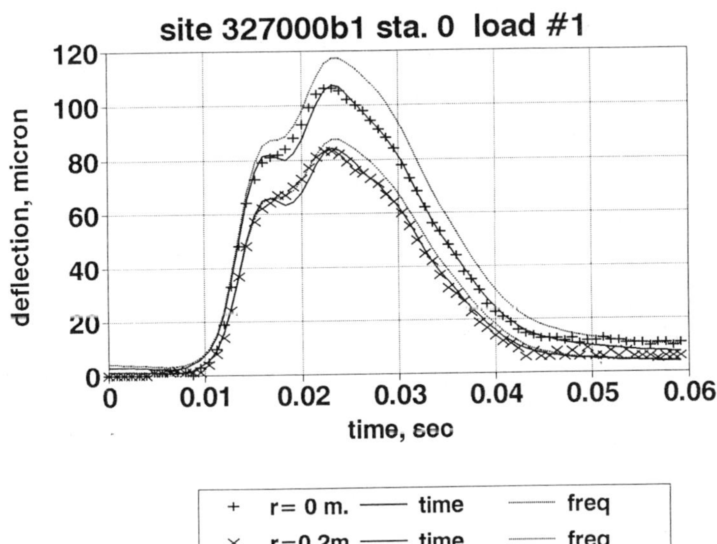

FIG. 10a--Comparison of Measured Deflection Histories with Computed
Ones Using Time and Frequency Domain Analysis, Site 327000b1, Sensors at
r=0 and 0.2m

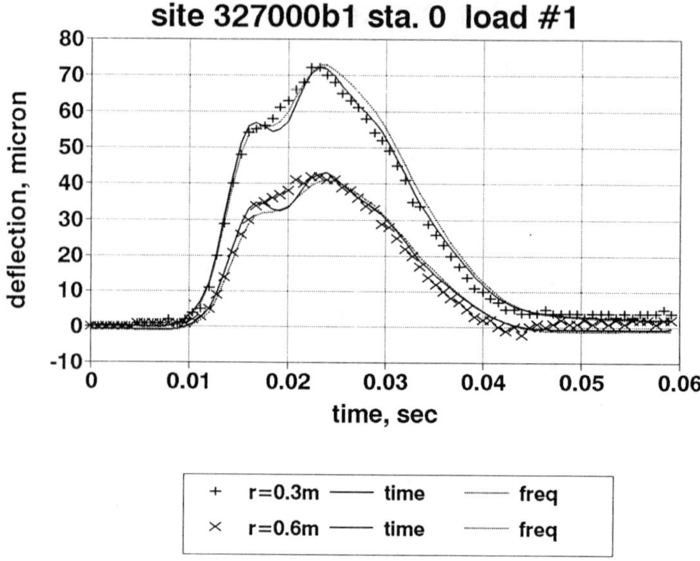

FIG. 10b--Comparison of Measured Deflection Histories with Computed Ones Using Time and Frequency Domain Analysis, Site 327000b1, Sensors at r=0.3 and 0.6 m

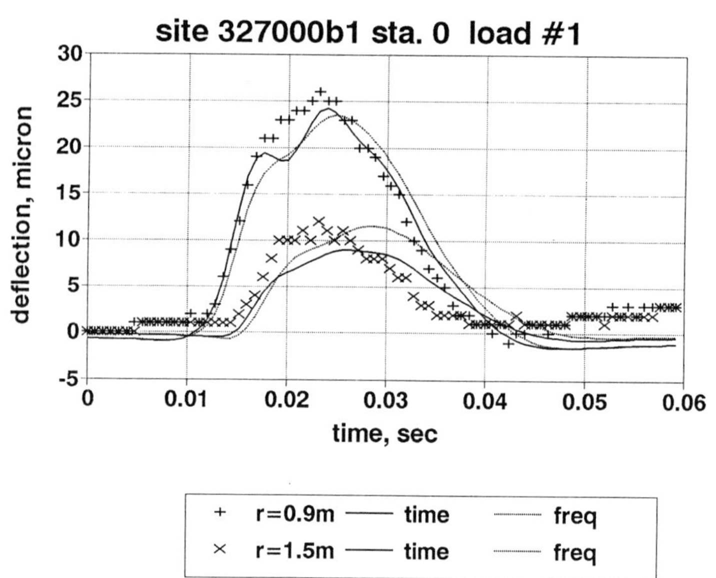

FIG. 10c--Comparison of Measured Deflection Histories with Computed Ones Using Time and Frequency Domain Analysis, Site 327000b1, Sensors at r=0.9 and 1.5 m

modulus of the base is lower and that of the subgrade is higher for the time domain analysis as compared to the frequency domain analysis. Such a difference is acceptable from an engineering point of view. It is noted that the time domain analysis was made with all data points in the 0 to 60 msec window, thus the two analyses are made on the same data. Figures 10a-c show the computed deflection histories for both the time and frequency domain analyses, for 6 sensors. It is seen that (a) both analyses predict a tail, i.e., the deflection does not decay to zero within the 60 msec window. As discussed in the previous section on the choice of the backcalculation procedure, this non-zero deflection at time equal 60 msec may be due to delayed elasticity and/or to viscous deformation; (b) the fitting of the data is better achieved with the time domain analysis as compared with the frequency domain analysis, especially for the first sensor at r=0. This may be due to the correction of the tail that was made in the frequency domain analysis. In the above case, when all data points are used in the analysis, the two types of analysis seem to give similar results. However, when the time domain analysis is made with the data for up to 0.04 sec (two third of the data points and ignoring the tail) slightly different backcalculated results are obtained. The backcalculated parameters are shown in Table 5. When the number of data points is reduced further to the window from 0 to 0.03 sec, the slope m reduces to 0.11 (from 0.281 obtained with the window from 0 to 0.06 sec).

TABLE 5--<u>Results of dynamic backcalculation.</u>

SITE	AC LAYER (Eq. 6)			MODULUS IN MPa, (ksi) OF			
	$1/D_o$ MPa (ksi)	$1/D_1$ MPa (ksi)	m	Base	Sub-base	Sub-grade	Hard Lay--er
327000b1 (Frequency Domain Analysis)	717 (104000)	0.897 (130)	0.294	0.078 (11)	--- ---	0.366 (53)	--- ---
327000b1 (Time Domain Analysis with 0.06 sec window)	717 (104000)	1.172 (170)	0.281	0.045 (6.5)	--- ---	0.506 (73)	--- ---
327000b1 (Time Domain Analysis with 0.04 sec window)	717 (104000)	1.421 (206)	0.235	0.037 (5.3)	___ ---	0.631 (92)	--- ---
341030A3 (Time Domain Analysis with 0.06 sec window)	58.14 (8430)	0.586 (85)	0.169	0.150 (22)	0.20 (29)	0.06 (9)	0.51 (74)

The moduli computed using the dynamic backcalculation compare very well with those obtained using the static linear elastic backcalculation (comparing Tables 4 and 5). However with the dynamic backcalculation, all three parameters of the viscoelastic power law are estimated while with the linear elastic one, only the elastic modulus was backcalculated.

<u>Site 341030A3</u>
    The measured and predicted deflection bowls are shown in Figures
11.a-b, and the backcalculated parameters in Table 5. The effect of the
shallow bedrock is well illustrated in the measured response of the
pavement which shows oscillations of decaying amplitude in all sensors.
The computed response of the sensors near the load seems to fit quite
well the measured response. However, the calculated oscillations of the
first three sensors appear to be more damped than the measured response.
The response of the two sensors far from the load is not well predicted.
The last sensor at 1.524 m (60 in) radial distance shows a negative
deflection while the predicted one is positive. A possible explanation
of this discrepancy is non-uniformity of the section which may be
cracked. The backcalculated moduli of the base and subgrade using the
dynamic scheme are similar to those obtained using the static linear
elastic scheme. Only the subbase modulus is overpredicted in the
dynamic backcalculation.

Discussion of the Results

    The dynamic linear backcalculation procedure was applied for two
sites. The results show that the backcalculated moduli compare very
well (except for the subbase modulus at site 341030A3) with those
obtained using the linear elastic procedure. This may look strange for
the second site with shallow bedrock. However, the bedrock is very
close to the surface and does not seem to affect the deflection.
Similar results are presented by Chang et al (1991), who analyzed a
similar pavement (route 1 with 150 mm AC and 180 mm base) on top of a
subgrade of varying thickness. Their results show that the dynamic
effect is smaller as the depth of bedrock decreases toward zero or
increases toward infinity and reaches its maximum value at about 4.50
meters (15 ft).
    The time histories of the deflection provide additional
information used by the backcalculation procedure, for example the
relative position of the different response curves or the arrival time
of the wave at different distances from the source. This additional
information may be one reason for the small difference in the
backcalculation results of the linear elastic and dynamic schemes. Note
that the forward computation may be an additional reason for the
difference in the backcalculation results.

CONCLUSIONS

    The backcalculation procedures have been separated into
categories, and a recommendation on which one to use was given as
follows:

1. The dynamic linear scheme should be preferred over the static
    linear scheme. Subdividing the subgrade into two layers
    should be used to account partly for non-linearity of the
    material behavior.
2. When no bedrock exists (or not indication of bedrock within at
    least 6 meters (20 feet)) and when the subgrade is a stress
    sensitive material (such as sands and sandy materials), it is
    recommended to use the static non-linear scheme.
3. The linear elastic backcalculation may be used in addition to
    the above two cases or for routine analysis of pavement with
    stiff and thick AC layers with no bedrock within at least 6
    meters (20 feet) and clayey material subgrades.

    However, in order to implement the two schemes recommended - the
dynamic linear and static non-linear backcalculation schemes, it is
imperative to initiate a basic research in material characterization and
measurements evaluation. The non-linear material behavior is
represented by several parameters, not only one modulus. All parameters
cannot be reliably backcalculated only from deflection bowls. The
research in material characterization is therefore intended to provide

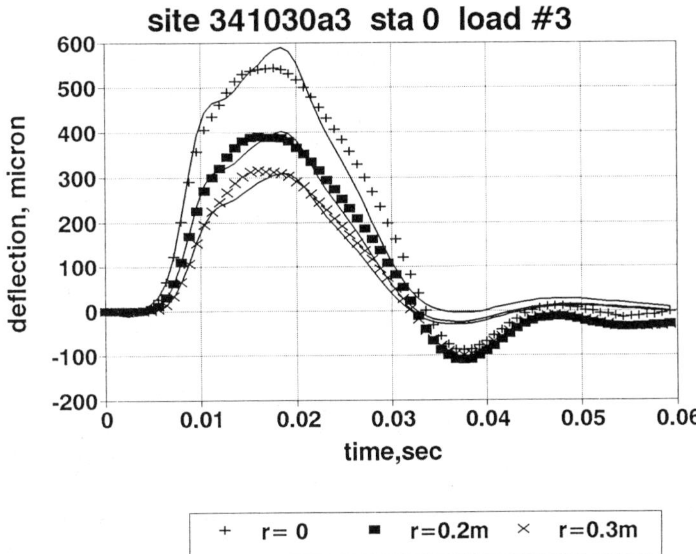

FIG. 11a--Comparison of Measured Deflection Histories with Computed Ones Using Time Domain Analysis, Site 341030A3, sensors at r=0, 0.2 and 0.3 m

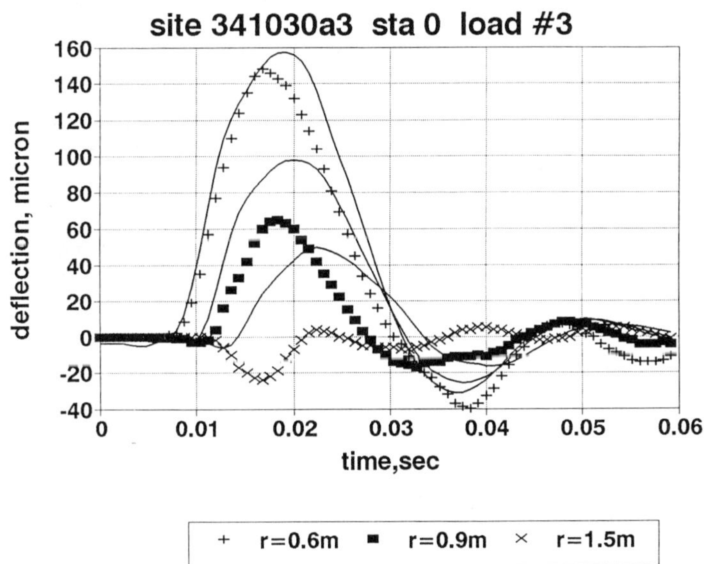

FIG. 11b--Comparison of Measured Deflection Histories with Computed Ones Using Time Domain Analysis, Site 341030A3, Sensors at r=0.6, 0.9 and 1.5 m

the parameters that cannot be reliably backcalculated. Examples of such parameters are: k_2 to k_5, damping ratios of granular and clayey materials for different density, moisture content and gradation conditions.

The accuracy of the tail shape in the deflection histories is questionable. This is a major concern in dynamic backcalculation. There is a need for basic research to improve the accuracy of the measurements, and at least evaluate the errors involved with each type of equipment - geophones or Linear Variable Differential Transformers (LVDT). An effort should be made to insure that the time histories are not shifted by a filtering technique (time shifting has an important effect on the backcalculated moduli) and that the tail of the signal can be used for evaluating depth of bedrock and viscoelastic properties.

Examples to illustrate the use of the static non-linear and dynamic linear backcalculation are presented in the paper. These examples show that: (a) in the case of non-linear materials, the distribution of the moduli in the layers is non-uniform. It would be difficult to represent it by an approximate scheme; (b) the dynamic backcalculation, by using more information than the static linear one, provides the possibility to derive viscoelastic parameters. In the examples presented, the parameters of the power law for the asphalt layer were evaluated. The scheme is being used in a SHRP project on performance models of flexible pavements. The results of the study will be published separately.

The backcalculated k_1-values are larger than those obtained from laboratory tests. This result seems to be due to different initial state of stress in the field and the laboratory tests. It suggests that verification of the backcalculation techniques must rely on instrumented pavement sections where deformation is measured at different locations inside the pavement as well as at the surface.

Additional research would be required to link the properties derived from laboratory tests to those backcalculated from the field on instrumented pavements.

ACKNOWLEDGEMENTS

The author would like to express his deep appreciation to Mr. Tom Scullion from Texas Transportation Institute for his review of the paper and his valuable comments.

REFERENCES

1.   Bentsen, R.A., Bush, A.J. and Harrison, A., 1989, "Evaluation of Non-Destructive Test Equipment for Airfield Pavements, Phase I, Calibration Test Results and Field Data Collection", _Technical Report_ GL-89-3, Department of the Army, Waterways Experiment Station, Vicksburg, Mississippi.

2.   "BISAR Users Manual, 1972, "Layered System Under Normal and Tangential Loads", Shell-Koninilijke/Shell Laboratorium, Amsterdam, The Netherlands.

3.   Brown, S.F., Tam, W.S. and Brunton, J.M., 1987, "Structural Evaluation and Overlay Design: Analysis and Implementation", _Proc._, Sixth International Conference on the Structural Design of Asphalt Pavements, University of Michigan, an Arbor, Michigan, pp. 1013-1028.

4.   Chang, T.Y., Ko, H.Y., Scott, R.F. and Westman, R.A., 1967, "An Integrated Approach to the Stress Analysis of Granular Materials", _Research Report_, California Institute of Technology, Pasadena, CA., p. 65.

5.   Chang, D.W., Kang, Y.V., Roesset, J.M., and Stokoe, K.H., 1991, "Effect of Depth of Bedrock on Deflection Basins Obtained with Dynaflect and FWD Tests", _Transportation Research Record_, 1355, pp. 8-17.

6.   Chen, W.F. and Mizuno, E., 1990, "Non-linear Analysis in Soil Mechanics, Theory and Implementation", in _Developments in Geotechnical Engineering_, 53, Elsevier.

7.   Chou, Y.J. and Lytton, R.L., 1991, "Accuracy and Consistency of Backcalculated Pavement Layer Moduli", _Transportation Research Record_, 1293, pp. 72-85.

8.   Davies, T.G. and Mamlouk, M.S., 1985, "Theoretical Response of Multilayer Pavement System to Dynamic Non-Destructive Testing," _Transportation Research Record_, 1022, pp. 1-7.

9.   Desai, C.S. and Siriwardane, H.J., 1984, _Constitutive Laws for Engineering Materials with Emphasis on Geologic Materials_, Rainbow Bridge Book Co.

10.   Eringen, A.C., 1962, _Non-Linear Theory of Continuous Media_, McGraw-Hill, New York.

11.   Irwin, L. and Szebenyi, T., 1991, "User's Guide to MODCOMP3, Version 3.2", CLRP _Report_ Number 91-4, Cornell University Local Roads Program, Ithaca, N.Y.

12.   Lytton, R.L., 1989, "Backcalculation of Pavement Layer Properties", _ASTM STP_ 1026, pp. 7-38.

13.   Lytton, R.L., German, F.P., Chou, Y.J. and Stoffels, S.M., 1990, "Determining Asphalt Concrete Pavement Structural Properties by Non-destructive Testing", NCHRP Report 327, _Transportation Research Board_, Washington, D.C., p.105

14.   Magnuson, A.H., 1988, "Computer Analysis of Falling Weight Deflectometer Data, Part I: Vertical Displacement Computations on the Surface of a Uniform (one-layer) Half-Space Due to an Oscillating Surface Pressure Distribution", _Research Report_ 1215-1F, Texas Transportation Institute, Texas A&M University, College Station, TX.

15.   Magnuson, A.H., Lytton, R.L. and Briggs, R.C., 1991, "A Comparison Study of Computer Predictions and Field Data for Dynamic Analysis of Falling Weight Deflectometer Data", <u>Transportation Research Record</u>, 1293, pp. 61-71.

16.   Michelow, J., 1963, "Analysis of Stresses and Displacements in an N-Layered Elastic System Under a Load Uniformly Distributed in a Circular Area", California Research Corp., Richmond, CA.

17.   PCS/LAW Engineering, 1990, "Analysis of Section Homogeneity, Non-Representative Test Pit and Section Data, and Structural Capacity, FWDCHECK Version 1.00", Volume I, <u>Technical Report</u>, prepared for SHRP by P-001B Technical Advisory Staff.

18.   Peutz, M.G.F., Van Kempen, H.P.M. and Jones, A., 1968, "Layered Systems Under Normal Surface Loads", <u>Highway Research Record</u>, 228, pp. 34-45.

19.   Press, W.H., Flannery, B.P., Teukolsky, S.A. and Vetterling, W.T., 1989, <u>Numerical Recipes, The Art in Scientific Computing,</u> Cambridge University Press.

20.   Rada, G.R., Richter, C.A. and Stephanos, P.J., 1992, "Layer Moduli from Deflection Measurements: Software Selection and Development of SHRP's Procedure for Flexible Pavements," <u>Transportation Research Record</u>, 1377, pp. 77-87.

21.   Roesset, J., 1987, "Computer Program UTFWIBM", The University of Texas at Austin, Texas.

22.   Rohde, G.T., 1990, "The Mechanistic Analysis of FWD Deflection Data on Sections with Changing Subgrade Stiffness with Depth", <u>Ph.D. Dissertation</u>, Texas A&M University, College Station, TX., p. 228.

23.   Shapery, R., 1987, "Lectures in Viscoelasticity", <u>Notes</u> from Course at Texas A&M Univ., College Station, TX.

24.   Torpunuri, V.S., 1990, "A Methodology to Identify Material Properties in Layered Viscoelastic Half Spaces", <u>M.Sc</u> <u>Thesis</u>, Texas A&M Univ., College Station, TX., p. 112.

25.   Ullidtz, P., 1977, "Overlay and Stage by Stage Design", <u>Proc</u>. Fourth International Conference on the Structural Design of Asphalt Pavements, Ann Arbor, Michigan, Volume 1, pp. 722-735.

26.   Uzan J., 1985, "Granular Material Characterization", <u>Transportation Research Record</u>, 1022, pp. 52-59.

27.   Uzan, J., Scullion, T., Michalek, C.H., Paredes, M. and Lytton, R.L., 1988, "A Microcomputer Based Procedure for Backcalculating Layer Moduli from FWD Data", <u>Report</u> No. FWHA/TX-88-1123-1, Texas Transportation Institute, Texas A&M University, College Station, TX., p. 82.

28.   Uzan J., Lytton, R.L. and German, F.P., 1989, "General Procedure for Backcalculating Layer Moduli", <u>ASTM STP</u> 1026, pp. 217-22.

29.   Uzan, J. and Scullion, T., 1990, "Verification of Backcalculation Procedures", <u>Proc</u>. Third International Conference on Bearing Capacity of Roads and Airfields, Trondheim, Norway, Volume 1, pp. 447-458.

30.   Uzan, J., Witczak, M.W., Scullion, T. and Lytton, R.L., 1992, "Development and Validation of Realistic Pavement Response

Models", <u>Proc.</u> Seventh International Conference on Asphalt Pavements, Nottingham, U.K., Vol. 1, pp. 334-350.

31.   Uzan, J., 1992, "Resilient Characterization of Pavement Materials", Short communication, <u>Int. Journal for Numerical & Analytical Methods in Geomechanics</u>, Vol. 16, pp. 453-459.

32.   Uzan, J., 1992, "Rigid Pavement Evaluation Using NDT – A Case Study", <u>Journal of Transportation Engineering</u>, ASCE, Vol. 118, No. 4, July-August, pp. 527-539.

33.   Uzan, J., Briggs, R.C. and Scullion, T., 1993, "Backcalculation of Design Parameters for Rigid Pavements", <u>Transportation Research Record</u>, 1377, pp. 107-114.

34.   Uzan, J., 1993, "Dynamic Linear Backcalculation of Pavement Material Properties", <u>Journal of Transportation Engineering</u>, ASCE.

35.   Van Cauwelaert, F.J., Alexander, D.R., White, T.D. and Barker, W.R., 1989, "Multilayer Elastic Program for Backcalculating Layer Moduli in Pavement Evaluation", <u>ASTM</u> <u>STP</u> 1026, pp. 171-188.

36.   Witczak, M.W. and Uzan J., 1988, "The Universal Airport Pavement Design System; <u>Report</u> I of IV: Granular Material Characterization", University of Maryland, College Park, MD.

37.   Wolf, J.P., 1985, <u>Dynamic Soil Structure Interaction</u>. Prentice-Hall, Inc.

Gonzalo R. Rada[1], Cheryl Allen Richter[2], and Peter Jordahl[3]

SHRP'S LAYER MODULI BACKCALCULATION PROCEDURE

REFERENCE: Rada, G. R., Richter, C. A., and Jordahl, P., "SHRP's Layer Moduli Backcalculation Procedure," Nondestructive Testing of Pavements and Backcalculation of Moduli (Second Volume), ASTM STP 1198, Harold L. Von Quintas, Albert J. Bush, III, and Gilbert Y. Baladi, Eds., American Society for Testing and Materials, Philadelphia, 1994.

ABSTRACT: Deflection basin measurements for the purpose of structural capacity evaluation are a key component of the SHRP's LTPP monitoring program. In the near term, SHRP will apply a backcalculation procedure to these data in order to estimate the in situ elastic moduli of the pavement layer materials. Because a standard method for evaluating the structural capacity of pavements from deflection data does not presently exist, SHRP has undertaken a study to develop a layer moduli backcalculation procedure for use in the initial analysis of the SHRP deflection data. The procedure covers not only the software but also the rules and guidelines used in applying the program. This paper focuses on the standard procedure used to ensure that the LTPP deflection data analysis is as consistent, productive, and straight-forward as possible.

KEYWORDS: backcalculation procedure, backcalculation rules and guidelines, layer moduli backcalculation, deflection testing

INTRODUCTION

Since the Spring of 1988, SHRP has completed an initial round of deflection testing on nearly 800 in-service pavement test sections, and has begun a second round. Although the raw deflection data is the primary data to be stored for use by pavement researchers, the initial Long-Term Pavement Performance (LTPP) data analyses require that SHRP derive estimates of the in-situ elastic moduli of the pavement layers from the deflection data. In order to do so, SHRP has developed a backcalculation procedure, consisting of an existing backcalculation program and a series of application "rules".

The development process for the SHRP backcalculation procedure involved four phases: (1) a literature review to identify backcalculation programs which might be used in the procedure; (2) the selection of a limited number of programs for detailed evaluation; (3) a detailed evaluation of those programs; and (4) the development of a procedure around the selected program. The first three stages of this endeavor, which resulted in the selection of the MODULUS backcalculation

[1]Principal engineer at PCS/Law Engineering, 12240 Indian Creek Ct, Suite 120, Beltsville, MD 20705; [2]Research highway engineer at FHWA Long Term Pavement Performance Division, 6300 Georgetown Pike, McLean, VA 22101-2296; [3]Systems analyst at Brent Rauhut Engineering, Inc., 8240 Mopac Expressway, Suite 220, Austin, TX 78759.

program, are discussed in detail elsewhere (Strategic Highway Research
Program 1991). This paper focuses on the standard backcalculation
procedure developed around the selected program.

In general, backcalculation is a laborious process, requiring a
high degree of skill, and the results are known to be moderately to
highly dependent on the individual doing the backcalculation. This
comes about for a number of reasons, including the lack of a consensus
standard addressing all aspects of the backcalculation process. In
order to ensure that the backcalculation process applied in the SHRP
data analysis is as consistent, productive, and straight forward as
possible, the SHRP backcalculation procedure combines an existing
backcalculation program with a rigorous set of application rules. In
addition, the initial backcalculation has been automated to a high
degree, to reduce opportunities for "operator" error and between user
inconsistencies.

Thus, in comparison to other procedures, the SHRP backcalculation
procedure is unique from a standpoint of data availability,
standardization, and automation. It is also unique in that a group of
experts -- the SHRP Expert Task Group (ETG) on Deflection Testing and
Backcalculation, chaired by Dr. A.J. Bush, III -- was instrumental in
its development, both in the software selection and rules development.
Despite all of this, it is anticipated that the procedure will be
refined as more is learned about its strengths, weaknesses and
requirements.

BACKCALCULATION RULES

The SHRP backcalculation procedure is best illustrated by
reference to Figure 1. The procedure relies on the wealth of
information stored in the LTPP data base -- deflection, pavement
structure and materials, and surface layer temperature data. -- to
generate the input for the backcalculation program. More specifically,
data base queries and rules are used to generate data files for input
into the MODULUS program, and additional rules address the subsequent
evaluation of the backcalculation results.

The SHRP backcalculation rules address three major areas. The
first group of rules focuses on the definition of the moduli ranges
required to run the MODULUS program, the second set of rules addresses
the modeling of the pavement structure for purposes of backcalculation,
and the third and final set of rules focuses on the evaluation of the
backcalculation results. In addition, new rules or modifications to the
existing ones based on preliminary LTPP data analysis results are
discussed in a later section.

Before proceeding with the discussion of the backcalculation
rules, it is worthwhile noting two key points related to the deflection
data to be used in the analysis. First, at any given SHRP pavement test
location and for each FWD drop height, four repeat drops are performed
and the variance of these drops is checked to ensure that it is within a
specified tolerance. Second, the deflection data associated with these
four repeat drops are averaged for purposes of the backcalculation
analysis to minimize the effects of random errors. Further details on
the SHRP deflection testing program are given elsewhere (Strategic
Highway Research Program 1993).

<u>Definition of Layer Moduli Ranges</u>

The MODULUS program requires that an estimate of the "expected"
range of moduli be specified for each pavement layer, except the
subgrade where only an estimate of the initial modulus is required. In
the SHRP backcalculation procedure, predictive equations that rely on
material property and field temperature data stored in the LTPP data
base are used to establish the moduli range for asphaltic concrete (AC)
layers -- the specific algorithm used depends on the available

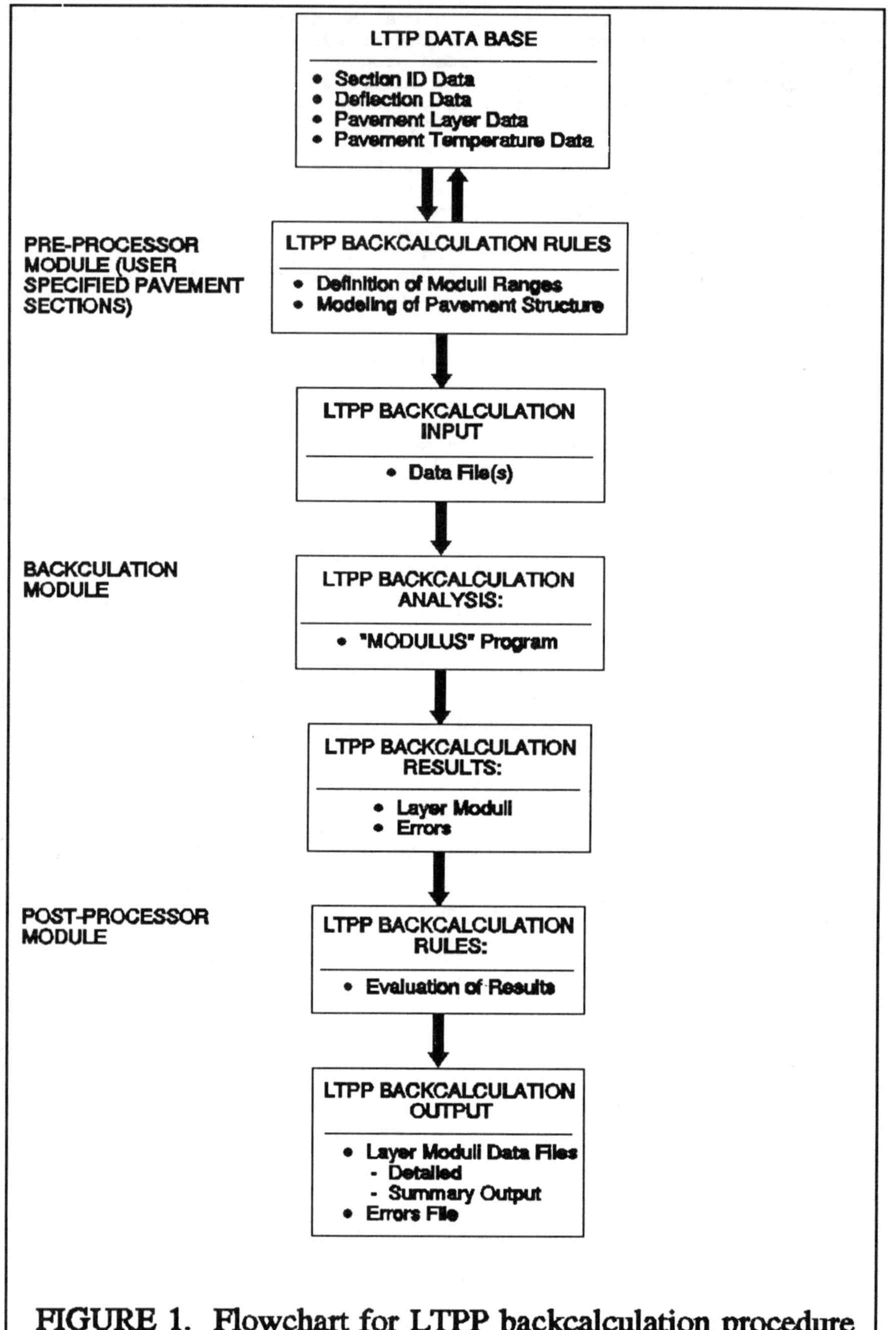

FIGURE 1. Flowchart for LTPP backcalculation procedure

information. Moduli ranges for portland cement concrete (PCC) layers and other stabilized materials are determined based on available laboratory test results, or assumed. Similarly, moduli ranges for unbound granular base and subbase layers are estimated on the basis of material type. Outer deflection readings and Boussinesq's one-layer deflection equation are used to estimate the initial subgrade modulus.

<u>Asphalt concrete layers</u>--The following rules are used to arrive at the modulus range for asphalt concrete layers:

1. Determine mid-depth temperature of AC layer(s)--Using the surface layer temperature gradient versus time data stored in the LTPP data base, the mid-depth temperature for each AC layer in the pavement structure at the time of testing is determined (extrapolated or interpolated).

2. Compute initial modulus of AC layer(s)--If mix data -- aggregate grading, maximum and bulk specific gravity of mix, and asphalt content -- are available from the LTPP data base, the following equation is used to estimate the initial modulus of AC layers (Witczak 1989):

$$
\begin{aligned}
\log_{10}[E^*] = {} & 5+2.250053-0.091756*V_{bc}-0.027949*V_a-0.096881*p_{200}+ \\
& 0.250094*p_{abs}-0.006447*t_p+0.060612*f-0.00007404*t_p{}^2+ \\
& 0.00191539*V_{bc}{}^2+0.0082813*p_{200}{}^2-0.0010225*p_{3/4}{}^2+ \\
& 0.0001909*p_{3/8}{}^2-0.0801155*p_{abs}{}^2+0.0148592*\eta_{70,10^6}{}^2- \\
& 0.0024159*f^2+0.00094015*p_{3/8}*V_{bc}+0.00084534*p_{3/4}*V_{bc}+ \\
& 0.0004965*p_{3/4}*p_4-0.00034328*p_{3/8}*p_4-0.00316297*p_{3/8}*p_{abs} \quad (1)
\end{aligned}
$$

where E^* = AC modulus, in psi; V_{bc} = effective asphalt content, by volume percentage; V_a = percent air voids in mix; p_{200} = percent aggregate weight <u>passing</u> the No. 200 sieve; p_{abs} = percent asphalt absorption, by weight of aggregate; f = test frequency of load wave, in Hz (assume 16 Hz in all cases); t_p = test temperature, in Fahrenheit (from Step No. 1); p_4, $p_{3/8}$, and $p_{3/4}$ = percent aggregate weight <u>retained</u> in the No. 4, 3/8" and 3/4" sieves, respectively; and $\eta_{70,10^6}$ = asphalt viscosity at 70°F (21°C), in 10^6 P.

The effective asphalt content (V_{bc}), by volume percentage, is determined by means of the following equation (Asphalt Institute 1988):

$$
V_{be} = \frac{\left(P_{ac} - P_{abs} - \dfrac{P_{abs} * P_{ac}}{100} \right) * G_{mb}}{G_b} \tag{2}
$$

where p_{ac} = percent asphalt content by weight of mix; p_{abs} = percent asphalt absorption by weight of aggregate; G_{mb} = maximum specific gravity of mix; and G_b = specific gravity of bitumen (assume a value of 1.010 if not stored in the LTPP data base).

If aggregate (effective and bulk) and bitumen specific gravities are stored in the LTPP data base, the following equation is used to determine the percent asphalt absorption (p_{abs}) by weight of aggregate (Asphalt Institute 1988):

$$
P_{abs} = 100 * \left[\frac{G_{se} - G_{sb}}{G_{sb}G_{se}} \right] G_b \tag{3}
$$

where G_{se} = effective specific gravity of aggregate; G_{sb} = bulk specific gravity of aggregate; and, G_b = specific gravity of asphalt. Otherwise, it is assumed that p_{abs} = 0.5% for crushed stone, gravel and sand mixtures and 1.5% for slag.

The percentage of voids in the mix, V_a, is determined from the following relationship (Asphalt Institute 1988):

$$V_a = 100 * \left[\frac{G_{mm} - G_{mb}}{G_{mm}} \right]$$

$$(4)$$

where G_{mm} = maximum specific gravity of compacted mix and G_{mb} = bulk
specific gravity of compacted mix.

The asphalt viscosity at 70°F (21°C), $\eta_{70,10^6}$, can be determined in
one of three ways. If measured absolute (140°F (60°C, in P) and
kinematic (275°F (135°C), in cSt) viscosities are stored in the LTPP data
base, a log(log(viscosity)) versus log(temperature) correlation is first
established and then extrapolated to 70°F (21°C) (Witczak 1989). Figure
2 graphically illustrates the computation of $\eta_{70,10^6}$ from known viscosity
and temperature data. When using this procedure, special care must be
taken to ensure that viscosity data have been converted into <u>centipoise</u>
and temperatures into degrees <u>Rankine,</u> prior to the development of the
correlation.

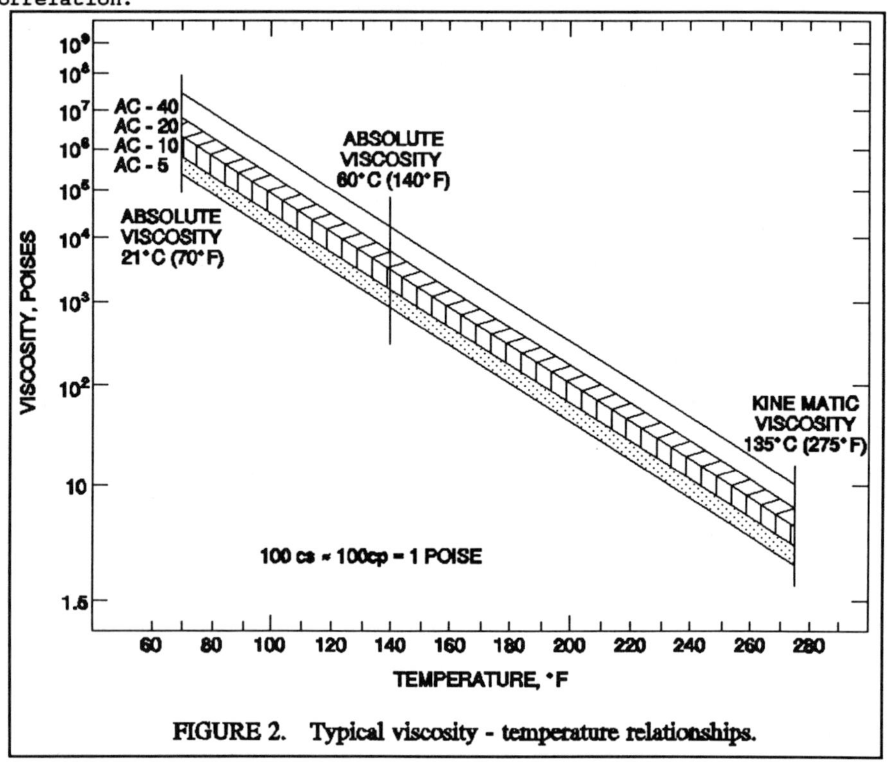

FIGURE 2.   Typical viscosity - temperature relationships.

If viscosity data is not available but a penetration value at 77°F
(25°C) or Pen$_{77}$ is known, the following relationship between asphalt
viscosity at 70°F (21°C) and penetration at 77°F (25°C) is used (Asphalt
Institute 1988):

$$\eta_{70,10^6} = 475,300 * \text{Pen}_{77}^{-2.93}$$

$$(5)$$

Finally, if the only information known about the asphalt
consistency is the general grade, either viscosity or penetration grade,

the values shown in Table 1 are used. In the event that asphalt
consistency data are not available, viscosity values are assumed on a
state-by-state basis; e.g., $2.5*10^6$ P (AC-20) for the State of Maryland.

Table 1 - Asphalt Viscosity at 70°F Based on Grade

Basis for Grade	Grade	Viscosity (70°F (21°C), 10^6 P)
Viscosity	AC-5	0.3
	AC-10	1.0
	AC-20	2.5
	AC-40	5.0
Penetration	60-70	2.5
	85-100	1.0
	100-120	0.5
	150-200	0.25
After Residue (AR)[1]	10	0.08
	20	0.3
	40	1.0
	80	2.5
	160	5.0

Note: [1]Viscosity values were established for viscosity and
penetration grade asphalt based on Witczak 1989; AASHTO M-226-80
correlations were used to establish values for the AR grades of
asphalt.

It has been assumed that a certain minimum amount of data --
aggregate grading, maximum and bulk specific gravity of mix, and asphalt
content -- are available for the computation of the initial modulus for
each AC layer in the pavement structure. In those cases where this
information is <u>not</u> available from the LTPP data base, the initial
modulus is computed using the following equation (Asphalt Institute
1988):

$$\log_{10}[E^*] = 5+0.553833+0.028829*p_{200}*f^{-0.17033}-0.03476*V_a+0.070377*\eta_{70,10^6}$$
$$+0.000005*[t_p^{(1.3+0.49825\log(f))}*p_{ac}^{0.5}]-0.00189[t_p^{(1.3+0.49825\log(f))}*p_{ac}^{0.5}*f^{-1.1}]+$$
$$0.931757f^{-0.02774}$$

$$(6)$$

where E^* = AC modulus, in psi; V_a = percent air voids in mix; f = test
frequency of load wave, in Hz (assume 16 Hz); t_p = test temperature; in
Fahrenheit (from Step No. 1); p_{200} = percent aggregate weight <u>passing</u> the
No. 200 sieve; p_{ac} = percent asphalt content by weight of mix; and, $\eta_{70,10^6}$
= asphalt viscosity at 70°F (21°C), in 10^6 P.
    If the information contained in the LTPP data base is not
sufficient to define one or more of the variables in Equation 6, the
following default values are used:

•   Percent air voids in mix, V_a: 4% for surface courses, 5% for
     binder courses, and 7% for base courses.

- Percent asphalt content by weight of mix, p_{ac}: 6% for surface courses, 5% for binder courses, and 4% for base courses; 8% for all sand asphalt mixtures.
- Percent aggregate weight passing the No. 200 sieve, p_{200}: 6% for surface courses, 5% for binder courses, and 4% for base courses; 6% for all sand asphalt mixtures.

The asphalt viscosity at 70°F (21°C), $\eta_{70,10^6}$, can be determined using any of the three procedures described earlier for the definition of this variable in Equation 1. If grade information is not present in the data base, viscosity values are assumed on a state-by-state basis.

3. Combine AC layers of same construction age--In general, backcalculation procedures are unable to handle individual AC construction lifts separately. As a consequence, in the SHRP backcalculation procedure, AC layers having the same construction age are combined into a single layer -- e.g., binder and surface course for an overlay or original surface layer are combined into one layer. The specific rules for combining AC layers are as follows:

- Add thicknesses of all AC layers having the same construction age, including any surface treatments:

$$h_{composite} \quad = \quad h(surface\ treatment)+h(surface\ course)+ \\ h(binder)+ \ldots +h_n \tag{7}$$

- Find initial composite modulus for the combination of AC layers having the same construction age:

$$E_{composite} = \sum \left[\left(\frac{h_i}{h_{composite}} \right) E_i^{\frac{1}{3}} \right] \tag{8}$$

where h_i = thickness of the "i"th layer; E_i = modulus of the "i"th layer (from Step No. 2); and i = 1 to n, where n is the number of AC layers having the same construction age. For example, if during construction, a 2 in. (51 mm) AC surface course with a modulus of 1,000,000 psi (6,900 MPa) is placed over a 3 in. (76 mm) AC binder course with a modulus of 500,000 psi (3,450 MPa), the composite modulus for the combined 5 in. (127 mm) AC layer is 673,000 psi (4,640 MPa). When surface treatments are present, their thickness should be included in the total surface thickness, but their presence should be ignored when determining the composite modulus value.

4. Define modulus range for AC layer(s)--Once the initial or composite modulus of each AC layer has been defined, the range of moduli is determined as follows:

$$Range \quad = \quad 0.25*E(initial\ or\ composite)\ to\ 3.00*E(initial\ or \\ composite) \tag{9}$$

The upper limit of the AC layer modulus range defined by the above relationship is not to exceed 3,000,000 psi (20,700 MPa).

<u>Portland cement concrete layers</u>--The procedure to define the modulus range for PCC layers is considerably simpler than that for AC layers. One reason for this is that more strength tests are being performed on PCC layer materials (static modulus, compressive strength, and splitting tensile strength). Most of this testing has been completed and the data is now stored in the LTPP data base. The other reason is that PCC moduli are not as temperature dependent as those for AC materials, thus the anticipated range of values can be more easily approximated in the absence of any information.

The specific set of rules used to define the range of moduli for PCC layers is as follows:

1. Determine initial modulus of PCC layer(s)--Depending on the type of laboratory strength data available, the initial PCC modulus is determined in the following priority order:

- If static modulus (E) test results are available, these values are used directly in the definition of the layer moduli range.
- If static modulus data are not available but compressive strength results are, the following equation is used to determine the initial modulus value (Hammitt 1974):

$$E = 57,000 * (f_c')^{0.5} \tag{10}$$

where E = PCC modulus in psi and f_c' = 28-day compressive strength in psi.

- If neither static modulus or compressive strength data are available but splitting tensile strength results are, the following equation is used (American Concrete Institute 1977):

$$f_c' = 12.53 * \text{Splitting tensile strength} - 1,275 \tag{11}$$

where the splitting tensile strength is in units of "psi". The resulting f_c' value is then entered into Equation 10 to estimate the initial PCC modulus value.

- If laboratory strength test data are <u>not</u> available, an initial modulus value of E = 4,000,000 psi (27,600 MPa) is used.

2. Define modulus range for PCC layer(s)--As with AC layers, once the initial modulus of each PCC layer has been established, the range of moduli is determined as follows:

$$\text{Range} = 0.25*E(\text{initial}) \text{ to } 3.00*E(\text{initial}) \tag{12}$$

Also, the upper limit of the PCC layer moduli range defined by the above relationship is not to exceed 9,000,000 psi (62,100 MPa).

<u>Base and subbase layers</u>--Although many models for estimating the modulus of unbound and stabilized base and subbase materials are available in the current literature, a somewhat simplistic approach is used in the SHRP backcalculation procedure to estimate the initial modulus and modulus ranges for these material types. One reason for taking this approach is that existing models do not cover the full range of material types encountered in the SHRP pavement test sections. Another reason is that many of the material parameters required by these models are not yet available from the LTPP data base, though eventually, many of them will be. For now, however, the following rules are used:

- For unbound granular base and subbase materials, the initial modulus and range of moduli are determined on the basis of material type as shown in Table 2. If the lower bound of the modulus range is greater than the initial subgrade modulus (discussed in the next section), use the latter as the lower bound instead of the above guidelines.
- For stabilized base and subbase layers, estimates of the initial modulus and range of moduli are based on unconfined compressive strength data, which are generally available from the LTPP data base. The recommended values are summarized in Table 3, according to the stabilizing agent used; if unconfined compressive strength data is lacking, a value of 400 psi (2,760 KPa) is assumed for lime stabilized layers, 700 psi (4,800 KPa) for asphalt stabilized layers, and 1000 psi (6,890 KPa) for cement stabilized materials.

Table 2 - Initial Modulus and Moduli Range for Unbound
Base and Subbase Materials

Material Type	Initial Modulus ksi (MPa)	Moduli Range ksi (MPa)
Crushed Stone, Gravel or Slag		
Bases	50.0 (345)	10.0 to 150.0 (70 to 1035)
Subbases	30.0 (205)	10.0 to 100.0 (70 to 690)
Gravel or Soil-Agg. Mix, Coarse		
Bases	30.0 (205)	10.0 to 100.0 (70 to 690)
Subbases	20.0 140)	5.0 to 80.0 (35 to 550)
Sand		
Bases	20.0 (140)	5.0 to 80.0 (35 to 550)
Subbases	15.0 (105)	5.0 to 60.0 (35 to 415)
Gravel or Soil-Agg. Mix, Fine		
Bases	20.0 (140)	5.0 to 80.0 (35 to 550)
Subbases	15.0 (105)	5.0 to 60.0 (35 to 415)

Subgrade layers--The initial subgrade modulus is estimated from
the composite moduli predicted for radial distances greater than the
effective radius, a_{3e}, of the stress bulb at the pavement-subgrade
interface; as indicated by the horizontal dashed line in Figure 3 for
linearly elastic subgrades or by the upward trend for non-linear (stress
dependent) subgrades. The composite modulus is a single value
representation of the overall pavement stiffness, at a given radial
distance, that combines the modulus of elasticity of all layers present
in the pavement.

The specific set of rules used in the SHRP backcalculation
procedure involves the following steps (Rada et.al. 1988):

- Calculate the composite modulus of the pavement at each radial
  distance beyond 5.91 in. (150 mm) -- i.e., 8, 12, 18, 24, 36, and
  60 in. (203, 305, 457, 610, 914 and 1524 mm) -- using the measured
  deflection data as input into Boussinesq's one-layer deflection
  equation:

$$E(comp) = \frac{p_c * a_c^2 * (1 - u^2) * C}{def * r} \tag{13}$$

where E(comp) = pavement composite modulus; p_c = contact pressure
applied by FWD; a_c = load plate radius (5.91 in. (150 mm)); u =
Poisson's ratio of subgrade, assume to be 0.4; def = measured
deflection at given radial distance "r" from center of load plate;
r = radial distance for deflection in question; and C = deflection
constant equal to:

$$C = 1.1 log\left(\frac{r}{a_c}\right) + 1.15 \tag{14}$$

Table 3 – Initial Modulus and Moduli Range for Stabilized
Base and Subbase Materials

Material Type	Unconf. Comp. Strength psi (KPa)	Initial Modulus ksi (MPa)	Modulus Range ksi (MPa)
Lime Stabilized	< 250	30.0 (205)	5.0 to 100.0 (35 to 690)
	250-500	50.0 (345)	10.0 to 150.0 (70 to 1035)
	> 500	70.0 (485)	15.0 to 200.0 (105 to 1380
Asphalt Stabilized	< 300	100.0 (690)	10.0 to 300.0 (70 to 2070)
	300-800	150.0 (1035)	25.0 to 800.0 (170 to 5515)
	> 800	200.0 (1380)	50.0 to 1500.0 (345 to 10350)
Cement Stabilized	< 750	400.0 (2760)	50.0 to 1500.0 (345 to 10350)
	750-1250	1000.0 (6900)	100.0 to 3000.0 (690 to 20685)
	> 1250	1500.0 (10350)	150.0 to 4000.0 (1035 to 27680)
Fractured PCC	-----	500.0 (345)	100.0 to 3000.0 (690 to 20685)
Others	-----	50.0 (35)	10.0 to 150.0 (90 to 1035)

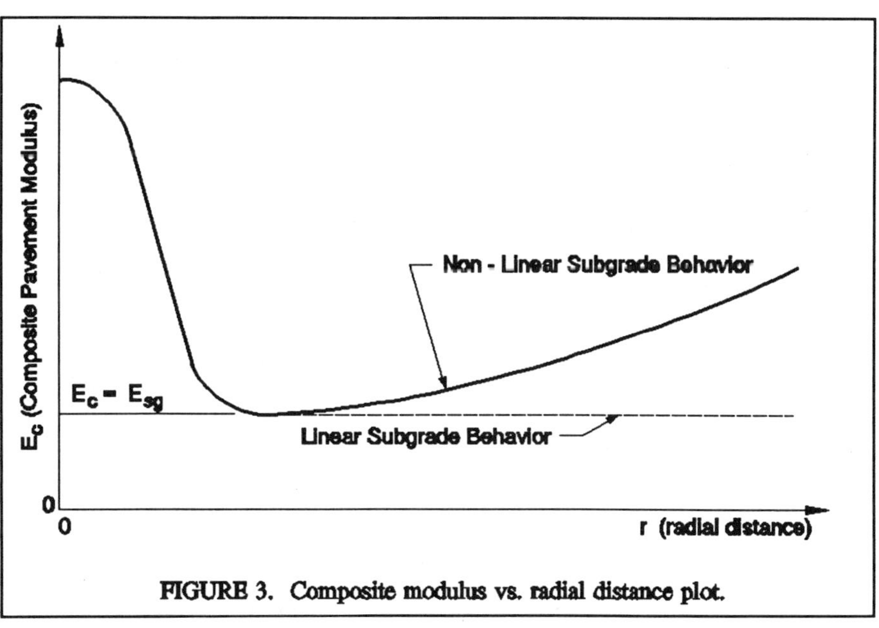

FIGURE 3. Composite modulus vs. radial distance plot.

- Assume that the initial subgrade modulus is equal to the minimum
  composite pavement modulus:

$$E(subgrade) \quad = \quad E(comp)_{minimum} \tag{15}$$

Note that the MODULUS program requires only an initial modulus value for
the subgrade, not a range of modulus.

<u>Modeling of Pavement Structure</u>

Along with known layer thicknesses, the layer moduli derived from
the SHRP backcalculation rules will provide much of the information
required to run the MODULUS program, but not all. Because the MODULUS
program is limited to a maximum of 4 unknown layers, prioritized
guidelines are required for combining two or more layers in pavement
structures with more than 4 layers. Likewise, rules for fixing layer
moduli in complex pavement structures, thin asphalt concrete layers, and
treated subgrade soils are also required. Another item that must be
covered by these rules is the assignment of a Poisson's ratio for each
pavement layer.
The specific set of rules used by the SHRP procedure for modeling
of pavement structures in backcalculation analyses is as follows:

<u>Subgrade layers</u>

- Lime, asphalt (mixed in place), or cement stabilized subgrade is
  treated as a subbase layer.
- If shoulder boring data or other similar information indicates
  that a rigid layer is present within 20 feet (6 m) of the surface,
  then the subgrade thickness is defined in accordance with this
  information. Otherwise, the MODULUS option to calculate the depth
  to an effective rigid layer is used; i.e., to look for rigid layer
  effects at depths of up to 50 feet[2] (15.2 m). If no rigid layer
  is found within this range, then the depth to rigid layer defaults
  to 50 feet (15.2 m).
- When analyzing a three-layer pavement system, the subgrade is
  modeled as two layers and the thickness of the top subgrade layer
  is assumed to be equal to 36 in. (914 mm). This is done to
  account for possible changes in subgrade modulus with depth due to
  such factors as the stress sensitivity of the subgrade soil,
  varying moisture conditions, etc. However, if the total subgrade
  thickness is less than 72 in. (1,829 mm)(due to presence of rigid
  layer) a single subgrade layer is used.

<u>Thin Layers</u>

- If the total thickness of an AC layer is less than 3 in., fix the
  modulus of this layer equal to that derived from Equation 1 or 6.
- If a thin layer ($\leq$ 2 in. (51 mm)) exists beneath PCC, neglect the
  modulus of this layer and combine its thickness with that of the
  underlying layer.

<u>Pavement Structure</u>--As indicated earlier, the maximum number of
layers (with known <u>or</u> unknown modulus) that can be modeled in the
MODULUS program is 4, exclusive of the effective rigid layer.   If more
than 4 layers are present, the prioritized list of rules given below are
used to reduce the number of layers included in the backcalculation
analysis.

---

[2]limit on the depth to effective rigid layer was changed from 25
feet (7.6 m) to 50 feet (15.2 m) in the SHRP backcalculation procedure.

1.   Combine adjacent granular base and subbase layers, <u>if</u> more than one is present <u>and</u> material types are similar (e.g., crushed stone base and crushed gravel subbase <u>not</u> crushed stone base and sand subbase). The total thickness for the composite layer is equal to the sum of the thicknesses for the adjacent layers, while the modulus range is defined by the combined range of the layers (i.e., largest maximum, smallest minimum).

2.   Combine adjacent AC layers of different construction dates, <u>if</u> more than one is present (e.g., overlay plus original surface). Use Equations 7, 8 and 9 to determine the total thickness and moduli range for the composite layer. If the total thickness is still less than 3 in. (76 mm), fix the modulus of this composite layer as that generated from Equation 8.

3.   Combine adjacent stabilized base and subbase layers, <u>if</u> more than one is present <u>and</u> material types are similar (e.g., cement stabilized base and subbase, <u>not</u> cement stabilized base and lime stabilized subgrade, which is treated as a subbase). The total thickness and modulus range for the composite layer is determined in the same fashion as in Item 1 above.

4.   Combine adjacent granular base and subbase layers, <u>if</u> more than one is present; material types <u>do not</u> have to be similar; e.g., crushed stone base and sand subbase. The total thickness and moduli range for the composite layer is determined in the same fashion as in Item 1 above.

5.   Combine adjacent subbase and subgrade layers, <u>if</u> material types are similar; e.g., sand subbase over sandy subgrade. If this done, the initial subgrade modulus should be used in the backcalculation analysis. The thickness of this combined layer will depend on whether or not a rigid layer (actual or effective) exists below the subgrade.

6.   Combine adjacent AC and asphalt treated layers, <u>if</u> more than one is present; e.g., original surface plus asphalt treated base. Use Equations 7, 8 and 9 to determine the total thickness and moduli range for the composite layer.

7.   Combine adjacent cement-stabilized and lime-stabilized base or subbase layers, <u>if</u> more than one is present; e.g., cement stabilized base and lime stabilized subgrade (treated as subbase). The total thickness and moduli range for the composite layer is determined in the same fashion as in Item 1 above.

<u>Poisson's Ratio</u>--Recommendations for assigning Poisson's ratios as a function of material type abound in the literature. Based on this information and recommendations by the Deflection Testing and Backcalculation ETG, the values shown in Table 4 have been selected for use in the SHRP backcalculation procedure.

Evaluation of Analysis Results

The third and final set of rules focus on the evaluation of the backcalculation results. Maximum allowable deflection matching error limits are established, both for the individual sensors as well as all sensors combined. Guidelines for checking the reasonableness of the results are also provided in these rules, along with procedures to be followed in case of bad or questionable data.

The specific rules for the evaluation of the results are as follows:

- All backcalculation results must be carefully reviewed by an engineer familiar with the backcalculation process.
- If the results fail the convexity test, the range of moduli must be widened (reduce lower bound by 50% and increase upper bound by 100%), and the backcalculation rerun. If the results are similar to those from the first run (within 10%), they should be accepted whether they pass the convexity test or not. If the results from the second run differ from those from the first run, but pass the

Table 4 - Poisson's Ratio as a Function of Material Type

Material Type	Poisson's Ratio

Asphalt Concrete
    E > 500 ksi (3,450 MPa)
    E < 500 ksi (3,450 MPa)
	
	0.30
	0.35

Portland Cement Concrete	0.15

Stabilized Base/Subbase
    Lime
    Cement
    Asphalt
    Other (stabilized subgrade)
    Other (fractured PCC)
	
	0.20
	0.20
	0.35
	0.35
	0.30

Granular Base/Subbase	0.35
Cohesive Subgrade	0.45
Cohesionless Subgrade	0.35

convexity test, they should be accepted. Otherwise, they are not
considered valid.
- Results having an average absolute arithmetic error[3] in excess of
  2% are not considered valid. This corresponds to a total sum of
  absolute error of 14% when all seven sensors are used in the
  backcalculation; 10% when only five sensors are used, and so on.
- Predicted moduli which hit the boundaries provided as input into
  the backcalculation are not considered valid.
- When the deflection errors fail to meet the 2% tolerance, the
  modulus results hit an upper or lower bound or, the results are
  considered "unreasonable" in the judgement of the reviewer, the
  engineer must look for obvious problems, by verifying the input
  data, comparing the results with laboratory data, and checking the
  distress film. In the absence of obvious errors, unacceptable
  results will be set aside for further evaluation.

Other Considerations

    Despite all of the above rules, the evolving nature of the science
(or art) of backcalculation makes it likely that early experience with
this procedure will bring to light areas where further refinement is
needed. Hence, it is anticipated that the initial release of the SHRP
backcalculation procedure will be followed up, as we learn more about
the strengths, weaknesses, and requirements of the process.
    While the initial analysis of the LTPP deflection data has not
been completed, it is anticipated that new rules will likely be added to
the existing SHRP backcalculation procedure and/or that existing ones
may be modified. For example, based on preliminary analysis results, it
is possible that the following rules will be implemented in the SHRP
backcalculation procedure:

- Using the same rules described earlier in the paper, fix the
  modulus of AC layers having thicknesses of 6 in. (152.4 mm) or
  less and constructed on portland cement concrete or other stiff
  materials (e.g., cement treated bases).

---

[3]average of absolute differences between measured and predicted
deflections for sensors used in backcalculation.

- In the case of portland cement concrete pavements, combine adjacent unbound base and subbase layers underneath the PCC slab. The total thickness for the composite layer is equal to the sum of the thicknesses for the combined layers, while the moduli range is defined by the combined range of the layers.
- In addition to AC layers, combine other "thin" material layers placed below PCC slabs with other adjacent base, subbase or subgrade layer.
- As more laboratory modulus test results become available, these data will be used to estimate the initial value and range of moduli for the various material types.

SUMMARY AND CONCLUSIONS

This paper focused on a standard procedure, developed around the MODULUS program , to ensure that the backcalculation process applied in the LTPP deflection data analysis is as consistent, productive, and straightforward as possible. The procedure consists of a rigorous set of application rules which rely on the wealth of information stored in the LTPP data base to generate the input -- modeling of pavement structure, layer moduli ranges, Poisson's ratios, etc. -- for the backcalculation program.

In conjunction with data base queries, the SHRP backcalculation rules are used to generate data files for direct input into the MODULUS program. As detailed in the paper, these rules are used to model the pavement structure and to establish initial moduli or moduli ranges for use in conjunction with measured deflections and loads in the analysis. Additional rules address the subsequent evaluation of the results.

The resulting SHRP backcalculation procedure is unique for several reasons: data availability, standardization (i.e., rules), and automation. It is also unique in that a group of experts (SHRP ETG on Deflection Testing and Backcalculation) was instrumental in the selection of the software and development of the rules.

Despite these rules and other unique aspects of the SHRP procedure, the evolving nature of the science (or art) of backcalculation makes it likely that early experience with this procedure will bring to light areas where further refinement is needed. Modifications and additions to the procedure presented herein will be made as necessary.

ACKNOWLEDGEMENTS

The work described in this paper was performed by PCS/Law Engineering under contract to the Strategic Highway Research Program, National Research Council. The publication of this article does not necessarily indicate approval or endorsement by the National Academy of Sciences, the United States Government or the American Association of State Highway and Transportation Officials or its member states of the findings, opinions, conclusions, or recommendations either inferred or specifically expressed herein. The authors gratefully acknowledge the cooperation and assistance of the SHRP Expert Task Group on Deflection Testing and Backcalculation, chaired by Dr. A.J. Bush, III and others that participated in the software evaluation and development of rules.

REFERENCES

ACI Committee 318, "Building Code Requirements for Reinforced Concrete (ACI-318-77)", American Concrete Institute, Detroit, Michigan, 1977.

Asphalt Institute, "Mix Design Methods for Asphalt Concrete and Other Hot-Mix Types", Manual Series No. 2, College Park, Maryland, 1988.

Hammitt, G.A., "Concrete Strength Relationships", Miscellaneous Paper S-74-30, U.S. Army Engineer Waterways Experiment Station, Vicksburg, Mississippi, December 1974.

Rada, G.R., Witczak, M.W. and Rabinow, S.D., "A Comparison of AASHTO Structural Evaluation Techniques using NDT Deflection Testing", TRB, Transportation Research Record 1207, Washington, D.C., 1988.

Strategic Highway Research Program, "SHRP Layer Moduli Backcalculation Procedure - Software Selection", SHRP Technical Report, July 1991.

Strategic Highway Research Program, "Manual for FWD Testing in the Long Term Pavement Performance Study: Operational Field Guidelines, Version 2.0", SHRP Manual, March 1993.

M.W. Witczak, "The Universal Airport Pavement Design System - Report II: Asphaltic Mixture Material Characterization", Department of Civil Engineering, University of Maryland, College Park, Maryland, May 1989.

Norris Stubbs[1], Vikram S. Torpunuri[2], Robert L. Lytton[3], and Allen H. Magnuson[4]

A METHODOLOGY TO IDENTIFY MATERIAL PROPERTIES IN PAVEMENTS MODELED AS LAYERED VISCOELASTIC HALFSPACES (THEORY)

REFERENCE: Stubbs, N., Torpunuri, V. S., Lytton, R. L., and Magnuson, A. H., "A Methodology to Identify Material Properties in Pavements Modeled as Layered Viscoelastic Halfspaces (Theory)," Nondestructive Testing of Pavements and Backcalculation of Moduli (Second Volume), ASTM STP 1198," Harold L. Von Quintas, Albert J. Bush, III, and Gilbert Y. Baladi, Eds., American Society for Testing and Materials, Philadelphia, 1994.

ABSTRACT: An accurate, rapid, and reliable methodology for identifying material properties of pavements modeled as layered viscoelastic halfspaces subjected to dynamic excitation is developed. The methodology is based on the forward model of system identification. The Modulus of elasticity of each layer, the damping in each layer, and the slope of the creep curve for the top layer are selected as the parameters to be identified. The validity and accuracy of the proposed methodology is demonstrated by several numerical experiments involving simulated changes in the selected pavement parameters. Iterative procedures developed to solve nonlinear problems which arise in estimating the pavement parameters are presented.

KEYWORDS: systems identification, viscoelasticity, pavements, multilayered, falling weight deflectometer

---

This paper addresses the general problem of nondestructively identifying the properties of multilayered pavement systems. An accurate nondestructive identification system for pavement properties could positively impact, among other things, the efficiency of pavement

---

[1]Associate Professor, Civil Engineering Department, Texas A&M University, College Station, Texas 77843.

[2]Research Associate, Texas Transportation Institute, Texas A&M University, College Station, TX 77843.

[3]Professor, Civil Engineering Department, Texas A&M University, College Station, TX 77843.

[4]Engineering Research Associate, Texas Transportation Institute, Texas A&M University, College Station, TX 77843.

maintenance, the allocation of future resources, and the safety of the public. During the past two decades, research intensity in the area of backcalculation of multilayered pavement properties has grown dramatically. Brown and Bush (1972) investigated the validity of applying linear elastic layered theory of pavements. Classen et al. (1976) derived structural properties of a pavement from the surface deflection and the shape of the deflection bowl under a test load. Vaswani (1977) used the deflection shape and size of the deflected basin to evaluate the modulus of two layers. Southgate et al. (1982) nondestructively evaluated the pavement subgrade modulus with a dynamic test and Nazarian et al. (1983) utilized spectral analysis of surface waves to determine the elastic properties of layered media. More recently, Roesset and Shao (1985) compared deflections obtained from dynamic and static tests and studied the implication of any differences. Practical curves and tables derived from Dynaflect measurements to evaluate pavement moduli were developed by Sebaaly et al. (1986).   Ali and Khosla (1987) conducted comparative studies of material properties identified by backcalculation using Falling Weight Deflectometer (FWD) data, and laboratory measurements. Most recently, Magnuson et al. (1991) demonstrated the extraction of certain engineering properties using dynamic analysis of FWD data; Parker (1991) conducted a study to develop methods for using FWD measurements to determine pavement in situ moduli; and Sivaneswaran et al. (1991) described a nonlinear least squares convergence method to back calculate both the layer moduli and the thickness.

An analysis of the foregoing sample of works suggests that no single approach simultaneously satisfies the following four conditions: a) the fact that FWD loads the pavement dynamically; b) that the dynamically-loaded deflection basin may differ significantly from the statically computed basin; c) the fact that pavement materials behave viscoelastically; and d) the method used a proven and systematic identification procedure. The objective of this paper is to demonstrate the feasibility of such a systems identification approach for layered viscoelastic pavement systems using dynamic measurements. The approach assumes a particular mathematical model of the pavement system and utilizes deflections obtained from a Falling Weight Deflectometer (FWD). The paper is organized into three main sections: A description of the mechanical model of the pavement system, a development of the system identification algorithm, and a presentation of several examples to illustrate the utilization of the algorithm.

DESCRIPTION OF PAVEMENT MODEL

A schematic of the generic multilayered halfspace model of a pavement system is shown in Figure 1. The structure includes three finite layers and a half space. The thickness of layer i is given by t_i and sensors, 0, 1...6 on the surface represent appropriatly spaced geophones associated with the FWD apparatus. In the general case, each layer i is assumed to be modeled as a linear viscoelastic material with complex modulus $E^*_i(\omega)$ where ω is the angular frequency. In this paper the following properties will be assigned to the layers. The creep behavior for the top asphaltic concrete layer is assumed to follow the two-parameter power law representation (Magnuson et al. 1991). The

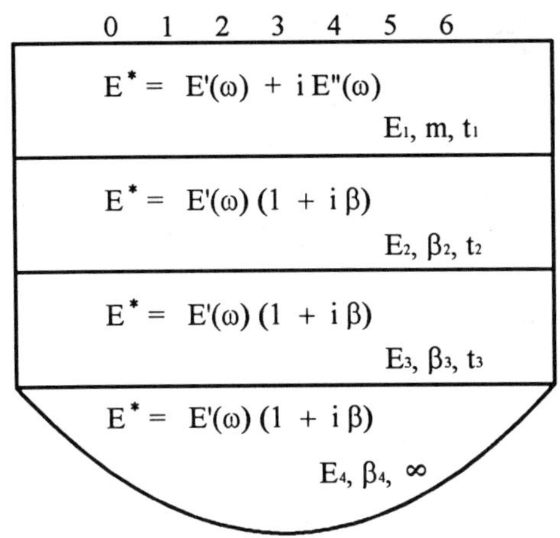

FIG. 1--Schematic of Multilayered Halfspace Model

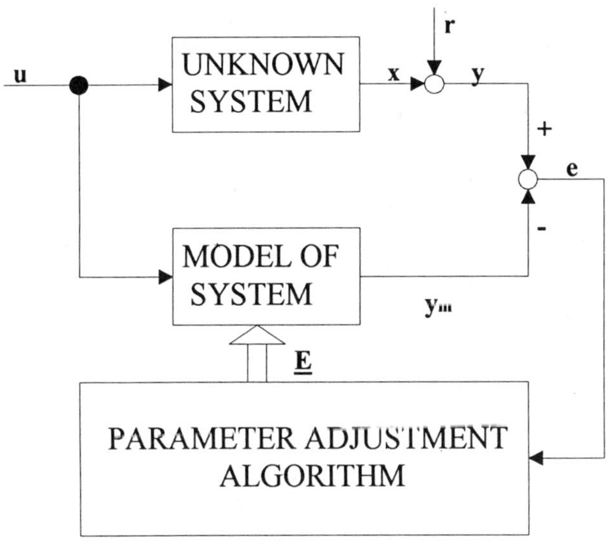

FIG. 2--Schematic of the Forward Identification Paradigm Showing the
Parameter Adjustment Algorithm

complex modulus for the top layer (i.e., the asphaltic concrete layer)
will be given by the model

$$E(\omega)^* = E'(\omega) + iE''(\omega) \tag{1}$$

where

$$E'(\omega) = E_1\Gamma(1-m)\,\omega^{-m}Cos(\Pi m/2), \tag{2}$$

$$E''(\omega) = E_1\Gamma(1-m)\,\omega^{-m}Sin(\Pi m/2), \tag{3}$$

and

$$2\beta = E''(\omega)/E'(\omega) = Tan(\Pi\frac{m}{2}) \tag{4}$$

Note that in equations (2) to (4), $i = \sqrt{-1}$, $\Gamma(\bullet)$ is the Gamma Function, m
is the slope of the creep compliance curve, and β is the frequency-
dependent damping ratio. For the remaining layers, the complex modulus
is given by

$$E^*(\omega) = E(1+i2\beta) \tag{5}$$

which assumes that the damping is frequency independent. Thus the
parameters to be determined are E_1 and m for the top layer, and E_i, β_i
($i = 2,3,4$) for the remaining layers.

THE SYSTEM IDENTIFICATION ALGORITHM

The Problem

The problem to be solved here can be stated as follows: Given a
real pavement system that is modeled with the idealized system given in
Figure 1 and given a set of data from an FWD apparatus, find the
parameters m_1, E_1, E_2, E_3, E_4, β_2, β_3, and β_4, assuming that the layer
thicknesses are known.

The Selected System's Identification (SID) Approach

The rationale behind the SID approach to be used here is summarized
in Figure 2. In the figure input, U, represents the physical excitation
from the FWD to the pavement. The measured outputs, y, from the
geophones include the true signal, x and the noise, r. The response of
the pavement model to the FWD excitation is Y_m and the difference
between the measured output y and the calculated output Y_m is denoted by
e. Specialized computer programs (e.g., PUNCH (Kausel, Roesset, and
Waas, 1974; Kausel, and Roesset, 1981; and Kausel, 1986), UTFWIBM
(Kausel, Roesset, and Waas, 1974; Kausel, and Roesset, 1981; and Kausel,
1986), or SCALPOT (Magnuson, 1988) are used to compute the model
response, given the input U. The Parameter Adjustment Algorithm accepts
the error, e, as input and generates updated values of parameters for
the model. The process continues until a convergence criterion on the
error is satisfied.

The Parameter Adjustment Algorithm

Let the mathematical model of some process be defined by N scalar parameters E_i ($i = 1, \ldots N$). Then any scalar function $f_k(E_1, E_2, \ldots E_n;$ $X_k, t_k)$, where X_k and t_k are, for example, independent spatial and temporal variables, may be expanded in a Taylor series such that

$$f_k(\tilde{E} + d\tilde{E}) = f_k(\tilde{E}) + (\nabla \cdot d\tilde{E}) f_k + \frac{1}{2} (\nabla \cdot d\tilde{E})^2 f_k + \ldots +$$

$$\frac{1}{N!} (\nabla \cdot d\tilde{E})^n f_k + R_n \tag{6}$$

Note that the parameters E_i have been collected into the vector E, ∇ is a vector with components $\partial/\partial E_1$, $\partial/\partial E_2$, $\ldots$ $\partial/\partial E_N$, and scalar R_n is the remainder. If we retain only first order terms, Equation (6) reduces to, in vectorial notation;

$$f_k(\tilde{E} + \tilde{d}E) \cong f_k(\tilde{E}) + \nabla f_k \cdot \tilde{d}E + 0^2(\tilde{d}E) \tag{7}$$

If we equate $f_k(E + dE)$ with the true (actual) output of a real system (e.g., the pavement) and $f_k(E)$ with the output of the mathematical model of the real system for most recent set of parameters, then the error, e_k, between the two outputs becomes (keeping only first order terms in a Taylor series expansion)

$$\tilde{e}_k = f_k(\tilde{E} + d\tilde{E}) - f_k(\tilde{E})$$

$$= \sum_{i=1}^{N} \frac{\partial f_k}{\partial E_i} dE_i \tag{8}$$

Note that e_k represents the difference between the actual performance of the system and the model performance when the independent variables take on values X_k and t_k. If the error is evaluated at $m \geq N$ values of the independent parameters, m equations in N unknowns may be written ($j = 1, \ldots m$)

$$e_j = \sum_{i=1}^{N} \frac{\partial f_j}{\partial E_i} dE_i \tag{9}$$

Equations (9) may be conveniently nondimensionalized by dividing both sides by $|f_k|$, the absolute value of each f_k, to give

$$\frac{e_k}{|f_k|} = \sum_{i=1}^{N} \frac{\partial f_k}{\partial E_i} \frac{|E_i|}{|f_k|} \frac{dE_i}{|E_i|} \tag{10}$$

On setting

$$z_k = e_k/|f_k|, \tag{11}$$

$$\alpha_i = dE_i/|E_i|, \tag{12}$$

and

$$f_{ki} = \frac{|E_i|}{|f_k|} \frac{\partial f_k}{\partial E_i} \tag{13}$$

the set of m equations (see Equation 10) (k = 1,.....,m) may be
rewritten as:

$$z_k = \sum_{i=1}^{N} f_{ki} \, \alpha_i \tag{14}$$

If we further define the load vector, z, the change vector, a, and the
sensitivity matrix, F, as $z = [z_1, z_2 \ldots z_m]^T$, $a = [\alpha_1, \alpha_2, \ldots \alpha_N]^T$, and
$F = (f_{ki})$ respectively, Equations (14) may be rewritten in matrix form
as:

$$z = Fa \tag{15}$$

Equation (15) forms the foundation for this study. The vector z is
completely determined from the outputs of the mathematical model and the
physical system (ie. pavement) under study. The sensitivity matrix F is
completely determined from the gradient of the scalar functions (f_k),
the current values of the parameters (E_i), and the value of the function
(f_k). Only the matrix 'a' in Equation (15) is unknown.

Considering Equation (15), the matrix of changes in parameters, a,
can be obtained by solving Equation (15) to give

$$a = F^{-1} z \tag{16}$$

where F^{-1} is the classical or true inverse of matrix F if m = N. In
the present case, since m is not necessarily equal to N, the true
inverse of F is replaced by the matrix F^{-I}, which is often called the
"Generalized Inverse" or "Pseudoinverse".

If F is real mxn matrix and m>n (i.e. an overdetermined system) the
pseudoinverse may be given by

$$F^{-I} = (F^T F)^{-1} F^T \tag{17}$$

Note that Equation (17) defines the left pseudoinverse of F, since

$$F^{-I}F = I \qquad (18)$$

If F is a real mxn matrix and m<n (i.e. the underdetermined system) the pseudoinverse may be given by

$$F^{-I} = F^T(FF^T)^{-1} \qquad (19)$$

while the solution to Equation (19) using this approach is not unique, the solution given by Equation (19) is the minimum norm solution. Equation (19) defines the right pseudoinverse of F since

$$FF^{-I} = I \qquad (20)$$

Once the α_i's have been obtained, the updated values of the parameters are given by

$$E_i^{New} \cong E_i^{Old}(1+\alpha_i) \quad (i=1,\ldots,N) \qquad (21)$$

where E_i^{Old} are the initial set of parameters and E_i^{New} are the updated set of parameters.

<u>Application of Methodology</u>

To identify a pavement system, six major problems must be solved: (1) Compute the output to an initial reference pavement model; (2) Compute a sensitivity matrix relative to the referenced pavement model; (3) Compute the load vector for the real pavement relative to the pavement model; (4) Solve for the apparant differences in parameters between the real pavement and the referenced pavement model; (5) Update the parameters of the reference pavement model; and (6) Check for convergence.

The procedure described below is used to solve the first problem (i.e., compute the output to the referenced system). First, we select a set of initial values of the parameters that define the model. These 'seed' values may be selected either on the basis of experience or from the results of other backcalculation procedures. Second, we model the input to the pavement system. This input will consist of the time history of the impulsive load from the Falling Weight Deflectometer. Third, we select the points on the model at which the output is to be computed to be coincident with the locations of the geophones in the FWD test. Fourth, we select a numerical procedure that is capable of handling the material description, the input loads, and computing the displacements, at the desired locations as a function of time. Here we select the program SCALPOT (Scalar Potential) developed by Magnuson (1991). The input by SCALPOT consists of the geometrical configuration of the FWD apparatus, the physical properties of each pavement layer and the time-dependent surface pressure distribution. The layer properties include thickness, density, viscoelastic parameters, Poisson's ratio and Young's modulus. The output to SCALPOT consist of the vertical surface deflections at the locations of the geophones resulting from the pressure distribution. Fifth, we compute the vertical displacements at the sensor locations. Finally, we express the output in a form appropriate to the current system's identification problem. Such a form, which guarantees that the physics of the system are accounted for, is the complex frequency response functions $H_{1i}(\omega)$ which relate the displacement response at geophone i (i = 1,...,7) to the impulse from the FWD apparatus at Location 1. Thus, for a given location, at each frequency we have two pieces of information corresponding to the real and imaginary part of the frequency response function. Here we select the magnitude and the phase angle, denoted as $h_{mi}{}^R(\omega)$ and $h_{\phi i}{}^R(\omega)$, respectively, as the desired output. In the latter sentence note that the superscript R denotes that the response is associated with the reference model. In this formulation we have eight parameters to identify; therefore, selecting as output the frequency response, at all seven locations, at a single frequency yields 14 pieces of information and thus provides us with an overdetermined system (See Equation 17).

The procedure described below is used to solve the second problem (i.e., compute the sensitivity matrix). Note from Equation (14) that if we perturb the jth parameters such that $\alpha_j = \alpha \neq 0$ and all other $\alpha_i = 0$ (i $\neq$ j), then it follows that

$$f_{kj} = Z_k/\alpha_j = Z_k/\alpha \qquad (22)$$

Therefore, if we begin with our reference pavement model, introduce a perturbation α_j, compute the corresponding relative sensitivity Z_k, then Equation (22) provides the sensitivities for the jth column of the sensitivity matrix F. If we repeat the procedure for each parameter we can, therefore, define the entire matrix. In the latter paragraph we explained how to obtain $h_{mi}{}^R(\omega_o)$ and $h_{\phi i}{}^R(\omega_o)$ where ω_o is a frequency to be selected. If we perturb only the jth parameter of the model, we can compute the outputs $h_{mi}{}^j(\omega_o)$ and $h_{\phi i}{}^j(\omega_o)$ where superscript j refers to the output with the jth parameter perturbed. For Location i we can now define the contribution to the load vector to be

$$a_i = (h_{mi}^j - h_{mi}^R)/h_{mi}^R \qquad (23)$$

$$b_i = (h_{\phi i}^j - h_{\phi i}^R)/h_{\phi i}^R \qquad (24)$$

The fourteen components Z_k associated with the seven sensors can now be collected into the matrix defined by

$$Z^j = [a_1 b_1 a_2 b_2 \ldots\ldots a_7 b_7]^T \qquad (25)$$

The procedure for solving the third problem (i.e., compute the load vector for the real pavement relative to the pavement model) is similar to the procedure used in the latter paragraph. First the FWD data is processed to provide field versions of the transfer functions. The magnitude and phase of these transfer functions at location i will be denoted by $h_{mi}{}^P(\omega_o)$ and $h_{\phi i}{}^P(\omega_o)$. The load vector is then defined as in Equations 23-25.

With the sensitivity matrix F and the load vector Z now defined Equation (16) can now be solved to yield the difference between the parameter values in the model and the pavement. Equation (21) can then be used to update the model.

The following procedure is used to check for convergence. First, the reference pavement model is updated using the latest parameter values. Next, an output from the update model is determined and a new load vector for the pavement relative to the updated model computed. If the size of the load vector is less than some small number, we stop the calculation and take the existing parameter values as the values for the real pavement. If the convergence criterion is not satisfied we repeat steps 1, 3, 4, and 5 (see the 1st paragraph of this subsection) and test again. Several such iterations may be necessary before convergence. If after using the latter strategy the problem still does not converge, the sensitivity matrix can be updated using the latest set of parameter values and the procedures in steps 1, 3, 4, and 5 repeated.

EXAMPLES

In this section three examples will be given to illustrate the proposed procedure. The examples have been designed to demonstrate the rate of convergence as a function of the nearness of the values of the initial parameters to the target values. The selected pavement section is shown in Figure 1 and the geometry and other physical properties which are assumed to be known are listed in Table 1. For each problem, values of all parameters needed to define the problem will be assumed to be known and SCALPOT will be used to simulate the response from a FWD test. Assuming only that the results from the FWD test are available we will use the proposed procedure to identify the following parameters: m_1, E_1, β_2, E_2, β_3, E_3, β_4, and E_4. In all problems the frequency response function will be evaluated at a frequency of 48.89 HZ, a value which falls within the practical spectrum of 0 ~ 150 HZ. During initial simulations of the response of the system described in Table 1 we found that the output at all the geophones were insensitive to changes in the model parameters β_2 and β_3, the damping in layers 2 and 3, respectively. Therefore, we ignored the latter two parameters and focused on identifying the six viscoelastic parameters given in the 6X1 matrix: $E = (E_1\ E_2\ E_3\ E_4\ m_1\ \beta_4)^T$.

<u>Example I</u>

Assume a pavement whose FWD response is given by a SCALPOT model with geometry and density given in Table 1 and viscoelastic properties given by $(0.8E_1^\circ\ 1.1E_2^\circ\ 0.85E_3^\circ\ 1.15E_4^\circ\ 0.8m_1\ 0.85\beta_4)^T$. Identify the viscoelastic parameters of the model in Figure 1 starting with a reference pavement with viscoelastic parameters given by the matrix $(E_1^\circ\ E_2^\circ\ E_3^\circ\ E_4^\circ\ m_1^\circ\ \beta_4^\circ)^T$ where the latter quantities are defined in Table 1. Note that in this problem the target values differ from the starting values by a maximum of ±20%.

Table 1--<u>Reference Pavement Section Characteristics</u>

Layer	Thickness (m)	Modulus GPa	Poisson Ratio	Wt. Density GN/m^3	Damping Percent
Asphaltic Concrete	0.229	$7.584(E_1^\circ)$	0.33	0.021	$20(m_1^\circ)$
Crushed Stone	0.178	$0.310(E_2^\circ)$	0.35	0.019	1.6
Lime-Treated Clay	0.254	$0.180(E_3^\circ)$	0.38	0.019	4.7
Subgrade-Clay	∞	$0.15(E_4^\circ)$	0.40	0.019	$7.5(\beta_4^\circ)$

To solve this problem, the FWD data was simulated and frequency response functions calculated. The viscoelastic parameters in Table 1 were taken as reference values. The solution procedure utilized here is equivalent to the Secant Method and can be summarized as follows:

Step 1.   The load vector, z and the sensitivity matrix, F, are computed from initial (reference) parameters and from simulated parameters;

Step 2.   The system of equations is solved using Equation (16) and updated using Equation (21);

Step 3.   Using the updated parameters in SCALPOT, load vectors are recomputed;

Step 4.   Using the same sensitivity matrix, F, and the recomputed load vectors the system of equations is solved and updated using Equation (21);

Step 5.   Steps 1 through 4 are repeated using the same sensitivity matrix, until the parameters converge.

The results are shown in Table 2. Note that the system converged after the 4th iteration. Note that superscript T refers to the target value of the parameter.

TABLE 2--<u>Convergence Rate for Example I</u>

	Parameter Ratios							
Iteration No.	m_1/m_1^T	E_1/E_1^T	β_2/β_2^T	E_2/E_2^T	β_3/β_3^T	E_3/E_3^T	β_4/β_4^T	E_4/E_4^T
Start	1.25	1.25	*	0.91	*	1.18	1.18	0.87
1	1.02	0.93	*	1.15	*	0.88	1.03	1.02
2	1.00	0.99	*	1.01	*	0.98	.94	.99
3	1.00	1.00	*	.98	*	1.01	.99	1.00
4	1.00	1.00	*	.99	*	1.00	1.00	1.00
5	1.00	1.00	*	1.00	*	1.00	1.00	1.00

Example II

Assuming the same starting point as the one considered for Example I, and given in Table I, identify the pavement whose FWD response is given by a SCALPOT Model with parameters given by the matrix $(0.5E^\circ_1\ E^\circ_2\ 1.25E^\circ_3\ 0.7E^\circ_4\ m^\circ_1\ 1.3\beta^\circ_4)^T$. Note that in this problem the parameters have been changed in the range ±50%.

The same procedure as in Example I was initially used to solve the problem, however, convergence was extremely slow. To speed up convergence, we solved the problem using a combination of the Secant and Tangent Method. The combined iterative procedure is as follows:

Step 1.   The load vector, z, is subdivided into r finite load vectors Δz_i;

$$z = \sum_{i=1}^{r} \Delta z_i$$

Step 2.   The sensitivity matrix, F, is computed using the reference parameters;

Step 3.   Using sensitivity matrix, F, and finite load vector, Δz_1, the system of equations is solved using Equation (16) for changes in parameters;

Step 4.   The parameters are updated using Equation (21);

Step 5.   Using the same sensitivity matrix, F, and updated parameters, steps 1 through 5 of Secant Method are repeated until parameters converge;

Step 6.   Using the converged parameters in SCALPOT the sensitivity matrix, F, is recomputed; and

Step 7.   Using the updated sensitivity matrix and the next finite load vector Δz_2 the above steps 3 through 6 are repeated for all finite load vectors. The parameters obtained using Δz_r are the converged parameters of the system.

The Results are shown in Table 3 in which it can be seen that the problem converged after 6 iterations.

TABLE 3--<u>Convergence Rate for Example II</u>

Iteration No.	Parameter Ratios							
	m_1/m_1^{T}	E_1/E_1^{T}	$\beta_2/\beta_2^{\mathsf{T}}$	E_2/E_2^{T}	$\beta_3/\beta_3^{\mathsf{T}}$	E_3/E_3^{T}	$\beta_4/\beta_4^{\mathsf{T}}$	E_4/E_4^{T}
Start	1.00	2.0	*	1.0	*	0.8	0.77	1.43
1	1.02	1.39	*	1.25	*	0.83	.23	1.27
3	0.998	1.02	*	0.71	*	0.97	.68	1.00
4	1.009	1.01	*	1.02	*	0.99	1.04	1.01
5**	0.998	.997	*	0.998	*	1.01	1.01	0.996
7	1.00	1.00	*	1.00	*	1.00	1.00	0.998

**Note: For iterations 1-4 ΔZ_1 = Z/3; after iteration 4, the sensitivity matrix was updated and $\Delta Z_2/$ = 2Z/3.

<u>Example III</u>

Assuming the same starting point as in Example II, identify the pavement whose FWD response is given by a SCALPOT Model with parameters given by the matrix ($0.25E_1^{\circ}$ $1.5E_2^{\circ}$ $0.5E_3^{\circ}$ $2E_4^{\circ}$ $0.5m_1^{\circ}$ $1.5\beta_4^{\circ\mathsf{T}}$). Note that in this problem the parameters have been changed in the range ±100% of the referenced values.

The solution procedure for this problem was identical to that of Example 2. The results are provided in Table 4 in which it can be seen that the problem has converged after 7 iterations.

TABLE 4--<u>Convergence Rate for Example III</u>

Iteration No.	m_1/m_1^{T}	E_1/E_1^{T}	$\beta_2/\beta_2^{\mathrm{T}}$	E_2/E_2^{T}	$\beta_3/\beta_3^{\mathrm{T}}$	E_3/E_3^{T}	$\beta_4/\beta_4^{\mathrm{T}}$	E_4/E_4^{T}
Start	2.0	4.0		0.67		2.0	0.67	0.5
1	2.27	1.48	*	1.47	*	0.79	1.16	0.73
3	1.75	1.23	*	1.08	*	1.03	.97	0.86
5	1.33	1.12	*	.78	*	1.24	1.39	.93
6**	1.08	1.02	*	.95	*	1.03	1.11	.98
8	1.01	1.00	*	.98	*	1.04	1.03	1.00

**Note: For iterations 1-4, $\Delta Z_1 = Z/4$; after iteration 4 sensitivity matrix was updated and $\Delta Z_2 = 3Z/4$.

Discussion of Results

The results of Example I confirm that small changes can be predicted by the proposed theory with a sufficient degree of accuracy. The predicted values approached the simulated values within a maximum error well below 1 percent.

The results of Example II and Example III indicate the limits of linear behavior. Since the fractional change was significant, the linear approximation of the sensitivity matrix is apparently inadequate and numerical techniques for nonlinear systems e.g. the Tangent Method had to be utilized. The only impact this addition had was to increase the number of iterations. The method continued to identify the parameters in systems that were quite different from the initial system.

Finally, we wish to make two points regarding the accuracy and the uniqueness of the results predicted by the algorithm. First, for all starting values the algorithm converged to the correct parameter values. Second, for each of the three examples presented here the algorithm yielded a single solution.

SUMMARY AND CONCLUSIONS

A multilayered halfspace was selected to model a pavement section. A System's Identification procedure developed to identify the parameters of a typical pavement system. A new program, SCALPOT, was selected as

the appropriate computer algorithm to numerically realize the model.
Simulated experiments were designed and the methodology to verify the
theory of parameter identification for multilayered halfspaces based on
these experiments was presented. Sensitivity matrices, load vectors,
and change vectors were generated for the simulated experiments. In the
process, the iterative procedures for linear and nonlinear problems were
implemented in estimating the pavement parameters. The proposed
approach for parameter identification for multilayered viscoelastic
halfspaces was validated for the simulated experiments.

References

Ali, N. A., and Khosla, N. P., 1987, "Determination of Layer Moduli
    Using a Falling Weight Deflectometer," _Transportation Research
    Record 1117_, Transportation Research Board, National Research
    Council, Washington, D.C., pp. 1-10.

Brown, S. F., and Bush, D. I., November 1972, "Dynamic Response of Model
    Pavement Structure," _Proceedings of the ASCE, Transportation
    Engineering Journal_.

Claessen, A. I. M., Valkering, C. P., Ditmarsch, R., 1976, "Pavement
    Evaluation with the Falling Weight Deflectometer," _Proceedings of
    the Association of Asphalt Paving Technologists Technical Sessions_,
    Vol. 45, New Orleans, Lousiana.

Kausel, E., 1982, "Wave Propagation in Anisotropic Layered Media",
    International Journal for Numerical Methods in Engineering, Vol. 71,
    pp. 1743-1761.

Kausel, E., and Roesset, J.M., 1982, "Stiffness Matrices for Layered
    Soils",_Bulletin of the Seismological Society of America_, Vol. 71,
    pp. 1743-1761.

Kausel, E., Roesset, J.M., and Waas, G., 1974, "Forced Vibrations of
    Circular Foundations on Layered Media, Report R74-11, Department of
    Civil Engineering, Massachusettes Institute of Technology,
    Cambridge, Massachusetts.

Magnuson, A. H., 1988, "Dynamic Analysis of Falling Weight Deflectometer
    Data, " _Part 1: Research Report_, No. 1215-1F, Texas Transportation
    Institute, College Station, Texas.

Magnuson, A. H., Lytton, R. L., and Briggs, R. C., 1991, "Comparison of
    Computer Predictions and Field Data for Dynamic Analysis of Falling
    Weight Deflectometer Data, " _Transportation Research Record 1293_,
    Transportation Research Board, National Research Council,
    Washington, D.C., pp. 61-71.

Nazarian, S., Stokoe II, K. H., and Hudson, W. R., 1983, "Use of
    Spectral Analysis of Surface Waves Method for Determination of
    Moduli and Thicknesses of Pavement Systems, " _Transportation
    Research Record 930_, Transportation Research Board, National
    Research Council, Washington, D.C., pp. 38-45.

Parker, Jr., F., 1991, "Estimation of Paving Materials Design Moduli from Falling Weight Deflectometer Measurements," _Transportation Research Record 1293_, Transportation Research Board, National Research Council, Washington, D.C., pp. 42-51.

Roesset, J. M., and Shao, K. Y., 1985, "Dynamic Interpretation of Dynaflect and Falling Weight Deflectometer Tests," _Transportation Research Record 1022_, Transportation Research Board, National Research Council, Washington, D.C., pp. 7-16.

Sebaaly, E. B., and Mamlouk, M. S., 1986, "Typical Curves for Evaluation of Pavement Stiffness from Dynaflect Measurements, " _Transportation Research Record 1070_, Transportation Research Board, National Research Council, Washington, D.C., pp. 42-52.

Sivaneswaran, N., Kramer, S. L., and Mahoney, J. P., 1991, "Advanced Backcalculation Using a Nonlinear Least Squares Optimization Technique, " _Transportation Research Record 1293_, Transportation Research Board, National Research Council, Washington, D.C., pp. 93-102.

Southgate, H. F., Sharpe, G. W., Deen, R. C., Havens, J. H., August, 1982, "Structural Capacity of In-Place Asphaltic Concrete Pavements from Dynamic Deflections," _Proceedings,_ Fifth International Conference on Structural Design of Asphalt Pavements, Vol. 1.

Torpunuri, V.S., 1990, "A Methodology to Identify Material Properties in Layered Viscoelastic Halfspaces," Master's Thesis, Department of Civil Engineering, Texas A&M University, College Station, Texas.

Vaswani, N. K., January 1977, "Determining Moduli of Materials from Deflections," _Transportation Engineering Journal_, No. 12673, pp. 125-141.

Ronald S. Harichandran,[1] Tariq Mahmood,[2] A. Robert Raab,[3]
and Gilbert Y. Baladi[4]

BACKCALCULATION OF PAVEMENT LAYER MODULI, THICKNESSES AND STIFF LAYER
DEPTH USING A MODIFIED NEWTON METHOD

REFERENCE: Harichandran, R. S. Mahmood, T., Raab, A. R., and Baladi,
G. Y., "Backcalculation of Pavement Layer Moduli, Thicknesses and Stiff
Layer Depth Using a Modified Newton Method," Nondestructive Testing of
Pavements and Backcalculation of Moduli (Second Volume), ASTM STP 1198,
Harold L. Von Quintas, Albert J. Bush, III, and Gilbert Y. Baladi, Eds.,
Amercian Society for Testing and Materials, Philadelphia, 1994

ABSTRACT: A modified Newton method for the backcalculation of pavement
layer moduli, thicknesses, and stiff layer depth from measured surface
deflections is presented. The method is an iterative one and can use any
mechanistic analysis program for forward calculations (presently an extended precision CHEVRON program is used for this purpose). This method
has been implemented in a new backcalculation program named MICHBACK.
Several examples of backcalculation based on theoretical deflection basins
generated by the extended precision CHEVRON program are presented, and the
effect of errors in input layer thicknesses and stiff layer depth on the
backcalculated moduli are studied. Moduli backcalculated by MICHBACK are
compared with results obtained by the EVERCALC 3.0 and MODULUS 4.0 programs.

KEYWORDS: backcalculation, pavement evaluation, elastic layer, Newton
method, gradient method, stiff layer, MICHBACK, EVERCALC, MODULUS

The backcalculation of layer properties from surface deflection measurements is of considerable importance for the accurate evaluation, design of overlays, and management of existing pavements. Most existing
methods predict only elastic layer moduli, but quite often the layer

---

[1]Associate Professor, Department of Civil and Environmental
Engineering, Michigan State University, East Lansing, MI 48824-1226.

[2]Graduate Student, Department of Civil and Environmental Engineering, Michigan State University, East Lansing, MI 48824-1226.

[3]Senior Staff Officer, NRC/Strategic Highway Research Program,
818 Connecticut Avenue N.W., Washington, DC 20006

[4]Professor, Department of Civil and Environmental Engineering,
Michigan State University, East Lansing, MI 48824-1226.

thicknesses and the depth to bedrock or stiff layer are known only approximately and may also need revision. The use of incorrect layer thicknesses
and/or stiff layer depth can yield poor estimates for the backcalculated
moduli, as illustrated by the numerical examples presented later.
There are three general classes of backcalculation methods:

1. Iterative methods that repeatedly adjust the layer moduli and
call a mechanistic analysis program until a suitable match between the
measured and calculated deflection basins is obtained [e.g., CHEVDEF/BIS-
DEF/ELSDEF series (Bush 1985), EVERCALC (Sivaneswaran et al. 1991)];

2. Methods that match the measured deflection basin with a database
of deflection basins computed in advance for a variety of layer moduli
[e.g., MODULUS (Uzan et al. 1989)]; and

3. Methods that use statistical regression equations [e.g., LOADRATE
(Chua and Lytton 1984)].

Iterative methods are usually slow since they require numerous calls to a
mechanistic analysis program, and sometimes the results are sensitive to
the initial seed moduli. Methods that use a database are fast, but the
database of deflection basins corresponding to the range of expected layer
properties must be established before backcalculation is performed, and
the results are usually sensitive to the seed moduli. Methods that use
statistical regression equations are very fast, but usually do not have
acceptable accuracy.

Almost all existing iterative methods estimate the layer moduli by
minimizing an objective function which is the weighted sum of squares of
the differences between calculated and measured surface deflections (Uzan
et al. 1989). One of the problems faced with this approach is that the
multi-dimensional surface represented by the objective function may have
many local minima, and as a result the minimum to which a numerical procedure converges may depend on the initial seed moduli supplied by the
analyst. Another problem is that convergence can be slow because numerous
calls to a mechanistic analysis program (i.e., forward calculations) are
required by most numerical minimization techniques. An efficient and general minimization method (Levenberg-Marquardt algorithm) has been implemented in EVERCALC, which makes it converge quickly with only a modest
number of calls to the mechanistic analysis program (CHEVRON). The EVER-
CALC program has also been used with theoretical deflection basins to successfully estimate the asphalt layer thickness in addition to the layer
moduli (Sivaneswaran et al. 1991).

BACKCALCULATION USING A MODIFIED NEWTON METHOD

A modified Newton method for the backcalculation of pavement layer
properties has been implemented in a program named MICHBACK. The details
of the modified Newton method for the backcalculation of layer moduli and
thicknesses has been presented elsewhere (Harichandran et al. 1993), and
is only summarized here. The initial formulation of the method for backcalculating layer moduli was conceived and first suggested to the research
team by Dr. A. Robert Raab of NRC/SHRP. A literature search has revealed
that the method was conceived previously and published by Hou (1977). The
method outlined by Harichandran et al. (1993) has been revised slightly
to improve its speed of convergence. The revision consists of a logarithmic transformation of the surface deflections and layer moduli (but not

layer thicknesses). The transformation reduces the curvature of the hyper-surfaces that characterize the variation of each surface displacement with layer moduli, and hence enables the Newton method to converge faster. In this method, the ith incremental corrections to the logarithm of the unknown moduli, and layer thicknesses, are obtained by computing the least-squares solution of the over-determined system of linear equations

$$[G]^i \left\{ \begin{array}{c} \{\Delta(\log E)\}^i \\ \{\Delta t\}^i \end{array} \right\} = \{\log w\} - \{\log \hat{w}\}^i \tag{1}$$

where

$$[G]^i = \left[\begin{array}{cccccc} \dfrac{\partial(\log w_1)}{\partial(\log E_1)} & \cdots & \dfrac{\partial(\log w_1)}{\partial(\log E_n)} & \dfrac{\partial(\log w_1)}{\partial t_1} & \cdots & \dfrac{\partial(\log w_1)}{\partial t_p} \\ \vdots & & \vdots & \vdots & & \vdots \\ \dfrac{\partial(\log w_m)}{\partial(\log E_1)} & \cdots & \dfrac{\partial(\log w_m)}{\partial(\log E_n)} & \dfrac{\partial(\log w_m)}{\partial t_1} & \cdots & \dfrac{\partial(\log w_m)}{\partial t_p} \end{array} \right]_{\{E\} = \{\hat{E}\}^i,\ \{t\} = \{\hat{t}\}^i}$$

= gradient matrix of partial derivatives of the logarithm of the m surface deflections, with respect to the logarithm of the n unknown moduli, and p unknown layer thicknesses; evaluated using the current moduli, $\{\hat{E}\}^i$, and thicknesses, $\{\hat{t}\}^i$;

$\{\Delta(\log E)\}^i$ = vector of corrections to the logarithm of the ith estimate of the moduli;

$\{\Delta t\}^i$ = vector of corrections to the ith estimate of the thicknesses;

$\{\log w\}$ = vector of logarithm of measured surface deflections; and

$\{\log \hat{w}\}^i$ = vector of logarithm of surface deflections computed by a mechanistic analysis program using the ith estimate of the moduli and thicknesses.

One technique for solving the least-squares problem is to solve the $n \times n$ normal equations

$$[G]^T[G] \left\{ \begin{array}{c} \{\Delta(\log E)\}^i \\ \{\Delta t\}^i \end{array} \right\} = [G]^T \{ \{\log w\} - \{\log \hat{w}\}^i \} \tag{2}$$

However, the condition number of the matrix $[G]^T[G]$ is the square of the condition number of $[G]$, and hence solving the normal equations can magnify the effect of errors in the elements of $[G]$, errors in $\{\log w\}$, and/or round-off errors that accumulate during calculations. Linear least-squares problems should be solved by using orthogonal factorizations or singular value decomposition (Kahaner et al. 1989).

The gradient matrix is computed numerically, and requires $(n + p + 1)$ calls to the mechanistic analysis program during each iteration. Eq. 2 is essentially the Newton method for the solution of a system of non-linear equations. By using a modified Newton method in which the gradient matrix is used for several iterations before it is revised, the number of calls to the mechanistic analysis program can be reduced (Mahmood 1993; Harichandran et al. 1993). The iteration is terminated when the changes in the layer moduli are sufficiently small. i.e.,

$$\frac{\hat{E}_k^{i+1} - \hat{E}_k^i}{\hat{E}_k^i} < \varepsilon, \qquad k = 1, 2, \ldots, n \tag{3}$$

In addition, if the computed and measured deflections match closely, the root-mean-square error defined by

$$\text{RMS error in deflections} = \sqrt{\frac{1}{m} \sum_{j=1}^{m} \left(\frac{\hat{w}_j^i - w_j}{w_j} \right)^2} \tag{4}$$

will also be small. It should be realized that for deflection basins measured in the field, it may not be possible to obtain an *arbitrarily* close match between the computed and measured deflections.

In principle the modified Newton method can be used to predict any layer property, including Poisson's ratio, as long as the number of unknown quantities does not exceed the number of sensors. All that is required is that the partial derivatives of the surface deflections with respect to the unknown quantities be calculated by mechanistic analysis. However, experience indicates that at times the iteration does not converge as the number of unknown quantities is increased.

With artificially generated deflection basins, the MICHBACK program is able to backcalculate all layer thicknesses accurately along with the layer moduli if the surface deflections are entered to full precision (i.e., to about six significant figures). When the surface deflections are rounded to the nearest hundredth of a mil, or with field measurements, the program often has difficulty backcalculating more than a single layer thickness along with the moduli. In order to ensure satisfactory results, the program currently allows only one unknown layer thickness to be backcalculated along with the layer moduli. If a stiff layer is suspected and its depth is unknown, then it is recommended that the thickness of the roadbed soil (i.e., depth to the stiff layer) be backcalculated, since this may be too deep to be determined by coring or radar-based methods. Further recommendations regarding the backcalculation when all layer thicknesses are unknown are provided in the MICHBACK User's Manual (Harichandran et al. 1994).

ESTIMATION OF STIFF LAYER DEPTH

In most situations a very stiff layer will be located at some depth. The stiff layer may represent actual bedrock, or may be an artificial one simulating the stress hardening behavior of the roadbed soil with depth. Some backcalculation programs do not allow for the presence of a stiff layer, some allow the stiff layer only at a known depth, and some have the capability to estimate the stiff layer depth. The ability to include a stiff layer, and accurate estimation of its depth, are very important for the accuracy of the backcalculation.

The EVERCALC 3.0 and MODULUS 4.0 programs enable the analyst to allow for the presence of the stiff layer, and use regression equations to estimate its depth from the surface deflections (Rodhe and Scullion 1990). The stiff layer modulus is specified by the analyst in EVERCALC and automatically set by the program in MODULUS. The estimation of the stiff layer depth solely based on regression equations can contribute to significant

TABLE 1--<u>Layer properties, seed moduli and moduli ranges</u>
<u>for three layer flexible pavements.</u>

Layer	Thickness mm	Modulus MPa	Poiss. Ratio	Seed Mod.[a] MPa	Modulus Range[b]
AC	127.0	3447.37	0.35	689.48	(689.48,5515.81)
Base	203.2	310.26	0.40	68.95	(48.26, 482.63)
Roadbed	Variable	51.71	0.45	6.89	48.26

[a] For MICHBACK and EVERCALC
[b] For MODULUS

error in the backcalculated layer moduli. The MICHBACK program allows the analyst to specify the stiff layer modulus, and is capable of accurately estimating the stiff layer depth for mechanistically calculated deflection basins. The stiff layer depth estimation has also been tested with numerous deflection basins obtained in the field, and although the actual depth to the stiff layer were unknown for these, the program yielded very reasonable results for most basins (Mahmood 1993). For stiff layer moduli greater than about 1 million psi, the exact value specified by the analyst is not important, since the stiff layer depth and the backcalculation results are insensitive to this.

In MICHBACK, the initial estimate of the stiff layer depth is obtained from the measured surface deflections using regression equations. The stiff layer depth, or equivalently the roadbed soil thickness, is then iteratively improved using the incremental approach described above. However, incorporating the layer moduli and the roadbed soil thickness simultaneously in Eq. 2 usually results in slow convergence. It has been found that the following sequence of steps results in good convergence:

1. Fix the stiff layer depth at the value predicted by the regression equation, and perform one iteration considering only the moduli as unknowns.

2. Perform the following two steps alternately until either of the user-specified convergence criteria is met:

(a) Consider the roadbed soil thickness as well as the moduli as unknowns and perform one iteration.

(b) Consider only the roadbed soil thickness as the unknown and perform one iteration.

NUMERICAL EXAMPLES AND COMPARISONS

Several examples of backcalculation using MICHBACK are given in this section for three layer pavements with and without a stiff layer, and a four layer composite pavement. The layer properties used for all the examples involving flexible pavements are given in Table 1. All the examples are based on theoretical deflection basins (i.e., deflections basins computed using the extended precision CHEVRON program) due to a wheel load of 40.034 kN (9000 lb) applied to a circular area of radius 150.1 mm (5.91 in.), and seven surface deflections calculated at radial distances of 0,

203.2, 304.8, 457.2, 609.6, 914.4 and 1524 mm (0, 8, 12, 18, 24, 36 and 60 inches) from the center of the loaded area. Surface deflections were rounded to the nearest hundredth of a mil (1 mil = 0.001 inch) before input to all programs. The moduli backcalculated by MICHBACK are compared with the values obtained from the EVERCALC 3.0 and MODULUS 4.0 programs. Careful attention was given to making the comparisons as fair and equitable as possible. Since EVERCALC uses the original uncorrected CHEVRON program for forward calculations, the deflection basins used with EVERCALC were generated by the original CHEVRON program to make the comparisons fair. The extended precision CHEVRON program gives essentially the same results as the WES5 programs used by MODULUS.

The MICHBACK and EVERCALC programs require the same types of input parameters. The convergence criterion for the moduli (Eq. 4) specified for these programs was $\varepsilon = 0.001$ (0.1%), and the seed moduli used for all examples are given in Table 1. In addition, upper and lower limits were placed on each layer modulus. These limits were immaterial for MICHBACK since it always converged, but as noted later, in the presence of a stiff layer EVERCALC occasionally diverged and encountered these limits.

The MODULUS program is somewhat different in its approach, does not allow the analyst to specify a convergence measure, and requires slightly different input parameters:

1. The most probable value of the roadbed soil modulus and lower and upper values indicating the range of all other layer moduli are required as input so that a database of deflection basins can be generated. It was found that the moduli backcalculated by MODULUS were quite sensitive to the initial value of the roadbed soil modulus, and somewhat sensitive to the moduli ranges. In order to give it a more than fair start, a roadbed soil seed modulus of 48.26 MPa (7,000 psi) was used for MODULUS, while a much poorer seed modulus of 6.89 MPa (1,000 psi) was used for the other two programs in all examples. The lower and upper moduli specified for the AC and base layers are shown in Table 1. In two analyses, the backcalculated AC modulus was found to be constrained by the upper value of the modulus range, and for these cases, the database of deflections were regenerated using 8273.7 MPa for the upper AC modulus range and the backcalculation was performed again.

2. MODULUS was allowed to automatically select appropriate weights to be applied to the readings of each sensor.

3. Use of a rigid stiff layer was suppressed for all examples except for those in which it was actually present. For the examples in which the stiff layer was present, MODULUS does not allow its modulus to be specified, but assigns it internally (Uzan et al. 1989).

4. The "RUN A FULL ANALYSIS" option was used for all examples, so that material types were not required as input.

<u>(a) Effect of Incorrect Thickness Specification</u>

Quite often, the layer thicknesses are known only approximately, and the thicknesses specified by the analyst in a backcalculation may be incorrect. At present MICHBACK is able to backcalculate all layer moduli and correct the thickness of a single layer. The analyst must identify the layer for which the thickness specification is possibly incorrect. The error in the thickness specification must also not exceed about 50% of the input value. Thickness correction appears to have been successfully performed using a developmental version of EVERCALC (Sivaneswaran et al. 1991), but is not available in the standard version. MODULUS does not perform thickness correction.

The effect of incorrectly specifying the AC or base thickness on the backcalculated layer moduli was investigated for the three layer pavement having the properties shown in Table 1. Either the AC or the base thickness was incorrectly specified, with the error ranging from -40% to +40%, and all three programs were used to backcalculate the moduli. Two types of backcalculations were performed with MICHBACK: one with automatic correction of the incorrectly specified thickness (MICHBACK1), and the other with the thickness held fixed at the incorrect value as done by the other programs (MICHBACK2). The percentage errors in the backcalculated AC and base moduli when the AC thickness was incorrectly specified are shown in Figs. 1 and 2, and the corresponding errors when the base thickness was incorrectly specified are shown in Figs. 3 and 4. In both cases, the errors in the roadbed soil modulus backcalculated by all programs was small and is not shown. Percentage errors were calculated as the ratio 100×(estimated modulus − actual modulus)/actual modulus, and therefore positive errors indicate that the estimates were larger than the actual moduli, while negative errors indicate that the estimates were smaller than the actual moduli. The results show that MICHBACK1 (i.e., MICHBACK with thickness correction enabled) is able to accurately backcalculate all layer moduli. Not correcting the error in the specified AC or base thickness causes significant error in the backcalculated AC and base modulus, but does not cause much error in the backcalculated roadbed soil modulus.

The effect of incorrectly specifying the AC thickness to be 10% higher than the actual thickness on the backcalculated moduli for "thin," "medium" and "thick" three layer pavements is presented in Table 2. The thin, medium and thick sections had AC thicknesses of 50.8, 127 and 228.6 mm (2, 5 and 9 inches), respectively, and all other properties were as specified in Table 1. Again, due to thickness correction, MICHBACK1 yields excellent results. The final AC thickness estimated by MICHBACK1 for the thin, medium and thick pavements was 51.2, 126.2 and 230.3 mm (2.02, 4.97 and 9.08 inches), respectively. When thickness correction is not performed, the error in the backcalculated AC modulus is larger for thinner pavements, while the error in the base modulus is larger for thicker pavements. Again MICHBACK2 and EVERCALC yield similar trends while MODULUS yields more erratic results. For the thick pavement MODULUS backcalculates a higher than actual AC modulus when the thickness is specified high, which is opposite to what is expected and what EVERCALC and MICHBACK2 yield. To compensate for the excess stiffness of the AC, the base modulus is grossly under-estimated by MODULUS.

<u>(b) Estimation of Stiff Layer Depth</u>

The depth of the stiff layer (from the pavement surface) estimated by the three computer programs, together with the percentage error in the backcalculated moduli, are presented in Table 3 for "shallow," "medium" and "deep" stiff layer locations. A three layer flexible pavement having the properties given in Table 1 was used. Again, the surface deflections input to the programs was calculated with the CHEVRON program. The presence of the stiff layer was indicated and the depth was specified as being unknown in all three backcalculation programs. Table 3 indicates that the regression equations used by EVERCALC and MODULUS for obtaining the stiff

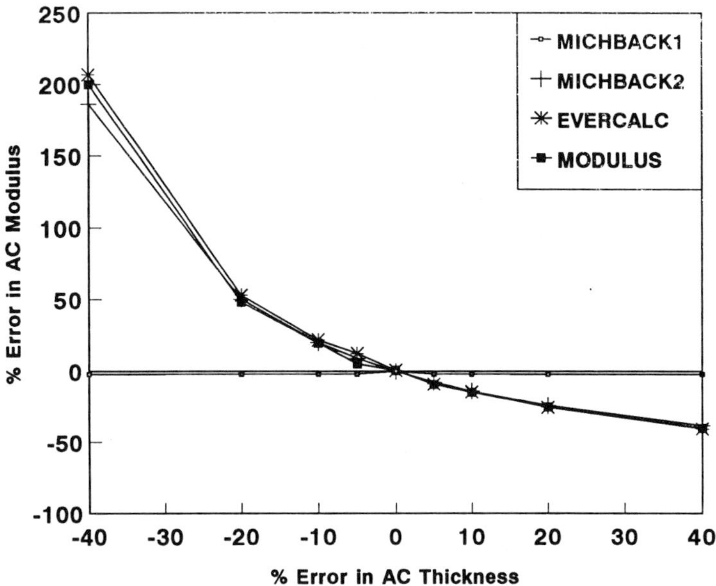

FIG. 1--Error in backcalculated AC modulus due to incorrect AC thickness specification.

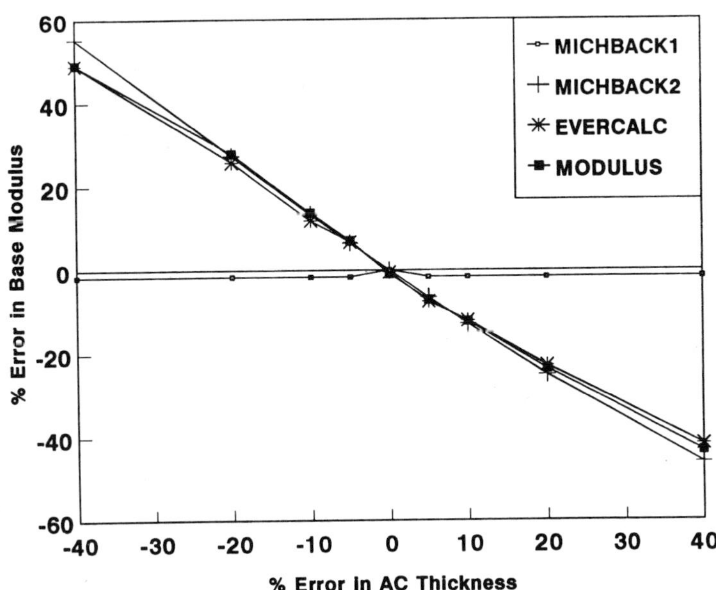

FIG. 2--Error in backcalculated base modulus due to incorrect AC thickness specification.

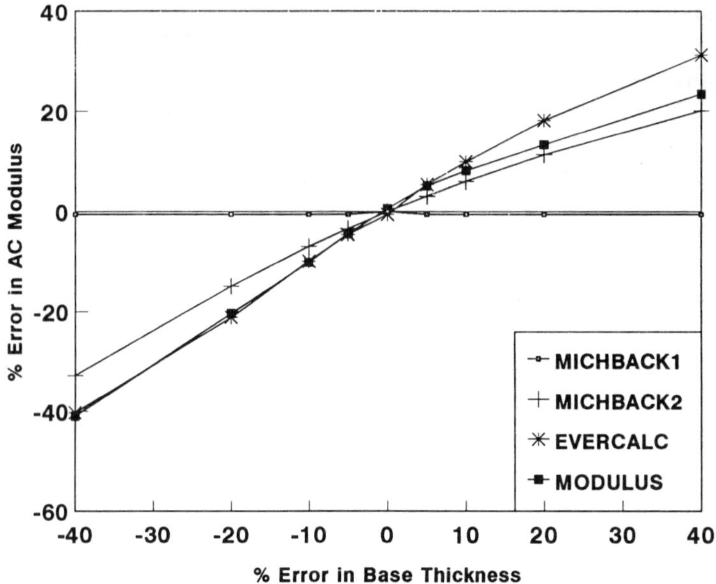

FIG. 3--Error in backcalculated AC modulus due to incorrect base thickness specification.

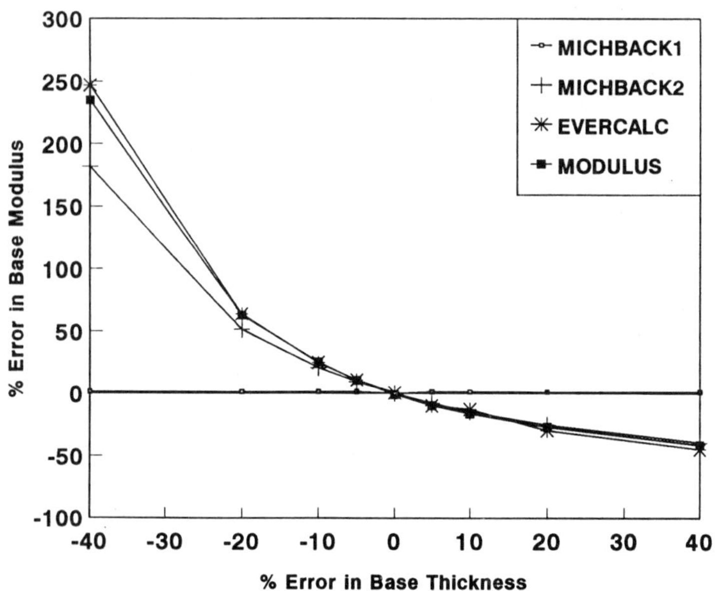

FIG. 4--Error in backcalculated base modulus due to incorrect base thickness specification.

TABLE 2--<u>Effect of +10% error in specified AC thickness on backcalculated moduli for thin, medium and thick flexible pavements.</u>

Pavement Type	Program	% Error in Backcalculated Moduli		
		AC	Base	Roadbed Soil
Thin	MICHBACK1	-3.2	-0.1	0.0
	MICHBACK2	-21.2	-2.1	0.1
	EVERCALC	-19.8	-2.4	0.0
	MODULUS	-15.5	-2.7	0.0
Medium	MICHBACK1	1.5	0.5	0.0
	MICHBACK2	-15.7	-15.8	0.2
	EVERCALC	-14.5	-12.0	0.0
	MODULUS	-3.2	-30.4	1.3
Thick	MICHBACK1	-0.9	-1.2	0.0
	MICHBACK2	-8.0	-37.4	0.4
	EVERCALC	-9.2	-32.9	0.0
	MODULUS	16.6	-78.0	4.0

TABLE 3--<u>Backcalculation of stiff layer depth and moduli for three layer flexible pavements.</u>

Stiff Layer Depth	Program	Percentage Error in Backcalculated Properties			
		Stiff Layer Depth	AC Modulus	Base Modulus	Roadbed Soil Modulus
Shallow (0.914 m)	MICHBACK	-0.2	0.5	0.7	-0.6
	EVERCALC	5.6	5.5	-13.7	5.6
	MODULUS	5.6	10.5	-20.4	21.3
Medium (3.658 m)	MICHBACK	-0.1	0.5	-0.6	0.1
	EVERCALC	-41.3	-29.9	122.2*	-41.9
	MODULUS	-41.6	-23.7	59.6	-32.0
Deep (6.096 m)	MICHBACK	0.3	1.0	-1.1	0.1
	EVERCALC	-56.7	-51.6	122.2*	-34.4
	MODULUS	-52.5	-39.6	118.9	-33.3

* EVERCALC restricted the base modulus at the upper limit of 6.89 MPa

layer depth give similar results and tend to under-estimate the depth, and
the error becomes progressively larger as the stiff layer depth increases.
As a result of this, the error in the backcalculated moduli also become
very large as the stiff layer depth increases. For the medium and deep
stiff layer cases, EVERCALC had difficulty converging, and the base modulus was restricted by the program to remain at the user-specified upper
limit of 6.89 MPa (100,000 psi). It should be noted that the divergence
in the base modulus also affects the results for the other moduli. MICH-
BACK, on the other hand, estimates the stiff layer depth to within 0.5%
of the actual depth. As a result, the moduli backcalculated by MICHBACK
are much more accurate than those estimated by the other two programs.

The error in the stiff layer depth predicted by the regression equations used in EVERCALC and MODULUS becomes much worse for composite pavements. The regression equations were probably developed for flexible
pavements and seem inapplicable to composite pavements. The properties,
input seed moduli and moduli ranges for a four layer composite pavement
used for comparison purposes are shown in Table 4. Table 5 shows the
percentage errors in the stiff layer depth and layer moduli backcalculated
by the three programs. The errors in the stiff layer depths backcalculated
by EVERCALC and MODULUS are now unacceptable, and as a result of this the
errors in the backcalculated moduli are also very large. For the medium
and deep stiff layer cases EVERCALC had difficulty converging, and the
slab modulus was restricted by the program to remain at the user-specified
upper limit of 41.37 MPa (6,000,000 psi). In addition, for the deep stiff
layer case, EVERCALC also restricted the base modulus at the user-specified upper limit of 6.89 MPa (100,000 psi). It is interesting to note
that EVERCALC and MODULUS no longer give similar stiff layer depths. In
general, the layer moduli of composite pavements are more difficult to
backcalculate than for flexible pavements (Harichandran et al. 1993; Mahmood 1993), and even MICHBACK yields slightly larger errors than for flexible pavements. However, the results obtained with MICHBACK are
substantially better than those obtained from the other two programs, and
are quite acceptable.

The effect of using an incorrect stiff layer location on the backcalculated moduli was studied by using all three programs to backcalculate
the moduli without attempting to improve the specified stiff layer depth.
A three layer flexible pavement with the stiff layer at a medium depth
(3.658 m) was used, and the error in the stiff layer depth specified to

TABLE 4--<u>Layer properties, seed moduli and moduli ranges
for four layer composite pavement.</u>

Layer	Thickness mm	Modulus MPa	Poiss. Ratio	Seed Mod.[a] MPa	Modulus Range[b]
AC	101.6	3447.37	0.35	689.48	(689.48,5515.81)
PCC slab	203.2	31026.41	0.25	3447.38	(13789.5, 41368.5)
Base	152.4	172.37	0.40	68.95	(48.26, 482.63)
Roadbed	Variable	51.71	0.45	6.89	48.26

[a] For MICHBACK and EVERCALC
[b] For MODULUS

TABLE 5--<u>Backcalculation of stiff layer depth and moduli for four-layer composite pavement.</u>

Stiff Layer Depth	Program	Percentage Error in Backcalculated Properties				
		Stiff Layer Depth	AC Modulus	Slab Modulus	Base Modulus	Roadbed Modulus
Shallow (1.219 m)	MICHBACK	−3.1	−1.6	1.9	−13.7	5.9
	EVERCALC	172.9	148.4	−59.2	−78.0	586.7
	MODULUS	172.3	−14.9	−24.5	36.0	353.3
Medium (3.658 m)	MICHBACK	0.5	−1.1	0.5	6.8	0.1
	EVERCALC	316.7	−40.3	33.3*	−78.0	−86.7
	MODULUS	108.3	27.0	−36.3	15.3	83.0
Deep (6.096 m)	MICHBACK	−0.7	0.2	−2.0	54.6	−1.0
	EVERCALC	150.0	−72.8	33.3*	300.0**	33.0
	MODULUS	25.1	17.4	−25.7	279.8	12.4

* EVERCALC restricted the slab modulus at the upper limit of 41.37 MPa
** EVERCALC restricted the base modulus at the upper limit of 6.89 MPa

the programs was varied from −40% to +40%. The errors in the backcalculated AC, base and roadbed soil moduli are shown in Figs. 5, 6 and 7, respectively. The results indicate that the AC and base moduli backcalculated by EVERCALC are most sensitive to errors in the stiff layer depth, while those obtained with MODULUS are slightly less sensitive than those obtained with MICHBACK. In general the backcalculated moduli of the AC and base layers interact with each other, and the over-estimation of one results in the under-estimation of the other, and *vice versa*. The roadbed soil modulus backcalculated by all three programs is also sensitive to errors in the stiff layer depth, since this error directly affects the roadbed soil thickness and hence its stiffness. That is, when the stiff layer depth is under- or over-estimated, all three programs compensate for the increase or decrease in roadbed soil stiffness by under- or over-estimating its modulus. These observations clearly indicate the importance of predicting the stiff layer depth accurately.

CONCLUSIONS

A modified Newton method for the backcalculation of pavement layer properties from measured surface deflections is summarized. The method is capable of backcalculating a layer thickness or the stiff layer depth in addition to the layer moduli, and has been implemented in a new backcalculation program named MICHBACK. Using deflection basins generated by an extended precision CHEVRON elastic layer analysis program, the moduli backcalculated by the MICHBACK, EVERCALC 3.0 and MODULUS 4.0 programs are compared when the AC or base thickness or the stiff layer depth is incor-

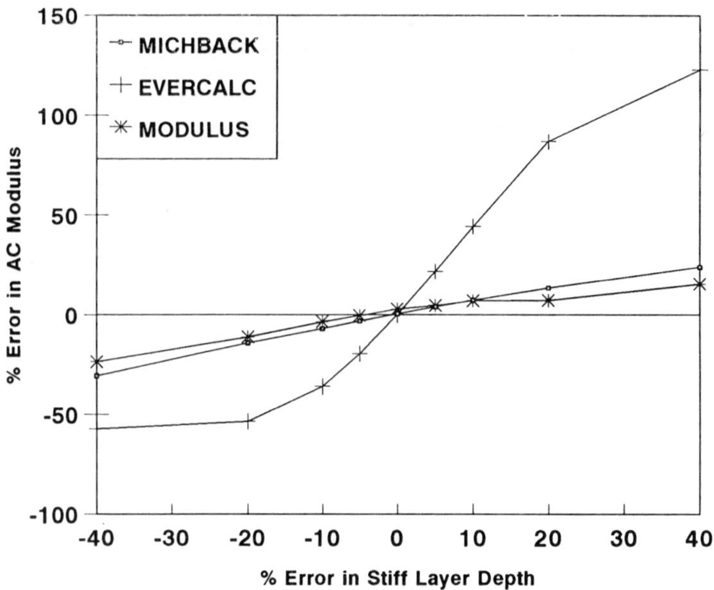

FIG. 5--Error in backcalculated AC modulus due to incorrect
stiff layer location.

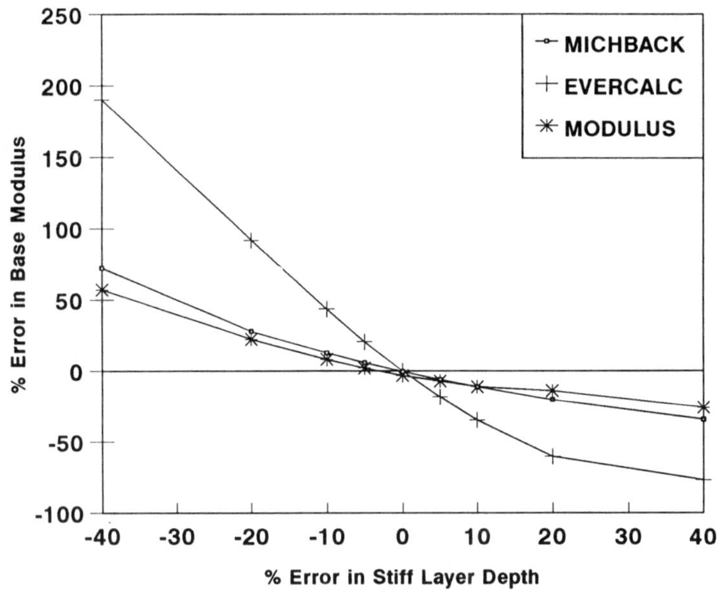

FIG. 6--Error in backcalculated base modulus due to incorrect
stiff layer location.

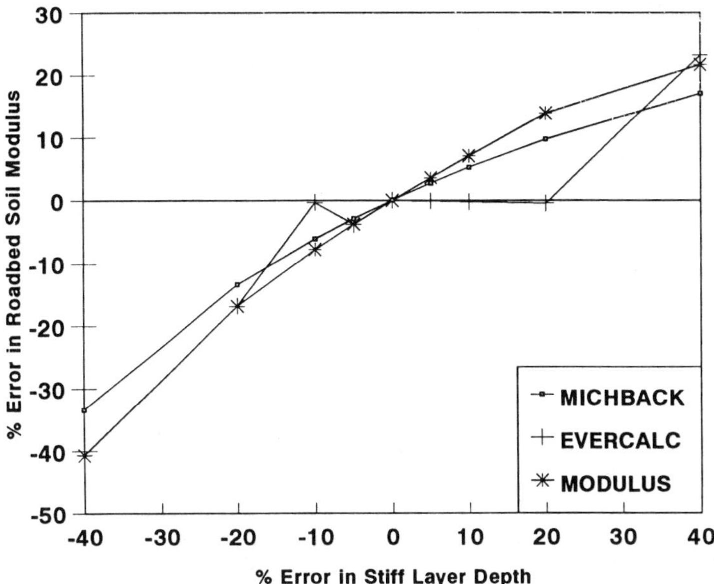

FIG. 7--Error in backcalculated roadbed soil modulus due to incorrect
stiff layer location.

rectly specified. Various three layer flexible pavements and a four layer
composite pavement are used for the comparisons.

An incorrect AC or base thickness causes significant error in *both*
the AC and base backcalculated moduli, but does not cause significant error in the backcalculated roadbed soil modulus. The ability of MICHBACK
to correct an incorrectly specified thickness allows it to backcalculate
the moduli much more accurately than either EVERCALC or MODULUS.

An incorrect stiff layer depth causes significant error in all the
layer moduli backcalculated by the programs. For a three layer flexible
pavement, the AC and base moduli backcalculated by EVERCALC appear to be
most sensitive to errors in the stiff layer depth. The regression equations used in EVERCALC and MODULUS produce relatively poor estimates of
the stiff layer depth, especially for flexible pavements with a deep stiff
layer and for composite pavements. On the other hand, if requested, MICH-
BACK is capable of automatically estimating the stiff layer depth very
accurately for flexible pavements, and reasonably accurately for composite
pavements, by using the Newton method. This is a major advance over other
backcalculation programs. As a result of its accurate stiff layer depth
estimation, MICHBACK is able to estimate all layer moduli much more accurately than either EVERCALC or MODULUS for an unknown stiff layer depth.

Although this paper provides results based only on artificially generated deflection basins, MICHBACK has been tested extensively using a
large number of field measurements obtained in Michigan, and its advanced
capabilities such as stiff layer depth estimation and thickness correction
have yielded very good results (Mahmood 1993).

ACKNOWLEDGMENTS

The authors would like to express their gratitude to the Michigan Department of Transportation and the University of Michigan Transportation Research Institute for providing the financial support for this work. We are also grateful to: Cynthia M. Ramon and Weijun Wang for their assistance in writing the MICHBACK program; Lynne Irwin for improving the quadrature in our version of the CHEVRON program and for his helpful suggestions; and Joe P. Mahoney, NCHRP and Robert L. Lytton for providing us copies of the EVERCALC 3.0 and MODULUS 4.0 programs for evaluation purposes.

REFERENCES

Bush III, A. J., November 1985, "Computer Program BISDEF," U. S. Army Corps of Engineers Waterways Experiment Station.

Chua, K.M. and Lytton, R. L., 1984, "Load Rating of Light Pavement Structures," _Transportation Research Record 1043_, Transportation Research Board, National Research Council, Washington, D.C.

Harichandran, R. S., Mahmood, T., Raab, A. R. and Baladi, G. Y., 1993, "Modified Newton Algorithm for Backcalculation of Pavement Layer Properties," _Transportation Research Record 1384_, Transportation Research Board, National Research Council, Washington, D.C., pp. 15-22.

Harichandran, R. S., Ramon, C. M., Mahmood, T. and Baladi, G. Y., 1994, "MICHBACK User's Manual," Michigan Department of Transportation, Lansing.

Hou, T. Y., 1977, "Evaluation of Layered Material Properties from Measured Surface Deflections," thesis presented to the University of Utah, Salt Lake City, in partial fulfillment of the requirements for the degree of Doctor of Philosophy.

Kahaner, D., Moler, C. and Nash, S., 1989, _Numerical Methods and Software_, Prentice-Hall, Englewood Cliffs, New Jersey.

Mahmood, T., 1993, "Backcalculation of Flexible Pavement Layer Properties from FWD Deflection Data," thesis presented to Michigan State University, East Lansing, in partial fulfillment of the requirements for the degree of Doctor of Philosophy.

Rodhe, G. T. and Scullion, T., 1990, "MODULUS 4.0: Expansion and Validation of the MODULUS Backcalculation System," _Research Report 1123-3_, Texas Transportation Institute, Texas A&M University System, College Station.

Sivaneswaran, N., Kramer, S. L. and Mahoney, J. P., 1991, "Advanced Backcalculation using a Nonlinear Least Squares Optimization Technique," _Paper No. 910362_, presented at the 70th Annual Meeting of the Transportation Research Board, Washington, D.C.

Uzan, J., Lytton, R. L. and Germann, F. P., 1989, "General Procedure for Backcalculating Layer Moduli," _Nondestructive Testing of Pavements and Backcalculation of Moduli, ASTM STP 1026_, A. J. Bush III and G. Y. Baladi, Eds., American Society for Testing and Materials, Philadelphia, pp. 217-228.

Kathleen T. Hall[1] and Michael I. Darter[2]

IMPROVED METHODS FOR ASPHALT-OVERLAID CONCRETE PAVEMENT BACKCALCULATION AND EVALUATION

REFERENCE: Hall, K. T., and Darter, M. I., "Improved Methods for Asphalt-Overlaid Concrete Pavement Backcalculation and Evaluation," Nondestructive Testing of Pavements and Backcalculation of Moduli (Second Volume) STP 1198, Harold L. Quintus, Albert J. Bush, III, and Gilbert Y. Baladi, Eds., Amercian Society for Testing and Materials, Philadelphia, 1994.

ABSTRACT: Structural evaluation is perhaps more difficult for asphalt-overlaid concrete (AC/PCC) pavement than for any other pavement type. This is partly due to the inadequacies of available tools to analyze the complex behavior of composite pavement structures, and partly due to the lack of guidance available for interpretation of structural analysis results. Guidelines for deflection data analysis and interpretation of backcalculation results for AC/PCC pavement are presented in this paper.

In recent research, a procedure has been developed for backcalculation of concrete slab and foundation moduli from deflections measured on asphalt-overlaid concrete (AC/PCC) pavement. This backcalculation method quickly and repeatably produces results consistent with other backcalculation methods and consistent with observed distress and the condition of cores, with efficiency and repeatability unmatched by iterative or database search methods.

Guidelines were also developed for practical interpretation of the backcalculation results, by this or any other reliable method. These guidelines were developed for evaluation of AC/PCC pavement but are also relevant to PCC pavement evaluation. Backcalculation can yield many useful results, including mean PCC slab and foundation moduli, variability in moduli, percentage of backcalculated moduli below a selected critical level, identification of specific areas with unusually high or low moduli, and the relationship of slab and foundation moduli to type, quantity, and severity of distress.

KEY WORDS: Asphalt, concrete, overlays, pavement, nondestructive deflection testing, backcalculation, structural evaluation.

---

[1]Post-Doctoral Research Associate, Department of Civil Engineering, University of Illinois at Urbana-Champaign, 1215 Newmark Civil Engineering Laboratory, 205 North Mathews Avenue, Urbana, IL 61801

[2]Professor, Department of Civil Engineering, University of Illinois at Urbana-Champaign, 1212 Newmark Civil Engineering Laboratory, 205 North Mathews Avenue, Urbana, IL 61801

INTRODUCTION

Concrete pavements are rehabilitated by asphalt resurfacing more commonly than by any other method. In Illinois, for example, more than 60 percent of the Interstate system (1750 two-way miles (2816 km) of concrete pavement) have been overlaid with asphalt, and nearly all of the Illinois Interstate is expected to be overlaid by the year 2000. Similar trends are seen in other states. As the mileage of bare concrete highway pavement decreases and the mileage of asphalt-overlaid concrete (AC/PCC) highway pavement increases, evaluation and rehabilitation of AC/PCC pavements become increasingly pressing concerns to state highway engineers.

Given this trend, it is a significant concern that to date the performance of AC overlays of PCC pavements has been very inconsistent and in many cases very poor. A large and growing mileage of AC/PCC pavements is deteriorating, some of it rapidly, for reasons which many pavement engineers do not fully understand. Despite the fact that AC/PCC pavements make up such a large percentage of the highway mileage of the United States, much less is known about their performance, evaluation, and rehabilitation than is known about other pavement types.

PROJECT-LEVEL EVALUATION OF AC/PCC PAVEMENT

The four key components of a project-level evaluation of an AC/PCC pavement are:

1. Functional evaluation,
2. Structural evaluation,
3. Drainage evaluation, and
4. AC material evaluation.

The condition, deflection, and materials testing data collected must be examined to determine whether or not the pavement has a deficiency in any of these four key areas. The deficiencies identified will play a major role in selection of appropriate rehabilitation alternatives. This paper addresses specifically the collection and analysis of condition, deflection and materials data for use in structural evaluation of AC/PCC pavement.

DATA COLLECTION

Distress Survey

The following distress types and quantities (per mile or per km) should be recorded during a distress survey of AC/PCC pavement:

1. Deteriorated reflection cracks,
2. Full-depth AC patches and expansion joints (except at bridges),
3. Localized failures (and punchouts, for AC/CRCP),
4. Mean rut depth,
5. Alligator cracking in wheelpaths and/or severe shoving of AC, and
6. Evidence of pumping of fines or water at cracks and pavement edge.

The primary objective of a distress survey is to obtain information needed to select and design rehabilitation alternatives and to prepare detailed plans, specifications, and bid estimates. The importance of surveying distress prior to rehabilitation cannot be overemphasized. Lack of data on PCC pavement condition prior to rehabilitation is the major hindrance to predicting the performance of AC/PCC pavements. Without distress surveys of AC/PCC pavements prior to second rehabilitation, it is just as difficult to predict the performance of second rehabilitation alternatives.

Nondestructive Deflection Testing

Deflection testing on the AC/PCC pavements evaluated in this study was conducted with a Dynatest Falling Weight Deflectometer (FWD), owned and operated by the Illinois Department of Transportation (IDOT). The procedures used for deflection testing of AC/PCC pavements in this study are described in this section. The available guidelines for deflection testing on bare PCC and AC pavements were modified as needed to establish procedures suitable for AC/PCC pavements.

Sensor positions -- The backcalculation procedure described in this paper requires deflections measured at the center of the load plate and at 12, 24, and 36 inches (30.5, 61, 91.5, and 122 cm) from the center of the load plate. The deflections measured at these four sensors are referred to here as d_0, d_{12}, d_{24}, and d_{36}. The backcalculation procedure described here is currently being modified to analyze deflection measurements at any sensor positions.

Test load -- For purposes of backcalculating highway pavement layer moduli, a target test load of 9000 pounds (40 kN) is typical. Deflections measured at actual applied loads within about 2000 pounds (9 kN) of the target level may be linearly scaled to 9000-pound (40 kN) deflections.

Number of load drops per test station -- In addition to an initial seating drop, it is common practice when testing AC pavements to apply multiple load drops for each load level at each station tested. Indeed, ASTM Test Method for Deflections with a Falling-Weight-Type Impulse Load Device (D 4694) recommends at least two load drops per load level, and making additional drops (up to five total) until deflections vary less than 5 percent. ASTM D 4694 further recommends that the first drop always be excluded from the deflection analysis. Multiple drops are considered necessary for asphalt pavement testing because deflections may decrease significantly between the first and second drop, due to compression of the pavement layers.

Multiple load drops are sometimes used when testing bare concrete pavements as well, for the purpose of averaging results from several drops and obtaining more accurate pavement moduli estimates. However, the variability between drops at a single point is not nearly as significant an issue in project-level evaluation as the variability in pavement moduli along the length of the project. Multiple load drops do not significantly increase the time required for deflection testing, but they do require storage and manipulation of a much larger quantity of deflection data, and increase the time required for data analysis.

To determine whether multiple load drops were necessary for testing AC/PCC pavements, two load drops were done at each station for two of the case study projects used in this study. The deflections measured at some 200 stations on each project were then subjected to paired t-tests at a confidence level of 95 percent. Maximum deflection (d_0) and deflection basin AREA computed from deflections at 0, 12, 24, and 36 inches (30.5, 61, 91.5, and 122 cm) are the two deflection basin parameters used to backcalculate pavement layer moduli for PCC and AC/PCC pavements according to the procedure presented here. The analyses did not show any significant difference in d_0 and AREA between the first and second drops. Based on these results, a single load drop was used for all of the remaining AC/PCC projects tested.

AC temperature measurement -- The resilient modulus of AC varies substantially with temperature. Therefore, in order to analyze deflections measured on AC pavement or AC/PCC pavement, it is essential to adjust the deflections to account for the variation in temperature of the AC mix which occurs during the duration of the deflection testing.

It is not uncommon for the AC mix temperature to vary by 40°F (22°C) during a typical day of FWD testing. This mcuh temperature variation could easily correspond to a range of 600 ksi (4137 MPa) in AC resilient modulus. Failure to account for temperature variation may result in considerable errors in backcalculated PCC slab and foundation moduli.

For the AC/PCC pavements tested for this study, the temperature of the AC layer was monitored during deflection testing by drilling a hole to the middepth of the AC overlay, inserting liquid and a temperature probe into the hole, and reading the AC mix temperature when it had stabilized. This process takes only a few minutes. This was done three or four times during each day of testing (before testing began in the morning, before and/or after lunch, and when testing was completed in the afternoon).

<u>Materials Sampling and Testing</u>

Cores are typically obtained from AC/PCC pavements for the following purposes:

1. Resilient modulus testing of AC, for use in backcalculation;
2. Split tensile testing of PCC, for comparison with backcalculated moduli;
3. Visual examination of PCC for evidence of "D" cracking;
4. Examination of AC/PCC interface (bonded or unbonded);
5. Confirmation of AC and PCC layer thicknesses; and
6. Recovery of stabilized subbase, if possible, for visual examination and resilient modulus testing.

BACKCALCULATION OF CONCRETE AND FOUNDATION MODULI

Analysis of AC/PCC pavement deflections measured at locations where the underlying PCC is severely deteriorated, as in the case of "D" cracking, will produce low backcalculated *in situ* PCC modulus values. These low modulus values should not be interpreted as the true stress/strain response of the PCC as a homogeneous elastic layer, but rather as an indication of the extent to which its behavior departs from that of a sound slab, i.e., the extent of the PCC's deterioration. The ability to diagnose the condition of the PCC from analysis of deflection measurements is particularly valuable in evaluation of AC/PCC pavements, since the extent of the deterioration of the PCC is often not fully evident from visible distress.

Structural evaluation using NDT data is perhaps more difficult for AC/PCC pavements than for all other pavement types. The computer programs available for backcalculation of pavement layer moduli possess a variety of theoretical and practical limitations which hinder their usefulness in AC/PCC pavement analysis. Valid and repeatable results are typically only obtained from even the best of these tools by very knowledgeable pavement engineers with considerable experience in backcalculation. A review of the capabilities and limitations of available backcalculation tools with respect to AC/PCC pavement analysis is presented in Reference 1.

Of the available methods, the most useful for concrete pavement analysis is the direct closed-form backcalculation method based on plate theory. This approach produces backcalculation results with efficiency and repeatability that cannot be matched by iterative or database backcalculation methods. Furthermore, only backcalculation methods based on plate theory can characterize the foundation by its k-value, which is an essential input to concrete pavement design and rehabilitation procedures.

In order to apply the backcalculation procedure described in the preceding section to an existing AC/PCC, deflections measured on the existing AC surface must be adjusted to account for the influence of the AC layer. The procedure for doing so is described in this paper. Some of the equations presented here appeared in slightly different form in Reference 1.

<u>AREA Concept in Backcalculation</u>

A simple two-parameter approach to backcalculation of surface and foundation moduli for a two-layer pavement system was proposed by Hoffman and Thompson for flexible pavements [2]. They proposed that the deflection basin could be characterized by its AREA as defined by the following equation:

$$AREA = 6 * \left[1 + 2 \left(\frac{d_{12}}{d_0} \right) + 2 \left(\frac{d_{24}}{d_0} \right) + \left(\frac{d_{36}}{d_0} \right) \right] \tag{1}$$

where d_0 = deflection at center of load plate, in (1 in = 2.54 cm)
       d_i = deflection at distance i from plate center, in

The AREA equation is derived from the trapezoidal rule, and the 6 in the equation is the sensor spacing of 12 inches divided by 2. The deflections are all normalized with respect to d_0 in order to remove the effect of different load levels and to restrict the range of values obtained. Thus, AREA has units of length rather than area. AREA and d_0 are independent parameters, from which the surface and foundation moduli in a two-layer pavement system may be determined. Hoffman and Thompson developed a nomograph for backcalculation of flexible pavement surface and subgrade moduli from d_0 and AREA.

The AREA concept was subsequently applied to backcalculation of PCC slab elastic modulus values and subgrade k-values. [3, 4] Further investigation of this concept [5, 6] has produced a forward solution procedure to replace the iterative and graphical procedures used previously. This solution is based on the fact that, for a given load radius and sensor arrangement, a unique relationship exists between AREA and the dense liquid radius of relative stiffness of the pavement system (ℓ), in which the subgrade is characterized by a k-value [7]:

$$\ell = \sqrt[4]{\frac{E\,h^3}{12\,(1 - \mu^2)\,k}} \tag{2}$$

where ℓ = dense liquid radius of relative stiffness, inches
      E = PCC elastic modulus, psi (1 psi = 6.895 kPa)
      h = PCC thickness, in (1 in = 2.54 cm)
      μ = PCC Poisson's ratio, typically 0.15
      k = effective k-value, psi/in (1 psi/in = 17.5 kPa/cm)

The following equation for ℓ as a function of AREA was developed by Hall [8]:

$$\ell = \left[\frac{ln \left(\frac{36 - AREA}{1812.279} \right)}{-2.559} \right]^{4.387} \tag{3}$$

Slab Size Correction

The above backcalculation procedure for k-value and concrete slab
E value employs Westergaard's equation for deflection of an infinite
plate on a dense liquid foundation. Recent research has shown that an
adjustment must be made to obtain appropriate k-values and concrete E
values for slabs which are not sufficiently large to approximate
infinite slab behavior. [9] If L/ℓ, the ratio of least slab dimension
(length or width) to radius of relative stiffness, is less than about 8,
incorrect k and E values will be backcalculated unless this adjustment
is made. The necessary adjustment involves the following steps:

1. Calculate AREA from Equation 1 above.
2. Estimate ℓ from Equation 3 above.
3. Calculate L/ℓ_{est}, where L = least slab dimension
4. Calculate adjustment factors for maximum deflection (d_0)
   and ℓ from the following equations:

$$AF_{d_0} = 1 - 1.06817\, e^{-0.66914\left(\frac{L}{\ell_{est}}\right)^{0.84408}} \qquad (4)$$

$$AF_{\ell} = 1 - 5.29875\, e^{-2.17612\left(\frac{L}{\ell_{est}}\right)^{0.49895}} \qquad (5)$$

5. Calculate adjusted d_0 = measured d_0 * AF_{d0}
6. Calculate adjusted ℓ = ℓ_{est} * AF_{ℓ}
7. Proceed with backcalculation of k-value and concrete E as
   described below, using adjusted d_0 and ℓ.

Backcalculated k-value

The backcalculated k-value may be obtained from Westergaard's
deflection equation:

$$k = \left(\frac{P}{8\, d_0\, \ell^2}\right)\left\{1 + \left(\frac{1}{2\,\pi}\right)\left[ln\left(\frac{a}{2\,\ell}\right) + \gamma - 1.25\right]\left(\frac{a}{\ell}\right)^2\right\} \qquad (6)$$

where d_0 = maximum deflection, inches
      P = load, pounds (1 pound = 4.448 N)
      γ = Euler's constant, 0.57721566490
      a = load radius, 5.9 in (15 cm) for the FWD

Figure 1 was developed from Equation 6 for a load P = 9000 pounds
(40 kN) and a load radius a = 5.9 inches (15 cm). It should be noted
that the backcalculated k-value is typically about twice the k-value
which would be obtained if a static plate bearing test were conducted on
the subgrade. Thus, the backcalculated k-value should be divided by 2
to obtain an estimate of the plate bearing k-value for use in any
overlay design procedure which requires a plate bearing k input.

Concrete Elastic Modulus

With the k-value known, the slab Eh^3 may be computed from the
definition of ℓ (Equation 2), and for a known or assumed slab thickness,
the concrete elastic modulus E may be determined. Figure 2 was
developed for determination of E, assuming a concrete Poisson's ratio
μ = 0.15 and a load radius a = 5.9 inches (15 cm).

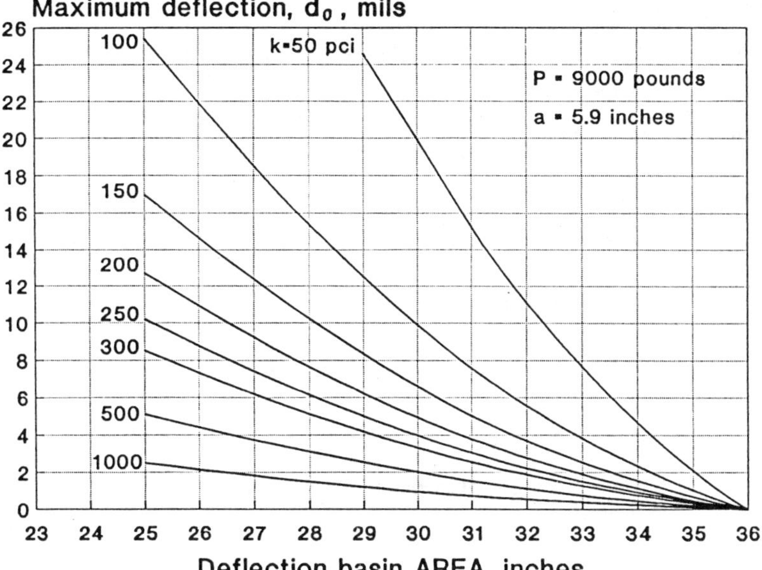

Figure 1. Effective dynamic k-value from d_0 and AREA.

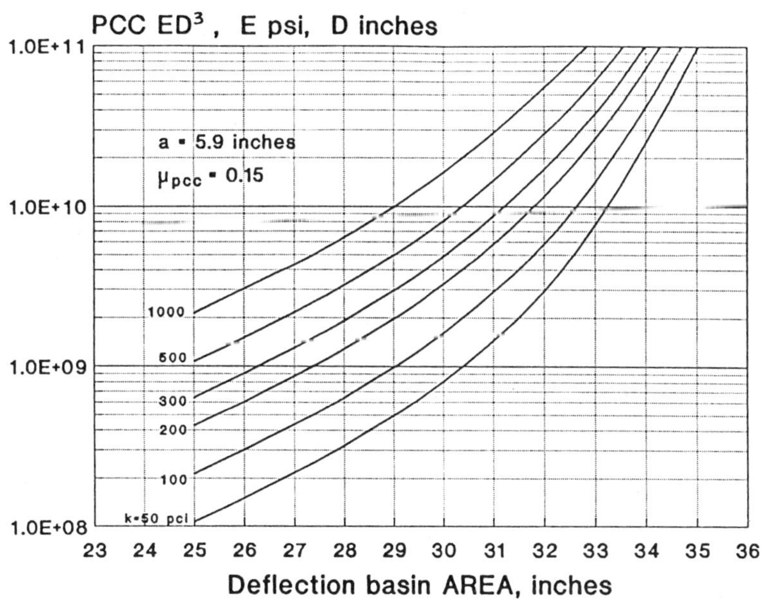

Figure 2. PCC modulus from k-value, AREA, and slab thickness.

AC Elastic Modulus

An existing AC/PCC pavement cannot properly be modelled as a slab on grade, since the AC overlay exhibits not only bending but also compression. To determine the amount of vertical compression that occurs in the AC overlay, the resilient modulus of the AC layer must be determined. Two methods for determining the AC elastic modulus as a function of mix temperature are presented here.

The first method uses the Asphalt Institute's equation for AC modulus as a function of mix parameters, mix temperature, and loading frequency. This equation, developed by Witczak for use in the Asphalt Institute's MS-1 Design Manual [10], is a refinement of work originally done for the Asphalt Institute by Kallas and Shook [11]. It is considered highly reliable for dense-graded AC mixes with gravel or crushed stone aggregates. [12]

$$\log E_{ac} = 5.553833 + 0.028829 \left(\frac{P_{200}}{F^{0.17033}} \right) - 0.03476\, V_v$$

$$+ 0.070377\, \eta + 0.000005\, t_p^{(1.3 + 0.49825 \log F)}\, P_{ac}^{0.5} \tag{7}$$

$$- \frac{0.00189}{F^{1.1}}\, t_p^{(1.3 + 0.49825 \log F)}\, P_{ac}^{0.5} + 0.931757 \left(\frac{1}{F^{0.02774}} \right)$$

$$
\begin{aligned}
\text{where}\quad E_{ac} &= \text{elastic modulus of AC, psi (1 psi = 6.895 kPa)} \\
P_{200} &= \text{percent aggregate passing the No. 200 sieve} \\
F &= \text{loading frequency, Hz} \\
V_v &= \text{air voids, percent} \\
P_{ac} &= \text{asphalt content, percent by weight of mix} \\
t_p &= \text{AC mix temperature, } ^{\circ}\text{F} \\
\eta &= \text{absolute viscosity at } 70^{\circ}\text{F, } 10^6 \text{ poise } (21^{\circ}\text{C, } 10^5 \text{ Pa}\cdot\text{s})
\end{aligned}
$$

This can be reduced to a relationship between AC modulus and AC mix temperature for a particular loading frequency by assuming typical values for the AC mix parameters P_{ac}, V_v, P_{200}, and η. For example, the following values are typical for AC overlay mixes placed by the Illinois DOT prior to 1984:

$$
\begin{aligned}
P_{200} &= 5 \text{ percent} \\
V_v &= 2 \text{ percent} \\
P_{ac} &= 5 \text{ percent} \\
\eta &= 1 \text{ for AC-10, 2 for AC-20}
\end{aligned}
$$

IDOT changed its AC overlay mix design in 1984 to address rutting problems attributed to gradation and low air voids. AC-20 was also required on Interstate-type pavements, whereas AC-10 had often been used in the past. These changes were implemented during the 1984 construction season. Therefore, the above mix parameters are appropriate for AC overlays placed prior to 1984, while for overlays placed in or after 1984, the following mix parameters are typical:

$$
\begin{aligned}
P_{200} &= 4 \text{ percent} \\
V_v &= 5 \text{ percent} \\
P_{ac} &= 5 \text{ percent} \\
\eta &= 2 \text{ for AC-20}
\end{aligned}
$$

The Dynatest FWD is an impulse loading device with a load duration of about 25 to 30 milliseconds. [13]   This corresponds to a loading frequency of approximately 18 Hz. For this frequency and the above mix parameters, the following equations were obtained from the Asphalt Institute equation for E_{ac}:

IDOT mix, AC-20, pre 1984:

$$\log E_{ac} = 6.712176 - 0.000164671\ t_p^{1.92544}$$

IDOT mix, AC-10, pre 1984:

$$\log E_{ac} = 6.641799 - 0.000164671\ t_p^{1.92544} \qquad (8)$$

IDOT mix, AC-20, 1984 and after:

$$\log E_{ac} = 6.451235 - 0.000164671\ t_p^{1.92544}$$

The Asphalt Institute's equation for AC modulus applies to new mixes. AC which has been in service for some years may have a higher modulus (due to asphalt hardening) or lower modulus (due to stripping or other deterioration of the AC) at any given temperature.

The second method for establishing a relationship between E_{ac} and mix temperature involves repeated-load indirect tension testing of AC cores taken from the in-service AC/PCC pavement, according to ASTM Test Method for Indirect Tension Test for Resilient Modulus of Bituminous Mixtures (D 4694). Testing at two or more temperatures is recommended to establish points for a curve of log E_{ac} versus temperature. AC modulus values at any temperature may be interpolated from the laboratory values obtained at any two temperatures, as shown below:

$$\log E_{ac\ t} = \left(\frac{\log E_{ac\ t_1} - \log E_{ac\ t_2}}{t_1 - t_2} \right) * (t - t_1) + \log E_{ac\ t_1} \qquad (9)$$

This straight-line interpolation method is suitable for assignment of moduli at temperatures which are not outside the range of t_1 to t_2 by more than about 10°F (5.5°C). Because the relationship of log AC modulus to temperature is not linear but rather S-shaped, extrapolation to much lower or much higher temperatures may produce unreasonably high or unreasonably low moduli, respectively. The AC modulus assignment procedure described in this paper is currently being modified to produce more accurate estimates of AC modulus over a wider range of temperatures than the laboratory testing range.

For purposes of interpreting NDT data, AC modulus values obtained from lab tests of cores must be adjusted for the difference between the loading frequency of the test apparatus (typically 1 to 2 Hz) and the loading frequency of the deflection testing device (18 Hz for the FWD). This adjustment is made by multiplying the laboratory-determined E_{ac} by a constant value which may be determined for each laboratory testing temperature using the Asphalt Institute's equation. Field-frequency E_{ac} values will typically be 2 to 2.5 times higher than lab-frequency values.

Some researchers have been able to correlate AC resilient modulus to split tensile strength for specific AC mixes compacted in the laboratory. However, AC resilient modulus and split tensile strength generally do not correlate as well for cores taken from in-service mixes. This is often true for samples taken from several different projects, and is sometimes true for samples taken from a single project. For this study, resilient modulus testing and split tensile testing was performed on a large number of cores from three different projects. Because of the poor correlation observed, development of a model for AC

resilient modulus versus split tensile strength for use with the
backcalculation procedure developed in this study was not pursued
further.

Correction to d_0

An elastic layer program (BISAR) was used to model AC/PCC pavement
structures over a broad range of parameters:

AC thickness:	3, 5 and 7 in (7.6, 12.7, 17.8 cm)
AC modulus:	250, 500, 750, 1000, 1250 ksi (1 725, 3 450, 5 175, 6 900, 8 625 MPa)
PCC thickness:	6, 9, and 12 in (15.2, 22.9, 30.5 cm)
PCC modulus:	3, 5, and 7 million psi (20 685, 34 475, 48 265 MPa)
Subgrade modulus:	6, 24, and 42 ksi (41, 165, 290 kPa)
AC/PCC interface:	Bonded and unbonded

A load magnitude of 9,000 pounds (40 kN) and a load radius of
5.9 in (15 cm) were used. Poisson's ratio values used for the AC, PCC
and subgrade were 0.35, 0.15, and 0.50 respectively. The PCC/subgrade
interface was modelled as unbonded.

Deflections were computed at the surface of the AC and the surface
of the PCC at radial offsets of 0, 12, 24, and 36 inches (30.5, 61,
91.5, and 122 cm). Vertical compression in the AC layer, as indicated
by the change in d_0 between the AC and PCC surfaces, often accounted for
a significant portion of the total deflection, depending primarily on
the thickness and modulus of the AC, and to a lesser extent on the
AC/PCC interface condition. For example, in systems with a thick AC
layer (7 in, 17.8 cm) and a low AC modulus (250 ksi, 1725 kPa), more
than 50 percent of the total deflection in the pavement occurred in the
AC layer.

The change in d_0 is significantly greater when the AC is not
bonded to the PCC than when it is bonded. For each interface bonding
condition, it was found that the change in d_0 could be predicted very
reliably as a function of the ratio of the AC thickness to AC modulus
(D_{ac}/E_{ac}). These relationships were found to be very insensitive to the
ranges of other parameters investigated. The following equations were
obtained for these relationships:

AC/PCC BONDED:

$$d_{0\ compress} = -0.0000328 + 121.5006 \left(\frac{D_{ac}}{E_{ac}} \right)^{1.0798}$$

AC/PCC UNBONDED:

$$d_{0\ compress} = -0.00002132 + 38.6872 \left(\frac{D_{ac}}{E_{ac}} \right)^{0.94551}$$

(10)

$d_{0\ compress}$ = AC compression at center of load, in
D_{ac} = AC thickness, in (1 in = 2.54 cm)
E_{ac} = AC elastic modulus, psi (1 psi = 6.895 kPa)

$$R^2 = 99.97 \text{ percent for both equations}$$
$$n = 180 \text{ for each equation}$$
$$\sigma_Y = 0.0350 \text{ mils (0.889 } \mu\text{m) for bonded AC/PCC}$$
$$\sigma_Y = 0.0434 \text{ mils (1.102 } \mu\text{m) for unbonded AC/PCC}$$

Using these equations, the d_0 of the PCC slab in the AC/PCC pavement under a 9000-pound (40 kN) load may be determined by subtracting the vertical compression which occurs in the AC surface from the d_0 measured at the AC surface.

The interface bond is a significant unknown in backcalculation. The AC/PCC interface is fully bonded when the AC layer is first placed, but how well that bond is retained is not known. Examination of cores taken at a later time may show that bond has been reduced or completely lost. This is particularly likely if stripping occurs at the AC/PCC interface. If the current interface bonding condition is not determined by coring, the bonding condition which is considered more representative of the project must be assumed.

In the elastic layer analyses conducted, only d_0 was found to change significantly between the AC and PCC layers. Changes in d_{12}, d_{24}, and d_{36} were very small over the entire range of parameters. The PCC d_0 values predicted by the above equation varied from the actual PCC d_0 values by -0.13 to +0.20 mils (-3.3 to +5.1 μm) when the AC and PCC were bonded, and by at most -0.13 to +0.10 mils (-3.3 to +2.5 μm) when the AC and PCC were unbonded.

Computed AREA of PCC

The AREA of the PCC slab may be computed from the following equation using the d_0 of the PCC slab determined as described above, and d_{12}, d_{24}, and d_{36} measured at the AC surface. The PCC d_0 and $AREA_{pcc}$ may then be used to determine the PCC elastic modulus and the k-value.

$$AREA_{pcc} = 6 * \left[1 + 2 \left(\frac{d_{12}}{d_{0\ pcc}} \right) + 2 \left(\frac{d_{24}}{d_{0\ pcc}} \right) + \left(\frac{d_{36}}{d_{0\ pcc}} \right) \right] \qquad (11)$$

The predicted PCC AREA values varied from the actual PCC AREA values by -0.75 to +0.45 inches (-1.91 to +1.14 cm) when the AC and PCC were bonded, and by -0.57 to +0.26 inches (-1.44 to +0.66 cm) when the AC and PCC were unbonded. However, the largest errors in the predicted PCC AREA corresponded to actual PCC AREA values greater than 33.5 inches (85 cm), which occur very rarely in the field.

STRUCTURAL EVALUATION OF AC/PCC PAVEMENT

A thorough structural evaluation requires investigation of the condition and load-carrying contribution of each of the layers of the pavement system: the AC surface, the PCC slab, the base, and the subgrade. The results of the structural evaluation are crucial to decisions which must be made for rehabilitation planning and design, including:

1. Division of the project into uniform sections,
2. Identification of areas requiring repair,
3. Determination of structural improvement needs,
4. Selection of an appropriate overlay type, and
5. Second overlay design.

The PCC elastic modulus and foundation k value or elastic modulus may be backcalculated from slab deflection measurements, using the procedure described in this paper, or a variety of other methods.

PCC Modulus and Strength

The average backcalculated concrete slab modulus over the length of a project is an important indicator of the pavement's structural capacity. Typical elastic modulus values for concrete of various types and conditions are summarized below, along with general estimates of the corresponding remaining life of the concrete.

Concrete Slab Condition	Typical Modulus (1 million psi = 6895 MPa)	Remaining Structural Life of Slab
Sound JRCP or JPCP	3 to 8 million psi	More than five years
Sound CRCP	2 to 8 million psi	More than five years
Concrete with significant "D" cracking	500,000 to 3 million psi	Three to five years
Concrete with severe "D" cracking	50,000 to 500,000 psi	Less than two years

For jointed plain or jointed reinforced pavement, the concrete modulus of rupture may be estimated from the concrete modulus backcalculated from deflection basins in uncracked areas, using the following equation [4]:

$$S'_c = 43.5 \left(\frac{E}{10^6} \right) + 488.5 \qquad (12)$$

where S'_c = third-point modulus of rupture, psi (1 psi = 6.895 kPa)
      E = backcalculated concrete slab modulus, psi
      R^2 = 71 percent
      σ_Y = 38.5 psi (265 kPa)

The modulus of rupture values used to develop Equation 12 were estimated from the indirect tensile strength of concrete cores using the following equation [14]:

$$S'_c = 1.02\, \sigma_t + 210 \qquad (13)$$

The backcalculated PCC elastic modulus may also be used to estimate the PCC strength in an AC-overlaid jointed pavement. However, unusually low values may be obtained at some basins if the underlying slab is cracked within the deflection basin, even if a reflection crack is not visible at the AC surface.

For CRC pavement, it is not advisable to estimate the PCC modulus of rupture from the backcalculated elastic modulus, because of the likelihood of shrinkage cracks within the deflection basin. Cores taken from CRCP may have much higher indirect tensile strengths, and thus higher flexural strengths, than the backcalculated modulus would suggest. This is even more true for AC/CRC pavements, since the AC obscures viewing of cracks in the CRCP.

The backcalculated PCC modulus may also be inconsistent with the strength of cores if the pavement has "D" cracking. Relatively good modulus values (e.g., 2 to 3 million psi, or about 14 000 to 21 000 MPa) may be backcalculated on "D"-cracked pavement from which it is difficult to obtain sound cores for testing. Conversely, strong cores may be obtained in some areas of a pavement which is severely "D"-cracked in other areas.

These inconsistencies raise the question of which parameter, the elastic modulus or the modulus of rupture, is the better measure of the pavement's structural capacity. It is better to think of the two as different measures, neither of which is better. The backcalculated elastic modulus is representative of the stiffness of the slab, exhibited within a radius of several feet of an applied load. This stiffness depends not only on the strength of the concrete material, but also on the homogeneity of the concrete, the contribution of the reinforcing steel, and the presence of visible cracks and microcracks, caused by shrinkage, fatigue, and durability (freeze-thaw or reactive aggregate) deterioration. The modulus of rupture is representative of the concrete material strength at the location of the core, and is not dependent on these other factors, with the exception of microcracking. It should not be surprising, therefore, that in some cases the two measures correlate well, while in other cases they do not.

Variability in PCC and Foundation Moduli

Backcalculated PCC and foundation moduli are not adequately described by their mean values alone, since these may vary considerably over the length of a project. It is important to have a sense of how much variability in backcalculated moduli is typical in order to recognize when a pavement exhibits unusually high variability. The variability is expressed by the coefficient of variation, which is the standard deviation of the values expressed as a percentage of the mean.

Caution should be exercised in ascribing significance to differences in PCC moduli or foundation moduli observed for different sections of a project, when in fact, due to the magnitude of variation, such differences may not be statistically significant. The topic of variability in backcalculation results for different pavement types and conditions deserves further study.

Frequency Distribution of Moduli

A cumulative frequency distribution of PCC modulus values is a useful means of determining the median value, which may be different than the mean value, particularly if some values are unusually high. The cumulative frequency distribution also illustrates the percentage of deflection basins with moduli below a critical low level.

Figure 3 shows the cumulative frequency distributions of E_{pcc} for 170 deflection basins tested in one lane of a section of I-74 near Mansfield, Illinois. The AC-overlaid 7-inch CRCP has extensive "D" cracking. E_{pcc} values greater than 9 million psi (62 055 MPa), most likely corresponding to locations of PCC patches greater than 7 in (10.16 cm) thick, are not shown. The frequency graph shows that the median E_{pcc} value is about 3.15 million psi (21 720 MPa), which is lower than the mean value of 3.87 million psi (26 684 MPa).

The frequency distribution also shows that E_{pcc} was below 2 million psi (13 790 MPa) for approximately 28 percent of the basins tested, and below 1 million psi (6 895 MPa) for 7 percent of the basins tested. The actual percentage of the total project area with E_{pcc} below each of these levels is certainly greater, since deflection testing was only done at locations with little or no surface distress.

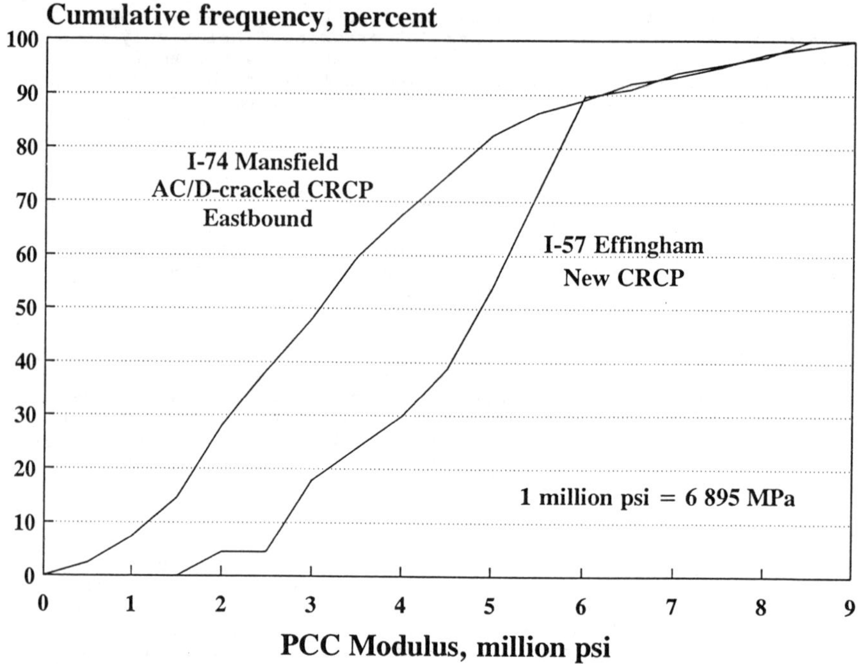

Figure 3. Comparison of PCC modulus frequency distributions.

The cumulative frequency distribution for E_{pcc} values for a recently constructed section of 10-inch CRCP on I-57 near Effingham, Illinois is also shown in Figure 3. The mean E_{pcc} value for this pavement was 4.71 million psi (32 475 MPa). This frequency distribution differs from that of the "D"-cracked pavement in several ways. The slope of the cumulative frequency line is steeper, indicating less variation in E_{pcc} values. (The coefficient of variation, excluding two values exceeding 9 million psi, or 62 055 MPa, is 31 percent.) The median value is much higher: 4.85 million psi (33 440 MPa). Less than 5 percent of the values are below 2 million psi (13 790 MPa), and none are below 1 million psi (6 895 MPa).

<u>Trends in PCC and Foundation Moduli</u>

A plot of PCC slab moduli and foundation moduli over the length of a project may reveal one or more significant shifts in the average modulus. What constitutes a significant shift must be judged in light of the variability in moduli. Using the mean and standard deviation of modulus values for two apparently dissimilar sections of the project, a statistical t-test may be applied to determine whether the mean values are truly different. If they are, the two sections may need to be considered separately in designing rehabilitation for the project.

If deflections differ markedly in one area from those in the rest of the project, the difference may be due to a change in the PCC slab thickness. For example, for one of the AC/PCC pavements studied, a dramatic drop in preoverlay deflections was observed for one 1000-foot section of the pavement. Subsequent conversations with IDOT District personnel revealed that the pavement in that section was actually 12-in (30.5 cm) CRCP, rather than 8-in (20.3 cm) CRCP with a 4-in (10.2 cm) cement-aggregate subbase. [15]

A dramatic change in deflections or backcalculated moduli in a particular area of a project will often coincide with a noticeable difference in the quantity and severity of distress in that area. Coring in the area may be warranted to determine whether a change in layer thicknesses or other cause is responsible for the difference.

Differences in Moduli by Direction

Nearly all of the AC/PCC pavement sections used as case studies showed differences in PCC moduli by direction which were sometimes minor but sometimes quite dramatic. Significant differences in slab moduli by direction usually coincided with significant differences in distress quantities and severities by direction, and differences in other condition measures (serviceability, roughness, etc.) as well. Foundation moduli were generally much more consistent by direction.

Figure 4 shows the PCC modulus frequency distributions for the I-74 Mansfield eastbound and westbound outer lanes. The eastbound distribution has a much lower median modulus and a larger percentage of low moduli. Patching records shed some light on the possible reason for the discrepancy seen for this project. Before the first AC overlay was placed in 1983, the eastbound lanes had more distress (although the reason for this is unknown.) About 2 percent of the eastbound traffic lane area was patched with AC prior to the 1983 overlay, while slightly less than 1 percent of the westbound area was patched. These percentages correspond to one AC patch about every 200 ft (61 m) eastbound and every 412 ft (126 m) westbound. The eastbound lanes continued to deteriorate more rapidly than the westbound lanes after the overlay, no doubt due in no small part to the continuity of the CRCP being disrupted by the closer spacing of AC patches. When the pavement was patched again in 1992 prior to a second AC overlay, the eastbound lanes again received more extensive patching. The total area patched in 1983 and 1992 was about 6 percent eastbound (an AC patch about every 67 ft, or 20 m), but only 2 percent westbound (an AC patch every 200 ft, or 61 m). Clearly, the two directions are in very different condition now, and even with more extensive patching, one must question whether the same overlay thickness is adequate for both directions. One must even question, for this particular example, whether the existing concrete, especially in the eastbound direction, can properly be considered a CRC slab anymore for purposes of overlay design.

It is inadvisable to base repair quantity estimates and rehabilitation designs on deflection data collected for only one direction of a project, particularly when distress or other condition factors (e.g., serviceability or roughness) indicate a clear difference in pavement condition by direction. Only when all condition indicators show that the two directions are very similar and the time available for deflection testing is very limited should only one direction be tested.

Unusually High or Low Modulus Values

Sometimes individual deflections are measured which are much higher or lower than the average values in the area tested. Unusually high deflections and low concrete modulus values usually indicate that the PCC slab is severely deteriorated. This may be true even at locations with little or no distress visible at the AC surface. High deflections also occur at locations where the underlying PCC has a full-depth AC patch, which may not necessarily be evident from reflection cracking at the surface.

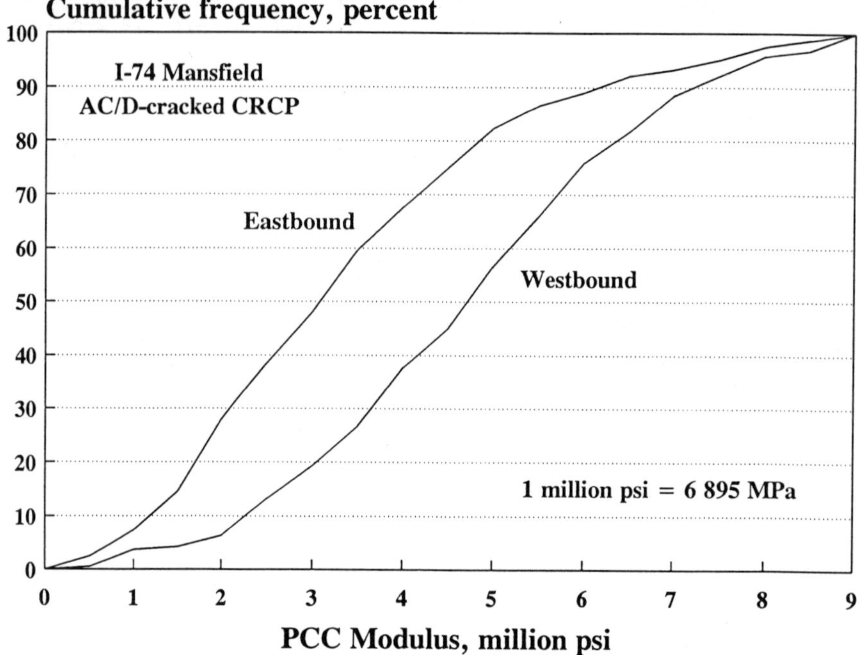

Figure 4.   PCC moduli distributions for one project, two directions.

Unusually low deflections (and unreasonably high PCC modulus values, e.g., greater than 9 million psi, or 62 055 MPa) are usually due to a localized change in slab thickness, such as a PCC repair which is thicker than the original slab. Occasionally a maintenance crew replaces both the concrete and underlying base with full-depth concrete when repairing a pavement. These repair locations often are not known in advance of deflection testing. These repairs do represent an improvement in the pavement condition, and thus should not be ignored. However, it may be unwise to include the backcalculated moduli at repairs in the calculation of the mean PCC modulus for the project, since a few very high values can change the mean enough to give a misleading impression of pavement condition in unrepaired areas.

A high backcalculated slab modulus may also be the result of bonding between the slab and a stabilized base. This generally did not occur on the projects tested for this study, since on the projects which did have stabilized (asphalt-treated) bases, the base was usually debonded from the slab and often very deteriorated.

<u>Relationship of PCC Modulus to Condition</u>

The most difficult aspect of AC/PCC pavement structural evaluation, the one which requires the most experience and expert judgement, is the assessment of the overall "condition" of the PCC slab. This requires consideration of the backcalculated PCC moduli along with the type, quantity, and severity of visible distress. This is made particularly difficult by the fact that the PCC modulus results obtained depend on the way in which deflection testing is conducted.

One possible option for relating PCC moduli to distress is to conduct deflection testing in both cracked and uncracked areas. An uncracked area is defined for the purpose of this discussion as an area in the interior of a PCC slab (away from slab joints or edges) without linear cracks or localized failures within the deflection basin. If deflection basins are measured in both cracked and uncracked areas, PCC moduli backcalculated from these deflections should be considered "effective" moduli, which represent not the true stress-strain behavior of the PCC, but the *condition of the pavement in its current state of cracking.*

An example of this approach is the work done by Rollings at the Waterways Experiment Station, in which a relationship was established between "E-ratio" (initial slab modulus versus cracked slab effective modulus) and Structural Condition Index (SCI, determined from cracking data). [16] FWD deflections were measured on full-size slabs when intact, and at several subsequent stages of cracking. Although the actual moduli obtained by Rollings are dependent on the backcalculation method used, the deflection data and trends in backcalculated moduli demonstrate that a decrease in SCI is accompanied by a decrease in effective slab modulus.

Testing PCC or AC/PCC pavements in cracked areas poses several practical difficulties, however. No detailed guidance is available for how to conduct deflection testing in cracked areas (i.e., what testing interval should be used, how many locations should be tested, where to position the load plate and sensors with respect to cracks, how close to joints should testing be done, etc). The deflection and backcalculation results are likely to be highly variable depending on exactly how the testing is done. For example, three very different deflection basins can be obtained at one transverse crack, depending on whether the crack is positioned between the d_{-12} and d_0 sensors, between the d_0 and d_{12} sensors, or at the d_0 sensor (directly beneath the load plate). The engineer is then faced with the difficulty of deciding which of the three PCC moduli backcalculated from these three basins best represents the condition of the slab in that area.

The second approach, which is the approach taken in this study, is to measure deflections in uncracked areas only. The PCC moduli backcalculated from these deflection basins represent the *condition of the pavement in uncracked areas,* separate from the type, quantity, and severity of visible distress.

A disadvantage of this approach is that the engineer must consider the backcalculation results and distress survey results separately in assessing the "condition" of the slab. Testing in uncracked areas has some significant advantages, however, over testing in cracked and uncracked areas. It has the practical advantage that it is much easier for an engineer or FWD operator to identify and avoid cracked areas during testing than to make the many decisions described above concerning how to test in cracked areas. Its other major advantage is that it is an excellent way to distinguish "D"-cracked PCC from sound PCC. This is particularly important in AC/PCC pavement structural evaluation, since the extent of "D" cracking deterioration in the PCC slab is very difficult to assess from visible distress alone.

If the PCC slab has "D" cracking or other severe deterioration, low backcalculated PCC modulus values will be obtained for deflection basins even in uncracked areas. These low modulus values will be reflected in the low mean and median values obtained, and also in the cumulative frequency distribution of modulus values.

Another of the case studies evaluated for this project, on I-70 near Marshall, Illinois, is an example of the relationship between

backcalculated PCC modulus and "D" cracking. This pavement is a
severely "D"-cracked CRCP with an AC overlay. It is characterized by
low median and mean PCC moduli, and large percentages of low PCC moduli
for deflection basins in uncracked areas. For example, in the westbound
direction, the mean PCC modulus is 3.6 million psi (24 822 MPa), and 40
percent of the basins have PCC moduli below 2 million psi (13 790 MPa).
At specific locations where high deflections were measured (and for
which low PCC moduli were backcalculated using the slab thickness of 8
in, or 20.3 cm), coring confirmed that the PCC was moderately to
severely deteriorated, as shown below. At locations where very low PCC
moduli (e.g., less than 200,000 psi, or 1379 MPa) were backcalculated,
as little as 1.5 in (3.8 in) of sound concrete was recovered by coring.

These examples demonstrate that extensive "D" cracking in an
AC/PCC pavement may be diagnosed from backcalculation results.
Indicators that the PCC slab is deteriorated are low mean or median PCC
moduli (e.g., less than about 4 million psi, or 27.6 MPa) or a high
percentage of low moduli (e.g., 15 percent or more less than 2 million
psi, or 13.8 MPa) obtained for deflection basins in uncracked areas.
Further analysis should be done using data for additional projects to
better establish critical levels for PCC modulus.

Deflection testing in both cracked and uncracked areas may become
a more useful structural analysis technique in the future if additional
field studies are conducted and detailed guidelines are developed for
testing and data analysis. For the present, deflection testing in
uncracked areas is recommended since the testing and analysis do not
require the level of expertise that testing in cracked areas requires.
Interpretation of backcalculated PCC moduli is an important subject
which certainly deserves further study.

CONCLUSIONS

Structural evaluation is perhaps more difficult for asphalt-
overlaid concrete pavement than for any other pavement type. This is
partly due to the inadequacies of available tools to analyze the complex
behavior of this composite pavement structure. A greater hindrance,
however, is the lack of guidance available for interpretation of
structural analysis results. Guidelines for deflection data analysis
and interpretation of backcalculation results for AC/PCC pavement are
presented in this paper.

The backcalculation procedure described produces results
consistent with observed pavement and core conditions, with efficiency
and repeatability unmatched by iterative or database search methods.
This procedure may easily be implemented in a spreadsheet or simple
computer program.

Backcalculation results are not used to their fullest advantage in
structural evaluation if nothing more is determined than the mean values
of concrete modulus and foundation modulus. A cumulative frequency
distribution of concrete moduli is very valuable in assessing the
variability in values and the percent of values below a level considered
critical for sound concrete (e.g., 2 million psi, or 13 790 MPa). The
backcalculation results together with visible distress and coring
results will indicate whether the concrete slab is sound and can be
maintained as a composite pavement in the future, or whether the
concrete is unsound and requires more substantial rehabilitation.

ACKNOWLEDGEMENT

The work described in this paper was conducted as part of the Illinois Cooperative Highway Research Program, Project 532 - "Rehabilitation of Asphalt-Overlaid Concrete Pavements," by the Department of Civil Engineering, University of Illinois at Urbana-Champaign, in cooperation with the Illinois Department of Transportation and the United States Department of Transportation, Federal Highway Administration. The views expressed in this paper are those of the authors, and do not necessarily reflect the official views or policies of the Illinois DOT or the FHWA.

The authors gratefully acknowledge Douglas Steele and Max Rexroad for their contributions to the research study described in this paper, and James Crovetti for sharing his experience on deflection testing and his suggestions for this paper.

REFERENCES

1.   Hall, K. T. and Mohseni, A., "Backcalculation of AC/PCC Pavement Layer Moduli," <u>Transportation Research Record</u> No. 1293, 1991.

2.   Hoffman, M. S. and Thompson, M. R., "Mechanistic Interpretation of Nondestructive Pavement Testing Deflections," Transportation Engineering Series No. 32, Illinois Cooperative Highway and Transportation Research Series No. 190, University of Illinois at Urbana-Champaign, 1981.

3.   ERES Consultants, Inc., "Nondestructive Structural Evaluation of Airfield Pavements," prepared for U.S. Army Engineer Waterways Experiment Station, Vicksburg, Mississippi, 1982.

4.   Foxworthy, P. T., "Concepts for the Development of a Nondestructive Testing and Evaluation System for Rigid Airfield Pavements," Ph.D. thesis, University of Illinois at Urbana-Champaign, 1985.

5.   Ioannides, A. M., "Dimensional Analysis in NDT Rigid Pavement Evaluation," <u>Transportation Engineering Journal</u>, American Society of Civil Engineers, Volume 116, No. TE1, 1990.

6.   Barenberg, E. J. and K. A. Petros, "Evaluation of Concrete Pavements Using NDT Results," Illinois Cooperative Highway Research Project IHR-512, University of Illinois and Illinois Department of Transportation, Report No. UILU-ENG-91-2006, 1991.

7.   Westergaard, H. M., "Stresses in Concrete Runways of Airports, <u>Proceedings</u>, Highway Research Board, Volume 19, 1939.

8.   Hall, K. T., "Performance, Evaluation, and Rehabilitation of Asphalt-Overlaid Concrete Pavements," Ph.D. thesis, University of Illinois at Urbana-Champaign, 1991.

9.   Crovetti, J. A. and Tirado-Crovetti, M. R., "Evaluation of Support Conditions Under Jointed Concrete Slabs," _Nondestructive Testing of Pavements and Backcalculation of Moduli_, Second Volume, ASTM STP 1198, H. L. Von Quintus, A. J. Bush, and G. Y. Biladi, Eds., American Society for Testing and Materials, Philadelphia, 1994.

10.   Asphalt Institute, "Research and Development of the Asphalt Institute's Thickness Design Manual (MS-1) Ninth Edition," Research Report 82-2, 1982.

11.   Kallas, B. F. and Shook, J. F., "Factors Influencing Dynamic Modulus of Asphalt Concrete," _Proceedings_, Association of Asphalt Paving Technologists, Volume 38, 1949.

12.   Miller, J. S., Uzan, J., and Witczak, M. W., "Modifications of the Asphalt Institute Bituminous Mix Modulus Predictive Equation," _Transportation Research Record_ No. 911, Transportation Research Board, 1983.

13.   Smith, R. E. and Lytton, R. L., "Synthesis Study of Nondestructive Deflection Testing Devices for Use in Overlay Design of Flexible Pavements," ERES Consultants, Inc., Federal Highway Administration Report No. FHWA/RD-83/097, 1984.

14.   Hammitt, G. M., II, "Concrete Strength Relationships," U. S. Army Engineer Waterways Experiment Station, Vicksburg, Mississippi, 1971.

15.   Barnett, T. L., Darter, M. I., and Laybourne, N. R. "Evaluation of Maintenance/Rehabilitation Alternatives for CRCP," Illinois Cooperative Highway Research Report No. IHR 901-3, Federal Highway Administration Report No. FHWA/IL/UI-185, 1981.

16.   Rollings, R. S., "Design of Rigid Overlays for Airfield Pavements," Ph.D. thesis, University of Maryland, 1987.

Anastasios M. Ioannides[1]

CONCRETE PAVEMENT BACKCALCULATION USING *ILLI-BACK 3.0*

REFERENCE: Ioannides, A. M., "Concrete Pavement Backcalculation Using ILLI-BACK3.0," <u>Nondestructive Testing of Pavements and Backcalculation of Moduli (Second Volume), ASTM STP 1198</u>, Harold L. Von Quintas, Albert J. Bush, III, and Gilbert Y. Baladi, Eds., American Society for Testing and Materials, Philadelphia, 1994.

ABSTRACT: Applications are presented of an improved closed-form backcalculation procedure for slab-on-grade concrete pavement systems. A choice is now possible from among several popular plate-sensor arrangements, involving the 300 or 450 mm dia. plate, with four or seven sensors, as well as the irregular sensor spacings arrangement used by the Strategic Highway Research Program. Based on experience from several projects, user guidelines are formulated.

KEYWORDS: backcalculation; k-value; concrete pavements; nondestructive testing; Falling Weight Deflectometer; data interpretation.

At the First International Symposium on Nondestructive Testing of Pavements and Backcalculation of Moduli, convened in 1988 in Baltimore, Maryland, an efficient and accurate closed-form backcalculation scheme for the determination of in situ Portland cement concrete (PCC) and foundation support moduli was proposed (Ioannides 1988). The method was subsequently documented in detail (Ioannides 1990), and was applied to the interpretation of a significant volume of data from in-service pavements (Ioannides et al. 1989). This closed-form backcalculation procedure is based on a consistent and theoretically rigorous approach utilizing the principles of dimensional analysis, and is applicable to two-layer, rigid pavement systems. A cardinal aspect of the process is the concept of the area of the deflection basin, first proposed by Hoffman and Thompson (1981).

Implementation of the method in a computer program or spreadsheet simplifies considerably the effort required in

---

[1]Engineering Consultant, 2 Clover Leaf Court, Savoy, IL 61874-9759.

interpreting nondestructive testing (NDT) data, such as those collected using the Falling Weight Deflectometer (FWD). Using software prepared for this purpose, the execution time for a single basin is usually less than 1 CPU sec. A unique feature of this approach is that in addition to yielding the required backcalculated parameters, it also allows an evaluation of the degree to which the in situ pavement system behaves as idealized by theory, i.e., as a medium-thick plate resting on a dense liquid (DL) or an elastic solid (ES) subgrade. Furthermore, the method provides an indication of possible equipment shortcomings as may arise in the field.

This Paper highlights an extension of the closed-form slab-on-grade backcalculation procedure, which allows a choice from among several popular plate-sensor arrangements, involving the 300 or 450 mm dia. plate, with four or seven sensors, as well as the irregular sensor spacings arrangement used by the Strategic Highway Research Program (SHRP). Application of the refined procedure to SHRP, Federal Highway Administration (FHWA), Illinois Department of Transportation (ILDOT) and other activities is discussed, and a number of user guidelines are formulated.

OVERVIEW OF CLOSED-FORM SLAB-ON-GRADE BACKCALCULATION METHOD

The backcalculation scheme described in this Paper employs two fundamental theoretical concepts. These are:

1. For any particular plate-sensor arrangement, a unique relationship exists between the deflection basin area, AREA, and the radius of relative stiffness, ℓ, of the slab-subgrade system (Ioannides 1990); and

2. Deflections in slab-on-grade pavements, expressed in a dimensionless form, are solely a function of the governing load size ratio, (a/ℓ), where a is the radius of applied load (Ioannides 1987).

All three of these quantities (AREA, ℓ, and a) are expressed in units of length. For sensors emanating from the center of the load plate at a uniform spacing Λ, the area of the deflection basin is calculated as follows:

$$AREA = \frac{\Lambda}{2D_0} \left(D_0 + 2 \left[D_1 + D_2 + \ldots + D_{n-1} \right] + D_n \right) \tag{1}$$

where D_i denotes the deflections recorded (i=0,n), and n is the number of sensors used, minus one. If the sensor spacing is nonuniform, a similar expression may be written using the trapezoidal rule.

Now, the radius of relative stiffness of the pavement-subgrade system is defined by:

For the DL foundation:
$$\ell = \ell_k = \sqrt[4]{\frac{E\,h^3}{12\,(1-\mu^2)\,k}} \tag{2}$$

For the ES foundation:

$$\ell = \ell_e = \sqrt[3]{\frac{E\,h^3\,(1-\mu_s^2)}{6\,(1-\mu^2)\,E_s}} \qquad (3)$$

where E is the slab Young's modulus; E_s is the soil Young's modulus; h is the slab thickness; μ is the slab Poisson ratio; μ_s is the soil Poisson ratio; and k is the modulus of subgrade reaction.

Application of dimensional analysis indicates that a unique relationship between AREA and ℓ exists and is valid for a particular plate size and sensor arrangement. Figure 1 shows the AREA vs. ℓ curves for four different loading and support conditions using four sensors at 12-in. spacing.

Inspection of the interior loading formula by Westergaard (1939) shows that the maximum deflection, D_0, may be rewritten in the following nondimensional form:

$$d_0 = \frac{D_0\,D}{P\ell^2} = \frac{D_0\,k\ell^2}{P} \qquad (4)$$

where P is the applied load, and D is the slab flexural stiffness, which is given by:

$$D = \frac{E\,h^3}{12\,(1-\mu^2)} \qquad (5)$$

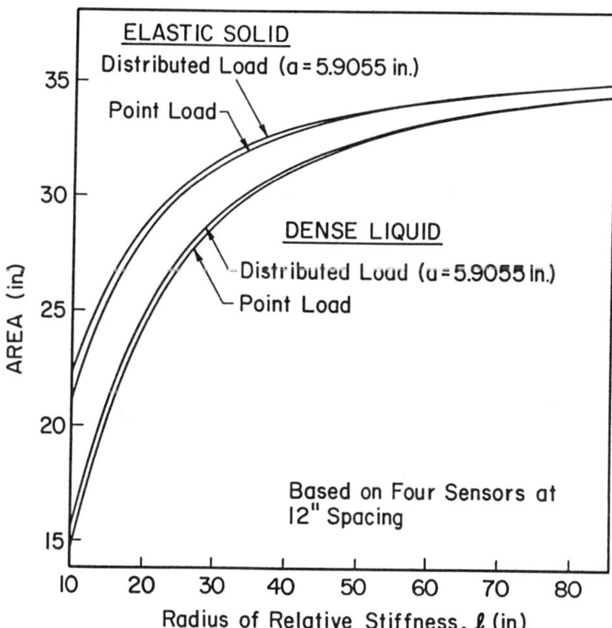

FIG. 1--Variation of AREA with ℓ (After Ioannides 1990)

Similar expressions can be derived for deflections at other sensor locations, i.e.,:

$$d_i = \frac{D_i\, D}{P\ell^2} = \frac{D_i\, k\ell^2}{P} \qquad (i=1,n) \tag{6}$$

where d_i denotes the nondimensional sensor deflections corresponding to the measured deflections, D_i. The nondimensional deflections are known functions of the ratio (a/ℓ) only. In the case of a constant plate load radius, they are uniquely defined by ℓ. Figure 2 shows the variation with ℓ of dimensionless deflections, d_i, corresponding to measured deflections using four sensors at 12-in. (305 mm) spacing and a circular load, radius a = 5.9055 in. (150 mm). The curves corresponding to d_0 are defined by Westergaard's interior loading maximum deflection formula for the DL (Ioannides et al 1985a) and by the corresponding ES equation by Losberg (1960), respectively. The remainder of the curves in Fig. 2 were derived more recently (Ioannides 1988; Ioannides et al 1989).

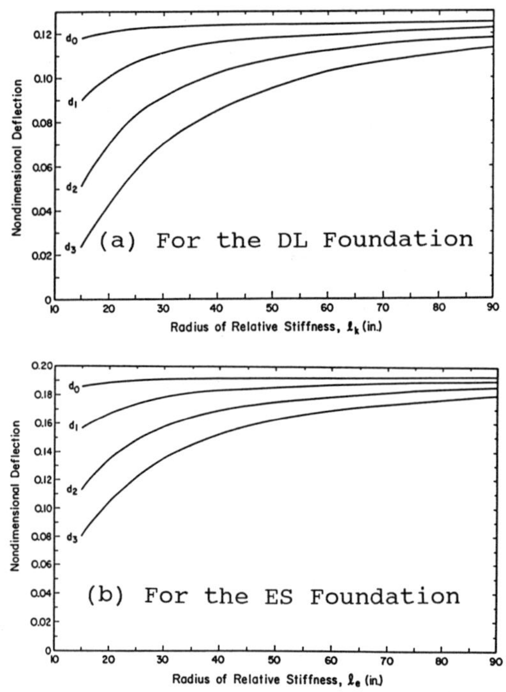

FIG. 2-- Variation of Dimensionless Deflections with ℓ (After Ioannides et al 1989)

On the basis of the two fundamental concepts noted above, the backcalculation procedure proceeds along the following lines:

1. Drop the weight, and record the applied load, P, as well as the resulting deflections, D_i.

2. Calculate the area of the deflection basin, AREA.

3. With this AREA-value, enter Fig. 1 or similar graph and pick the corresponding radius of relative stiffness value, ℓ.

4. With this ℓ-value, Enter Fig. 2 or similar graph and determine the corresponding dimensionless deflections, d_i.

5. Backcalculate the subgrade support values, as follows:

For the DL foundation:

$$k = \frac{d_i}{D_i} \frac{P}{\ell^2} \tag{7}$$

For the ES foundation:

$$C = \frac{d_i}{D_i} \frac{2P}{\ell} \tag{8}$$

For a chosen value of the subgrade Poisson's ratio, (say, μ_s = 0.4 to 0.5), Eq. (8) can be rewritten to yield the foundation elastic modulus, E_s:

$$E_s = (1-\mu_s^2) \frac{d_i}{D_i} \frac{2P}{\ell} \tag{9}$$

6. Backcalculate the slab flexural stiffness, D, using:

$$D = \frac{E h^3}{12 (1-\mu^2)} = \frac{d_i}{D_i} P\ell^2 \tag{10}$$

Thus, if the slab thickness, h, is known, the slab modulus can be calculated as follows:

$$E = \frac{12 (1-\mu^2)}{h^3} \frac{d_i}{D_i} P\ell^2 \tag{11}$$

Alternatively, if the slab modulus, E, is known, one can backcalculate thickness, h, from:

$$h = \sqrt[3]{\frac{12 (1-\mu^2)}{E} \frac{d_i}{D_i} P\ell^2} \tag{12}$$

For the slab Poisson ratio, μ , a value of 0.15 may be used. Note that using these backcalculation equations (Eqs. 7 through 12), n+1 determinations for each pavement system parameter (k, E_s, h or E) are possible, each corresponding to one measured deflection, D_i. This provides a control on the accuracy of individual sensor readings, as well as a measure of in situ material variability, and of the departure of the real system from the idealized conditions assumed in theory.

EXTENSION OF CLOSED-FORM BACKCALCULATION PROCEDURE

The main focus of the efforts described in this Paper has been the extension of the method's applicability to the interpretation of deflection data obtained using several popular plate-sensor arrangements. Recall that the curves in Fig. 1 apply only to data collected using a 300 mm diameter load plate, with four sensors located at 0, 12, 24, and 36 in. (0, 305, 610, 915 mm) from the center of the plate. This is one of the most widely used FWD configurations in the United States, particularly when highway pavements are evaluated. The chosen plate size of 300 mm corresponds to a 9000 lb (40 kN) circular wheel load applied at 80 psi (550 kPa) of pressure, often recognized as a "typical" highway pavement wheel load. The sensor locations adopted correspond to (r/a) ratios of 0, 2, 4, and 6, where r denotes the radial offset of the sensor, and a is the load plate radius.

In evaluating airfield pavements, however, it is sometimes considered more appropriate to use a larger load plate, to simulate the larger tire prints of conventional aircraft traffic. Some FWDs currently in use are equipped with a load plate of 450 mm in diameter, with which seven sensors located at 0, 12, 24, 36, 48, 60 and 72 in. (0, 305, 610, 915, 1220, 1525, 1830 mm) are used (Bentsen et al. 1989). This plate size corresponds to a 45,000 lbs (200 kN) wheel load applied at 180 psi (1250 kPa) contact pressure, which is typical of the wheel load applied by several modern aircraft types. Little information is available concerning any real advantages of a larger plate. According to one user, "a 1 to 2% difference in backcalculated parameters was observed when the 300 mm plate was replaced by the 450 mm plate on a 16-in. thick concrete slab" [2].

It is anticipated that in the near future, significant volumes of deflection data will become available from the database of the Long Term Pavement Performance (LTPP) program, recently transferred from SHRP to FHWA. This information was been collected using a 300 mm plate and seven sensors located at 0, 8, 12, 18, 24, 36, and 60 in. (0, 205, 305, 455, 610, 915, 1525 mm). The additional two sensors near the plate are intended to provide a better description of the deflection profile in this critical region. The outer sensor is probably used in conformity with testing asphalt concrete (AC) pavements, which sometimes requires such a distant measurement for the determination of the subgrade modulus of elasticity, E_s.

For these reasons, the closed-form slab-on-grade backcalculation procedure has been expanded to offer the user a choice among five plate-sensor arrangements, as shown in Table 1. In the software prepared for this purpose, an

---

[2]Crovetti, J.A. (1990), Personal communication.

TABLE 1--<u>Plate-Sensor Arrangement Options in *ILLI-BACK 3.0.*</u>

Arrangement 1:	300 mm dia. plate, 4 sensors at uniform 12 in. spacing (0, 12, 24, and 36 in.)
Arrangement 2:	300 mm dia. plate, 7 sensors at uniform 12 in. spacing
Arrangement 3:	450 mm dia. plate, 4 sensors at uniform 12 in. spacing
Arrangement 4:	450 mm dia. plate, 7 sensors at uniform 12 in. spacing
Arrangement 5:	300 mm dia. plate, 7 sensors at SHRP (non-uniform) spacings, i.e., at 0, 8, 12, 18, 24, 36, and 60 in.
Arrangement 6:	User specified, additional information required by program

<u>Note</u>: 1 in. = 25.4 mm

additional user specified plate-sensor arrangement may be used, provided some additional information required by the code is first generated. Another feature offered by the revised software is an improved set-up procedure, which may be used to initialize the Mode (Interactive or Batch), Units (English or Metric) and Plate-Sensor Arrangement to be used in the backcalculation process. The Interactive Mode is designed for use with a limited number of deflection basins, e.g., fewer than five. On the other hand, the Batch Mode is more convenient when a large number of deflection basins needs to be analyzed. In such cases, the user prepares an Input File which is read by the program directly, without resorting to on-screen questions. The Input File is formatted in such a way that in most cases files generated in situ by the FWD are acceptable with little or no modification. In the English Units Configuration, the software accepts inputs in the conventional units of pounds for plate load, and mils (1×10^{-3} in.) for sensor deflections. In the Metric Units Configuration, inputs are provided in the units of kilopascals for plate pressure, and microns (1×10^{-6} m) for sensor deflections. This configuration is primarily intended to accommodate data directly from the FWD. The user is reminded to take into account the selected plate radius when comparing inputs of plate pressure and plate load. Typically, execution times for the expanded software remain under 1 sec.

CLOSED-FORM BACKCALCULATION IN SHRP DATA ANALYSIS ACTIVITIES

The revised and expanded software, *ILLI-BACK 3.0,* was one of six backcalculation software packages selected for evaluation during a recent review conducted for SHRP's

Expert Task Group (ETG). The ultimate purpose of this undertaking was to choose "an existing backcalculation program as the best available for SHRP's purposes and to develop an evaluation procedure around it" (PCS/Law 1991). Only two of the programs reviewed were evaluated using data obtained from PCC surfaced pavements, ELCON (Dynatest 1990) and *ILLI-BACK 3.0*. The other four codes were evaluated using AC surfaced sections. The PCC test sections evaluated are shown in Fig. 3, and the results obtained for each of the three load levels considered are presented in Table 2.

To begin with, backcalculations were conducted using Arrangement 5 in Table 1 (SHRP configuration). Since a closed-form solution is only available for a two-layered system (i.e., a slab-on-grade), backcalculations for the more complex sections provided by SHRP-ETG were conducted using the top layer (PCC slab) thickness and ignoring the remainder of the layers. Thus, the moduli determined are in essence "effective moduli" E_{eff} and k_{eff}, and incorporate the effect of the presence of the base and subbase layers.

Traditionally, the sole effect of the base and subbase has been assumed to be an increased value of the subgrade modulus, k. The theoretical justification for this assumption is provided by Odemark's work using the Method of Equivalent Thicknesses (Ioannides et al. 1992). Recent investigations (Ioannides 1991; Darter et al. 1991; Barenberg et al. 1992) confirm that "top-of-the-base" k-values may be unrealistically high, and that a better transformation might result from an increased slab modulus value, E_{eff}, instead. In fact, in some cases, such as when thin slabs of low E are used, the bulk of the effect of the base is to increase E, leaving k almost unchanged. This observation can be verified easily by generating some deflection basins using a layered elastic program for a three-layered system, and then backcalculating the moduli as if this were a two-layered system. This phenomenon has important repercussions not only with respect to selecting a backcalculation program, but more generally on our understanding of effective k-values and their use in design.

Table 2 shows that in general, similar results are obtained for the three load levels, suggesting that there is no stress dependence effect involved when competent PCC pavements are considered. This confirms earlier findings using a three-dimensional finite element program (Ioannides and Donnelly 1988). Concerns, however, arise with respect to the deflections for Section K (a four-layer system incorporating a cement treated base), especially those for the lowest load level. For this section, large variations were observed in the values of the Coefficient of Variation (COV) for different load levels. Similar observations also apply to the PCC overlay Sections, O and P, which are "inverted" five- and four-layer sections, respectively. It would be advisable to use the highest load level available in such pavements because of deflection size and sensor

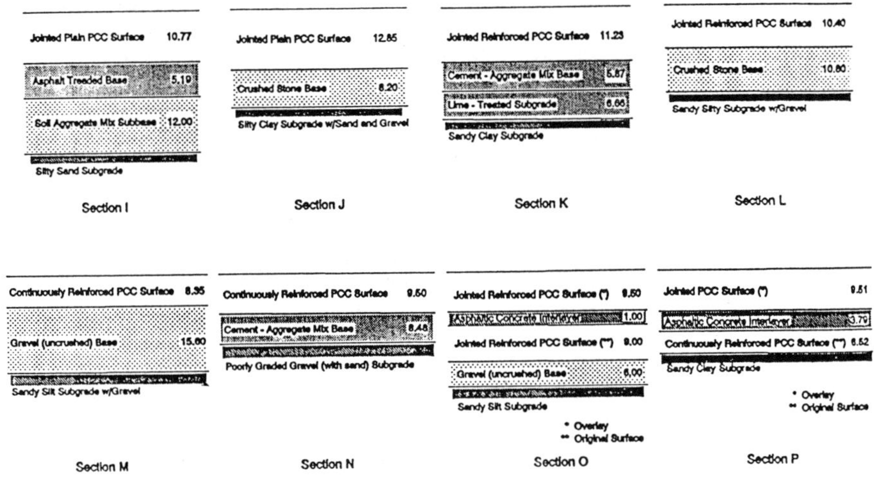

SUMMARY OF DEFLECTION DATA: PCC SURFACED PAVEMENTS

Section ID	Load (lbs)	Deflection (mils) @ r=0"	r=8"	r=12"	r=18"	r=24"	r=36"	r=60"
I	9098	1.86	1.75	1.68	1.54	1.34	1.11	0.82
	11966	2.57	2.37	2.28	2.04	1.84	1.50	1.12
	16302	3.47	3.19	3.09	2.79	2.52	2.06	1.50
J	9554	1.44	1.44	1.32	1.30	1.23	1.16	0.95
	12922	2.03	1.98	1.82	1.78	1.66	1.52	1.27
	16544	2.62	2.61	2.37	2.34	2.22	2.05	1.68
K	10116	1.62	1.65	1.60	1.56	1.50	1.39	1.13
	13258	2.49	2.26	2.19	2.09	2.02	1.84	1.50
	18238	3.02	2.86	2.80	2.68	2.59	2.37	1.91
L	9462	4.16	4.01	3.89	3.74	3.54	3.07	1.95
	12668	5.74	5.55	5.36	5.16	4.88	4.20	2.71
	16642	7.15	6.84	6.72	6.43	6.06	5.24	3.34
M	9736	4.77	4.27	4.01	3.60	3.20	2.57	1.56
	12998	6.45	5.82	5.46	4.93	4.40 .	3.50	2.16
	16938	7.90	7.12	6.70	6.04	5.40	4.35	2.72
N	9234	3.27	3.13	2.96	2.71	2.53	2.13	1.55
	11660	4.18	4.00	3.78	3.49	3.19	2.78	1.97
	15566	5.37	5.12	4.84	4.46	4.11	3.56	2.54
O	9740	2.09	2.05	1.99	1.91	1.85	1.70	1.39
	12760	2.61	2.60	2.51	2.42	2.34	2.18	1.77
	16554	3.36	3.34	3.11	2.87	2.82	2.70	2.20
P	10054	1.88	1.91	1.71	1.49	1.48	1.14	0.84
	13164	2.88	2.61	2.36	2.07	2.01	1.60	1.14
	17840	3.72	3.34	3.11	2.80	2.64	2.17	1.52

FIG. 3--SHRP-ETG PCC Surfaced Sections (After PCS/Law 1991)

TABLE 2--*ILLI-BACK 3.0* Backcalculation Results for SHRP-ETG.

SECT. ID	LOAD (lbs)	DENSE LIQUID E_{eff} (Mpsi)	k_{eff} (pci)	ELASTIC SOLID E_{eff} (Mpsi)	E_s (ksi)
I	9098	6.6	505	4.7	63
	11966	5.9	518	4.1	63
	16302	6.1	508	4.2	62
	16302	4.8	634	3.8	62
J	9554	14.9	252	12.2	49
	12922	10.5	328	8.3	55
	16544	12.5	272	10.1	50
	16544	8.1	407	7.1	58
K	10116	34.6	144	30.2	34
	13258	11.3	323	8.7	51
	18238	17.5	268	14.0	48
	18238	13.4	340	13.0	48
L	9462	5.8	143	4.5	22
	12668	5.6	139	4.3	22
	16642	5.9	149	4.4	23
	16642	5.9	145	5.2	20
M	9736	4.1	302	2.7	33
	12998	4.2	288	2.7	32
	16938	4.6	299	3.0	33
	16938	3.5	373	2.7	33
N	9234	6.9	241	5.0	33
	11660	6.9	233	5.0	32
	15566	7.1	244	5.2	33
	15566	5.2	308	4.5	33
O	9740	28.5	161	23.5	32
	12760	34.6	146	29.3	31
	16554	25.5	202	20.6	38
	16554	17.2	289	15.3	40
P	10054	11.8	501	8.5	65
	13164	6.9	618	4.6	70
	17840	8.0	590	5.4	69
	17840	5.8	775	4.4	73

Note: 1 lb = 4.448 N; 1 psi = 6.895 kPa; 1 pci = 0.271 MN/m^3. For each section, the first three drops use SHRP Plate-Sensor Arrangement, whereas the fourth uses Arrangement 1 in Table 1.

sensitivity.

In several cases, the calculated slab modulus is considerably higher than what would normally be expected. It would not be advisable to use these high moduli for stress calculation, since this would lead to greatly overestimated stresses. The primary reason for the apparently high slab moduli is the contribution of the base and subbase layers, especially when these are bonded to the concrete slab. Experience with numerical forward calculation methods shows that more complex pavement section profiles pose more difficult analytical problems. Examples of such "difficult" sections include those consisting of more than three layers, or exhibiting abrupt changes in layer moduli or even reversals of the trend of decreasing stiffness with depth ("inverted" sections), or having relatively thin layers. Such problems naturally become more pronounced in backcalculation efforts. On the other hand, when stiffer base and subbase layers are used, the recorded deflections will be smaller. This is especially true for the outer sensor (r=60 in. or 1525 mm). Thus, if sensors of the same, fixed sensitivity are used for all sensors (as is most common), it may be speculated that measurements from such pavements will be less reliable themselves. This reinforces an earlier recommendation that increasingly more sensitive sensors should be used as r increases (Ioannides et al 1989). Furthermore, since backcalculations may be performed on site immediately following an FWD drop, an effort should be made to verify excessively high slab moduli by performing additional drops, or even by destructive testing. It is not unlikely that other factors, such as equipment malfunction, plate seating errors, loss of slab support or continuity, etc., may also influence the reliability of backcalculation results. In addition, research efforts should be intensified toward obtaining forward and backcalculation solutions for multi-layered PCC pavement systems.

Since most phenomena in PCC pavements tend to be localized in nature, use of sensors at great r values should be avoided. In fact, for all SHRP cases considered, the moduli calculated on the basis of the outer sensor (r=60 in. or 1525 mm) were considerably different from those calculated from the other six sensors. The reason for placing a sensor at a large distance is related to a suggestion by Ullidtz (1977) that the soil modulus can be defined with reference to the outer sensor alone. There are two reasons, however, why this sensor should not be considered when using the closed-form backcalculation procedure. The first is that the soil modulus is not calculated from the outer sensor alone, but from every available sensor. The second is that Ullidtz's proposal is valid theoretically only as r tends to infinity. For real pavements "infinity" translates to "large enough," but how large "large enough" is depends on the stiffness of the pavement. For AC pavements, "large enough" may be 60 to 90 in. (1.5 to 2.25 m). For PCC pavements, however, "large enough" must be expected to be much larger, certainly well in excess of 60 in. (1.5 m). With deflections being as small as they are in PCC pavements, extremely sensitive sensors are necessary if measurements at such large r-values are to be meaningful.

To illustrate the effect of ignoring sensors at large r, the highest load level drop for each section was reanalyzed using Arrangement 1 in Table 1 (i.e., using four sensors with the 300 mm plate). Backcalculated moduli values are recorded in the fourth row for each section in Table 2. It is observed that in general the k-values are somewhat higher, and the slab moduli are correspondingly lower. It is believed that these moduli are more reliable "effective" values than the parameters estimated using the seven sensors. Once again, this suggests the need to reconsider prevailing practices which require more sensors.

The slab modulus values, E_{eff}, backcalculated on the basis of the plate-on-DL idealization are not the same as the corresponding moduli determined using the plate-on-ES assumption. This need not be surprising, however. It is a direct consequence of the fact that real subgrade soils are neither dense liquids nor elastic solids. The position of a real soil within the spectrum defined by these two extreme conventional idealizations may be inferred by examining the values of the COV of parameters backcalculated from each of the sensors.

CLOSED-FORM BACKCALCULATION IN FHWA DATA ANALYSIS ACTIVITIES

An example indicating that the closed-form slab-on-grade backcalculation procedure may be used with confidence in evaluating PCC pavements incorporating base and subbase layers is provided by a recently completed study sponsored by FHWA, and conducted by ERES Consultants, Inc. (Smith et al. 1990). Ninety-five PCC pavement systems were evaluated, which included a wide variety of both plain and reinforced section profiles, located in all four climatic zones of the continental United States and Canada. Table 3a presents a list of the sections evaluated, together with pertinent structural information and the mean values of the backcalculated effective parameters, E_{eff} and k_{eff}.

The sections considered in this FHWA study involved a wide variety of PCC slab thicknesses, base and subbase types and thicknesses, and subgrade types. Yet, very few of the reported backcalculation results are the cause of any concern. It should be noted that in this study, Arrangement 1 in Table 1 was used, i.e., a 300 mm load plate with four sensors. A target plate load of 9,000 lbs (40 kN) was adopted, and ten slabs were tested at each test section location. Such testing is considered necessary to ensure the repeatability and representative nature of the

TABLE 3a--<u>Backcalculation Results from FHWA Study by ERES (After Smith et al. 1990)</u>.

CLIMATE ZONE	PROJECT LOCATION	No. OF SECTIONS	PVT TYPE	SLAB h in.	BASE TYPE	h in.	SUBBASE TY-PE	h in.	SOIL TYPE	SLAB E_{eff} Mpsi	k_{eff} pci
DRY	I-94 Rothsay, MN	12	JRCP	8-9	1,2,3	5-6	0	0	A-6	6.7-9.4	172-314
FREEZE	I-94 Rothsay, MN	1	JRCP	9	1	3	0	0	A-6	7.6	156
	I-90 Albert Lea, MN	4	JR,PCP	8-9	1	5-6	0	0	A-2-7	6.6-8.0	127-178
	I-90 Austin, MN	1	JRCP	9	1	4	0	0	A-4	8.8	256
	TH 15 New Ulm, MN	1	JPCP	7.5	1	5	0	0	A-2-6	6.3	222
	TH 15 Truman, MN	1	JRCP	8.8	4	4	0	0	A-2-4	6.6	199
DRY	RT 360 Phoenix, AZ	6	JPCP	9-13	0,3,5	0-6	0,1	0-4	A-4,6	3.1-3.7	344-621
NO	I-10 Phoenix, AZ	1	JPCP	10	5	5	0	0	A-6	5.6	174
FREEZE	I-5 Tracy, CA	5	JPCP	8-11	3,5	5	1	24	A-1-a	5.2-7.0	232-433
	I-5 Sacramento, CA	1	JPCP	10	3	5	8	5	A-2-4	6.3	326
	I-210 Los Angeles, CA	2	JPCP	8	3,6	5	1	3-6	A-4	5.0-7.0	572-1423
	US 101 1000 Oaks, CA	1	JPCP	10	7	5	1	9	A-7	6.4	339
	RT 14 Solemint	1	JPCP	9	5	4	7	2	A-2-4	6.7	294
WET	US 10 Clare, MI	8	JR,PCP	9	1,4	4	1	10	A-2-4	5.3-6.3	300-502
FREEZE	I-69 Charlotte, MI	2	JRCP	10	9	4	1	3	A-2-4	4.4	186
	I-94 Marshall, MI	1	JRCP	9	1	4	1	10	A-4	4.5-4.8	189-283
	I-94 Paw Paw, MI	1	JRCP	10	9	4	1	21	A-2-4	4.5	233
	RT 23 Catskill, NY	6	JR,PCP	9	1,2	3-6	1	8	A-2-4	3.8-4.1	503-619
	I-88 Otego, NY	4	JR,PCP	9	1	4-6	0,1	0,8	A-1-a	5.1-6.1	273-471
	RT 23 Chillicothe, OH	7	JRCP	9	1,2	4-8	0	0	A-4,6	3.4-5.3	340-525
	SR 2 Vermillion, OH	2	JPCP	15	0	0	0	0	A-4	-	-
	HWY 3N Ruthven, ONT	4	JPCP	7-12	0,4,5	4-5	0	0	A-7-6	-	-
	HWY 427 Toronto, ONT	1	JPCP	9	3	6	0	0	-	-	-
	RT 422 Kittanning, PA	5	JRCP	10	1,3,4	5-13	0,1	0-8	A-4	3.2-4.5	538-1040
	RT 130 Yardville, NJ	1	JRCP	10	1	5	1	7	A-4	6.7	234
	RT 676 Camden, NJ	2	JRCP	9	4,10	4	11	4	A-2-4	5.3-5.4	210-356
WET	US 101 Guyserville, CA	3	JPCP	9	3	5	1	6	A-4	3.5-4.2	286-397
NO	I-95 Rocky Mount, NC	8	JR,PCP	8-9	1,2,3	4-6	0	0	A-2-4	3.9-5.5	128-672
FREEZE	I-85 Greensboro, NC	1	JPCP	11	5	5	0	0	A-4	5.9	293
	I-75 Tampa, FL (Hill.)	1	JPCP	13	1	6	0	0	A-3	5.6	378
	I-75 Tampa, FL (Man.)	1	JPCP	9	5	6	0	0	A-3	4.2	529
	TOTAL	95									

Base and Subbase Types: 0 NONE; 1 AGG; 2 ATB; 3 CTB; 4 PATB; 5 LCB; 6 PCTB; 7 HMAC; 8 LTSG; 9 PAGG; 10 NSOG; 11 LFAS

TABLE 3b--<u>Sections With E_{eff}-values in Excess of 7 Mpsi (After Smith et al. 1990)</u>.

CLIMATE ZONE	PROJECT LOCATION	SECTION No.	PVT TYPE	SLAB h in.	BASE TYPE	h in.	SUBBASE TYPE	h in.	SOIL TYPE	SLAB E_{eff} Mpsi	k_{eff} pci
DF	I-94 Rothsay, MN	1-6	JRCP	8	2	5	0	0	A-6	9.4	314
DF	I-94 Rothsay, MN	1-5	JRCP	8	2	5	0	0	A-6	9.1	304
DF	I-90 Austin, MN	3	JRCP	9	1	4	1	10	A-4	8.8	256
DF	I-94 Rothsay, MN	1-7	JRCP	9	2	5	0	0	A-6	8.3	287
DF	I-94 Rothsay, MN	1-11	JRCP	8	3	5	0	0	A-6	8.0	245
DF	I-90 Albert Lea, MN	2-2	JPCP	8	2	6	0	0	A-2-6	8.0	127
DF	I-94 Rothsay, MN	1-8	JRCP	9	2	5	0	0	A-6	7.9	278
DF	I-94 Rothsay, MN	1-12	JRCP	8	3	5	0	0	A-6	7.8	239
DF	I-94 Rothsay, MN	1-2	JRCP	9	1	6	0	0	A-6	7.8	172
DF	I-94 Rothsay, MN	5	JRCP	9	1	3	0	0	A-6	7.6	156
DF	I-90 Albert Lea, MN	2-3	JRCP	9	2	5	0	0	A-2-6	7.3	162
DF	I-94 Rothsay, MN	1-1	JRCP	9	1	6	0	0	A-6	7.1	191
DNF	I-210 Los Angeles, CA	2-2	JPCP	8.4	6	5.4	1	3	A-4	7.0	1423

<u>Note</u>: 1 lb = 4.448 N; 1 psi = 6.895 kPa; 1 pci = 0.271 MN/m^3.

backcalculated parameters. In many cases, additional
destructive testing was used to provide more information
about the section considered and the materials it comprised.
Thus, modulus of rupture, M_R, values were obtained at many
of the sites, by testing core samples taken from the PCC
slabs. The cores also served the very crucial function of
verifying the actual slab thickness in the field, in view of
the great impact small departures from the design thickness
can have on the backcalculated parameters. In addition,
some of the test sections considered were designated as
control sections.

Considering in particular the backcalculated mean
concrete modulus values, it is observed in Table 3b that
very few fall outside SHRP's "range of reasonableness" of 3
to 7 Mpsi. Darter et al. (1991) attribute these results to
slab age (mean= 15 years). A closer examination of these
"exceptional" cases suggests that the climatic zone and
presence of reinforcement may also be contributing factors.

CLOSED-FORM BACKCALCULATION IN ILDOT DATA ANALYSIS ACTIVITIES

An innovative approach to extending the applicability
of the closed-form slab-on-grade backcalculation approach to
more complex pavement systems was recently presented by Hall
(1991). During a project sponsored by ILDOT, PCC pavements
that had been overlaid with a thin AC layer were evaluated.
The primary departure of such pavements from the slab-on-
grade system is the through-the-thickness compression
experienced by the AC layer. Comparisons with results
obtained using layered elastic theory -program BISAR (Peutz
et al. 1968)- suggest that sufficiently accurate
backcalculated moduli are obtained if the maximum
deflection, D_0, is reduced by the amount of AC compression,
δ_0, before using the measured deflection basin in the slab-
on-grade procedure as usual. The following expressions have
been suggested for δ_0, in inches (Hall 1991):

(a) Bonded AC overlay:
$$\delta_0 = -0.0000328 + 121.5006 \left(\frac{h_{AC}}{E_{AC}}\right)^{1.0798} \qquad (13)$$

(b) Unbonded AC overlay:
$$\delta_0 = -0.00002132 + 38.6872 \left(\frac{h_{AC}}{E_{AC}}\right)^{0.94551} \qquad (14)$$

where E_{AC} and h_{AC} are the elastic modulus in psi and
thickness in inches, of the AC overlay, respectively. Note
that relatively smaller through-the-thickness compressions
occur at other FWD sensors, as well, but these may be
neglected.

A large volume of data were analyzed by Hall (1991)
using this extended procedure. For one hundred deflection
basins obtained on I-74 near Mansfield, IL, on a 7-in.

continuously reinforced concrete pavement section overlaid
with a 3-in. AC overlay, backcalculated moduli were compared
to results from BISDEF (Bush and Alexander 1985), a robust
backcalculation program using layered elastic analysis
program BISAR. Figure 4 presents a comparison between the
concrete and subgrade moduli obtained using the two
procedures. It is interesting to observe that the BISDEF
concrete moduli are consistently higher than those from the
extended procedure, while the BISDEF subgrade moduli were
correspondingly lower. Hall (1991) ascribes the scatter
observed to "inherent differences between the two theories,"
i.e., characterization of the concrete slab as a plate in
one compared to an elastic layer in the other. She also
notes that "the outputs of the BISDEF backcalculation are
less repeatable than those of the plate theory
backcalculation method. In fact, the PCC moduli obtained
from BISDEF may be varied by several hundred thousand psi by
manipulating the seed moduli, moduli limits, and tolerance
limits." No seed moduli are required by the original or the
extended closed-form backcalculation procedures.

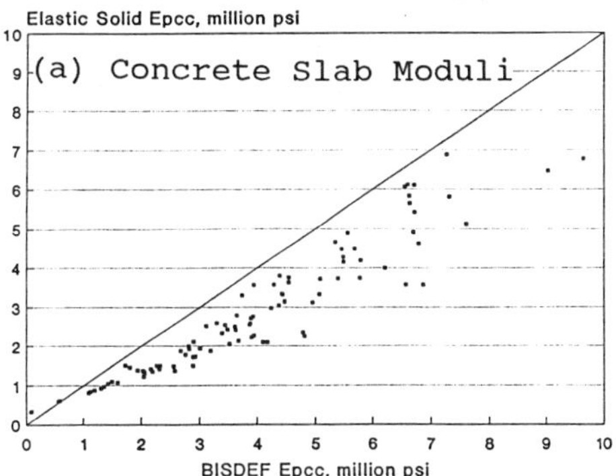

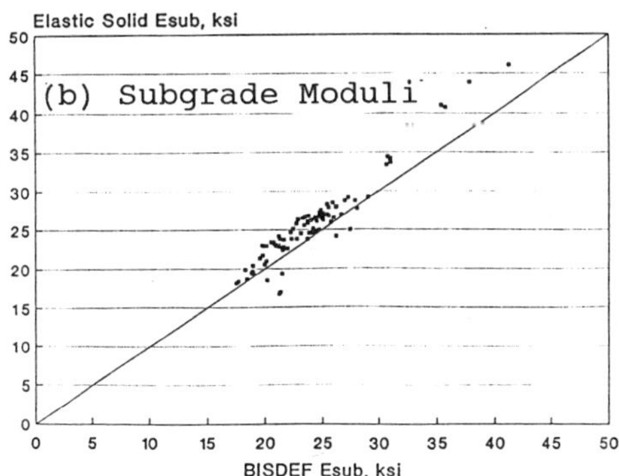

FIG. 4--Comparison of Backcalculated Parameters Using Plate
and Layered Elastic Theories (After Hall 1991)

The effect of slab size on the value of the subgrade modulus, k, backcalculated using the closed-form procedure, which was originally developed based on the infinite-plate assumption, may be accounted for using another corrective approach developed by Crovetti and Tirado-Crovetti (1994). This involves multiplying the measured maximum deflection, D_0, as well as a first estimate of the radius of relative stiffness, ℓ_{est}, (obtained, for example, from an initial backcalculation, which assumes infinite slab conditions) by appropriate correction factors, as follows:

$$CF = 1 - k_1\, e^{-k_2\left(\frac{L}{\ell_{est}}\right)^{k_3}} \qquad (15)$$

in which L is the actual slab size. Three regression constants, k_i, are required for each of the two correction factors. These are: $k_1=5.29875$, $k_2=2.17612$ and $k_3=0.49895$ for the ℓ-value correction; and $k_1=1.06817$, $k_2=0.66914$ and $k_3=0.84408$ for the D_0-value correction. These factors were established by analyzing deflection basins generated using the finite element computer program ILLI-SLAB (Ioannides et al. 1985b) and comparing backcalculated k-values to those provided as inputs to the forward calculation. Figure 5 shows that this corrective approach is more critical as the governing (L/ℓ) ratio decreases below about 5, e.g., for cracked slabs.

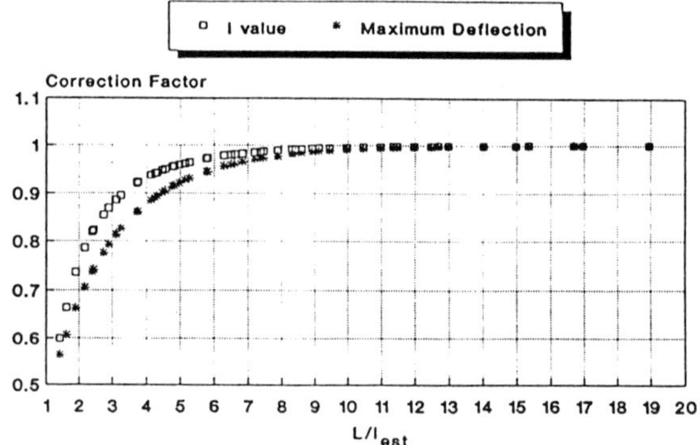

FIG. 5--Correction Factors for Slab Size Effect (After Crovetti and Tirado-Crovetti 1994)

BEYOND THE PLATE LOAD TEST

An important motivation for developing a backcalculation procedure specifically for concrete pavement systems is the desire to obtain in situ estimates of the modulus of subgrade reaction, k. The conventional method for determining the subgrade k-value has been the Plate Load Test (PLT), conducted on the unprotected subgrade. This test is well suited for first design exercises, which are relatively insensitive to the value of k. In fact, in view of this insensitivity and the cumbersome nature of the PLT, it is quite adequate in many cases to obtain estimates of k from correlations with soil classification (Terzaghi 1955). As a result, the parameter thus obtained approximates rather crudely the support provided to the pavement slab, resting as it is on a sequence of layers of differing thickness and stiffness characteristics, and being subjected to a seemingly endless variety of tire loads applied at arbitrary locations on the pavement surface.

In contrast, evaluation of in-service PCC pavements and rehabilitation design calculations are more sensitive to the k-value obtained. Furthermore, even though a multi-layer concrete pavement system may be reduced to a slab-on-grade for the purposes of first design, such a simplification may not be adequate in an evaluation scheme aimed at characterizing in situ all materials present in the pavement system. Thirdly, conducting a PLT in such cases would necessitate the removal of the slab, base and subbase materials to expose the subgrade surface. Note that unrealistically high k-values may result from testing on top of the base or subbase, primarily due to the departure of such a system from the assumed dense liquid behavior. Therefore, the k-value estimated must reflect as far as possible the actual support provided by the subgrade under the action of the actual load.

The need to determine the modulus of subgrade reaction, k, as it is actually mobilized under field conditions stems from the fact that this parameter is not an intrinsic soil property, but a pavement system characteristic. As such it may be expected to be sensitive to changes in any of the three primary pavement components, namely of the constructed layers, of the natural subgrade and of the geometry of the applied loads. Thus, the conventional 30-in. diameter plate used in the PLT is not relevant when testing in situ pavement systems, as opposed to unprotected subgrades. In addition, loads of a frequency close to that of pavement traffic need to be applied. How then should the k-value be determined?

Teller and Sutherland (1943) identified "at least three methods or procedures by which the load sustaining ability of the soil can be measured under field conditions." The first of these is the PLT, conducted on the unprotected subgrade itself. The other two methods involve testing on the slab surface. In the so-called "volumetric approach," k is determined as the ratio of the total applied plate load to the volume of the deflection basin measured. In the third approach, k is obtained "from the maximum slab deflection under the applied load by means of the deflection formulas given by Westergaard... if the elastic modulus of the concrete in the slab is known" (Teller and Sutherland 1943). The latter method may be recognized as a rudimentary form of the modern-day backcalculation approach, and, interestingly enough, was Westergaard's own preferred method for estimating the k-value: "The modulus k may be determined empirically for a given type of subgrade by comparing the deflections found by tests of full-sized slabs with the deflections given by the formulas" (Westergaard 1926). Why then did the PLT become the conventional method for determining the k-value?

Although anyone who has actually been involved in performing a PLT probably knows how cumbersome a test that is, it was until a few years ago the most convenient one to use among the three options available. The primary difficulty with both approaches which involve testing on top of in situ slabs is the very small magnitude of the deflections produced, or conversely the very high load levels that should be achieved if deflections were to be measured with sufficient accuracy. In fact, since the volumetric approach theoretically demands integrating the deflection basin from the center of the load to infinity, this difficulty all but precluded its application to actual field situations. It is precisely the availability of sensitive, inexpensive and easy to use deflection sensors that has provided more recently the impetus for backcalculation studies. A second, less important, difficulty presented by the test-on-slab approaches is that a slab is necessary before a k-value is determined, but a k-value is necessary before a slab may be designed and built. In view of the availability of comparable sections nearby and the insensitivity of predicted responses (particularly of bending stresses) to the k-value assumed, as discussed earlier, this need not be an insurmountable obstacle even when considering first design.

If backcalculation were to replace the PLT as the conventional method for determining the in situ subgrade modulus, it appears desirable to limit field measurements to a region reasonably close to the center of the applied load, in recognition of the sensitivity and limitations of deflection sensors currently available. In this region, deflections assume their larger values and may, therefore, be expected to be most reliable. This argument, however, needs to be balanced with the requirement to obtain representative estimates of the k-value, reflecting the natural variability of support conditions over an adequately broad area. Although the final choice is probably best left to the judgment and experience of local engineers, it appears that the four sensor arrangement spaced at 12-in. (305 mm) centers (i.e., extending to 36 in. (915 mm) from the center of the plate) may be a suitable choice for many projects. Positioning an additional sensor at 8 in. (205 mm) from the center of the load, as required by the SHRP arrangement, may also prove to be beneficial. The practice of using a distant sensor for the determination of the subgrade modulus -in SHRP's case at r=60 in. (1525 mm)- may lead to questionable results because of the very small magnitude of the deflection measured, especially when concrete pavements are considered.

CONCLUSION

A discussion of issues related to the backcalculation
of concrete pavement system parameters is presented,
particularly in light of recent experiences using a closed-
form slab-on-grade approach. Through a computerized
implementation of the procedure, it is possible to
backcalculate k and E_s from each of the sensors;
backcalculate either E or h; perform statistical analysis of
backcalculated parameters; use the interactive or batch mode
of execution; analyze metric or U.S. customary input data;
and maintain trivial execution times (less than 1 CPU sec.
per basin); etc. Enhancements described in this Paper allow
the engineer a choice from among several popular plate-
sensor arrangements, involving the 300 or 450 mm dia. plate,
with four or seven sensors, as well as the irregular sensor
spacings used by SHRP.
Factors influencing the selection of the plate-sensor
arrangement are outlined, and the need to limit the extent
of the deflection basin only to a region large enough to be
considered representative is emphasized. Therefore, use of
the conventional 300 mm plate with four uniformly spaced
sensors is encouraged, whereas use of an "outer sensor",
e.g., at r=60 in. (1525 mm) in the SHRP arrangement, is not
recommended. It is considered desirable to adopt
backcalculation as the conventional method for estimating
the subgrade modulus. Such a change is hampered at the
present time by the lack of a forward calculation procedure
for multi-layered concrete pavement systems. The prospect
of formulating such a procedure has recently been
considerably improved, mainly as a result of a study by Van
Cauwelaert (1990) that extended Burmister's multi-layer
theory to the DL foundation.

REFERENCES

Barenberg, E. J., Lane, R. and Goodman, G. R., 1992,
    "Pavement Design for Heavy Aircraft Loading,"
    Proceedings, Third International Symposium on Heavy
    Vehicle Weights and Dimensions, Queens' College
    Cambridge, U.K., pp. 284-290.

Bentsen, R. A., Bush III, A. J. and Harrison, J. A., 1989,
    "Evaluation of Nondestructive Test Equipment for
    Airfield Pavements - Phase I: Calibration Test Results
    and Field Data Collection," Technical Report GL-89-3,
    U.S. Army Engineer Waterways Experiment Station,
    Vicksburg, MS.

Bush III, A. J. and Alexander, D. R., 1985, "Pavement
    Evaluation Using Deflection Basin Measurements and

Layered Theory," _Transportation Research Record 1022_, Transportation Research Board, National Research Council, Washington, D.C., pp. 16-29.

Crovetti, J. A. and Tirado-Crovetti, M. R., 1994, "Evaluation of Support Conditions Under Jointed Concrete Pavement Slabs," _Nondestructive Testing of Pavements and Backcalculation of Moduli (Second Volume)_, ASTM STP 1198, Harold L. Von Quintas, Albert J. Bush, and Gilbert Y.Baladi, Eds., American Society for Testing Materials, Philadelphia, PA.

Darter, M. I., Smith, K. D. and Hall, K. T., 1991, "Concrete Pavement Backcalculation Results from Field Tests," _Transportation Research Record 1377_, Transportation Research Board, National Research Council, Washington, D.C., pp. 7-16.

Dynatest Consulting, Inc., 1990, "_ELMOD and ELCON: Evaluation of Layer Moduli and Overlay Design - User's Manual_," Ojai, CA.

Hall, K. T., 1991, "_Performance, Evaluation, and Rehabilitation of Asphalt-Overlaid Concrete Pavements_," thesis presented to the University of Illinois, at Urbana, IL, in partial fulfillment of the requirements for the degree of Doctor of Philosophy.

Hoffman, M. S. and Thompson, M. R., 1981, "Mechanistic Interpretation of Nondestructive Testing Deflections," _Civil Engineering Studies, Transportation Engineering Series No. 32, Illinois Cooperative Highway and Transportation Research Program Series No. 190_, University of Illinois, Urbana, IL.

Ioannides, A. M., 1987, Discussion of "Response and Performance of Alternate Launch and Recovery Surfaces (ALRS) Containing Stabilized-Material Layers," by R.R. Costigan and M.R. Thompson, _Transportation Research Record 1095_, Transportation Research Board, National Research Council, Washington D.C., pp. 70-71.

Ioannides, A. M., 1988, "A Closed-Form Backcalculation Procedure for Concrete Pavements," presented at _First International Symposium on Nondestructive Testing of Pavements and Backcalculation of Moduli_, Workshop 2, ASTM, Baltimore, MD.

Ioannides, A. M., 1990, "Dimensional Analysis in NDT Rigid Pavement Evaluation," _Journal of Transportation Engineering_, ASCE, Vol. 116, No. 1, pp. 23-36.

Ioannides, A. M., 1991, "Subgrade Characterization for

Portland Cement Concrete Pavements," <u>Proceedings</u>, International Conference on Geotechnical Engineering for Coastal Development -Theory and Practice-, Port and Harbour Institute, Ministry of Transport, Yokohama, Japan, pp. 809-814.

Ioannides, A. M. and Donnelly, J. P., 1988, "Three-Dimensional Analysis of Slab on Stress Dependent Foundation," <u>Transportation Research Record 1196</u>, Transportation Research Board, National Research Council, Washington D.C., pp. 72-84.

Ioannides, A. M., Thompson, M. R., and Barenberg, E. J., 1985a, "Westergaard Solutions Reconsidered," <u>Transportation Research Record 1043</u>, Transportation Research Board, National Research Council, Washington, D.C., pp. 13-23.

Ioannides, A. M., Thompson, M. R., and Barenberg, E. J., 1985b, "Finite Element Analysis of Slabs-On-Grade using a Variety of Support Models," <u>Proceedings</u>, Third International Conference on Concrete Pavement Design and Rehabilitation, Purdue University, W. Lafayette, IN, pp. 309-324.

Ioannides, A. M., Barenberg, E. J. and Lary, J. A., 1989, "Interpretation of Falling Weight Deflectometer Results Using Principles of Dimensional Analysis," <u>Proceedings</u>, Fourth International Conference on Concrete Pavement Design and Rehabilitation, Purdue University, W. Lafayette, IN, pp. 231-247.

Ioannides, A. M., Khazanovich L., and Becque, J. L., 1992, "Structural Evaluation of Base Layers in Concrete Pavement Systems," <u>Transportation Research Record 1370</u>, Transportation Research Board, National Research Council, Washington D.C., pp. 20-28.

Losberg, A., 1960, "Structurally Reinforced Concrete Pavements," <u>Doktorsavhandlingar Vid</u> Chalmers Tekniska <u>Högskola</u>, Göteborg, Sweden.

PCS/Law Engineering, 1991, "SHRP's Moduli Backcalculation Procedure: Software Selection," <u>Report under Contract No. SHRP-90-P-001B</u>, Beltsville, MD.

Peutz, M. G. F., Van Kempen, H. P. M. and Jones, A., 1968, "Layered Systems Under Normal Surface Loads," <u>Highway Research Record No. 228</u>, Highway Research Board, National Research Council, Washington, D.C., pp. 34-45.

Smith, K. D., Peshkin, D. G., Darter, M. I., Mueller, A. L. and Carpenter, S. H., 1990, "Performance of Jointed

Concrete Pavements - Volume 4: Appendix A - Project
Summary Reports and Summary Tables," <u>Report No. FHWA-RD-89-139</u>, ERES Consultants, Inc., Savoy, IL.

Teller, T. W. and Sutherland, E. C., 1943, "The Structural
Design of Concrete Pavements: Part 5," <u>Public Roads</u>,
Vol. 23, No. 8, pp. 167-212.

Terzaghi, K., 1955, "Evaluation of Coefficients of Subgrade
Reaction," <u>Géotechnique</u>, Vol. 5, No. 4, London,
England, pp. 297-326.

Ullidtz, P., 1977, "Overlay and Stage by Stage Design,"
<u>Proceedings</u>, Fourth International Conference on the
Structural Design of Asphalt Pavements, Vol. 1,
University of Michigan, Ann Arbor, MI, pp. 722-735.

Van Cauwelaert, F., 1990, "'Westergaard's Equations' for
Thick Elastic Plates," <u>Proceedings</u>, Second
International Workshop on the Design and Rehabilitation
of Concrete Pavements, Siyüenza, Spain, pp. 165-175.

Westergaard, H. M., 1926, "Computation of Stresses in
Concrete Roads," <u>Proceedings</u>, 5th Annual Meeting of the
Highway Research Board, Part I, pp. 90-112.

Westergaard, H. M., 1939, "Stresses in Concrete Runways of
Airports," <u>Proceedings</u>, 19th Annual Meeting of the
Highway Research Board, Washington, D.C., pp. 197-202.

Sameh M. Zaghloul[1], Thomas D. White[1], Vincent P. Drnevich[1], and Brian Coree[2]

DYNAMIC ANALYSIS OF FWD LOADING AND PAVEMENT RESPONSE USING A THREE-DIMENSIONAL DYNAMIC FINITE ELEMENT PROGRAM

REFERENCE: Zaghloul, S. M., White, T. D., Drnevich, V. P., and Coree, B., "Dynamic Analysis of FWD Loading and Pavement Response Using a Three-Dimensional <u>Nondestructive Testing of Payments and Backcalculation of Moduli (Second Volume), ASTM STP 1198</u>, Harold L. Von Quintas, Albert J. Bush, III, and Gilbert Y. Baladi, Eds., American Society for Testing and Materials, Philadelphia, 1994.

ABSTRACT: Multi-layer analyses are commonly used to analyze falling weight deflectometer (FWD) measurements and back calculate pavement layer moduli. This type of analysis assumes static loading conditions and linear elastic material properties. The FWD loading cycle cannot be considered in any sense as a static load. It is a dynamic load with a duration in the range of 30 to 40 msec. Also, paving materials and subgrades are not linear elastic materials and their response to static loads is different than that to dynamic loads, such as FWD loading. The difference between the multi-layer analysis assumptions and actual loading and material conditions is significant.

In this paper a three-dimensional dynamic finite element program (3D-DFEM), ABAQUS, is used to conduct a non-linear dynamic analysis of FWD tests on a flexible pavement section. Verification studies have been conducted of the 3D-DFEM to verify its static and dynamic analysis of both rigid and flexible pavements and no significant difference was found between the predicted pavement response using the 3D-DFEM and the field measured pavement response. A design of experiment was developed to study the effect of layer thicknesses and moduli on pavement surface deflections at various offset distances. Analysis of variance (ANOVA) and regression analysis were conducted to develop statistical models which can be used to predict pavement surface deflection at different offset distances as a function of layer thicknesses and moduli.

KEYWORDS: three-dimensional finite element analysis, dynamic analysis, nondestructive testing, FWD, backcalculation layer moduli, non-linear material models.

FWD LOADING CYCLE

A data set for a full depth asphalt section shown in Figure 1 is used in the analysis that follows. A typical load history from that data

---

[1]Post-Doctoral research engineer, professor, professor and head, respectively, School of Civil Engineering, Purdue University, West Lafayette, IN 47907.

[2]Pavement engineer, Division of Research, Indiana Department of Transportation, West Lafayette, IN 47906.

set is shown in Figure 2, and the associated surface deflection histories at the various offset distances are shown in Figure 3. As can be seen from those figures, the loading duration is about 30 msec and the maximum deflection at the various offset distances occurred after the peak load (i.e. there is a phase lag between load and deflection).

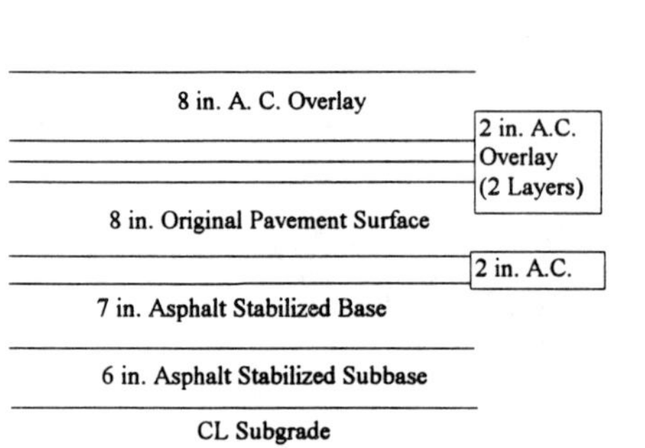

Figure 1 Structure of the Full Depth Asphalt Section

Figure 2 FWD Loading Cycle (1 psi = 6896 Pa)

3D-DFEM ANALYSIS

Finite Element Mesh Geometry

A finite element mesh (FEM) was configured, as shown in Figure 4, to represent the pavement structure as three layers. Bed rock was initially assumed at a depth of 95 in (241.3 cm) from the pavement surface. Figure 5 shows a horizontal plane of the mesh used in the analysis. With the exception of approximating the area of the load plate, a three-dimensional (3-D), 8-node element was used to fill this mesh.

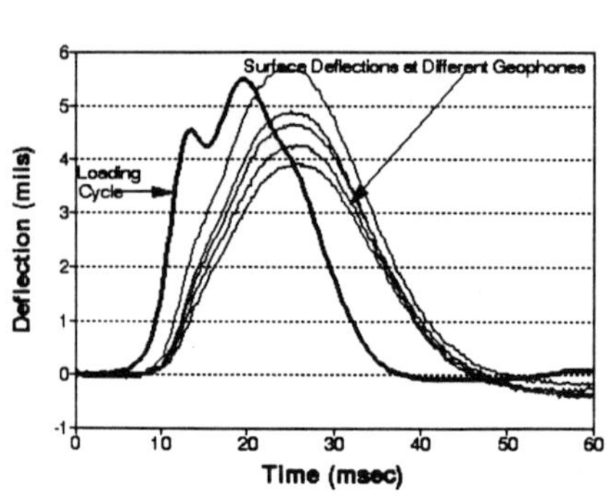

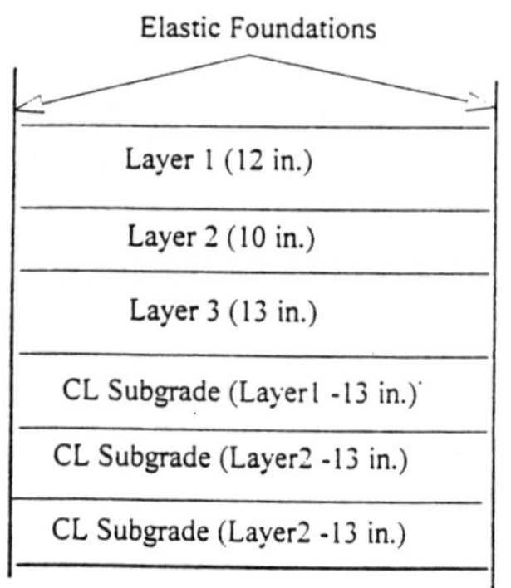

Figure 3 Deflection Histories at Various Offset Distances (1 in. = 25.4 mm)

Figure 4 Structure of the Modeled Pavement Section (1 in. = 25.4 mm)

In FWD tests, the load is applied to the pavement surface through a circular plate. In the FEM modeling, two cases of loading were considered:

1.   Approximating the circular load by a point load, assuming that the dimension of the loaded plate, 5.9 in (14.99 cm) radius, is relatively small compared to the pavement width of more than 288 in

(731.52 cm).

2.   A distributed load on an approximation of a circle, as shown in
     Figure 6.

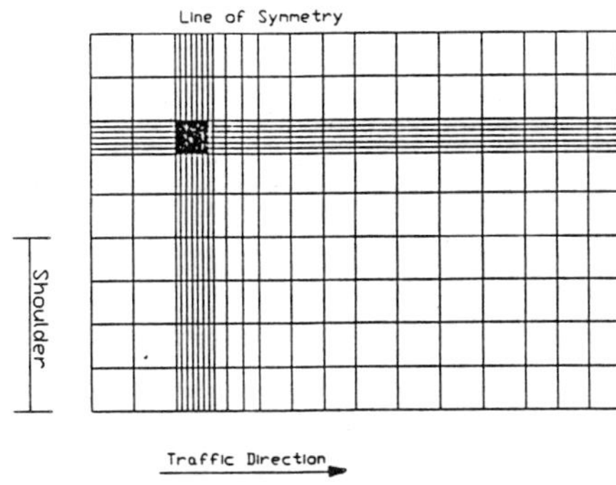

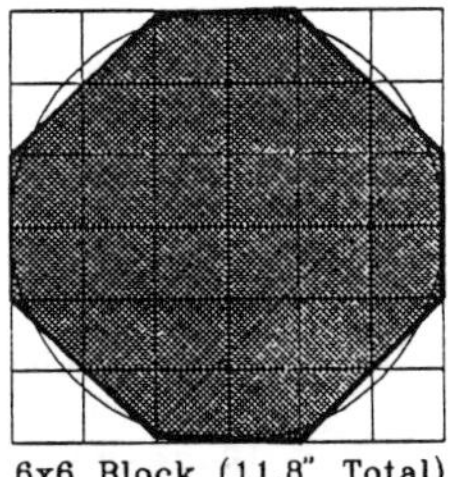

Figure 5 One Quarter Symmetry of the   **Figure 6** The 3D-DFEM Loading Area
        3D- Finite Element Mesh

     In the second case a set of 3-D, 6-node triangle elements was used
to approximate the loaded area. A comparison between the surface
deflection under the point load and under the distributed load is shown in
Figure 7. Although the difference between the predicted deflections for
the two cases at zero offset distance was not considered to be
significant, a decision was made to use the distributed load in subsequent
analysis.

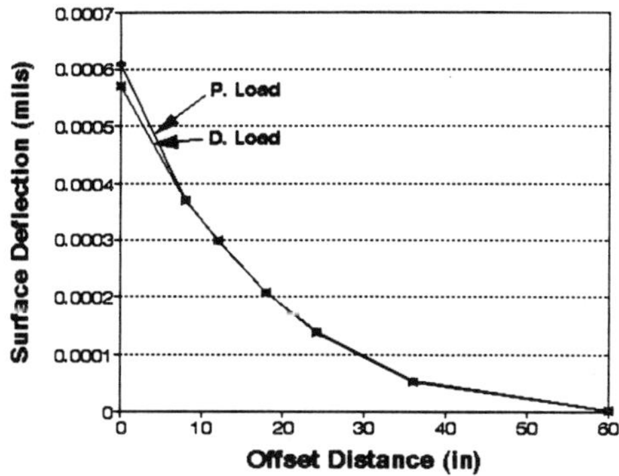

Figure 7 Comparison Between Point and Distributed Loads
              (1 mils = 25.4e-3 mm)

Material Properties

     In multi-layer analysis, such as with Bitumen Structures Analysis in
Roads (BISAR) [1], assumptions are made that the paving materials and
subgrades are linear elastic. In the 3D-DFEM analysis, the paving
materials and subgrade were categorized into three groups: asphalt
aggregate mixtures, granular materials and fine-grained, cohesive soils.
     The asphalt mixture is modeled as a visco-elastic material and the
time dependent properties are represented by instantaneous shear modulus
and long-term shear modulus [2]. Instantaneous shear modulus was selected
at a loading time of 10 msec, while long term shear modulus was selected

at a loading time of 1 sec. Figure 8 shows the results of a series of
laboratory creep tests in which the loading time and temperature were
varied [3]. In these tests the mix stiffness (E_{asp}) was defined as:

$$E_{asp} = \frac{\sigma_t}{\epsilon_t}$$

Where

ϵ_t = total strain at any time measured from a constant stress
creep test.

$$G_{asp} = \frac{E_{asp}}{2(1+\nu)}$$

Where

G_{asp} = Shear modulus.

ν = Possion's ratio.

In previous studies [4 and 5] a sensitivity analysis was conducted of the
effect of different material properties on predicted pavement response. In
these studies, ranges of values of instantaneous and long-term shear
moduli of asphalt mixtures were utilized. With this background and the
information available in the literature, reasonable values were assumed in
this current study for visco-elastic model input parameters.

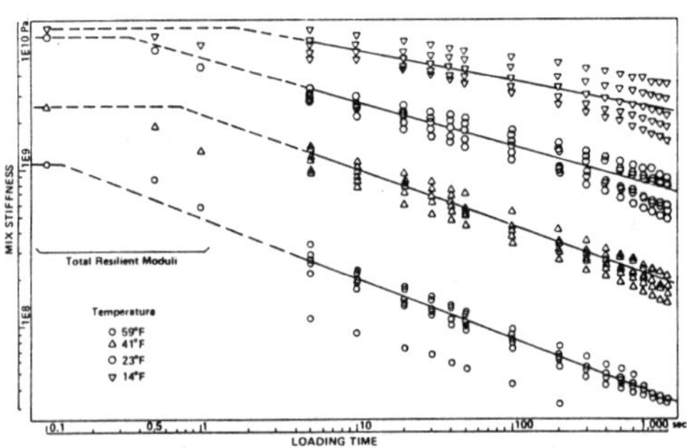

Figure 8 Effect of Loading Time and Temperature on Asphalt Mixtures
Stiffness [3]

Granular materials, base and subbase courses and subgrade in some
cases, can be modeled using the extended Drucker-Prager model [6 and 2].
This model assumes that the material will behave as an elastic material
for low stress levels. When the stress level reaches a certain limit,
yield stress, the material will start to behave as an elastic-plastic
material. The assumed stress-strain curve for a granular material is shown
in Figure 9.

The extended Cam-Clay model [7, 8 and 2] can be used to model fine-
grained, cohesive soils. This model uses a strain rate decomposition in
which the rate of deformation of the clay is decomposed additively into an
elastic and plastic part. Figure 10 shows the assumed soil response in
pure compression.

Also, damping coefficients for all layers are included for the
dynamic analysis. Other material properties required in the analysis
include modulus of elasticity, Poisson's ratio and bulk density. Values of
these parameters were obtained from the information available in the
literature and were evaluated and verified in previous sensitivity studies

conducted and reported by Zaghloul and White 1993a. Table 1 shows a sample of the material and model characteristics used in the analysis.

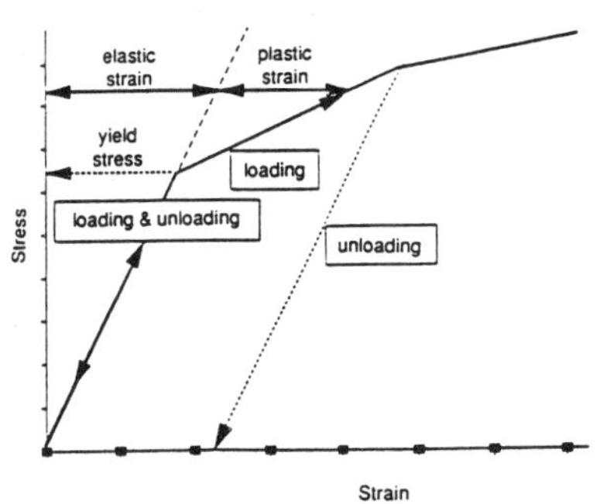

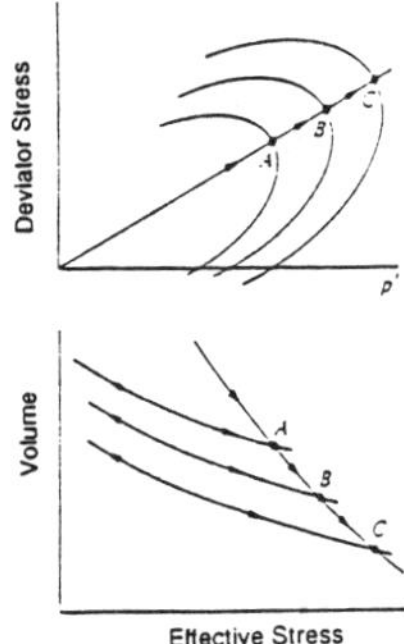

Figure 9 Drucker-Prager Model for Granular Materials [2]

Figure 10 Cam-Clay Model for Clays [13]

Table 1 An Example of the Material and Model Characteristics Used in the Analysis

Material	E (psi)	Possion's Ratio	G-Ratio	Damping Coeff.	Density (pcf)	Phi (degree)	Cohesion (psf)
Asphalt Mixtures	600,000	0.3	0.85	0.05	150	-	-
Granular Base	60,000	0.35	-	0.05	135	38	-
CL Subgrade	5,000	0.4	-	0.05	115	0	500

Loading Cycle

An FWD loading cycle is shown in Figure 11. This loading cycle was modeled by the straight line segments shown in the same figure.

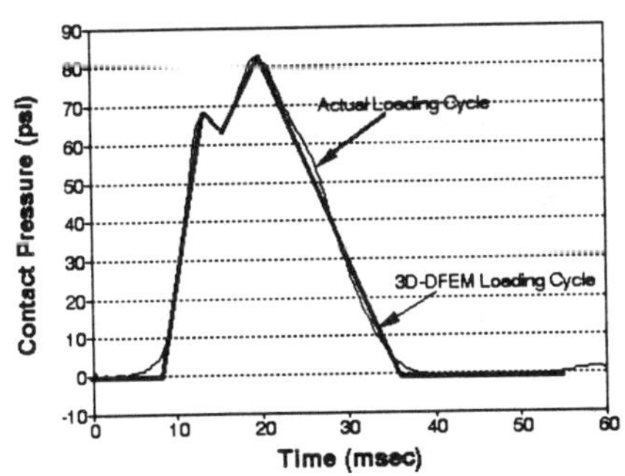

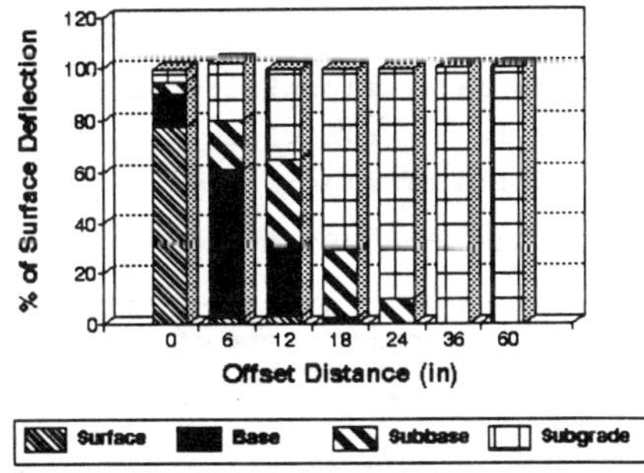

Figure 11 The 3D-DFEM FWD Loading Cycle (1 psi = 6896 Pa)

Figure 12 Pavement Deflection Distribution by Layer (3D-DFEM Predictions - 1 in. = 25.4 mm)

COMPARISON BETWEEN THE MEASURED AND PREDICTED DEFLECTIONS

Prior to initiating the process of matching the field deflections of the given pavement section, an evaluation was made of the contribution of each layer to the surface deflection at various offset distances. The full depth asphalt section was modeled as three layers on a CL subgrade soil. The FWD load cycle was applied. Peak surface deflections as well as peak deflections at the top of each layer were predicted and are shown in Figure 12. As expected, at the center of the loaded area all layers contribute with varying ratios to the surface deflection, while far from the load, most of the surface deflection comes from the subgrade. Most importantly, as suggested by Figure 12, the 3D-DFEM provides reasonable definition for the dynamic case of the contribution of each layer to the surface deflection at various offsets.

The same case was analyzed for no bedrock and bedrock at 95 inches with BISAR using layer moduli obtained from BISDEF. Results of these analysis are shown in Figure 13. Bisar predicts a significantly different distribution of deformation in the layers compared to the 3D-DFEM. This particular case is interesting and indicates a limitation of the elastic multi-layer analyses.

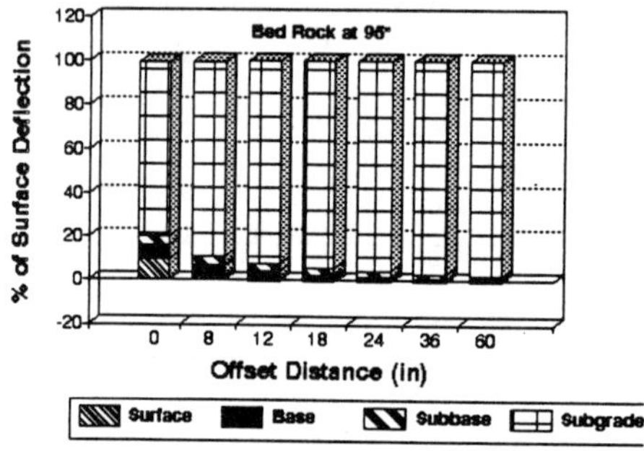

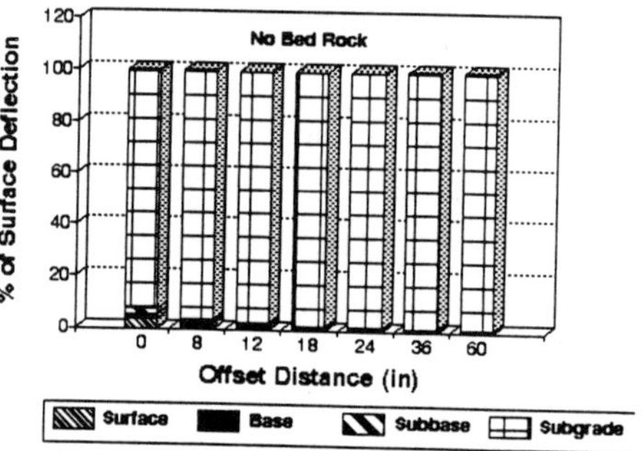

Figure 13A Pavement Deflection Distribution by Layer (BISAR Bed Rock @ 95 in. - 1 in. = 25.4 mm)

Figure 13B Pavement Deflection Distribution by Layer (BISAR No Bed Rock - 1 in. = 25.4 mm)

To study this point in more detail, a partial factorial design of experiment (DOE) was developed for the full depth asphalt section. Four factors at three levels each were considered:

- Modulus of elasticity of the first layer (E_s).
- Modulus of elasticity of the second layer (E_b).
- Modulus of elasticity of the third layer (E_{sb}).
- Modulus of elasticity of the CL subgrade (E_{sg}).

An analysis of variance (ANOVA) was conducted on the 3D-DFEM results to test the significance of different layer moduli for explaining the offset surface deflections. It was found that at the furthest geophone, 60 in (152.4 cm) offset, the subgrade modulus is the only significant factor. At geophone number 6, 36 in (91.44 cm) offset, both the subgrade and the subbase moduli are significant, while for the other geophones all layer moduli are significant.

A regression analysis was conducted to develop statistical models to predict the surface deflection from FWD loading as a function of the significant layer moduli. Seven statistical models were developed, the models for geophones 1 to 5 are functions of all layer moduli. The models for geophones 6 and 7 are functions of the subbase and subgrade moduli only.

1) $DF1 = a*E_s + b*E_b + c*E_{sb} + d*E_{sg}$

$$2)\ DF2 = a*E_s + b*E_b + c*E_{sb} + d*E_{sg}$$

$$3)\ DF_3 = a + b*E_s + c*E_b + d*E_{sb} + e*E_{sg}$$

$$4)\ DF_4 = a + b*E_s + c*E_b + d*E_{sb} + e*E_{sg}$$

$$5)\ DF_5 = a + b*E_s + c*E_b + d*E_{sb} + e*E_{sg}$$

$$6)\ DF_6 = a + b*E_{sb} + c*E_{sg} + d*E_{sb}*E_{sg}$$

$$7)\ DF_7 = a + b*E_{sb} + c*E_{sg} + d*E_{sb}*E_{sg}$$

where

DF_i = peak deflection at geophone i, and

a, b, c, d and e are regression constants and their values   are
shown in Table 2.

Table 2 Summary of the Regression Models of the Full Depth
Asphalt Section

Constant	X- Offset Distance - in						
	0	8	12	18	24	36	60
a	13.3679	11.586	10.8927	10.0778	9.288	11.0043	6.4566
b	-0.089	-0.047	-0.0578	-0.05427	-0.0499	-0.3022	-0.13611
c	-0.0896	-0.0849	-0.05687	-0.04583	-0.03644	-2.288	-0.10622
d	-0.1139	-0.1113	-0.1029	-0.0938	-0.08423	0.097	0.04233
e	-0.6193	-0.5928	-0.5523	-0.5205	-0.4913		
R-square	0.98	0.95	0.98	0.98	0.98	0.92	0.94

The above models and measured deflections were used to predict the
full depth asphalt section layer moduli. The first step is to solve
equations for models 6 and 7 simultaneously to obtain the subgrade and
subbase moduli. Surface and base layer moduli were then predicted solving
the equations for models number 1 and 2, simultaneously, using the
previously calculated subbase and subgrade moduli. The remaining three
models were used to evaluate the predicted layer moduli. Surface
deflections at the various geophone offset distances were predicted using
the appropriate models. It was found that the maximum difference for any
geophone between the predicted deflections using the regression models and
the measured deflections is 3.43%, while the absolute sum of difference
for all geophones is 6.17% (Table 3).

COMPARISON BETWEEN DIFFERENT BACKCALCULATION METHODS

In addition to the 3D-DFEM analysis three other methods were used to
match the measured peak surface deflections and backcalculate layer moduli
from the FWD testing of the full depth asphalt pavement section. The four
methods considered are:

1. 3D-DFEM
2. BISDEF [9]
3. WESDEF [9]
4. WESDEF with varying depth to bedrock

The first three methods were used to analyze the pavement section without
any constraints on the moduli. In the fourth method the depth to bedrock
was varied to force better agreement of the predicted surface deflections
with the measured peak surface deflections. For this case, good agreement
with the measured peak surface deflections was obtained with a depth to
bedrock of 95 in. (241.3 cm). Figure 14 shows the predicted and measured

Table 3 Comparison Between Different Backcalculation Methods

a. Deflection Basins

Position	Offset (in)	Measured Deflection	3D-DFEM				BISDEF		
			Computed Deflection	Difference (M - C)	%Difference (M-C)/M		Computed Deflection	Difference (M - c)	%Difference (M-C)/M
1	0	5.66	5.66	0	0.00		4.9	0.76	13.43
2	8	4.9	4.9	0	0.00		4.3	0.6	12.24
3	12	4.64	4.698	-0.058	-1.02		4.1	0.54	11.64
4	18	4.25	4.418	-0.168	-3.43		3.8	0.45	10.59
5	24	3.97	3.89	0.08	1.72		3.6	0.37	9.32
6	36	3.22	3.2	0	0.00		3.2	0.02	0.62
7	60	2.04	2.04	0	0.00		2.5	-0.46	-22.55
				Sum	6.17				80.39

Position	Offset (in)	Measured Deflection	WESDEF			WESDEF	BED Rock @ 95"		
			Computed Deflection	Difference (M - C)	%Difference (M-C)/M	Computed Deflection	Difference (M - C)	%Difference (M-C)/M	
1	0	5.66	6.5	-0.84	-14.84	5.65	0.01	0.18	
2	8	4.9	4.7	0.2	3.50	4.97	-0.07	-1.43	
3	12	4.64	4.2	0.44	8.10	4.66	-0.02	-0.43	
4	18	4.25	3.8	0.45		4.25	0	0.00	
5	24	3.97	3.5	0.47	12.20	3.87	0.1	2.52	
6	36	3.22	3	0.22	5.80	3.17	0.05	1.55	
7	60	2.04	2.2	-0.16	-8.30	2.03	0.01	0.49	
				Sum	52.74			6.60	

b. Layer Moduli (ksi)

Prediction Method	LAYER #1	LAYER #2	LAYER #3	SUBGRAD
BISDEF	1000	200	200	2
WESDEF	306	200	200	3
3D-DFEM Regression Models	302	221	159	7.5
WITH BED ROCK @95"	960	85	48.3	4.9

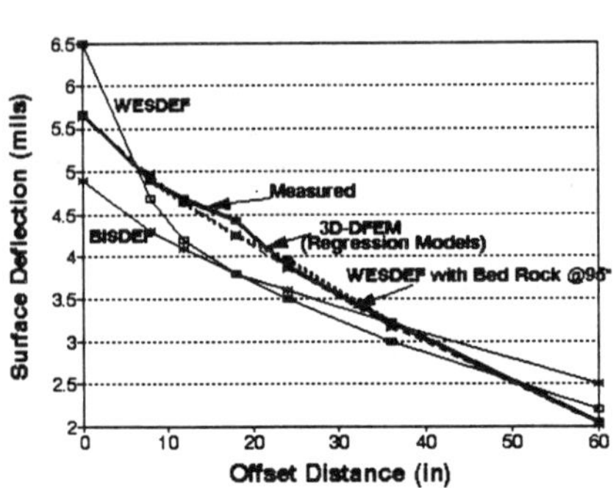

Figure 14 Comparison between Different Backcalculation Methods (1 in. = 25.4 mm, 1 mils = 25.4E-3 mm)

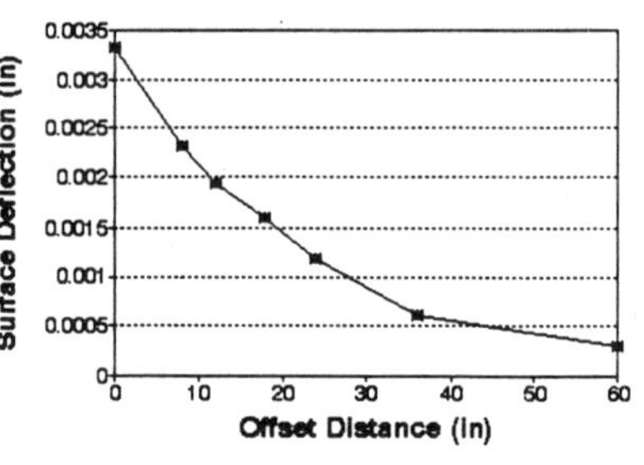

Figure 15 Effect of Bed Rock Location on Surface Deflection (3D-DFEM - 1 in. = 25.4 mm, 1 mils = 25.4E-3 mm)

pavement deflections. The corresponding layer moduli for each method are shown in Table 3.

Results of the application of these four methods were significantly different. The 3D-DFEM matched the measured peak deflection basin with the lowest sum of differences between the predicted and measured peak deflections. At the same time, the associated moduli for the full depth asphalt pavement were reasonable.

Deflections predicted by both BISDEF and WESDEF were not in good agreement with measured peak deflections. The sum of the difference between the measured and predicted deflections was %80.39 and %63.33 for BISDEF and WESDEF, respectively. Moduli as shown in Table 3 did not agree with each other and agree marginally with the moduli obtained by use of the 3D-DFEM regression models. Although good agreement was obtained between the measured and predicted deflections when bedrock was arbitrarily limited to 95 in. (241.3 cm), the resulting moduli were not reasonable. For instance, although the first and second layers are hot mix asphalt the backcalculated moduli were 960 ksi (6.62 E09 Pa) and 85 ksi (0.586 E09 Pa), respectively.

The effect of the depth to bedrock on the predicted surface deflections was examined in more detail. Construction records indicated that the pavement section had been cored to 72 in. (182.88 cm) and no bedrock was encountered. A sensitivity analysis was conducted using the 3D-DFEM and BISAR in which the depth to bedrock was assumed at depths of 95 in. (241.3 cm), 105 in.(266.7 cm), and 140 in. (355.6 cm). Results shown in Figure 15 indicated that the bedrock location did not affect predicted deflections for the 3D-DFEM. However, the depth to bedrock does affect the deflections predicted with BISAR as shown in Figure 16.

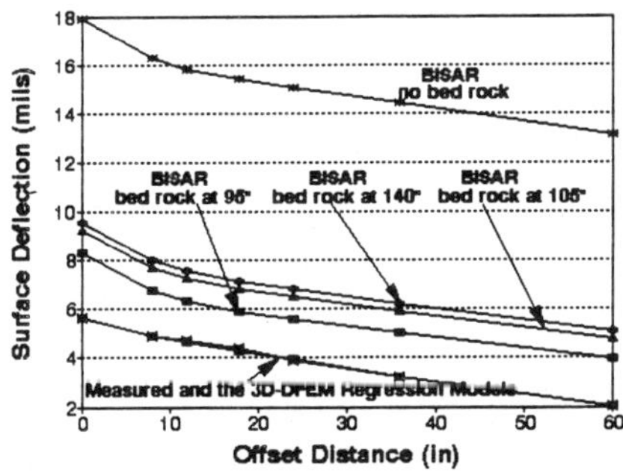

Figure 16 Effect of Bed Rock Location on Surface Deflection (BISAR - 1 in. = 25.4 mm, 1 mils = 25.4E-3 mm)

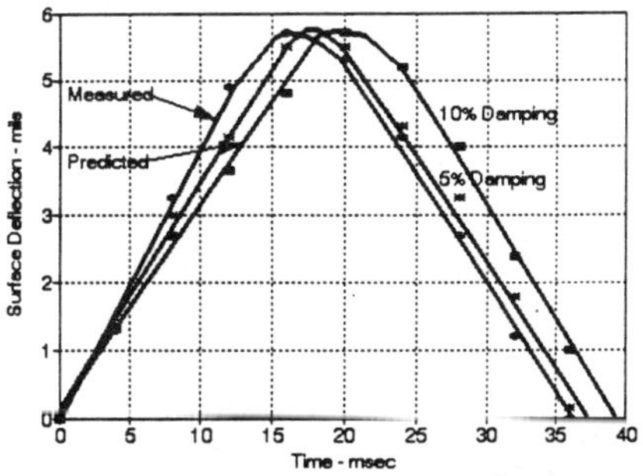

Figure 17 Surface Deflection History (1 mils = 25.4 E-3 mm)

This question of the effect of depth to bedrock underscores, as did the above analysis of accumulated layer deformation at offsets, the benefit of using a 3D-DFEM analysis for pavements. At other than very shallow depths to bedrock there should be no effect of the short time (30 msec) loading function from a bedrock boundary. On the other hand, layered elastic analysis such as BISAR are mathematically constrained by the depth to bedrock or integration limit. A layered elastic model does not realistically represent dynamic loading of pavements and it is not surprising that backcalculation procedures predict different moduli combinations when so many factors, real and assumed, affect the results.

FWD SIMULATION

A sensitivity study was conducted of 3D-DFEM simulation of the FWD test. In this step, the predicted layer moduli were used to predict not only the peak surface deflections, but also the deflections with time at the various offsets. In a previous study, Sebaaly [10] recommended a value of 5% for pavement layer damping coefficients.

A range of damping coefficients was considered in this current analysis. It was found that the time shift between the deflection histories decreases as the damping coefficients decrease to 5%. Damping coefficients below 5% produced no further reduction in the time shift. Therefore, a 5% damping coefficient was used in the subsequent analysis.

Figure 17 shows the variation of measured and predicted pavement deflection with time at an offset distance equal to zero (geophone 1). As can be seen from this figure, there is a 2 msec difference between the peaks of the measured and predicted deflections with 5% damping. Careful examination also shows some difference in the slope between the predicted and measured deflection curves.

EXPANDED STUDY OF DEFLECTION PREDICTION

In the previous design of experiment (DOE), only one thickness for each layer was included. In the expanded or main design of experiment (MDOE) a range of pavement layer thicknesses and moduli were included. The factors included in the MDOE are as follow:

1. Modulus of elasticity of the surface layer (E_s), three levels.
2. Modulus of elasticity of the base course (E_b), three levels.
3. Modulus of elasticity of the subbase course (E_{sb}), three levels.
4. Modulus of elasticity of the subgrade (E_{sg}), three levels.
5. Thickness of the surface layer (t_s), two levels.
6. Thickness of the base course (t_b), two levels.
7. Thickness of the subbase course (t_{sb}), two levels.

Table 4 shows the different factor levels. The values of other material and model characteristics (i.e., damping) which were used in the verification study and resulted in the best agreement with the FWD deflections were used in this MDOE as well. A fine-grained subgrade (CL) was assumed for this part of the study.

A partial factorial design was used to reduce the number of sections that needed to be analyzed. Each problem requires 8 to 12 hrs. on a Sun SPARC Workstation I with 16 Mbytes of memory and a 207 Mbyte hard drive. In this partial factorial design the number of sections analyzed using the 3D-DFEM was adequate to consider main effects, second order effects and two-way interactions. An analysis of variance (ANOVA) was conducted to determine the significance of the various factors in predicting the deflection at different offset distances, DF1 at 0 in. (0 cm), DF2 at 8 in. (20.32 cm), DF3 at 12 in. (30.48 cm), DF4 at 18 in. (45.72 cm), DF5 at 24 in. (60.96 cm), DF6 at 36 in. (91.44 cm) and DF7 at 60 in. (152.4 cm). Based on the ANOVA study, an individual regression model was developed for each offset distance to predict the surface deflection as a function of the significant factors only.

Two approaches were considered in developing the regression models. One approach was to develop regression models having terms in the form $(1/E_i)$ similar to that used to express the surface deflection of a half space subjected to a load-on a rigid circular plate [11]. Reasonable R^2 values were obtained for sensors 1 though 5 (between 87% to 92%). However, for sensors 6 and 7 the inverse of layer moduli was found to be insignificant and therefore regression equations could not be developed. The second approach was to develop best fit regression models. High R^2 values were obtained as shown in Table 5. The regression equations are:

$$1)\ DF1 = \frac{a}{E_s} + b * E_b + c * E_{sg} + d * t_s + e * t_b + f * E_b * t_b + g * E_{sb} * t_{sb}$$

$$2)\ DF_2 = a + \frac{b}{E_s} + c*E_b + d*E_{sb} + e*E_{sg} + f*t_s + g*t_b + h*t_{sb}$$

$$3)\ DF_3 = a + b*E_s + c*E_b + d*E_{sb} + e*E_{sg} + f*t_s + g*t_b + h*t_{sb} + i*t_s*E_{sg} + j*E_s*E_b$$

$$4)\ DF_4 = a*E_b + b*E_{sg} + c*t_s + d*E_b*t_s + e*t_s*E_{sb}$$

$$5)\ DF_5 = a*E_{sb} + b*E_{sg} + c*t_s + d*t_{sb} + e*E_{sb}*t_{sb}$$

$$6)\ DF_6 = a*E_{sb} + b*E_{sg} + c*E_{sb}*E_{sg} + d*t_s + e*t_b + f*t_s*E_{sb} + g*t_b*E_{sb}$$

$$7)\ DF_7 = a*E_{sb} + b*E_{sg} + c*E_{sg}*E_{sb} + d*t_s$$

Where

DF_i = peak deflection at geophone i.

a, b, ..., j are regression constants and their values are given in Table 5.

Table 4 Factor levels Included in the Main Design of Experiment

Factor	Level		
	1	2	3
Modulus of the Surface Layer (ksi)	200	400	600
Modulus of the Base Course (ksi)	30	45	60
Modulus of the Subbase Course (ksi)	10	20	30
Modulus of the Subgrade (ksi)	2.5	10	20
Thickness of the Surface Layer (in)	4	8	-
Thickness of the Base Course (in)	6	10	-
Thickness of the Subbase Course (in)	6	10	-

Table 5 Summary of the Regression Models for the Main Design of Experiment

Constant	X - Offset Distance - in						
	0	8	12	18	24	36	60
a	1091.8115	36.989	39.426	0.316999	0.4511	0.1917	0.0403
b	0.65523	353.753	-0.0149	-0.16947	-0.117	0.0364	0.0423
c	-0.214	-0.097	-0.1711	2.363	-0.198	-0.0039	-0.00227
d	-1.1899	-0.1909	-0.1569	-0.05278	1.4257	0.245	0.0749
e	3.605	-0.2484	-0.3755	-0.0168	-0.0629	0.226	
f	-0.0847	-0.2484	-1.1679		0	-0.0096	
g	-0.02372	-1.0196	-0.4888		0	-0.0106	
h		-0.522	-0.28486				
i		-0.2818	-0.309				
j			-0.00029264				
R-square	0.96	0.9	0.93	0.97	0.975	0.994	0.96

As expected, the elastic moduli of the subgrade and subbase were found to be significant at all locations, while the elastic moduli of the surface and base layers were found to be significant only at locations which are close to the loading point. The seven regression equations can be used to predict layer moduli. Knowing different layer thicknesses and material type the regression models will be functions of layer moduli only.

Models 6 and 7 with the measured deflections can be solved simultaneously to predict the subgrade and subbase moduli. Model 5 can be used as a check of the accuracy of the subgrade and subbase moduli. Knowing the subgrade and subbase moduli, model number 4 can be used to determine the base course modulus, then model number 1 can be used to determine the surface layer modulus. Models number 2 and 3 can be used to evaluate the predicted layer moduli. This procedure was applied to the FWD test results of a flexible pavement section located in Indiana.

FWD DATA RECORD

X -in (cm)	0	8	12	18	24	36	60
	(0)	(20.32)	(30.48)	(45.72)	(60.96)	(91.44)	(152.4)
DEFLECTION	27.89	23.46	19.59	15.78	11.59	4.72	0.80
mils (mm)	0.708	0.6	0.498	0.401	0.294	0.12	0.02

Indiana Department of Transportation (INDOT) records indicate that the pavement cross section for this site consists of:

- 4 in. (10.16 cm) asphalt layer,
- 6 in. (15.24 cm) granular base course, and
- 6 in. (15.24 cm) granular subbase course.

The subgrade of this section is low plasticity clay (CL). Using models 6 and 7, the subgrade and subbase moduli were found to be 10,000 psi (68.96 E06 Pa) and 35,000 psi (241.36 E06 Pa), respectively. Using model 4 the base course modulus was found to be 90,000 psi (620.64 E06 Pa). The surface modulus was determined from model 1 and found to be 350,000 psi (2 413.6 E06 Pa). Computers should be used to apply this method because numerical accuracy is important. The predicted deflections are compared with the measured deflections in Figure 18.

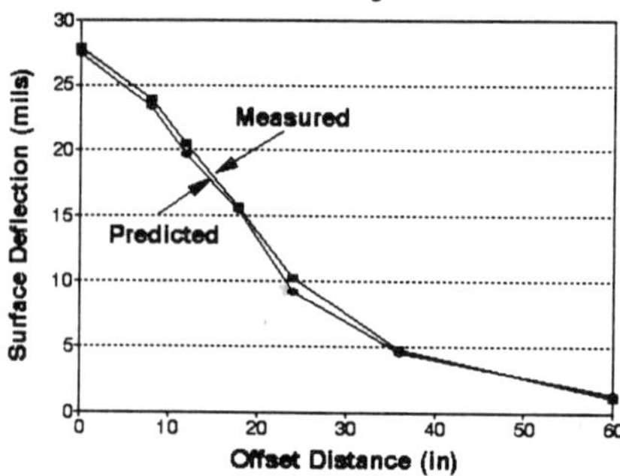

Figure 18 Comparison Between Measured and Predicted Pavement Deflection (1 in. = 25.4 mm, 1 mils = 25.4E-3 mm)

SUMMARY AND CONCLUSIONS

Multi-layer analysis is commonly used to analyze falling weight deflectometer (FWD) data and backcalculate layer moduli. This analysis assumes a static loading and linear elastic material properties. However, actual FWD loading is dynamic and the resulting pavement response is time dependent, as is evidenced by the time shift in the measured surface deflections at various offset distances from the point of loading. The difference between the multi-layer assumptions and actual FWD loading and paving material properties is significant and can lead to unrealistic back calculated layer moduli.

A three-dimensional dynamic finite element program (3D-DFEM) was used to simulate a FWD loading cycle and actual material characteristics. A study was conducted to verify the dynamic analysis capabilities of the 3D-DFEM using a FWD data set for a full depth asphalt section located in Indiana. The 3D-DFEM was used to model the pavement section which was loaded with a loading cycle similar to that of the FWD. Predicted peak deflections were found to match the measured ones. The same section was analyzed using BISDEF and WESDEF backcalculation programs. Moduli backcalculated using BISDEF and WESDEF varied significantly and the absolute sum of differences between the measured peak deflections and predicted peak deflections at various offset distances using the programs was 80% and 63%, respectively. The absolute sum of the differences using the 3D-DFEM was %6.17. Forcing the predicted deflections using WESDEF to closer agreement with the measured values by adjusting the depth to bed rock resulted in unrealistic layer moduli.

Variation of bed rock location below the subgrade had no effect on the predicted peak surface deflection with the 3D-DFEM. On the other hand, the depth of the bed rock was found to have a significant effect on multi-layer analysis results (BISAR) because these analysis are mathematically constrained by the depth to bed rock or integration limits.

The 3D-DFEM analysis predicts a reasonable contribution to the surface deflection from various layers for dynamic FWD loading. However, the predicted layer contribution to surface deflection using multi-layer analysis (BISAR) is not reasonable.

Deflection time histories at different offsets due to a FWD loading was predicted using the 3D-DFEM. Time difference between the measured and predicted deflection histories approached a minimum value at 5% damping.

A partial factorial design of experiment was developed to study the effect of layer moduli and thickness on the surface deflection at different offsets. Regression models were developed to predict the surface deflection as a function of the significant layer moduli. The subgrade and subbase moduli were found to be the only significant moduli at the furthest three geophones (60 in., 36 in., and 24 in.). All layer moduli were significant at the other geophone locations. These models were used to develop a layer moduli prediction method. These moduli predictions are based on a loading cycle similar to that of the FWD loading cycle and on dynamic analysis.

REFERENCES

[1]   "Bitumen Structures Analysis in Roads (BISAR), Computer Program," Koninlilijke/Shell - Laboratorium, Amsterdam, July 1972.

[2]   "ABAQUS, Finite Element Computer Program," Version 4.9, Hibbitt, Karlsson and Sorensen, Inc., 1989.

[3]   Roque, R. Tia, M. and Ruth, B., "Asphalt Rheology to Define the Properties of Asphalt Concrete Mixtures and Performance of Pavements," Asphalt Rheology: Relationship to Mixture, <u>ASTM STP 941, O. E. Brisco, Ed., American Society for Testing Materials</u>, Philadelphia, 1987, pp. 3-27.

[4]   Zaghloul, S. and White, T., "Use of a Three Dimensional-Dynamic Finite Element Program for Analysis of Flexible Pavement," Presented at the 72nd Annual Meeting of the Transportation Research Board, Washington, D.C., January 1993a.

[5]   Zaghloul, S. and White, T., "Non-Linear Dynamic Analysis of Concrete Pavements, <u>Proceedings of the 5th International Conference on Concrete Pavement Design and Rehabilitation</u>, Purdue University, W. Lafayette, Indiana, April 1993b.

[6]   Drucker, D. C., and W. Prager, "Soil Mechanics and Plastic Analysis or Limit Design," Quarterly of Applied Mathematics, Vol. 10,pp. 157-165, 1952.

[7]   Schofield, A., and C. P. Worth, "Critical State Soil Mechanics," McGraw Hill, New York, 1968.

[8]   Parry, R. H., Editor, "Stress-Strain Behavior of Soils," G. T. Foulis and Co., Henley, England, 1972.

[9]   Van Cauwelaert, F. J., Alexander, D. R., White, T. D., and Barker, W. R., "Multilayer Elastic Program for Backcalculating Layer Moduli in Pavement Evaluation," <u>Nondestructive Testing of Pavements and Backcalculation of Moduli, ASTM STP 1026,</u> A. J. Bush III and G. Y. Baladi, Eds., American Society for Testing and Materials, Philadelphia, 1989, pp. 171-188.

[10]   Sebaaly, B., Davies, T. and Mamlouk, M., "Dynamics of Falling Weight Deflectometer," American Society of Civil Engineers (ASKEW) Journal of Transportation Engineering, Vol. 111, No. 6, November, 1985.

[11]   Selvadurai, A., "Elastic Analysis of Soil-Foundation Interaction," Elsevie Scientific Pub. Co., 1979.

[12]   George, K. P., "Resilient Testing of Soils Using Gyratory Testing Machine," Transportation Research Board, 71 st Annual Meeting, Washington, D.C., January 12-16, 1992.

[13]   Kim, D. and Stokoe, K. H., "Characterization of Resilient Modulus of Compacted Subgrade Soils Using Resonant Column and Torsional Shear Test," Transportation Research Board, 71 st Annual Meeting, Washington, D.C., January 12-16, 1992.

[14]   Wood, D. M., "Soil Behaviour and Critical State Soil Mechanics," Cambridge University Press, 1990.

[15]   Derucher, K. and Korfiatis, G., "Materials for Civil and Highway Engineers,: Second Edition, Prentice Hall, New Jersey, 1988.

Measurements and Calculation Techniques in the Field and Laboratory

Per Ullidtz[1], Jørgen Krarup[2] and Thomas Wahlman[3]

VERIFICATION OF PAVEMENT RESPONSE MODELS

REFERENCE: Ullidtz, P., Krarup, J., and Wahlman, T., "Verification of Pavement Response Models," Nondestructive Testing of Pavements and Backcalculation of Moduli (Second Volume), ASTM STP 1198, Harold L. Von Quintas, Albert J. Bush, III, and Gilbert Y. Baladi, Eds., American Society for Testing and Materials, Philadelphia, 1994.

ABSTRACT: When backcalculating layer moduli from Falling Weight Deflectometer (FWD) tests and when forward calculating critical stresses and strains, a number of assumptions have to be made. Some of the important assumptions normally are that

1)   the system is in equilibrium,
2)   the materials are continuous and remain continuous under deformation, and
3)   the materials are elastic, isotropic and homogeneous.

None of these assumptions are valid for FWD testing of real pavement structures. A number of mathematical models exist for describing the pavement response, but they are all based on idealization of real pavement structures and materials. The only way of determining whether a given mathematical model is acceptable for pavement design or evaluation, is by comparing the response predicted by the model to that measured in real pavements.

Many attempts at doing this have been done over the years, but none of them have let to the conclusive verification of a specific mathematical model. This paper presents the results of measurements carried out on four instrumented test sections, one in the Danish Road Testing Machine and three on experimental sections in Sweden. On all sections the deflections were measured with a FWD, the layer moduli were backcalculated using different methods, and the stresses and strains in different materials, caused by FWD loading or by traffic loading, were calculated and compared to the stresses and strains measured at the same locations with pressure and strain gauges.

KEYWORDS: pavement response, analytical methods, verification, instrumentation, full scale testing, falling weight deflectometer

Mathematical models for calculating the pavement response, i.e. stresses, strains and deflections, are needed for evaluation or design of pavement structures, when using the analytical-empirical method. The models are needed both for back calculating layer moduli

[1] Dr.techn, The Technical University of Denmark, Lyngby, Denmark

[2] M.Sc., The Danish Road Directorate, Roskilde, Denmark

[3] M.Sc., Gatukontoret Malmö, Sweden

from Falling Weight Deflectometer (FWD) tests and for forward calculating the critical stresses and/or strains in the structure. Because of the complexity of pavement structures, these models are always based on simplifications, such as assuming the materials to be linear elastic, with no viscous, visco-elastic or plastic deformations, disregarding dynamic effects, assuming compatibility or continuity (even for granular materials) etc. Because of the many simplifications the models need to be verified by comparing the response predicted by the models, to the response measured on real pavements. This is the only way of determining whether a given mathematical model is acceptable or not.

This paper describes an attempt at verifying three different response models:

     1) based on the Odemark-Boussinesq approach (OB),
     2) using elastic layer theory (MODULUS, WES5), and
     3) using the Finite Element method (FE)

The procedure used was the following:

     - first ordinary Falling Weight Deflectometer (FWD) tests were carried out on instrumented test sections (measuring the deflection basins). The instrumented sections were in the Danish Road Testing Machine (RTM) and on three in situ sections in southern Sweden,

     - secondly FWD testing was done with a hydraulic pad (to ensure a uniform stress distribution under the loading plate) and the stresses and strains in the pavement structures were recorded. On the in situ sections the stresses and strains were also measured under an 11.5 tons axle load at different speeds,

     - from the deflection basins the moduli of the pavement layers were then calculated using the three models, and

     - finally the stresses and strains in the pavement were calculated with the corresponding moduli and models, and compared to the measured values.

Measuring the stresses and strains in a pavement structure is, unfortunately, rather difficult. The instruments used in these experiments have been developed over 20 years, and are believed to be reasonably accurate. The strain gauges in the asphalt layer have a low stiffness, but could possibly underregister by about 10-20 μstrain (10^{-6} m/m). The strain cells in the unbound layers have practically no stiffness and are built into the layers with very little disturbance to the materials. The presence of a pressure cell will cause a change in the stress distribution, but the pressure cells were recalibrated after completion of the pavement structures and are believed to be correct within approximately ten percent. The recalibration was done by measuring the stress at different distances from the FWD and calculating the total force at the level of the gauge (assuming axial symmetry). As the force exerted by the FWD is known a calibration factor may be determined. The gauges used for the instrumentation were described by Ullidtz and Ertman (1989).

MODELS USED

The first two models (Odemark-Boussinesq and WES5 used in MODULUS) are based on three fundamental principles:

    1) Equilibrium,
    2) Compatibility (continuity), and
    3) Elasticity

If in addition the materials are assumed to be homogeneous, isotropic and linear elastic, these three assumptions lead to a series of fourth order differential equations that can be solved, when the boundary conditions and the loads are known. For a semi--infinite half-space loaded by a point load a closed form solution was obtained by Boussinesq in 1885. To solve the problem for a layered system a number of programs have been developed. The program used in this analysis is the WES5 program developed for the USAE Waterways Experiment Station by Frans van Cauwelaert. The theoretical background is described in detail by Fr van Cauwelaert (1989). This program is fast and is believed to provide exact solutions to the mathematical problem.

It may be worthwhile noting once more that none of the basic assumptions are fulfilled for actual pavement structures. The loads (rolling wheel or FWD) are dynamic, many materials are not continuous, some are even particulate, deformations are not only elastic but also plastic, viscous, and visco-elastic, and they are seldom proportional to the stresses (Hook's law). In addition to this the materials are seldom homogeneous but vary considerably with space (and time), and they often show some degree of anisotropy. None of the models therefore provide "exact" solutions to the physical problem. With Finite Element programs it is possible to use more realistic assumptions (non-linearity, dynamic loading), although the correct treatment of particulate materials will probably require Distinct (or Discrete) Element methods.

Poisson´s ratio was not measured and a standard value of 0.35 was used for all layers with all models.

<u>Odemark-Boussinesq (OB)</u>

Several programs for backcalculating layer moduli from FWD testing make use of the Odemark-Boussinesq approach. A detailed description of this approach and its application in the ELMOD program was given by Ullidtz (1987). The main features are described below. Odemark assumed that the stresses, strains and deflections below a layer will depend on the stiffness of the layer, only. The stiffness of a layer is:

$$\frac{E \times I}{1 - \mu^2}$$

where I is the moment of inertia of the layer (proportional to the cube of the thickness), E is the modulus and μ is Poisson's ratio.

A layer may, therefore, be transformed to an "equivalent" layer with a different modulus, by multiplying the thickness by the cube root of the modular ratio (assuming Poisson's ratio to be the same for the two materials). By using Odemark's assumption (successively) a layered system can be transformed into a semi-infinite half-space, for which Boussinesq's equations for stresses, strains and displacements can be used.

A simple non-linearity, of the type often encountered in subgrade soils,

$$E = C * \left(\frac{\sigma_1}{\sigma_{ref}} \right)^n$$

may be incorporated with the method. C and n are constants (n is negative), σ_1 is the major principal stress, and σ_{ref} is a reference stress. Calculations with a Finite Element program have shown that this type of non-linearity has little influence on the stress distribution. It can, therefore, be incorporated directly into the OB method. This is extremely important for the correct determination of the moduli, when the subgrade is non-linear.

It is also possible to determine the approximate depth to a rigid layer. The (equivalent) depth is found as the distance to the point where the surface deflection would be 0 (zero). This is done by extrapolating from the measured deflections using Boussinesq's equation for surface deflection, d(r):

$$d(r) = \frac{(1-\mu^2) * P}{\pi * E * r}$$

where P is the load, E is the modulus of the subgrade and r is the distance from the loading centre.

Elastic Layer Theory (MODULUS)

To determine the layer moduli from elastic layer theory MODULUS 4.0 (Uzan, Scullion, Michalak, Parades and Lytton, 1988) was used. The program generates a database of deflection bowls using the WES5 linear elastic program. The number of deflection bowls generated depends on the user supplied limits on the acceptable moduli for each layer. Once the database is generated, a pattern search routine is used to fit measured and calculated bowls. MODULUS also contains a routine for determining the depth to a rigid layer, but is, in the version used, restricted to linear elastic subgrades.

Finite Element (FE)

The Finite Element program used is a modified version of the axial symmetric program developed by Wilson at University of California, Berkely (Duncan, Monismith and Wilson, 1968). The procedure used was similar to that used with MODULUS. First a database of deflection bowls was generated with the FE program, and then the bowls matching the measured deflections were found.

On data from the RTM a fixed depth of 1700 mm (corresponding to the rigid concrete bottom) was used and the maximum radius of the mesh was 2500 mm. Because the OB analysis had shown the subgrade to be linear elastic, only linear elastic materials were used.

For the three in situ sites infinite depth was simulated and different degrees of non-linearity were used for the subgrade. The load was transferred via a 20 mm thick plate with a modulus of 1000 MPa, rather than as a uniformly distributed load.

LAYER MODULI

When allowed to search for a stiff layer at some depth (bedrock), both the OB method and MODULUS locate the concrete bottom of the RTM quite close to the actual depth. This is shown in Figure 1, where both the actual physical depth to the concrete bottom and the "equivalent" depth are shown. From meter 8 to 9 the depth is decreasing, as also indicated to some extent in the calculations. From meter 0 to 8 OB gives a mean depth of 1928 mm and MODULUS a mean depth of 1872 mm. The actual depth is 1714 mm and the equivalent depth approximately 1870 mm.

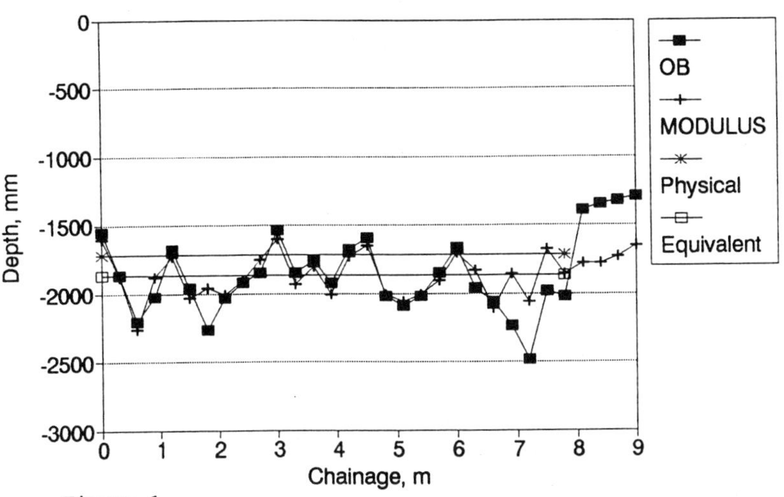

Figure 1

The pavement in the RTM had a thin asphalt layer of 64 mm on a 130 mm thick granular base course on 390 mm of granular subbase on a sandy moraine subgrade.

For the three in situ sections the thicknesses were:

	Sect. I	Sect. II	Sect. III
Asphalt	135 mm	160 mm	190 mm
Macadam	120 mm	120 mm	600 mm
Subbase	560 mm	540 mm	--

During testing in the RTM the asphalt temperature was about 8 °C and for the in situ testing it was about 0 °C. The macadam layer in section III was split into two 300 mm thick layers for the evaluation. The mean moduli of the layers are given below. In calculating the mean values it was assumed that the logarithms of the moduli were normally distributed, so that the mean value given is 10 raised to the mean of the logarithms. The moduli are given in MPa. Co is the constant in the non-linear surface modulus relationship and n is the power.

```
RTM                 OB   MODULUS        FE

Asphalt            7276      6583      3608
Base course         423       271       321
Subbase             141       148       166
Subgrade            121       133        89


Section I           OB   MODULUS        FE
Asphalt           15290     19644     17905
Macadam             949       577       454
Basecourse          225       148       113
Subgrade            125        69       122
Co                   84        --        67
n                 -0.25        --     -0.29


Section II          OB   MODULUS        FE
Asphalt           11187     15750     13614
Macadam             934       471       453
Basecourse          242        55        80
Subgrade            114       290       134
Co                   59        --        71
n                 -0.37        --     -0.31


Section III         OB   MODULUS        FE
Asphalt            9004     11692     10830
Macadam 1           509       200       192
Macadam 2           229        63        90
Subgrade            116       200       119
Co                   72        --        66
n                 -0.27        --     -0.27
```

An example of the moduli calculated for each point is shown in
Figures 2 to 4, for section II. The distance between consecutive
points is only 300 mm (0.3 m).

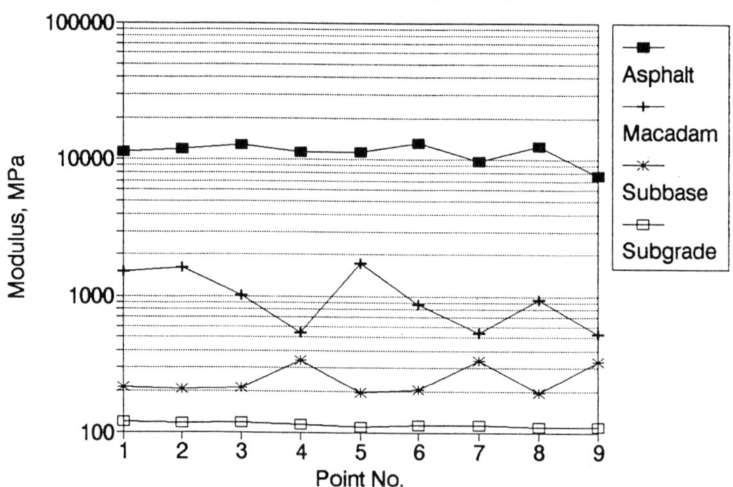

Figure 2

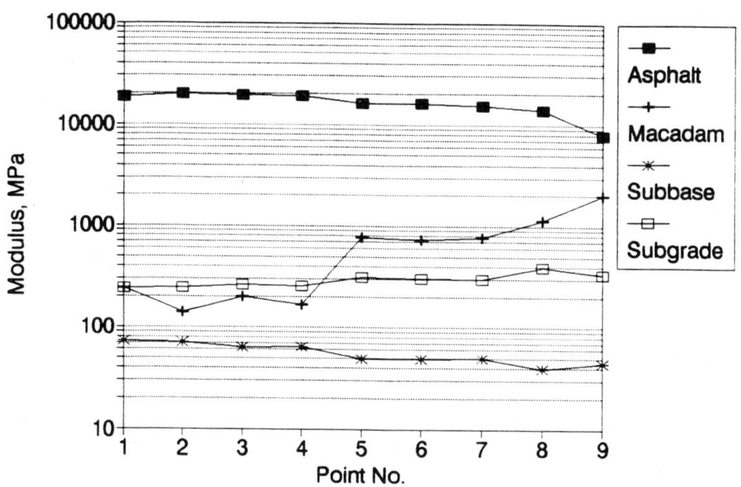

Figure 3

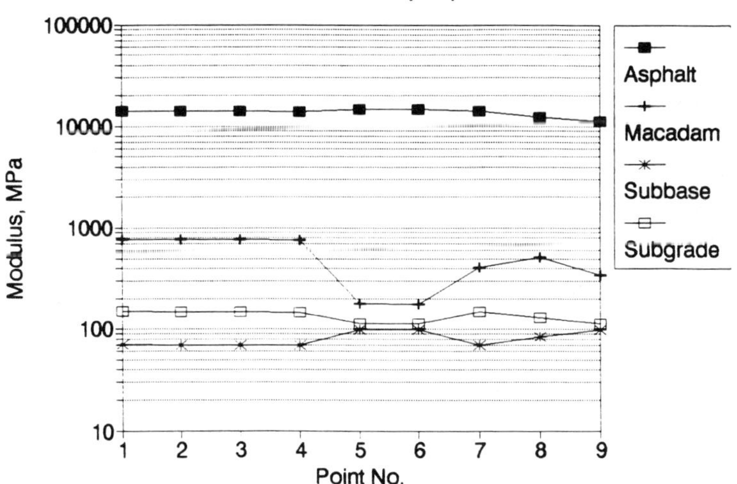

Figure 4

For the asphalt layer the agreement between the three methods is generally good although there is a difference of a factor two for the RTM, where the layer thickness was low and the evaluation, therefore, less reliable.

For the intermediate layers and the subgrade the agreement is rather poor for the in situ sections. The main difference is caused by the non-linearity of the subgrade. With MODULUS this is treated by introducing a rigid layer at some depth, but the result of this is obviously very different for the three sections. The very high moduli at section II and III are probably incorrect.

The variation in moduli is often very large, particularly with MODULUS and FE, indicating that the number of deflection basins in the databases may not have been sufficient. For MODULUS the non-linearity of the subgrade may also play a role. The variation in the modulus of the macadam in section II, from 120 MPa to 2000 MPa within 2.4 m, appears to be excessive.

MEASURED AND CALCULATED RESPONSE

In the RTM the response reported was measured under a FWD with a hydraulic loading plate, to ensure a uniform stress distribution. The horizontal strain at the bottom of the asphalt layer (ϵasph) was measured at ten positions, or points in the structure, the vertical stress on the top of the base course (σbc) at four positions, the vertical strain in the lower half of the base course (ϵbc), the vertical stress on top of the subbase (σsb1), the vertical strain in the upper part of the subbase (ϵsb1), the vertical stress in the middle of the subbase (σsb2), the vertical strain in the lower half of the subbase (ϵsb2), and the vertical stress on top of the subgrade (σsg) were all measured at three positions, and the vertical strain in the upper part of the subgrade (ϵsg) at two positions. The results are presented in Figures 5 to 13 and in the table below. Strains are in μstrain and stresses in kPa:

RTM

	OB	MODULUS	FE	Measured
ϵasph	196	189	215	155
σbc	297	266	329	352
ϵbc	445	706	710	1420
σsb1	113	111	132	188
ϵsb1	518	529	505	564
σsb2	43	43	41	65
ϵsb2	231	248	229	350
σsg	31	28	24	42
ϵsg	150	183	248	287

The best comparison between measured and calculated values is obtained from the graphs with all the details, rather than from the mean values.

Strain at bottom of asphalt
Road Testing Machine, Jan 92

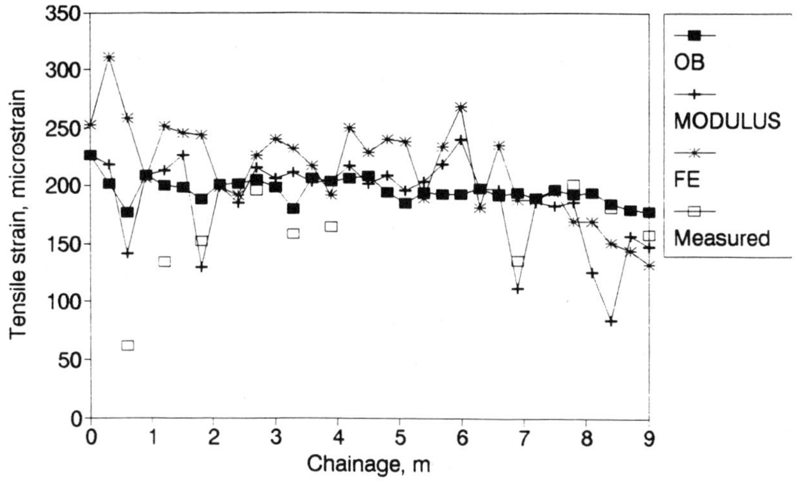

Figure 5

Stress at top of basecourse
Road Testing Machine, Jan 92

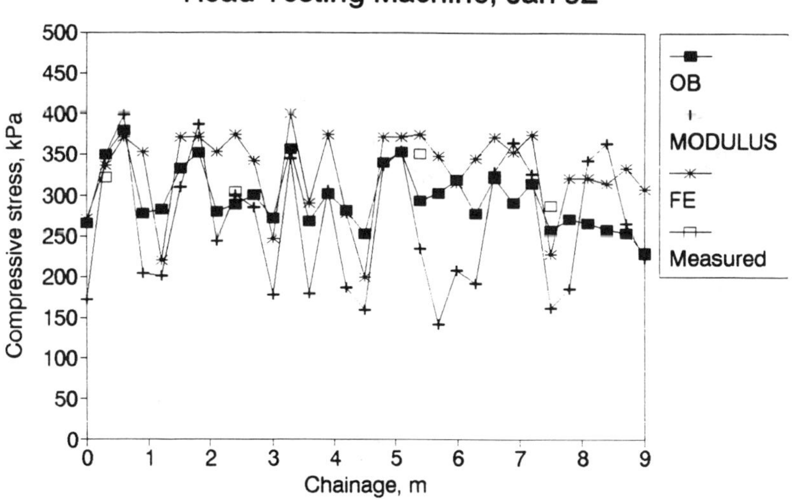

Figure 6

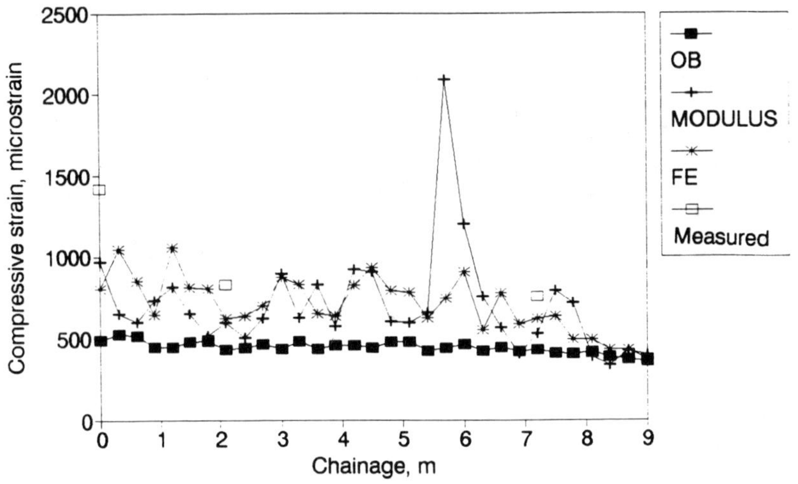

Figure 7

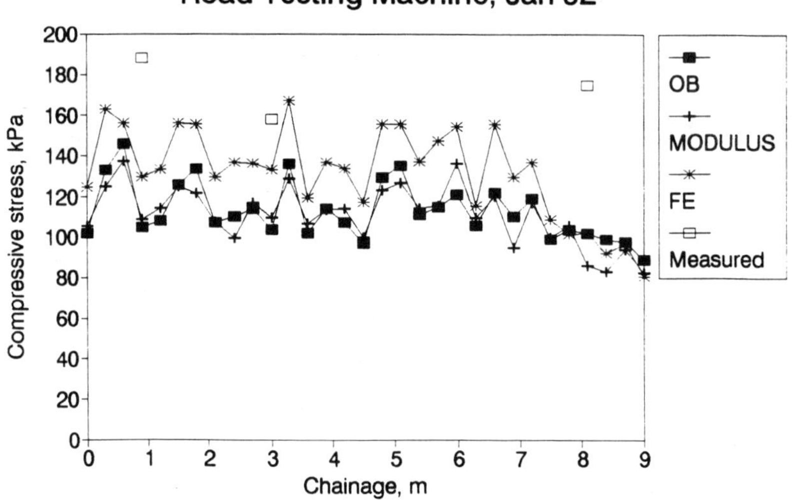

Figure 8

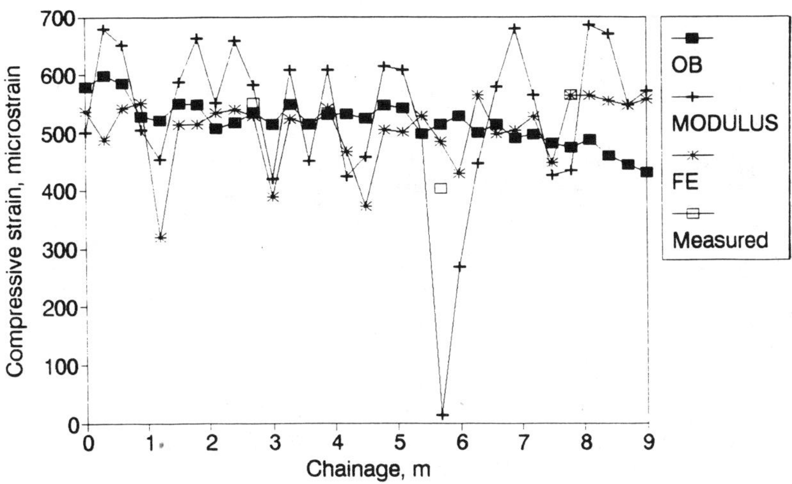

Figure 9

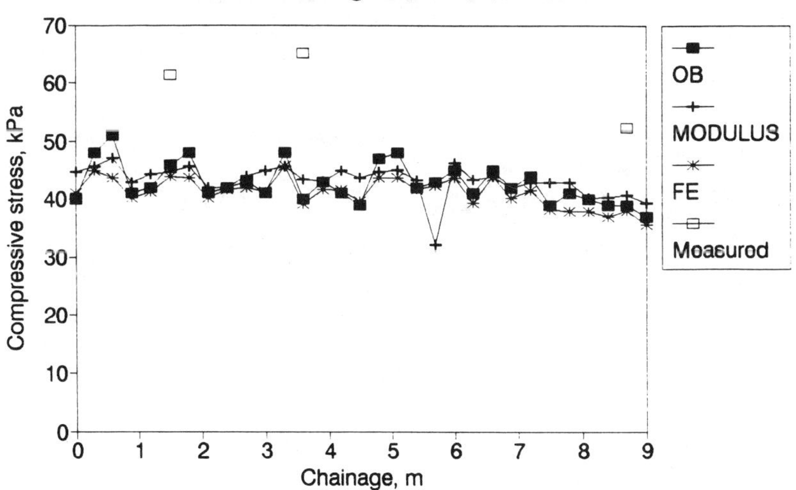

Figure 10

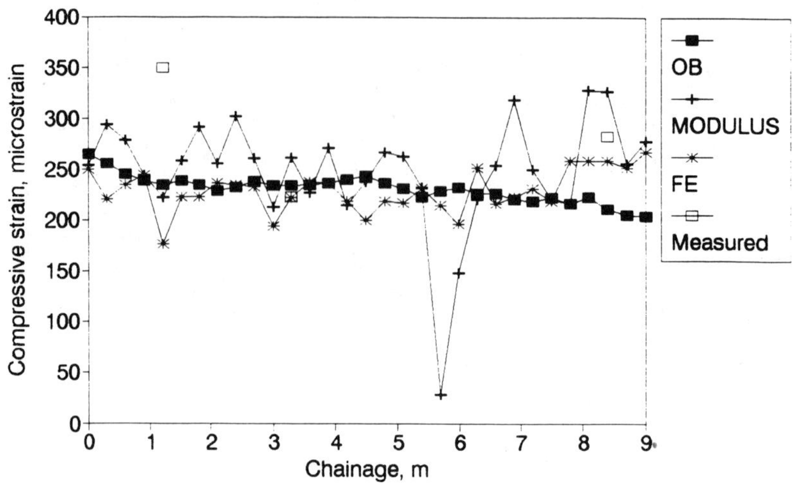

Figure 11

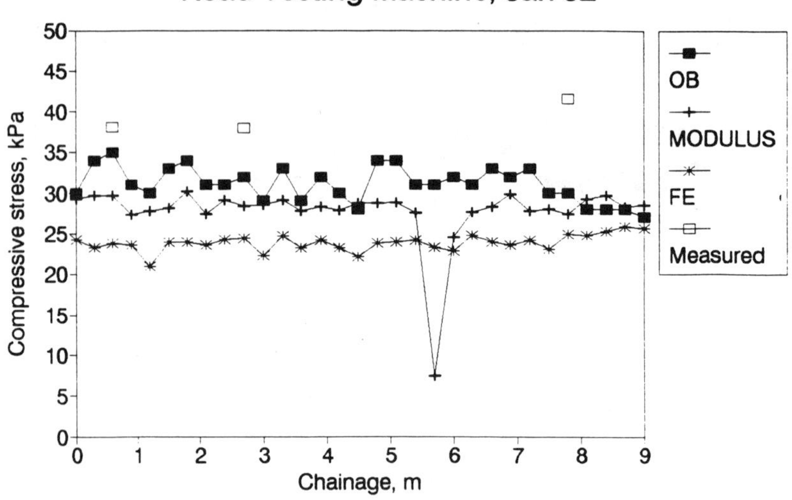

Figure 12

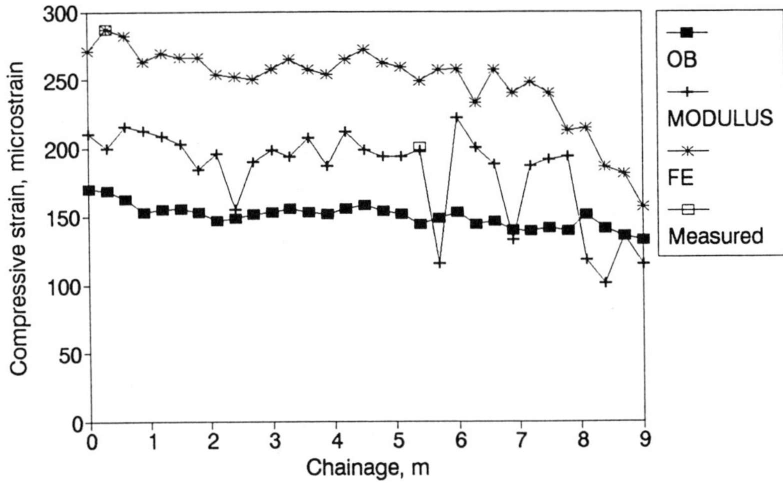

Figure 13

On the three in situ sections the response was measured under a
rolling wheel load. A dual wheel was used with an axle load of 11.5
tons. The results reported here were obtained with a driving speed of
70 km/h. The modulus of the asphalt (E_{rwl}) was calculated from the
value determined with the FWD (E_{fwd}) using:

$$E_{rwl}=0.368*t^{-.277}*E_{fwd}$$

where the loading time corresponding to the rolling wheel load was
found from:

$$t_{rwl}=\frac{d+h}{v}$$

where d is the distance between the wheels in the dual wheel (340
mm), h is the thickness of the asphalt in mm and v is the driving
speed in mm/sec. The response measured was the strain at the bottom
of the asphalt layer, ϵasph, the stress on the subgrade, σsg, and the
strain in the upper part of the subgrade, ϵsg. Strains are in μstrain
and stresses in kPa:

Section I

	OB	MODULUS	FE	Measured
ϵasph	96.8	80.8	82.1	46.0
σsg	17.7	12.2	12.6	13.5
ϵsg	198	172	159	126

Section II

	OB	MODULUS	FE Measured	
ϵasph	90.0	82.2	86.4	81.2
σsg	16.6	16.5	12.4	15.8
ϵsg	241	48	144	228

Section III

	OB	MODULUS	FE Measured	
ϵasph	109.3	89.8	86.0	58.2
σsg	16.0	17.2	11.9	13.7
ϵsg	201	75	145	176

The standard deviations are given below (in the same units):

Section I

	OB	MODULUS	FE Measured	
ϵasph	6.4	10.7	4.6	3.0
σsg	1.0	0.2	0.8	0.3
ϵsg	26.1	26.1	22.2	80.0

Section II

	OB	MODULUS	FE Measured	
ϵasph	5.9	2.8	7.1	5.9
σsg	0.4	0.4	0.5	0.7
ϵsg	20.0	9.2	28.4	21.5

Section III

	OB	MODULUS	FE Measured	
ϵasph	12.1	12.0	9.9	2.6
σsg	0.7	1.0	0.7	0.2
ϵsg	17.4	10.2	20.0	14.5

In Figure 14 all of the calculated values are shown versus the
measured values, in a log-log plot. The two dotted lines corresp
to twice and half the measured value, respectively.

A regression analysis on the logarithms leads to:

	OB	MODULUS	FE
Slope	0.976	0.945	0.973
Standard error of estimate	0.200	0.212	0.151
$10^{St.e.e}$	1.8	1.63	1.42
R squared	0.83	0.82	0.92

Measured and calculated response
RTM and in situ sections

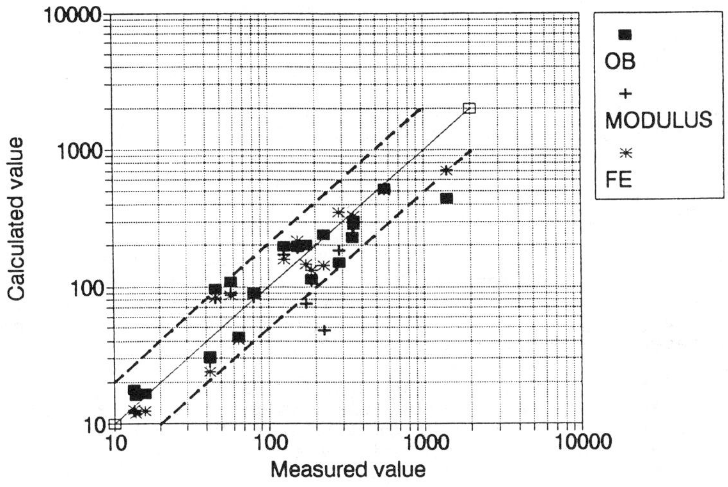

Figure 14

CONCLUSION

Although the present study was limited to only a few structures, some
interesting points may be noted:

- Both measured and calculated values often show a very high
variability. Coefficients of variation of 20% or more are not
unusual.

- When the subgrade is non-linear elastic a linear elastic method
should not be used. In section II the subgrade strain determined with
MODULUS is only one fifth of the measured value.

- The best overall agreement is obtained with the Finite Element
method. The improvement over the Odemark Boussinesq approach comes,
however, at a cost of a computing time which is between ten thousand
and a hundred thousand times longer, with a non-linear subgrade.

- The agreement between measured and calculated values is far from
satisfactory. There is an urgent need for more comprehensive studies
involving other mathematical models and improved instrumentation
technology.

ACKNOWLEDGEMENT

This research was sponsored by the Municipality of Malmö, Sweden,the
Danish Road Research Laboratory and the Technical University of
Denmark

References

Cauwelaert, Fr van, 1989, "Les Bases Essentielles des Modéles de Dimensionnement", <u>Journée Technique LAVOC</u>, Ecole Polytechnique Fédérale de Lausanne.

Duncan, J.M., Monismith, C.L. & Wilson, E.L., 1968, Finite Element Analyses of Pavements", <u>Highway Research Record 224</u>.

Scullion, T. & Michalak, C., Texas 1991, "MODULUS 4.0, User's Manual", <u>Research Report 1123-4</u>, Texas Transportation Institute, The Texas A&M University System, College Station.

Ullidtz, P., Elsevier 1987, "Pavement Analysis", <u>Developments in Civil Engineering</u>.

Ullidtz, P. & Ertman Larsen, H.J., 1989, "State-of-the-art Stress, Strain and Deflection Measurements", State of the Art of Pavement Response Monitoring Systems for Roads and Airfields, <u>Special Report 89-23</u>, US Army Corps of Engineers, Cold Regions Research & Engineering Laboratory.

Uzan, J., Scullion, T., Michalak, C.H., Parades, M. & Lytton, R.L., Texas 1988, "A Microcomputer Based Procedure for Backcalculating Layer Moduli from FWD Data", <u>Research Report 1123-1</u>, Texas Transportation Institute, The Texas A&M University System, College Station.

Vikram S. Torpunuri[1], Norris Stubbs[2], Robert L. Lytton[3], and Allen H. Magnuson[4]

FIELD VALIDATION OF A METHODOLOGY TO IDENTIFY MATERIAL PROPERTIES IN PAVEMENTS MODELED AS LAYERED VISCOELASTIC HALFSPACES

REFERENCE: Torpunuri, V. S., Stubbs, N., Lytton, R. L., and Magnuson, A. H., "Field Validation of a Methodology to Identify Material Properties in Pavements Modeled as Layered Viscoelastic Halfspaces," Nondestructive Testing of Pavements and Backcalculation of Moduli (Second Volume), ASTM STP 1198, Harold L. Vonquintas, Albert J. Bush, III, and Gilbert Y. Baladi, Eds., American Society for Testing and Materials, Philadelphia, 1994.

Abstract: The methodology developed in the accompanying paper is verified using Falling-Weight Deflectometer (FWD) data from in-service highways. The procedure to transform Falling-Weight Deflectometer data from the time domain to the frequency domain and its analysis in the frequency domain is presented. Frequency response curves based on back-calculated parameters obtained by other independent methods, parameters identified using the proposed methodology, and the actual Falling-Weight Deflectometer data collected from the field, are compared to demonstrate the validity of the approach.

KEY WORDS: field validation, systems identification, falling weight deflectometer, viscoelastic, layered media

The feasibility of a methodology to identify viscoelastic properties of a multilayered media has been established in the accompanying paper (Stubbs et al., 1994). The objective of this paper is to demonstrate the practicality of the proposed system identification procedure. The practicality of the approach will be demonstrated by comparing the values of the identified parameters with values expected in practice, comparing the frequency response of the identified system with the frequency response derived directly from the Falling Weight

---

[1]Research Associate, Texas Transportation Institute, Texas A&M University, College Station, TX 77843.

[2]Associate Professor, Civil Engineering Department, Texas A&M University, College Station, TX 77843.

[3]Professor, Civil Engineering Department, Texas A&M University, College Station, TX 77843.

[4]Engineering Research Associate, Texas Transportation Institute, Texas A&M University, College Station, TX 77843.

Deflectometer results, and comparing the accuracy of the proposed
predictions to the predictions of other backcalculated results.

EXPERIMENTAL VERIFICATION

Comparison of Predicted Pavement Properties with Expected Properties

Experimental verification of the Parameter Evaluation Algorithm
(Stubbs et al 1994) will be based on FWD data gathered on in-service
Highway, SH 19 & 24, in Texas. FWD data were gathered in the Fall of
1988. Typical input load-time history and the displacement-time history
at the seven sensors for the pavement is provided in Figure 1.
Frequency response functions, $H_{1j}(\omega)$, between the point of impact at
location 1 and sensor j were determined by first, taking separately the
Fast Fourier Transform (FFT) of the load and the appropriate deflection,
then dividing the FFT of the deflection by the FFT of the load. The
procedure was repeated for each sensor to give a total of seven
frequency response curves (Torpunuri, 1990). Typical frequency response
curves expressed in terms of magnitude and phase for the SH 19 & 24
pavement are shown in Figure 2.

The pavement structure at the test site is a four-layer system with
properties listed in Table 1 (Torpunuri, 1990). The properties in
parentheses are assumed values used to initiate the analysis. Note that
the section corresponds to the example pavement used in the accompanying
paper (Stubbs et al., 1994). The eight model parameters to be
identified are m_1, E_1, E_2, β_2, E_3, β_3, E_4, and β_4, where m_1 is the slope
of the creep compliance curve for the top asphaltic concrete layer, E_1 -
E_4 are the moduli of the four layers, and β_2 - β_4 are the damping ratios
of layers 2-4.

Table 1--<u>Reference Properties for Pavement Section at SH 19 & 24</u>

Layer	Thickness (m)	Modulus GPa	Poisson Ratio	Wt. Density GN/m^3	Damping Percent
Asphaltic Concrete*	0.229	7.584(E_1^R)	0.33	0.021	20(m_1^R)
Crushed Stone	0.178	0.310(E_2^R)	0.35	0.019	1.6
Lime-Treated Clay	0.254	0.180(E_3^R)	0.38	0.019	4.7
Subgrade-Clay	∞	0.15(E_4^R)	0.40	0.019	7.5(β_4^R)

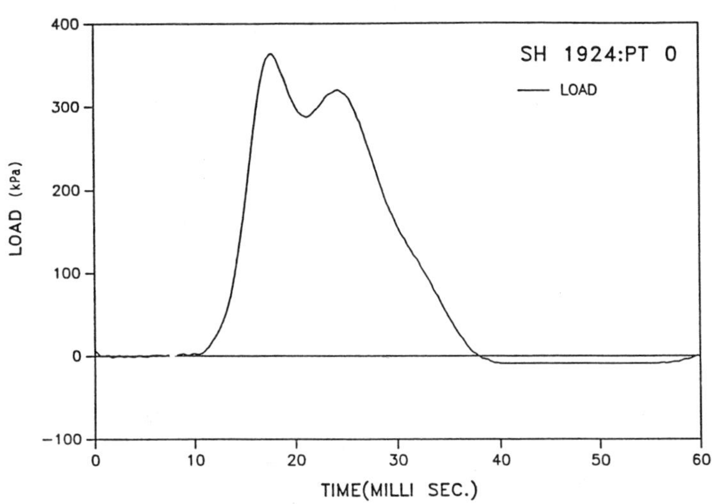

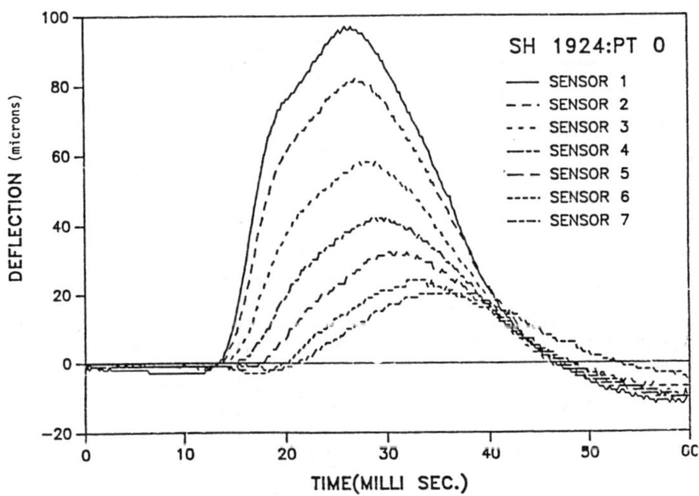

FIG. 1--Typical FWD Pulse, Lowest Load (SH 19 & 24).
(1 in. = 0.0254m)

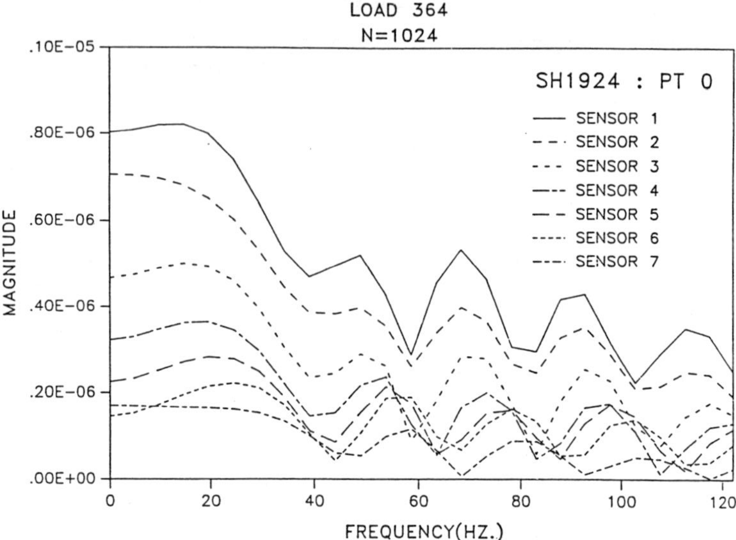

FIG. 2--Frequency Response from FWD Data

Since we have shown that the theoretical model is insensitive to
changes in the parameters β_2 and β_3, the parameters to be identified
here become E_1, E_2, E_3, E_4, $\bar{m}_1$, and β_4 (Stubbs et al., 1994). Because in
the theoretical paper (Stubbs et al., 1994) we found that it was
sufficient to utilize only one frequency in the output spectrum, here we
will again base our analysis on the response of all sensors at a
frequency of 48.89 Hz (See Torpunuri, 1990, for justification). Thus we
have seven load vectors and six parameters to be identified.
Consequently the sensitivity matrix has a size of 14X6.

We first generate the sensitivity matrix F (using Equations 22-25 in
the accompanying paper) with the viscoelastic properties in Table 1
serving as a seed. We also generate the load vector Z (i.e. the
fractional change in the magnitude or phase of the transfer function of
the pavement of SH 19 & 24 relative to the equivalent values for the
pavement modeled by SCALPOT with properties given in Table 1.) The
viscoelastic layer properties were identified as in Example II and III
in the accompanying paper and the rate of convergence is demonstrated in
Table 2. Note that all values are nondimensionalized with respect to
the reference value. The actual identified values and the range of
expected values developed on the basis of known properties and
engineering judgment, are listed in Table 3.

<u>Table 2--Rate of Convergence For SH 19 & 24 Pavement</u>

Current Parameters Value/Reference Parameter Value

Iteration No.	E_1/E_1^R	E_2/E_2^R	E_3/E_3^R	E_4/E_4^R	m_1/m_1^R	β_4/β_4^R
Start	1.00	1.00	1.00	1.00	1.00	1.00
1	1.05	0.59	1.58	0.81	0.79	0.68
3	1.09	0.44	0.30	0.62	0.57	0.29
5	1.10	0.49	0.45	0.54	0.45	0.08
6^*	0.99	0.81	0.55	0.46	0.34	0.08
8	0.95	1.20	0.50	0.44	0.29	0.08
10	0.93	1.32	0.46	0.44	0.27	0.08
11	0.93	1.36	0.45	0.44	0.27	0.079

Note: For iterations 1-5 ΔZ_1 = Z/4; sensitivity matrix updated at iteration 6 and ΔZ_2 = 3Z/4.

Table 3--<u>Comparison of Identified Parameters with Expected Engineering Values</u>

Parameters	Predicted Value (GPa)	Range of Engineering Value (GPa)		
E_1	7.025	2.068	-	7.184
E_2	0.422	0.241	-	0.517
E_3	0.083	0.068	-	0.206
E_4	0.067	0.034	-	0.172
m	0.060	0.05	-	0.10
β_4	0.0059	0.05	-	0.1

Comparison of Field-Determined Frequency Responses with Predicted Frequency Responses

The field-determined response functions $H_{1j}^{F}(\omega)$ ($j = 1,...7$) computed directly from the FWD data were described in the last section and are shown in Figure 2. SCALPOT provides the frequency response functions, $H_{1j}^{S}(\omega)$, directly. Thus the desired frequency response was obtained from a SCALPOT simulation with the viscoelastic parameters identified in Table 3. A comparison of the predicted frequency responses and the field-determined frequency responses is provided in Figures 3-6. Also provided in the figures are the results of a trial and error backcalculation procedure utilized by Magnuson (1988).

Discussion of Results

It is apparent from Table 3 that, except for β_4 the proposed approach predicted material parameters that are realistic and are in the expected engineering range (Torpunuri, 1990). The moduli for layers 1, 2, 3, and 4 and slope of the creep curve for the top layer, m, were all well within the common engineering range for those pavement materials. The damping ratio for halfspace is predicted to be an order of magnitude smaller than expected. However, the expected magnitude of damping in that case was small to begin with.

Several general comments can be made on reviewing Figures 3-6. First, the System Identification (SID) generated results agree closely not only for the calibration frequency of 48.89 Hz but also for the entire spectrum of interest (0-110 Hz). The agreement for the magnitude is generally better than that for the phase. The fit for the magnitude of the response is excellent for frequencies greater than approximately 20 Hz. The predicted response consistently overshoots the magnitude for frequencies less than 20 Hz. On the other hand, the phase comparisons are good for frequencies below approximately 40 Hz but the quality of the comparisons deteriorates the frequency increases. In fact, it appears that the measured phases become quite erratic suggesting low signal/noise ratio in the FWD data.

The second general comment relates to the comparative performance of the proposed SID procedure. A review of Figures 3-6 indicates that the proposed SID procedure is more accurate (i.e., in terms of bias of error and deviation of error) than the back calculated results. Furthermore, the error distribution in the SID results appears, as it should be, to be balanced. On the other hand, the error distribution in the back-calculated results is biased.

It should be emphasized that this excellent agreement between observation and prediction has been achieved as a result of calibrating the pavement model at a single frequency. Certainly better results could be obtained if calibrations were performed at two or more frequencies. Such calibrations would have the effect of increasing the overdeterminedness of the system. Of course, another avenue to improve the agreement would be to improve the quality of the viscoelastic

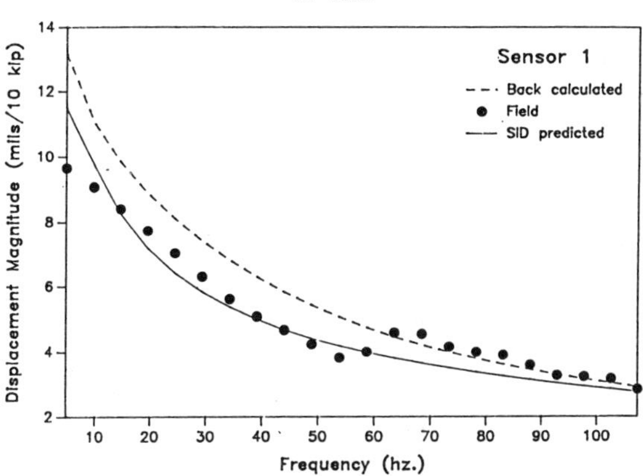

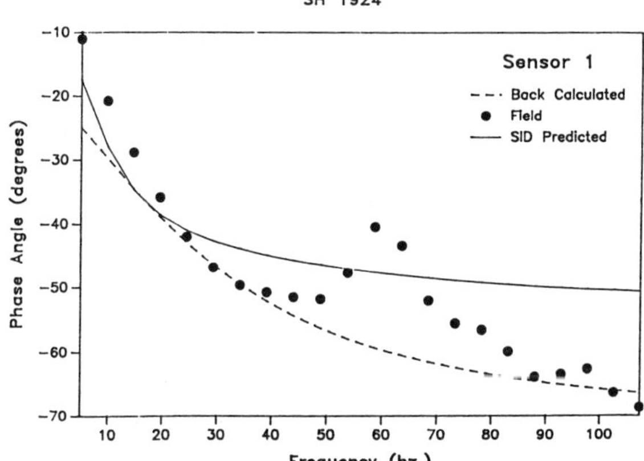

FIG. 3--Magnitude and Phase of Frequency Response at Sensor 1.
(1 in. = 0.0254m; 1 Kip = 4448.2N)

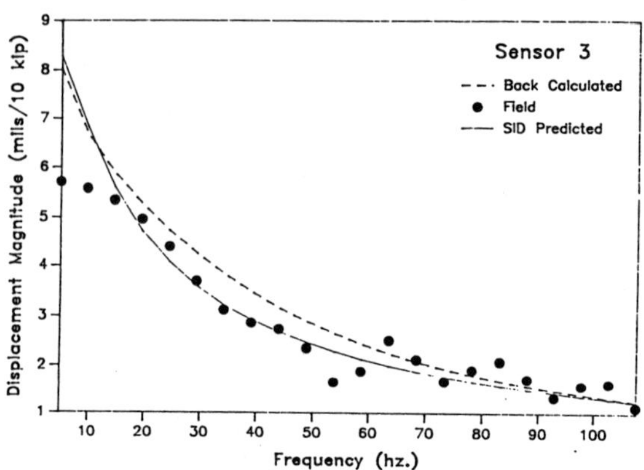

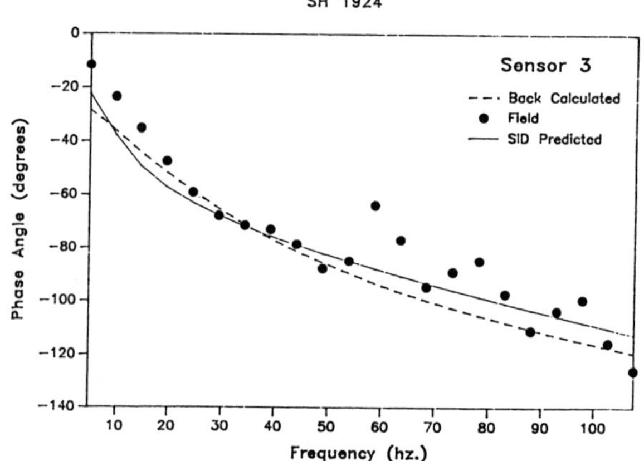

FIG. 4--Magnitude and Phase of Frequency Response at Sensor 3.
(1 in. = 0.0254m; 1 Kip = 4448.2N)

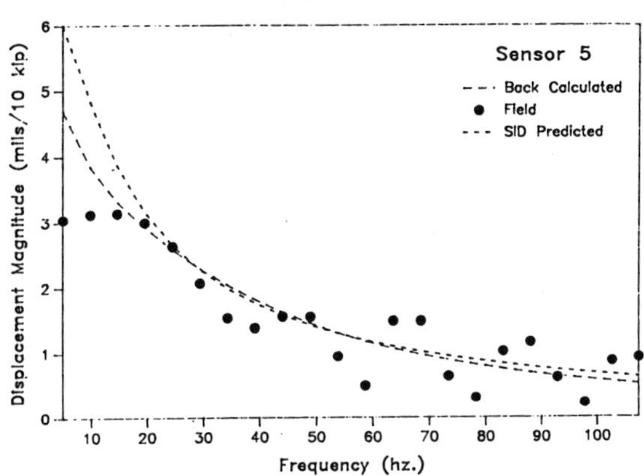

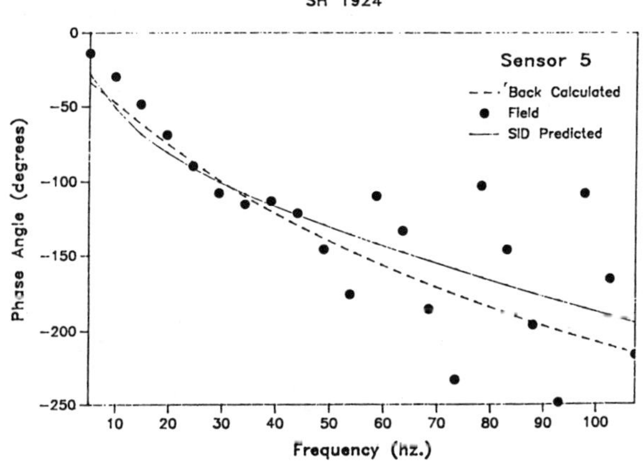

FIG. 5--Magnitude and Phase of Frequency Response at Sensor 5.
(1 in. = 0.0254m; 1 Kip = 4448.2N)

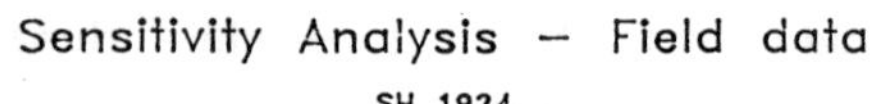

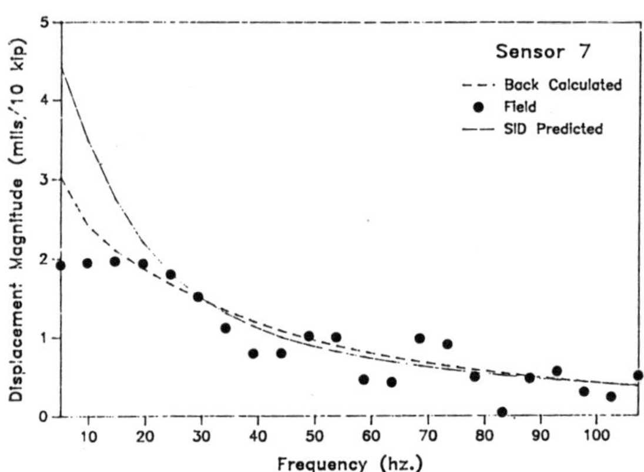

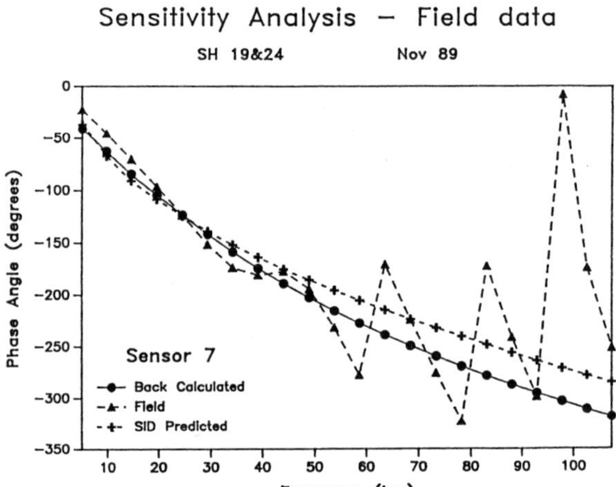

FIG. 6--Magnitude and Phase of Frequency Response at Sensor 7.
(1 in. = 0.0254m; 1 Kip = 4448.2N)

models. For example, a three parameter model for the top layer might improve the fit in the magnitude for frequencies below 20 Hz _a priori_.

SUMMARY AND CONCLUSION

The objective of this paper was to demonstrate the practicality of a systems identification approach described in an accompanying paper (Stubbs et al., 1994). We proposed to demonstrate the practicality of the approach by comparing values of the identified parameter with expected values, comparing the field-determined frequency response function with frequency response functions for the identified model of the pavement, and comparing the accuracy of the SID results with the results of other backcalculation techniques. Data to evaluate the procedure were obtained from FWD test on an in-service highway in Texas. First, we found that identified values for the pavement model were in, or close to, the expected range for such materials. Second, we found excellent agreement between the field-determined frequency responses and the frequency responses of the identified model of the pavement. Finally, inspection of the results indicate that the proposed SID approach provided a more accurate prediction of the true pavement frequency response functions than the frequency response functions developed on the basis of trial and error. These findings establish the practicality of the approach. The main implication of this work is the potential of systematically evaluating the viscoelastic properties of real pavements using an automated version of the system identification approach presented here along with the numerical modeling capabilities of a program like SCALPOT and the appropriate results from FWD tests.

REFERENCES

Magnuson, A. H., 1988, "Dynamic Analysis of Falling Weight Deflectometer Data, " _Part 1: Research Report_, No. 1215-1F, Texas Transportation Institute, College Station, Texas.

Stubbs, N. Torpunuri, V.S., Lytton, R.L., and Magnuson, A.H., 1994, "A Methodology to Identify Material Properties in Pavements Modeled as Layered Viscoelastic Halfspaces," _Nondestructive Testing of Pavements and Backcalculation of Moduli (Second Volume), ASTM STP 1198_, Harold L. Von Quintas, Albert J. Bush, and Gilbert Y. Baladi, Eds., American Society for Testing and Materials, Philadelphia.

Torpunuri, V.S., 1990, "A Methodology to Identify Material Properties in Layered Viscoelastic Halfspaces," Master's Thesis, Department of Civil Engineering, Texas A&M University, College Station, Texas.

Tayyeb Akram[1], Tom Scullion[2], & Roger E. Smith[3]

COMPARING LABORATORY AND BACKCALCULATED LAYER MODULI ON INSTRUMENTED PAVEMENT SECTIONS

REFERENCE: Akram, T., Scullion, T., and Smith, R. E., "Comparing Laboratory and Backcalculated Layer Moduli on Instrumented Pavement Sections," <u>Nondestructive Testing of Pavements and Backcalculation of Moduli (Second Volume), ASTM STP 1198</u>, Harold L. Von Quintas, Albert J. Bush III, and Gilbert Y. Baladi, Eds., American Society for Testing and Materials, Philadelphia, 1994.

ABSTRACT: This paper discusses the analysis of deflection data on thick and thin asphaltic concrete pavement sections instrumented with Multidepth Deflectometers (MDD's). Surface and depth deflection data were collected under both Falling Weight Deflectometer (FWD) and truck loadings. Linear elastic analyses were performed to compare layer moduli backcalculation from;

- FWD surface deflections only; and
- FWD surface and depth deflections only.

Backcalculated moduli values were compared with those measured in the laboratory. The moduli calculated from FWD loads were also used to predict pavement response under known truck loads. These predictions were compared to actual measured responses. In general it was found that the FWD characterization overestimates the strength of the subgrades and underpredicts subgrade strains induced by the truck by 18%.

KEYWORDS: Modulus, Backcalculation, Instrumentation, FWD, Truck, MDD, Laboratory Testing

[1]Graduate Research Assistant, Texas Transportation Institute, Texas A&M University, College Station, Texas 77843-3135.

[2]Assistant Research Engineer, Texas Transportation Institute, Texas A&M University, College Station, Texas 77843-3135.

[3]Associate Research Engineer, Texas Transportation Institute, Texas A&M University, College Station, Texas 77843-3135.

LAYOUT AND CROSS-SECTION OF TEST PAVEMENT SECTIONS

This study was conducted on test sites located on Farm to Market Road 2818 (Section I (Thin)) near Bryan, Texas and the outer west bound lane of State Highway 21 (Section II (Thick)) between Bryan and Caldwell, Texas. MDDs with four LVDT modules each were installed at each site. To determine the transverse position of the truck tires relative to the MDD location, a grid was painted on the pavement surface next to the MDD hole. As the test vehicle passed over the MDD, the lateral position of the outer tires relative to the MDD position was recorded by a video camera. The cross-sections of the test pavements showing the locations of MDD sensors are shown in Fig. 1.

OVERVIEW OF FIELD TESTING AND DATA ANALYSES

Field testing included surface deflection testing under FWD loadings with simultaneous recording of both surface and depth deflections. Depth deflections were also measured in the pavement under controlled truck loadings. Truck testing was completed immediately after FWD testing. A water tanker as shown in Fig. 2, was used for the testing.

Typical MDD response from Section II (Thick) under the test vehicle loading, is shown in Fig. 3a. From the information shown in this figure, the average vertical compressive strains within the pavement layers can be computed. The average vertical compressive strain at top of the subgrade layer was calculated as the deflection measured by MDD 2 minus MDD 3 divided by the spacing between them for Section I (Thin), and MDD 3 minus MDD 4 divided by the spacing between them for Section II (Thick). A typical multidepth strain profile computed for truck loading on Section II (Thick) is shown in Fig. 3b.

In the analysis phase attempts are made to match both surface and depth deflections using a linear elastic model. The moduli backcalculated were compared with those from laboratory testing. In further analysis the moduli backcalculated from the FWD loads were used to characterize the pavement structure. These were then used to predict the vertical compressive strains at top of the subgrade layer that should be induced by a known tire load. The predicted subgrade strains were compared with those measured under the actual truck loads.

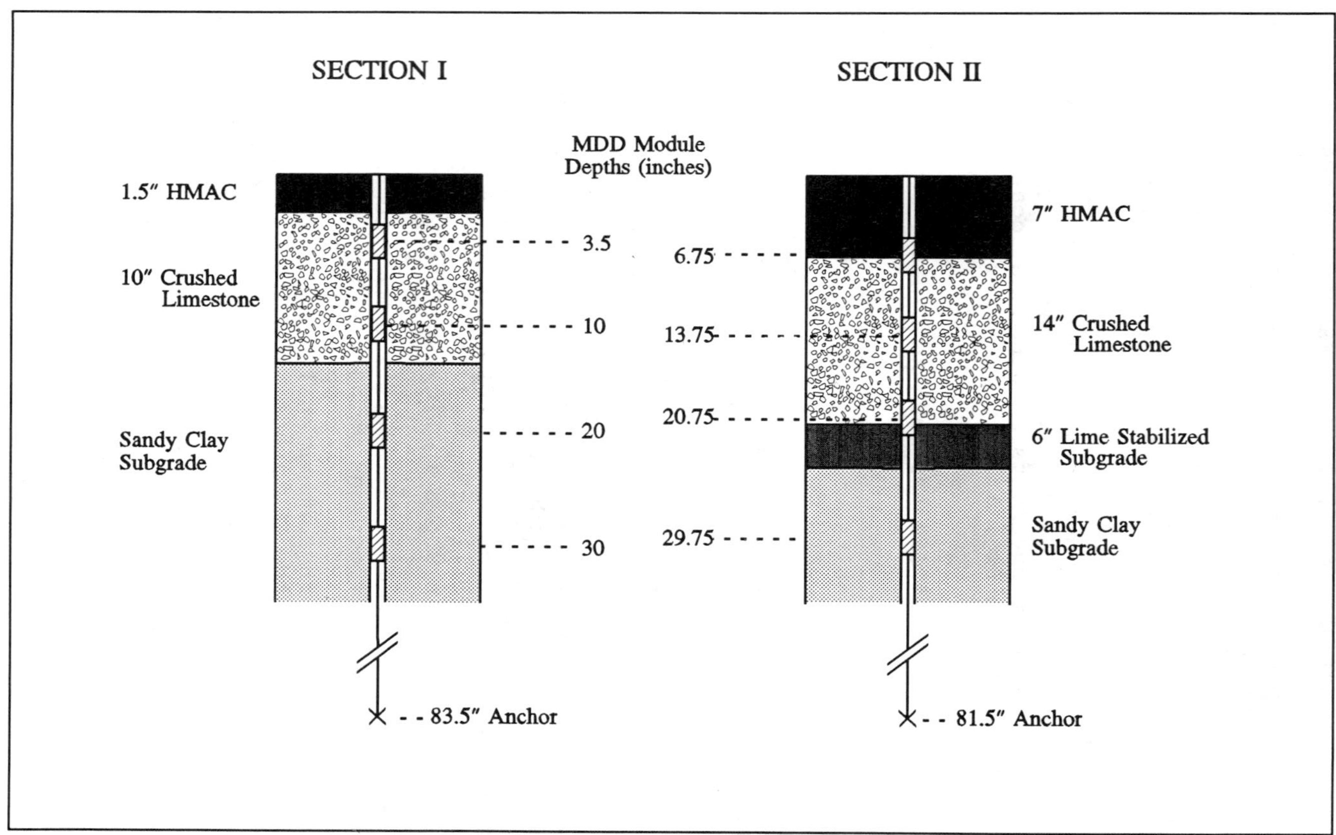

FIG. 1--MDD locations in test pavements.

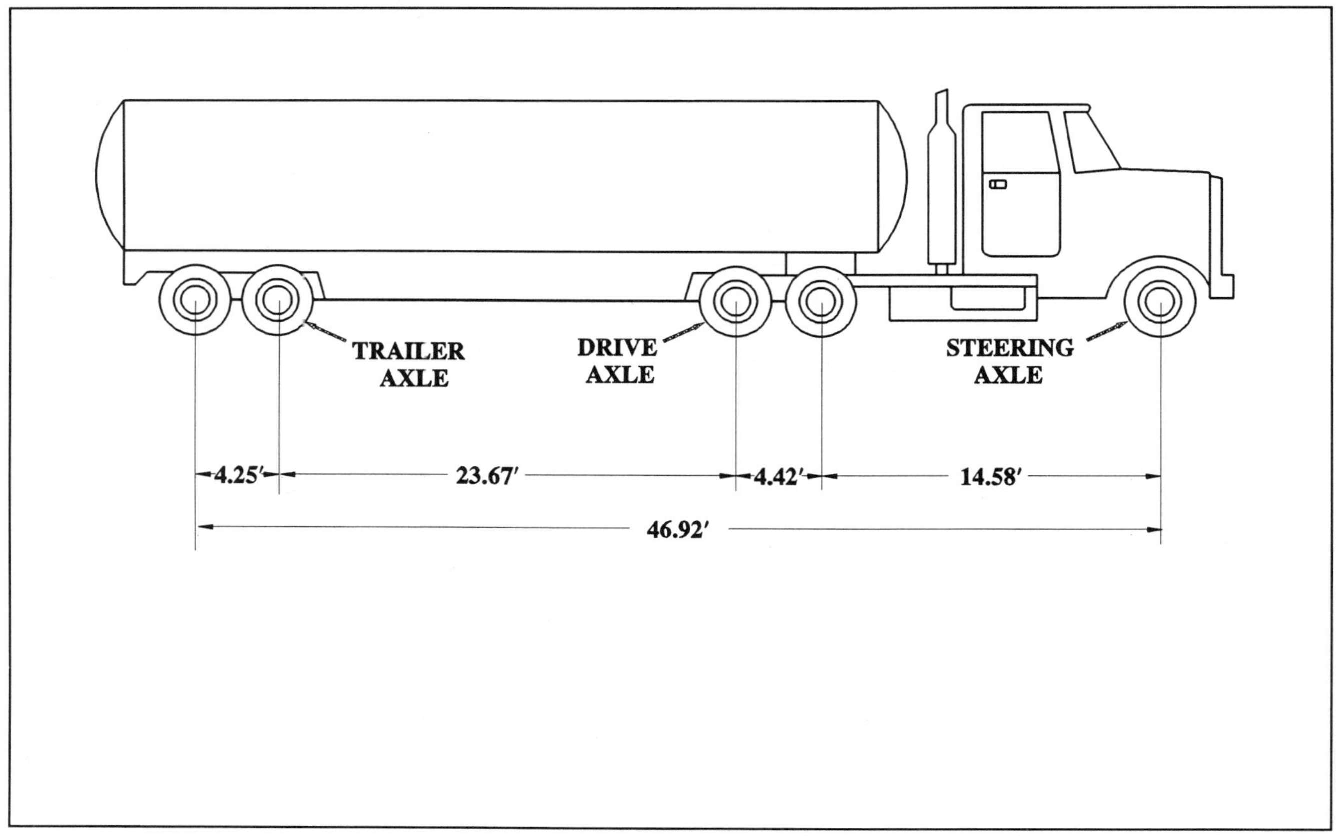

FIG. 2--3S2 water tanker used for testing.

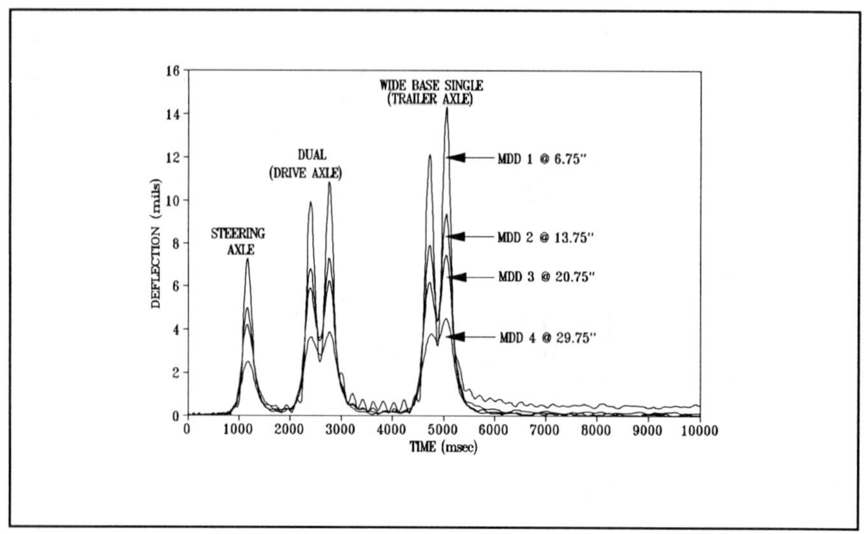

FIG. 3a--Typical depth deflections profile measured by MDD on Section II (thick) under the test vehicle (5 axle) loading.

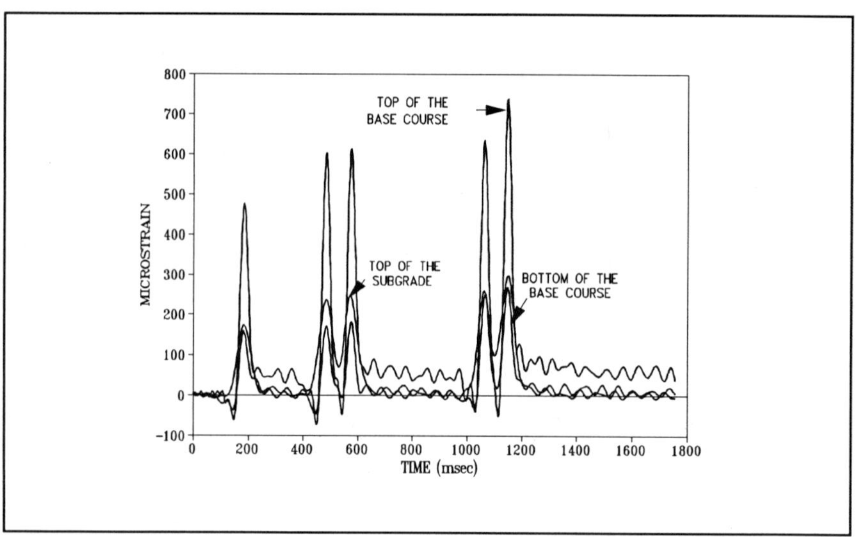

FIG. 3b--Typical multidepth strain profile measured by MDD on Section II (thick) under the test vehicle (5 axle) loading.

ANALYSIS OF SURFACE DEFLECTION DATA UNDER FWD LOADING

On both test sections, deflection measurements were obtained using an FWD at two different load levels with 4 replicate drops. Peak load and peak deflections were used in the analysis. Using the layered elastic backcalculation program MODULUS 4.0, the deflection data were analyzed, and the moduli values were backcalculated.

The Analysis of Section I (Thin)

In the surface deflection analysis for Section I (Thin), the 1.5 inches (38 mm) thick asphalt concrete surface layer modulus was fixed at 293 ksi (2020 MPa), based on asphalt concrete layer temperature at the time of testing.

The summary results of surface deflection analysis for Section I (Thin), at different load levels are shown in Table 1. The average base modulus value at the higher load level is slightly higher than at the nominal 8,000 lbs (35.6 KN) load level. The subgrade modulus decreases slightly with the increase in load.

The Analysis of Section II (Thick)

The pavement was considered a three layer system. Based on cone penetration test results, the lime stabilized layer was considered as part of the subgrade layer. A summary of the backcalculation results for different load levels along with averaged moduli values and absolute error per sensor are tabulated in Table 2. The granular base layer moduli values increased slightly with the increase in load level. Similar to Section I (Thin), the base layer on Section II (Thick) also demonstrated non-linear behavior. The subgrade modulus did not change significantly under different load levels.

ANALYSIS OF SURFACE AND DEPTH DEFLECTIONS UNDER FWD LOADINGS

On the two test sections, simultaneous surface and depth deflection measurements were obtained from FWD loadings at three different load levels and at three offsets from the MDD location. The detailed description of MDD installation techniques, testing procedures and measurement of anchor movements are described in detail elsewhere [1]. Peak load, surface deflections, and depth deflections were used in the analysis. The deflection results were analyzed using a generalized linear elastic backcalculation procedure developed by Uzan [2].

TABLE 1--Summary results of the surface deflection analysis at different load levels of Section I (thick).

TTI MODULUS ANALYSIS SYSTEM (SUMMARY REPORT)   (Version 4.0)

		Thickness(in)	MODULI RANGE(psi) Minimum	Maximum
District:	1			
County:	1			
Highway/Road: 2818	Pavement:	1.50	292,971	293,029
	Base:	10.00	10,000	150,000
	Subbase:	0.00	0	0
	Subgrade:	288.50	10,000	

Station	Load (lbs)	Measured Deflection (mils): R1	R2	R3	R4	R5	R6	R7	Calculated Moduli values (ksi): SURF(E1)	BASE(E2)	SUBB(E3)	SUBG(E4)	Absolute ERROR/Sens.	Depth to Bedrock
1.000	8,383	34.12	20.15	9.61	5.88	4.16	2.94	2.41	293.	31.7	0.0	8.9	4.71	292.66
1.000	8,343	33.57	19.81	9.53	5.84	3.99	3.10	2.41	293.	32.4	0.0	8.9	5.11	293.83
1.000	8,375	34.04	20.15	9.65	5.96	4.12	3.06	2.41	293.	32.2	0.0	8.8	4.20	300.00
1.000	8,295	34.28	20.36	9.74	6.04	4.16	3.22	2.45	293.	31.9	0.0	8.6	4.97	300.00
Mean:		34.00	20.12	9.63	5.93	4.11	3.08	2.42	293.	32.0	0.0	8.8	4.75	300.00
Std. Dev:		0.31	0.23	0.09	0.09	0.08	0.12	0.02	0.	0.3	0.0	0.2	0.40	16.03
Var Coeff(%):		0.90	1.13	0.91	1.50	1.96	3.75	0.83	0.	1.0	0.0	1.7	8.40	5.34
1.000	14,703	55.05	33.99	17.38	10.86	7.37	5.43	4.30	293.	41.7	0.0	8.2	5.61	274.24
1.000	14,071	55.21	34.20	17.51	10.78	7.41	5.39	4.34	293.	39.3	0.0	7.8	5.64	300.00
1.000	14,758	55.29	34.33	17.59	10.42	7.45	5.55	4.38	293.	41.2	0.0	8.2	6.61	208.82
Mean:		55.18	34.17	17.49	10.69	7.41	5.46	4.34	293.	40.7	0.0	8.1	5.95	260.48
Std. Dev:		0.12	0.17	0.11	0.23	0.04	0.08	0.04	0.	1.3	0.0	0.2	0.57	48.17
Var Coeff(%):		0.22	0.50	0.61	2.19	0.54	1.53	0.92	0.	3.1	0.0	2.7	9.53	18.49

TABLE 2--Summary results of the surface deflection analysis (second model) as different load levels for Section II (thick).

TTI MODULUS ANALYSIS SYSTEM (SUMMARY REPORT)   (Version 4.0)

District:	1		Thickness(in)	MODULI RANGE(psi) Minimum	Maximum
County:	1				
Highway/Road:	21	Pavement:	7.00	100,000	400,000
		Base:	14.00	10,000	100,000
		Subbase:	0.00	0	0
		Subgrade:	83.40	16,900	

Station	Load (lbs)	Measured Deflection (mils): R1	R2	R3	R4	R5	R6	R7	Calculated Moduli values (ksi): SURF(E1)	BASE(E2)	SUBB(E3)	SUBG(E4)	Absolute ERROR/Sens.	Depth to Bedrock
1.000	8,839	13.23	7.32	3.35	2.09	1.54	1.25	1.05	169.	38.3	0.0	17.3	4.69	111.75
1.000	8,863	13.12	7.28	3.31	2.05	1.50	1.25	1.05	174.	37.8	0.0	17.7	4.69	106.42
1.000	8,839	13.04	7.28	3.35	2.05	1.50	1.25	1.05	181.	36.9	0.0	17.6	4.30	100.17
1.000	8,783	13.04	7.32	3.35	2.05	1.46	1.25	1.05	181.	36.3	0.0	17.5	4.43	100.00
Mean:		13.11	7.30	3.34	2.06	1.50	1.25	1.05	176.	37.3	0.0	17.5	4.53	104.36
Std. Dev:		0.09	0.02	0.02	0.02	0.03	0.00	0.00	5.	0.9	0.0	0.2	0.20	4.78
Var Coeff(%):		0.69	0.32	0.60	0.97	2.18	0.00	0.00	3.	2.5	0.0	0.9	4.37	4.58
1.000	10,799	14.57	8.72	4.09	2.45	1.79	1.49	1.33	231.	37.5	0.0	17.3	4.26	90.39
1.000	10,791	15.51	8.80	4.13	2.57	1.87	1.57	1.37	181.	41.3	0.0	16.4	4.55	110.74
1.000	10,823	15.47	8.80	4.13	2.53	1.87	1.57	1.33	186.	40.1	0.0	16.7	4.31	101.22
Mean:		15.18	8.77	4.12	2.52	1.84	1.54	1.34	199.	39.7	0.0	16.8	4.37	100.09
Std. Dev:		0.53	0.05	0.02	0.06	0.05	0.05	0.02	27.	1.9	0.0	0.4	0.16	8.35
Var Coeff(%):		3.50	0.53	0.56	2.43	2.51	2.99	1.72	14.	4.9	0.0	2.6	3.64	8.34
1.000	15,783	20.77	12.02	5.89	3.63	2.66	2.21	1.93	218.	43.2	0.0	17.9	3.57	105.50
1.000	15,775	20.65	12.02	5.85	3.63	2.66	2.21	1.93	218.	43.5	0.0	17.9	3.85	109.31
1.000	15,751	20.69	12.02	5.89	3.63	2.66	2.21	1.93	219.	43.2	0.0	17.8	3.60	105.48
Mean:		20.70	12.02	5.88	3.63	2.66	2.21	1.93	218.	43.3	0.0	17.8	3.67	106.73
Std. Dev:		0.06	0.00	0.02	0.00	0.00	0.00	0.00	1.	0.2	0.0	0.0	0.15	1.78
Var Coeff(%):		0.30	0.00	0.39	0.00	0.00	0.00	0.00	0.	0.4	0.0	0.2	4.18	1.67

The analyses includes simultaneously matching both the
theoretical surface and depth deflections to those measured
in the field.

<u>The Analysis of Section I (Thin) Deflection Data</u>

Three sets of surface and depth deflection data were
collected at offsets of 2.25, 8.5 and 17.5 inches (57, 215
and 444 mm) from the center of the FWD loading plate to the
middle of the MDD hole. An offset distance of 2.25 inches
(57 mm) from the center of the FWD load plate to the middle
of the MDD hole was used to prevent the center geophone from
sitting directly on the MDD top cap. FWD drops were made at
each offset. The resulting surface and depth deflections
were measured simultaneously. When the FWD plate was not
over the MDD hole, the movement of the anchor was monitored
by measuring the movement on the center rod of the MDD
system. This was achieved by placing one of the FWD outer
geophones (the geophone at 60 inches (1524 mm)) on top of a
pedestal mounted on the center core and recording the core
movement via the FWD system.

The averaged measured data for Section I (Thin)
normalized to 9,000 lbs (40 KN) load are plotted in Fig. 4.
It includes average FWD surface deflections for all offsets
combined, and averages of MDD depth deflections for each of
three offset. The depth deflections measured at the top and
the bottom of the granular base layer by MDD 1 and MDD 2, at
an offset of 8.5 and 17.5 inches (215 and 444 mm) from the
load, were larger than those measured at the top of the
asphalt concrete layer. These deflection values suggest
that some dilation or extension is taking place in the
granular base layer. To address these effects, it appears
that non-linear finite element backcalculation models
incorporating dilation effects may be needed to model this
behavior for the granular base material more realistically
[3].

For the purpose of linear analysis, Section I (Thin)
was modelled as a three layer system. The thin asphalt
layer modulus value was fixed at 293 ksi (2020 MPa) based on
the temperature at the time of testing, and the base and
subgrade layer moduli were backcalculated. In the
backcalculation analysis, the average of all FWD surface
deflections measured at all three offsets and the average of
the MDD depth deflections for individual offset (4 depths x
3 offsets from load) were included in the analyses. The
measured and calculated surface and depth deflections as
well as the backcalculated moduli are listed in Table 3.
The best match base modulus was 24.6 ksi (170 MPa) with a
subgrade modulus of 7.6 ksi (52 MPa). Figs. 5a and 5b
graphically illustrates the results of deflection analyses.

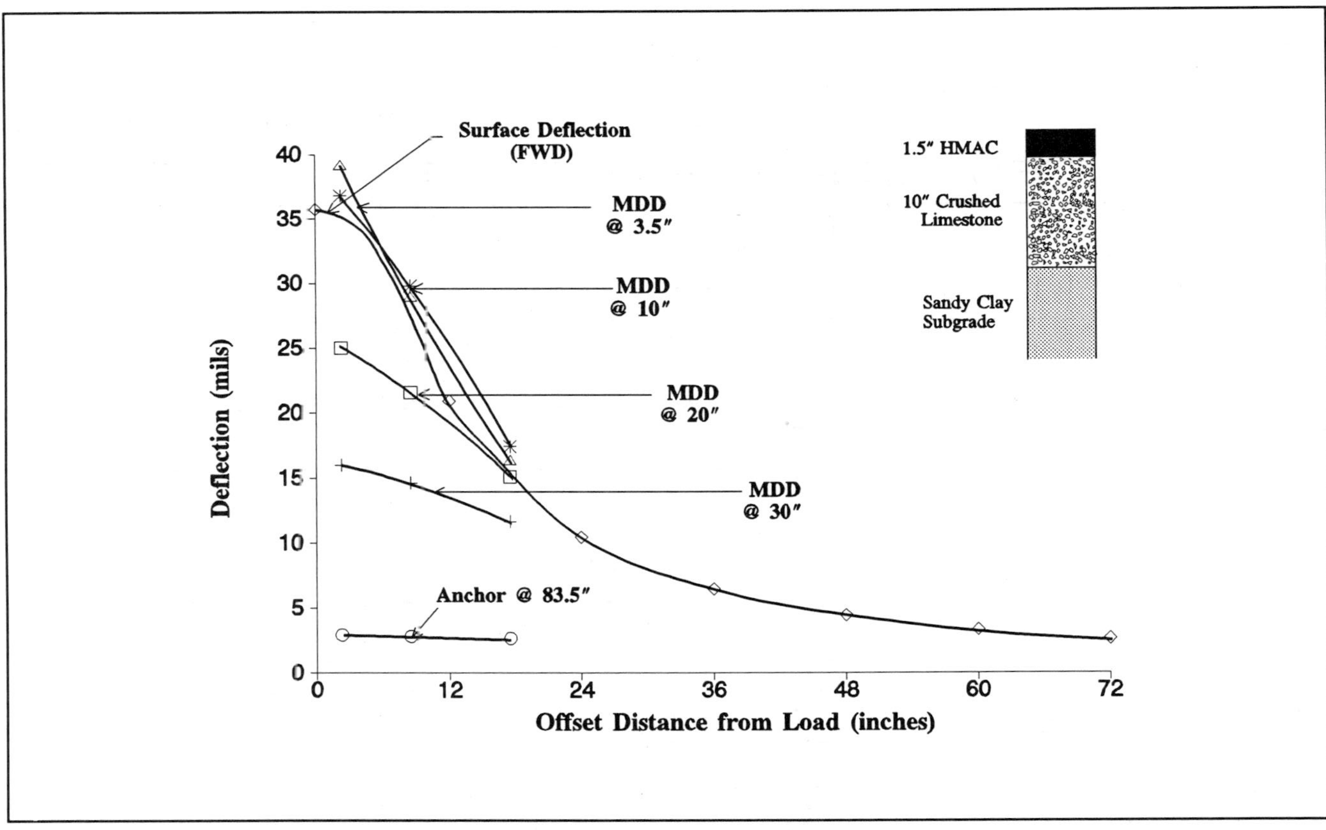

FIG. 4--Measured surface and depth deflections under FWD loading on Section I (thin).

TABLE 3--Results of the surface and depth deflection analysis for Section I (thin) under FWD loadings.

Surface Deflections

Offset r (inches)	0.00	12.00	24.00	36.00	48.00	60.00	72.00
Measured Deflection (mil)	35.75	20.92	10.46	6.47	4.45	3.33	2.64
Predicted Deflection (mil)	45.78	21.51	11.91	7.44	5.02	3.58	2.67
Error (percent)	-25.06	-2.80	-13.87	-14.97	-12.75	-7.66	-1.16
Absolute Error (mil)	10.03	0.59	1.45	0.97	0.57	0.25	0.03

Backcalculated Moduli

Asphalt Concrete (psi)	293,000
Granular Layer (psi)	24,580
Subgrade (psi)	7,588

MDD Deflections

Offset from Load to MDD Hole (inches)	2.25	8.5	17.5

LVDT at a Depth of 3.5 inches (Top of the Base)

Measured Deflection (mil)	39.10	28.87	16.27
Predicted Deflection (mil)	39.87	27.35	16.10
Error (percent)	-1.97	5.28	1.05
Absolute Error (mil)	0.77	-1.52	-0.17

LVDT at a Depth of 10 inches (Bottom of the Base)

Measured Deflection (mil)	36.82	29.81	17.41
Predicted Deflection (mil)	30.72	24.72	15.81
Error (percent)	16.56	17.06	9.18
Absolute Error (mil)	-6.10	-5.09	-1.6

Table 3--Continued.

LVDT at a Depth of 20 inches (8.5 inches into the Subgrade)			
Measured Deflection (mil)	25.02	21.59	15.09
Predicted Deflection (mil)	19.83	18.01	13.85
Error (percent)	20.73	16.59	8.24
Absolute Error (mil)	-5.19	-3.58	-1.24
LVDT at a Depth of 30 inches (18.5 inches into the Subgrade)			
Measured Deflection (mil)	15.99	14.62	11.65
Predicted Deflection (mil)	14.20	13.47	11.45
Error (percent)	11.18	7.86	1.70
Absolute Error (mil)	-1.79	-1.15	-0.20

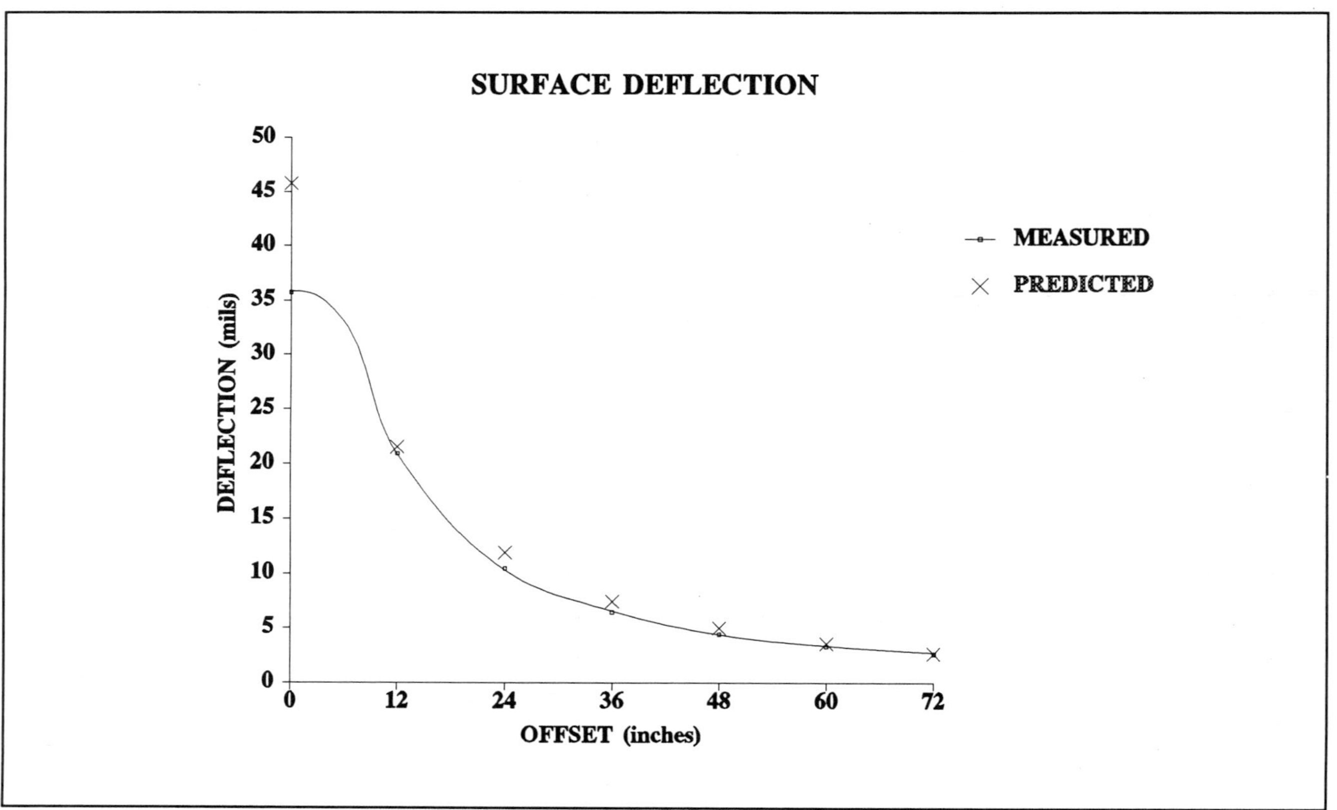

FIG. 5a--Measured and Predicted Surface Deflections Under FWD Loading on Section I (Thin).

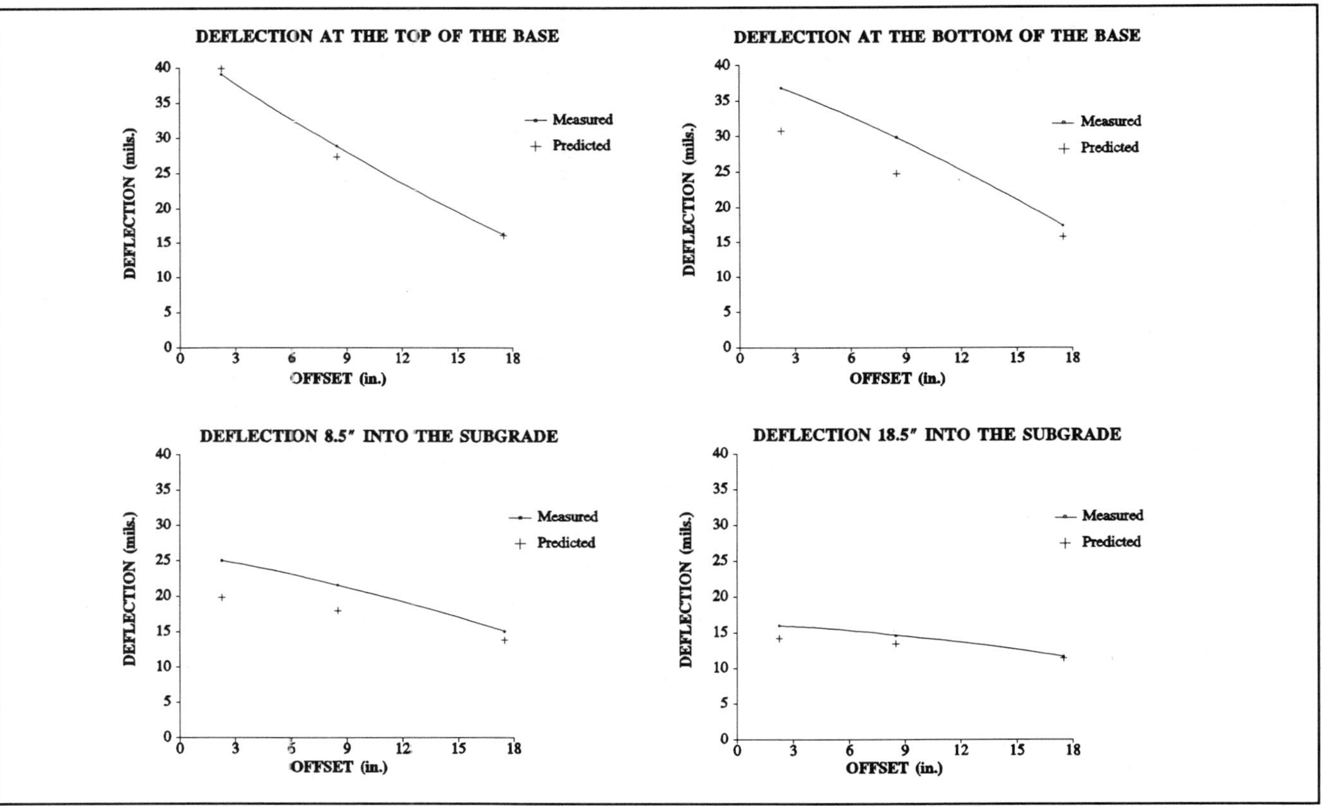

FIG. 5b--Measured and predicted depth deflections under FWD loading on Section I (thin).

From studying the results, for nineteen deflections (seven surface plus twelve depth) the best match linear elastic model resulted in an average error of about 10.5% per sensor. The error value is quite reasonable for a thin section analyzed with an assumption of linearity they are of the same magnitude as those presented by Uzan and Scullion [3]. The surface deflections as shown in Fig. 5a are matching well for all sensors except the one under the loading plate, which is over predicted by about 28%. For depth deflections, the match is good for MDD 1 (top of the base layer) and for MDD 4 (18.5 inches (470 mm) into the subgrade layer), as shown in Fig. 5b. The largest error in this system occurs at MDD 2 (bottom of the base) and at MDD 3 (8.5 inches (216 mm) into the subgrade). The deflections were under predicted at MDD 2 and MDD 3 for all offsets. The subgrade modulus appears to be slightly overpredicted.

The Analyses of Section II (Thick) Deflection Data

The pavement was modelled as a three layer system. The lime stabilized layer was considered part of the subgrade layer. Three data sets were collected at offsets of 2.5, 8.75 and 14.5 inches (63, 222 and 368 mm) from the center of the FWD loading plate to the middle of the MDD hole. The data collection procedure was similar to that explained for Section I (Thin).

The measured data on Section II (Thick) normalized to 9,000 lbs (40 KN) are plotted in Fig. 6. It shows that the depth deflection measured at the top of the granular base layer (MDD 1), at an offset of 14.5 inches (368 mm) from the load, is larger than that measured at the surface of the pavement. Similar to Section I (Thin). The measured and calculated surface and depth deflections as well as the backcalculated moduli are listed in Table 4. Figs. 7a and 7b graphically illustrate the results of the deflection analyses.

From studying the results, the linear elastic backcalculation model resulted in an average error of about 7% per sensor over sixteen sensors. The surface deflections as shown in Fig. 7a are matching well for all the sensors. For depth deflections, the match was good at MDD 1 (top of the base layer) and MDD 2 (middle of the base layer), as shown in Figs. 7b. The largest errors in this system (10-16% error) occur at MDD 3 (bottom of the base layer) and MDD 4 (8.75 inches (222 mm) into the subgrade). The deflections were under predicted at MDD 3 and MDD 4 for all offsets. The subgrade backcalculated modulus value appears to be a slight over estimate of the subgrade support.

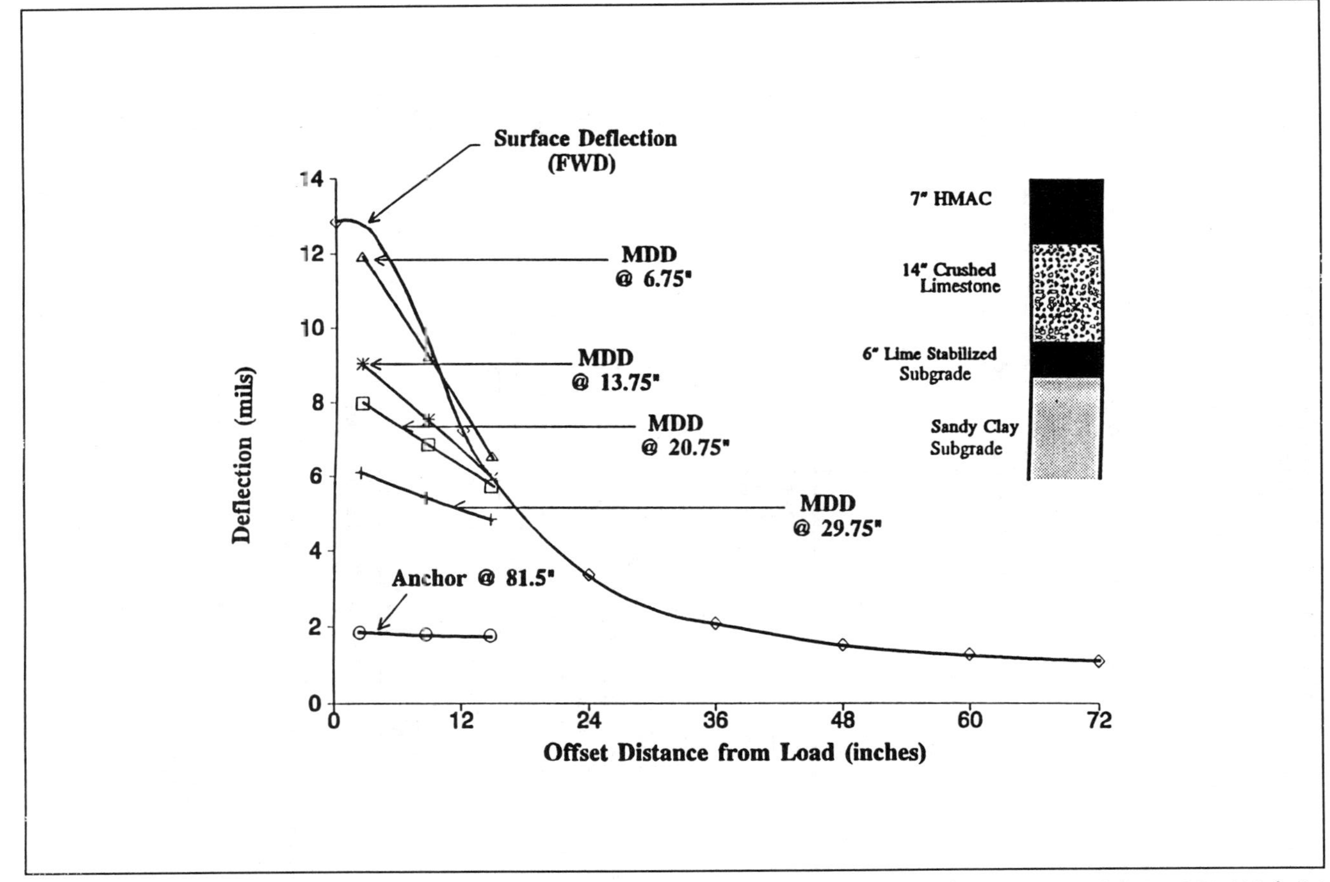

Fig. 6--Measured surface and depth deflection under FWD loading on Section II (thick).

TABLE 4--Results of the surface and depth deflection analysis for section II (thick) under FWD loadings (four surface plus twelve depth sensors).

Surface Deflections				
Offset r (inch)	0.00	12.00	24.00	48.00
Measured Deflection (mil)	12.85	7.24	3.36	2.09
Predicted Deflection (mil)	14.55	7.30	4.00	2.32
Error (percent)	13.22	0.90	18.34	10.90
Absolute Error (mil)	1.70	0.06	0.64	0.23

Backcalculated Moduli	
Asphalt Concrete (psi)	135,7
Granular Base (psi)	37
Subgrade (psi)	44,69
	8
	15,19
	9

MDD Deflections			
Offset from Load to MDD Hole (inches)	2.5	8.75	14.5

LVDT at a Depth of 6.75 inches (Top of the Base)			
Measured Deflection (mil)	11.88	9.21	6.50
Predicted Deflection (mil)	11.97	8.92	6.45
Error (percent)	0.80	-3.14	-0.75
Absolute Error (mil)	0.09	-0.29	-0.05

LVDT at a Depth of 13.75 inches (Middle of the Base)			
Measured Deflection (mil)	9.04	7.55	5.93
Predicted Deflection (mil)	8.96	7.68	6.11
Error (percent)	-0.85	1.74	3.04
Absolute Error (mil)	-0.08	0.13	0.18

LVDT at a Depth of 20.75 inches (Bottom of the Base)			
Measured Deflection (mil)	7.97	6.85	5.72
Predicted Deflection (mil)	7.26	6.52	5.46
Error (percent)	-8.88	-4.81	-4.46
Absolute Error (mil)	-0.71	-0.33	-0.26

LVDT at a Depth of 29.75 inches (8.75 inches into the Subgrade)			
Measured Deflection (mil)	6.09	5.42	4.83
Predicted Deflection (mil)	5.11	4.78	4.25
Error (percent)	-	-	-
Absolute Error (mil)	16.15	11.87	12.10
	-0.98	-0.64	-0.58

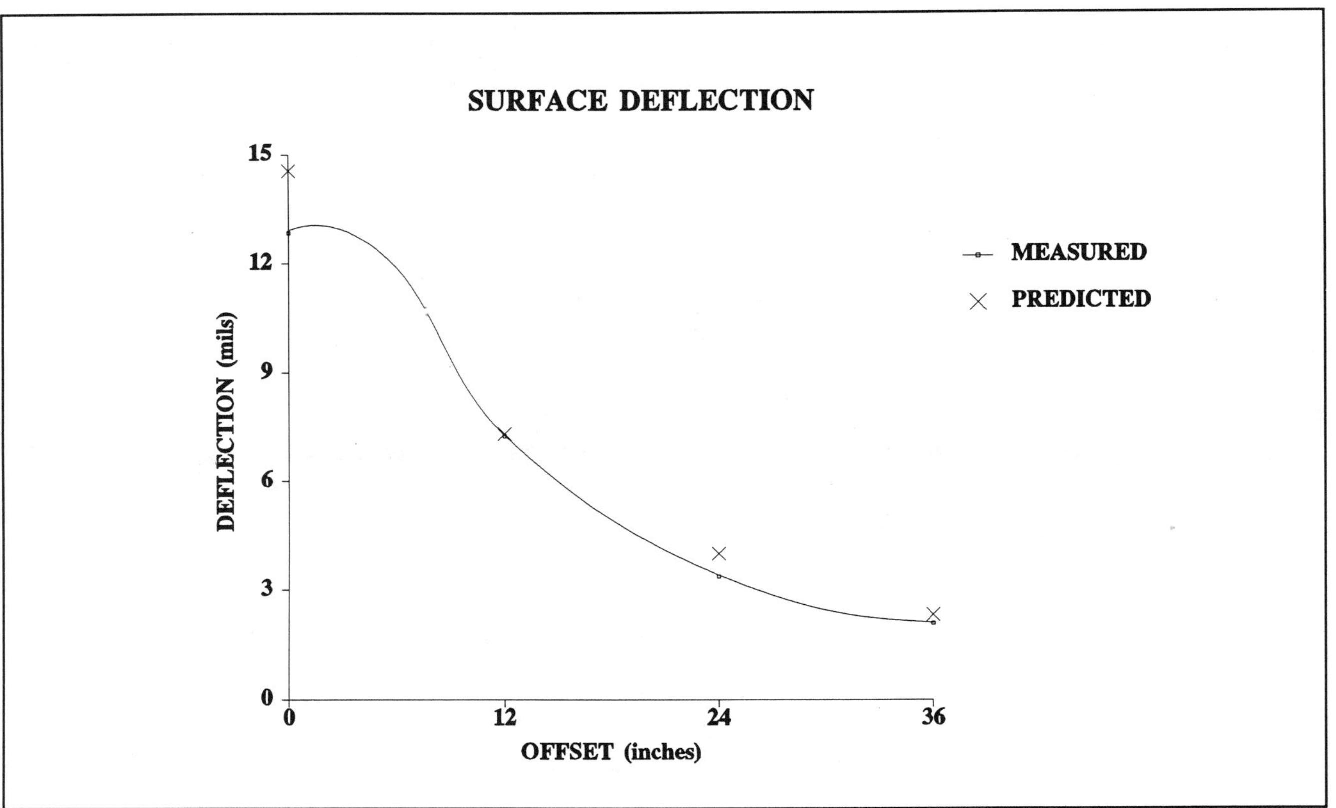

FIG. 7a--Measured and predicted surface deflections (first backcalculation model for four FWD surface sensors) under FWD loading on Section II (thick).

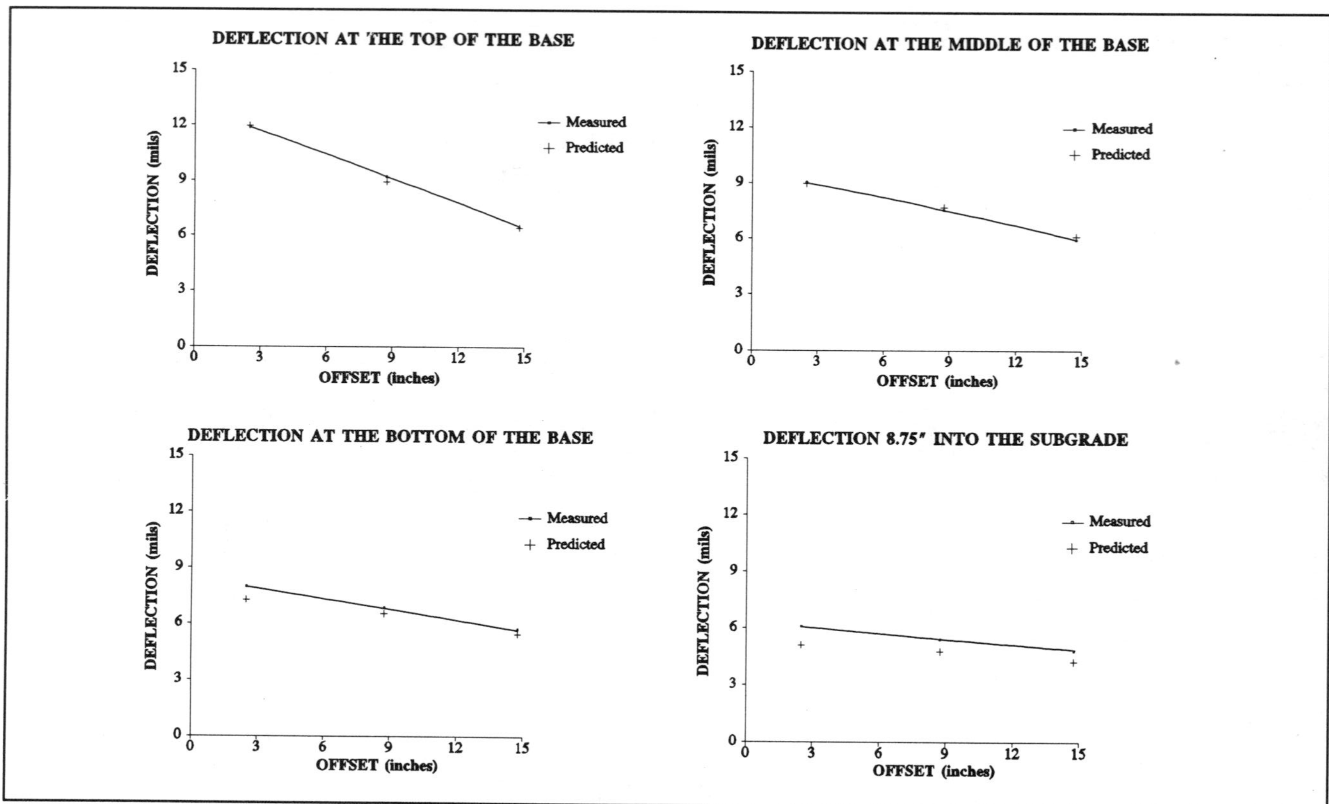

FIG. 7b--Measured and Predicted Depth Deflections Under FWD Loading (First Backcalculation Model for Twelve MDD Depth Deflections) on Section II (Thick)

COMPARISON BETWEEN FWD AND TRUCK LOADING CONDITIONS

The FWD applies approximately a haversine shaped impulse with a duration of about 28 msec. Moving wheel deflection signal durations reported by Bohn [4], Hoffman [5], and Wiman et al. [6] were much longer than FWD deflection signal durations. These researchers found the durations caused by moving trucks to be three to five times longer than those caused by FWD loadings.

The deflection pulse times under moving truck and FWD loadings were measured at MDD sensor locations on Section I (Thin) and Section II (Thick). Typical measured MDD responses in the subgrade under FWD, single and tandum axle loads are shown in Figures 8a and 8b. A complete set of results is presented in Akram [7]. The results show that the FWD deflection pulse duration is more or less constant with depth on both test sections (24 msec to 27 msec). These results were similar to the past findings by Bohn [4] and Wiman [6].

The duration of the pulse measured under the truck travelling at 55 mph (88 kph) was substantially longer than that measured under the FWD. The ratio of FWD pulse duration to single to tandum axles was 1 to 2.6 to 5.8 on the thin pavement and 1 to 3.25 to 6.75 on the thick pavement. The effect of vehicle speed on the measured pulse width is shown in Figures 8a) and b). As the speed increased from 5 to 55 mph (8 to 88 kph) the pulse width measure at the top of the subgrade decreased by 80% (800 msec to 158 msec).

The manner in which the granular layer performed under the moving vehicular loading is illustrated in Figs. 9a and 9b. Both figures show dilation/extension in the granular layer. It appears that the moving wheel is compressing the granular material directly under it and pushing (extending) the material ahead. This phenomenon is not observed under the FWD loading.

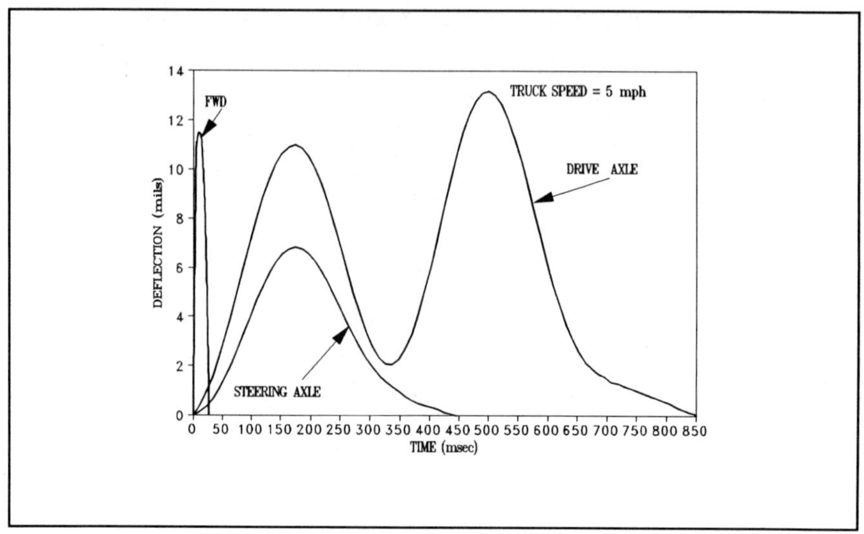

FIG. 8a--Deflection pulse duration measured by MDD sensor at a depth of 29.75 inches under moving truck (5 mph) and FWD loading on Section II (thick).

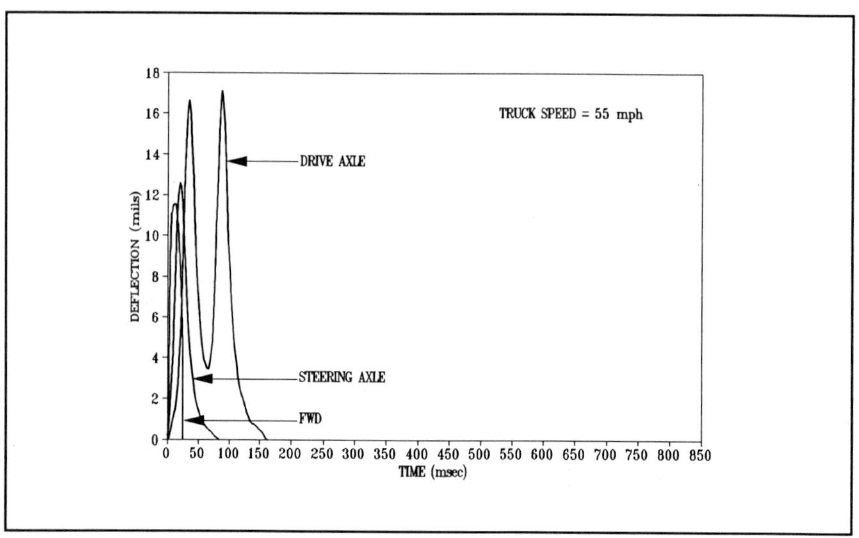

FIG. 8b--Deflection Pulse Duration Measured by MDD Sensor at a Depth of 29.75 inches Under Moving Truck (55 mph) and FWD Loading on Section II (Thick).

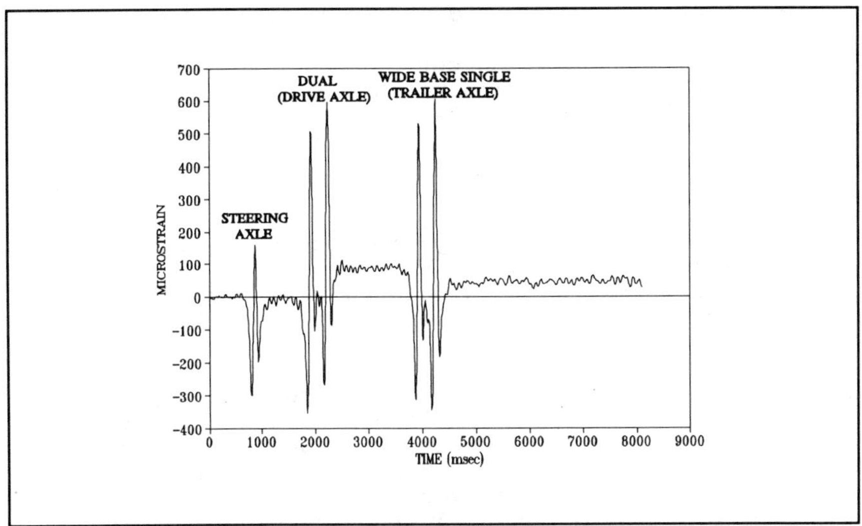

FIG. 9a--Vertical strain profile for granular base layer in Section I (thin) showing material dilation at a speed of 10 mph.

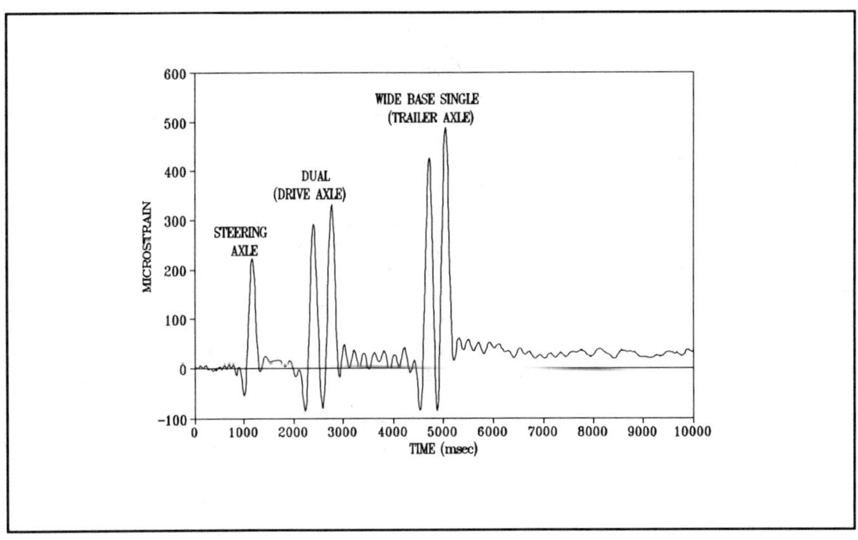

FIG. 9b--Vertical strain profile for granular base layer in Section II (Thick) showing material dilation at a speed of 35 mph.

COMPARISON OF BACKCALCULATION RESULTS TO THE LABORATORY DATA

The laboratory testing consisted of indirect tension tests on asphalt concrete cores and resilient modulus tests on remolded samples of the base and subgrade materials, performed at 0.4, 5, and 10 Hz loading frequencies. The laboratory and backcalculated moduli are illustrated in Figs. 10a and 10b.

When comparing laboratory and backcalculated moduli, perfect agreement should not be expected. The laboratory tests are performed under simulated stress conditions expected in the pavements under repeated vehicular loadings. The material samples are disturbed by the extraction process, and granular base materials were remolded. The results from the backcalculation, on the other hand, are model properties rather than true material properties [8]. By using the linear elastostatic approach a single layer stiffness is obtained for each layer. This is only an apparent stiffness for the whole layer. Actually, the stiffness of each pavement layer changes vertically and horizontally due to material changes and stress sensitivity. As a result, the backcalculated moduli do not match perfectly with the laboratory results.

<u>Asphaltic Concrete</u>

The stiffness of asphalt concrete is influenced by the temperature and loading frequency. In Table 6, the laboratory results from the indirect tension test are tabulated for various temperatures and loading frequencies. A halfsine loading and rest period (10 time pulse duration) were used.

On Section II (Thick), the backcalculated surface moduli from both of the backcalculation procedures are tabulated in Table 5. The backcalculated asphalt concrete moduli were considerably less than the laboratory results. The asphalt temperature at the time of conducting the FWD survey was approximately 85°F. The lab results indicate that the modulus should be considerably higher than the 135 to 176 ksi (930 to 1225 MPa) backcalculated. The only major factor not included in the analysis is surface cracking. Wheelpath cracking was observed close to the test area.

<u>Base Course</u>

Standard triaxial tests were made on both base and subgrade materials. The measured moduli were related to the confining conditions by a regression equation. The stress conditions in the pavement were then computed and the corresponding layer moduli calculated using the developed regression equations. As shown in Fig. 10a, no significant trend in loading frequency was observed in the laboratory

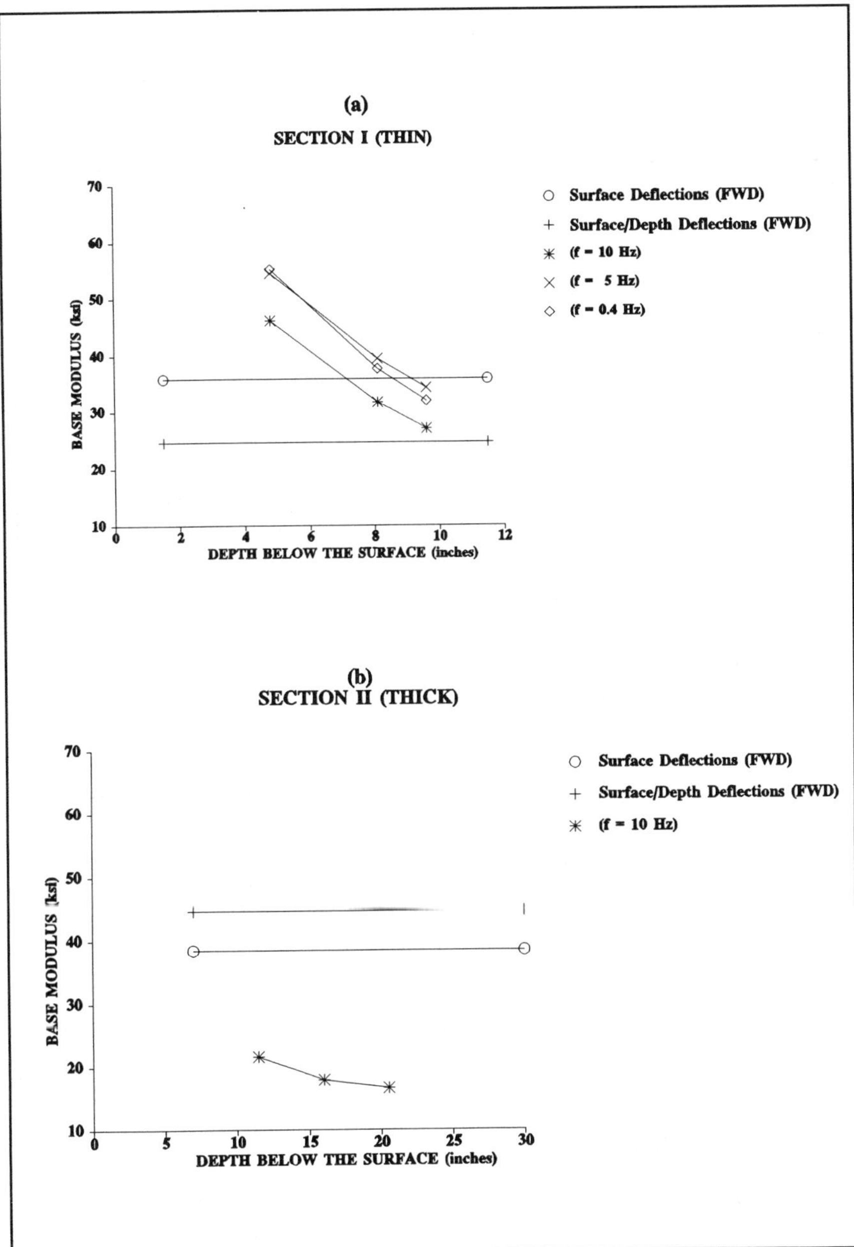

FIG. 10a--Comparison between backcaluated and laboratory results for the base of Section I (thin) and Section II (thick).

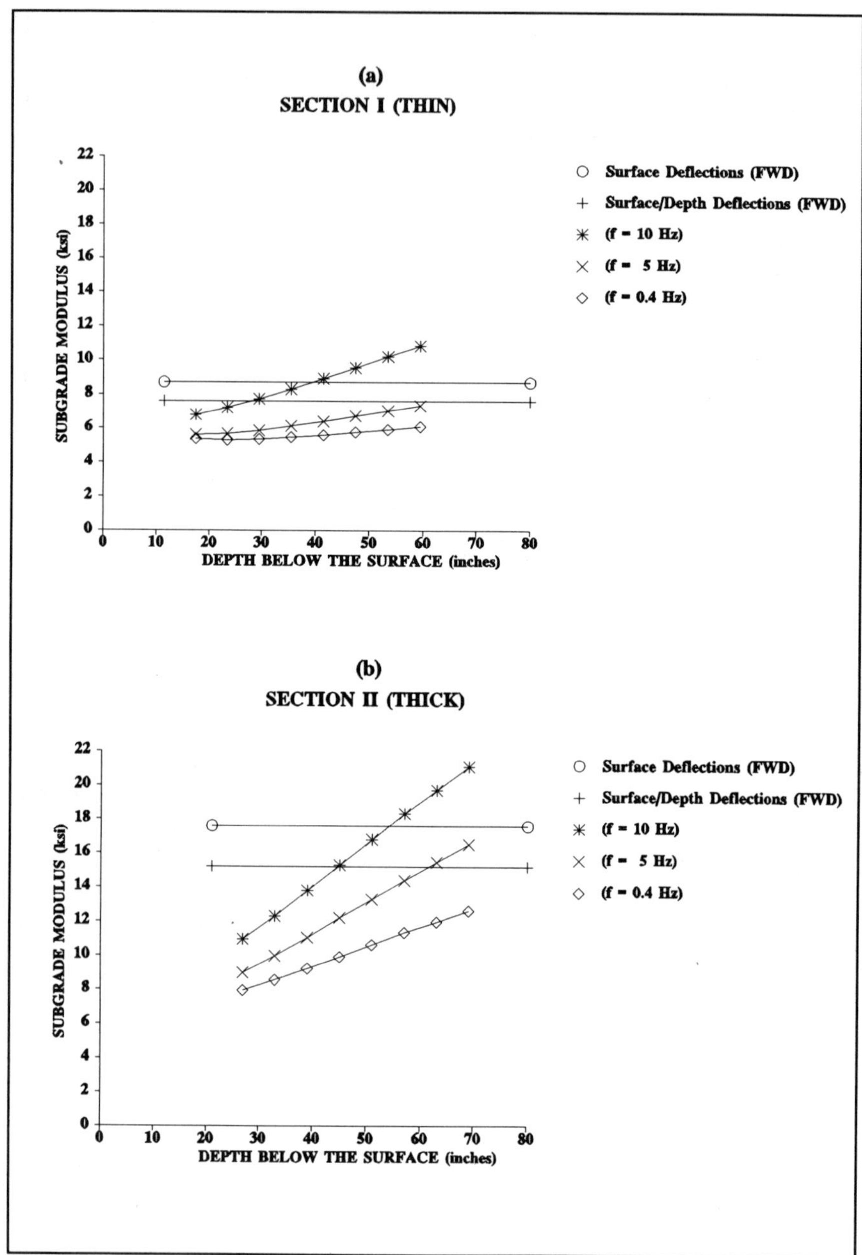

FIG. 10b--Comparison between backcaluated and laboratory results for the subgrade of Section I (thin) and Section II (thick).

TABLE 5--Backcalculated layer moduli Under FWD loadings by using different deflection data set on section I (Thin) and section II (Thick).

Backcalculation Procedure	Backcalculated Moduli (psi)	
	Section I (Thin)	Section II (Thick)
Surface Deflections (FWD)	293,000[*] 32,000[$] 8,800[**]	176,000[*] 37,300[$] 17,500[**]
Surface and Depth Deflections (FWD)	293,000 24,580 7,588	135,000 44,698 15,199
[*]Asphalt Concrete [$]Granular Base [**]Subgrade		

TABLE 6--The laboratory results for asphalt concrete layer samples at different temperatures and loading frequencies.

Test Section	Temperature (°F)	Frequency (Hz)	Modulus (ksi)
I (Thin)	77	10	743
		5	547
		0.4	310
	104	10	325
		5	210
		0.4	162
II (Thick)	77	10	591
		5	253
		0.4	149
	104	10	347
		5	84
		0.4	55

data. However, the laboratory results show a decreasing trend in the base course modulus with depth.In the linear elastic backcalculation a single number is generated to represent the average value of the base course this is shown as the horizontal lines in Figure 10. The stress conditions vary from the top to the bottom of the base and hence a range of moduli are computed from the lab data. The agreement between the laboratory and the backcalculated results in the lower half of the base layer was reasonably good. As shown in Fig. 10a, the laboratory results over predicted the modulus value for the upper half of the base layer in Section I (Thin). The granular material on Section II (Thick) was tested for only one loading frequency. As shown in Fig. 11a, the laboratory results show a decreasing trend in the modulus value with depth; however, compared to Section I (Thin), it is a relatively small change. Compared with the backcalculated base moduli, the laboratory results under predicted the backcalculated base modulus.

<u>Subgrade</u>

The laboratory results indicate the subgrade modulus on both the test sections to be frequency sensitive. The laboratory modulus increased with an increase in loading frequency and confining pressure.

On Section I (Thin), the agreement between the laboratory data and the backcalculated moduli over a range of frequencies is good. The agreement between the subgrade moduli backcalculated from surface deflections under FWD and from combined surface and depth deflections under FWD is reasonably good with laboratory results at high frequency (10 Hz). The laboratory results at low frequencies (0.4 and 5 Hz) match well with the subgrade moduli backcalculated from depth deflections under FWD and truck loadings. On Section II (Thick), the backcalculated subgrade modulus was found to be between 15 and 17.5 ksi (103 and 120 MPa). This was found to be representative of a stiffness at a depth of between 40 and 50 inches (1013 and 1267 mm) below the surface.

COMPARISON BETWEEN THE MEASURED AND PREDICTED SUBGRADE STRAINS UNDER TRUCK LOADINGS

One of the main purposes of backcalculation is to determine layer moduli that can be used in a forward calculation model to predict truck strains within the pavement. In this study the theoretical predictions under truck loadings were made using the BISAR computer program [9] using the backcalculated layer moduli and static axle loads. The calculated vertical compressive subgrade strains were compared with measured ones under the truck loadings.

The comparison for Section I (Thin) and Section II (Thick) along with the percentage errors are tabulated in Table 7. The measured strains are computed by subtracting adjacent depth deflections and dividing by the gauge separation, as described previously. The computed strains are predicted at the midpoint of the two MDD's.

On Section I (Thin), the errors between the measured subgrade strain and the subgrade strain predicted by the theoretical model was relatively high (18%). However, the strains predicted by using the backcalculated moduli from surface and depth deflections under FWD loadings matched very well with the measured strain value. As shown in Table 7, the lowest percentage difference of 2% was obtained when moduli values, backcalculated using both surface and depth deflections under FWD loading, were used to make theoretical strain prediction. This shows that on a thin pavement section, despite all the material non-linearities, subgrade strain predictions made using a linear layered elastic model were reasonably good. However, using moduli obtained from FWD surface deflections only, the strains predicted in the subgrade were under estimated by 18%.

On Section II (Thick), the match between the measured strains and those predicted by the theoretical model using different backcalculated moduli were found to be between 6% and 10%. From Table 7 it is clear that the set of moduli values backcalculated using both surface and depth deflections under FWD loading produced the best comparison between computed and measured subgrade strains.

CONCLUSION

Both surface and multidepth deflection data were collected under FWD and truck loadings on Section I (Thin) and Section II (Thick). Using a linear, elastostatic backcalculation technique, the layer moduli were backcalculated from different sets of deflection data and are summarized in Table 5.

These trends were evident from the backcalculated layer moduli under different conditions:

• on Section I (Thin), using the FWD surface deflections, a reasonably low mean square error of 4.75 percent per sensor was found, the base and subgrade modului values exhibited slight non-linear behavior with an increase in load;

TABLE 7--Comparison between the vertical compressive subgrade strain values measured under truck loadings and those predicted by the linear elastic program BISAR for dual tires on tandem axles for Section I (thin) and Section II (thick).

Backcalculation Procedure	Test Section	Backcalculated Moduli (psi)	Subgrade Strain (μstrain)		Error (percent)
			Measured	BISAR	
Surface Deflections (FWD)	I	293,000[*] 32,000[$] 8,800[**]	1047	857	-18
Surface and Depth Deflections (FWD)	I	293,000 24,580 7,588	1047	1030	-2
Surface Deflections (FWD)	II	176,000 37,300 17,500	242	209	-14
Surface and Depth Deflections (FWD)	II	135,000 44,698 15,199	242	228	-6

[*]Asphalt Concrete　　[$]Granular Base　　[**]Subgrade

• on Section II (Thick), the surface deflection analyses resulted in an average mean square error of 4.53 percent per sensor, similar to Section I (Thin), the base and subgrade moduli values demonstrated a non-linear behavior with an increase in load;

• the measured surface plus depth deflection data on both test sections showed material non-linearity; however, the backcalculated moduli were reasonably close to those predicted from the surface deflections;

The evaluation of the pulse durations between the FWD and truck loadings showed that the FWD pulse duration stayed almost constant with depth, whereas, the deflection pulse under truck loading changed with speed and depth. The pulse time measured at the bottom MDD sensor under a truck moving at a speed of 55 mph (88 kph) was about 6 to 7 times the FWD pulse time for tandem axle and about 3 times for the steering axle.

The backcalculated moduli did not match well with the laboratory data. The laboratory tests are conducted on remolded samples under simulated stress conditions. However, they give an indication of the material behavior. Laboratory data showed higher moduli with an increase in the loading frequency for subgrade materials which indicates that the deeper pavement layers are also affected by the loading frequency. Similar findings were reported by Wiman [6]. The backcalculated asphalt concrete stiffnesses at lower temperature and frequency compared well with those found in the laboratory. The base course and subgrade layer values were generally closer for Section I (Thin) than Section II (Thick).

As shown in Table 7, by using the different combinations of deflection measurement (surface or depth) it is possible to backcalculate layer moduli. However, the main purpose of backcalculation is to determine layer moduli that can be used in a forward calculation mode to predict strains within the pavement. The set of moduli values backcalculated by matching surface plus depth deflections under FWD loading on both thin and thick pavement sections produced the best match between the subgrade strains. Using FWD moduli only resulted in an overprediction of subgrade modulus and an underestimation of the truck induced subgrade vertical compressive strains by 15 to 18%.

REFERENCES

[1] Scullion, T., Uzan, J., Yazdani, J, I., and Chan, P. (1988). "Field evaluation of multidepth deflectometers." *Research Report 1123-2*, Texas Transportation Institute, The Texas A&M University, College Station, Texas.

[2] Uzan, J., Lytton, R. L., and Germann, F. P. (1988a). "General procedures for backcalculating layer moduli." *First Symposium on NDT of Pavements and Backcalculation of Moduli*, ASTM, Baltimore, Maryland.

[3] Uzan, J., and Scullion T. (1990). "Verification of backcalculation procedures." *Proceedings of the Third International Conference on the Bearing Capacity of Roads and Airfields*, Trondheim, Norway.

[4] Bohn, A., Ullidtz, P., Stubstad, R., and Sorensen, A. (1972). "Danish Experiments with the French falling weight deflectometer." *Proceedings 3rd International Conference on the Structural Design of Asphalt Pavements*, London, England.

[5] Hoffman, M. S., and Thompson, M. R. (1981). "Mechanistic interpretation of nondestructive pavement testing deflections." *Transportation Engineering Series No. 32*, Illinois Cooperative Highway and Transportation Research Program, Series No. 190, University of Illinois, Urbana-Champaign, Illinois.

[6] Wiman, L. G., Jansson, H. (1990). "A Norwegian/Swedish in-depth pavement deflection study (2) - seasonal variations and effect of loading type." *Third International Conference on Bearing Capacity of Roads and Airfields*, The Norwegian Institute of Technology, Trondheim, Norway.

[7]   Akram, T. "In situ Flexible Pavement Materials Characterization and Estimation of Distress under Dual and Wide Based Tires." Ph.D. Thesis, Texas A&M University, Dec. 1992.

[8]   Rohde, G. T. (1990). "The mechanistic analysis of FWD deflection data on sections with changing subgrade stiffness with depth." PhD Dissertation, The Texas A&M University, College Station, Texas.

[9]   *BISAR*. (1978). "Bitumen structural analysis in roads user's manual." Koninklijke/Shell - Laboratorium, Shell Research, Amsterdam, Holland.

K.P. George[1] and Waheed Uddin[2]

IN SITU AND LABORATORY CHARACTERIZATION OF NONLINEAR
PAVEMENT LAYER MODULI

REFERENCE: George, K. P., and Uddin, W., **"In Situ and Laboratory Characterization of Nonlinear Pavement Layer Moduli,"** <u>Nondestructive Testing of Pavements and Backcalculation of Moduli (Second Volume), ASTM STP 1198</u>, Harold L. Von Quintas, Albert J. Bush, III, and Gilbert Y. Baladi, Eds., American Society for Testing and Materials, Philadelphia, 1994.

ABSTRACT: The accuracy of five different backcalculation procedures for in situ characterization of granular/subgrade layers is studied by utilizing deflection data, both Dynaflect and Falling Weight Deflectometer. ELSDEF, BISDEF, MODCOMP2, MODULUS V.4 and FPEDD1/RPEDD1 are the programs evaluated with the authors' data as well as those from the literature. Included in this study is the gyratory resilient moduli of laboratory compacted samples from five sites in Mississippi. Another aspect of backcalculation procedure, lacking total consensus, is how to account for nonlinear response of granular and subgrade materials in a backcalculation algorithm. Patterned after the results of two decades of soil dynamic studies, the authors propose strain-softening models to effect correction of nonlinear behavior. The correction curve, otherwise known as shear strain attenuation curve is implemented in the FPEDD1/RPEDD1 programs. Backcalculated moduli with the FPEDD1 program show satisfactory agreement with laboratory moduli, validating the reasonableness of the strain-softening correction in light load devices.

KEYWORDS: resilient moduli, backcalculation, nonlinear, laboratory, deflection, pavement, gyratory testing

---

CHARACTERIZATION OF GRANULAR/SUBGRADE SOILS
     Resilient modulus, M_r, is defined as the ratio of the repeated axial deviator stress simulating traffic loading to

---

[1]Professor, Civil Engineering Department, The University of Mississippi, University, MS 38677
[2]Assistant Professor, Civil Engineering Department, The University of Mississippi, University, MS 38677

the recoverable axial strain. Methods for the determination
of M_r are described in AASHTO T274-82 (AASHTO 1982). Because
results are mandated at numerous stress combinations, and
since for each stress level 200 pre-conditioning cyclic
loadings are required, evaluation of a single soil takes
several hours.

SHRP Protocol P46 (SHRP 1991) includes modifications
over the AASHTO procedure. Proposed to be included in the
Protocol are standardized procedures for measuring load and
deformation; however, the basic deficiencies of the AASHTO
test procedure are not adequately addressed. The testing
time is somewhat reduced by revising the pre-conditioning
cycles from 200 to 100.

In view of the complexity of triaxial testing, some
attempts have been made to find substitute tests. In one
such study, the conventional repeated load test (RLT) was
compared to another test, namely, diametral repeated load
procedure (Montalvo et al. 1984). This study did not
succeed in establishing a correlation between the diametral
moduli and the RLT moduli. The senior author has proposed
the use of U.S. Corps of Engineers Gyratory Testing Machine
(GTM) in lieu of the RLT test. The GTM, a combination
kneading compaction, "dynamic consolidation", and shear
testing machine, is a rather realistic simulator of abrasion
effects caused by repetitive stress and intergranular
movement owing to moving wheel loads (US Army 1962). The
material (granular base/subbase or subgrade soil) while
confined in the soil mold, is subjected to repeated load,
simulating the wheel load repetitions and is also programmed
to undergo shear stress reversal, the stress state
associated with the passage of moving wheel (George, 1992).
A comparative analysis by George (1992) shows that the
kneading gyratory moduli, M_{rk}, are consistently lower than
the M_r values.

Many research studies have been conducted to
investigate the sensitivity of various factors affecting
resilient modulus (George 1992; Thompson et al. 1976).
These factors include material type, sample preparation
method, stress state, and the condition of the samples. The
same factors more or less affect the gyratory resilient
modulus as well. For the first time, the gyratory study was
able to show that shear stress reversal strongly influences
the resilient modulus (George 1992). Note that shear stress
reversal can be introduced by gyrating the sample at some
prescribed angle.

The question arises as to how closely the repeated load
test can simulate the field conditions, and, in turn, the
load bearing characteristics of the material. A number of
studies show that the in situ modulus of granular sublayers
is strongly correlated to the stress state and thickness of

the upper layers (or the pavement structure). Confirming
this philosophy, May and Witczak (1981), asserted that "the
subgrade moduli may be a function of its intended use".
Based on field deflection studies, they concluded that
current laboratory methods for granular material
characterization appear to be inadequate for modeling in
situ behavior, regardless of the measuring device. This
discrepancy is attributed to the difference in the shear
strain (which is mobilized under the triaxial simulation
device) compared to that induced in the field. The
laboratory modulus, therefore, should be viewed critically
with respect to the shear strain mobilized during the test
procedure.

Recognizing that there is lack of agreement between the
resilient modulus of granular and soil materials determined
in the laboratory and those calculated from in situ testing,
an investigation is initiated to review all of the factors
responsible for this discrepancy. A major factor
responsible for this difference is the strain-sensitivity of
the material. A procedure to apply an appropriate
correction to in situ moduli (derived from nondestructive
testing, NDT), thereby to deduce laboratory equivalent
modulus, will be discussed in the paper. To validate the
correction procedures, the authors compare the
backcalculated moduli, corrected for strain sensitivity,
with the laboratory values.

COMPARISON OF LABORATORY RESILIENT MODULUS WITH BACKCALCULATED VALUES

Because base/subgrade characterization plays a crucial
role in the rehabilitation design, interest is focused on
evaluating in situ modulus of these layers. Backcalculation
of moduli from a deflection basin is a cost-effective
approach for estimating in situ moduli. Several
backcalculation procedures have been used in the past; the
basic tenet of these procedures is that the combination of
layer moduli providing the best match of theoretical and
measured deflection basins yields the effective in situ
characterization of pavement layers. How close are the
laboratory moduli (generally used for design) to in situ
values is a question that confused the pavement engineer.
The following investigation is planned to shed light on this
issue.

Test Plan

With the objective of comparing in situ modulus with
laboratory value, five subgrade soils were sampled from the
roadbed of actual pavement projects during construction.
Table 1 represents the basic properties of the test soils.
Replicate subgrade soil samples were compacted at AASHTO T99
density and moisture content. After equilibrating the

samples, they were tested in accordance with the AASHTO T 274 procedure with one exception that the load duration, by necessity, was set at 0.3 seconds as opposed to 0.1 seconds, recommended in AASHTO procedure. Another set of replicate samples was tested for gyratory resilient modulus, M_{rk}, at 69 kPa/138 kPa cyclic pressure, 1 second load duration with the gyratory angle set at 0.1 degree.

Dynaflect- and FWD-deflections were obtained on these pavements at various stages making it possible to backcalculate in situ modulus of each layer. Dynaflect deflections were obtained in 1991, a few months after the pavement construction, and FWD deflection in 1992, approximately one year after the Dynaflect tests.

Comparison of Various Backcalculation Programs

In order to make a meaningful comparison of laboratory versus in situ moduli, it is essential that suitable backcalculation procedure(s) be selected from a list of algorithms developed during the last two decades. A preliminary screening was conducted by backcalculating moduli of three pavements for which FWD deflection data was adapted from Zhou et al. (1989). Five backcalculation programs were investigated: they are BISDEF, ELSDEF, MODCOM2, MODULUS 4.0 and FPEDD1/RPEDD1. Calculations using the first three programs were conducted by Zhou et al., while analyses with the last two programs were performed in this study. Average modulus values along with the standard deviations of five locations in each pavement site are listed in Table 2. Substantially low aggregate base moduli are predicted with ELSDEF, MODCOMP2 and MODULUS 4.0 in the three flexible pavements, a reason for not pursuing those programs further. The subgrade moduli calculated by all of the five programs are in satisfactory agreement, however. In backcalculating the moduli for Mississippi sections, FPEDD1 and MODULUS 4.0 will be used.

Moduli Results

From five road sites in Mississippi, disturbed subgrade soil samples were collected for laboratory testing. Conducted also were parallel deflection tests, the results of which afforded in situ modulus determination by backcalculation procedure. When comparing the laboratory resilient modulus with the corresponding FWD backcalculated moduli, it is noted that the laboratory moduli are consistently lower than that of the in situ counterpart (George 1992). Similar results have been reported by several researchers. Mamlouk et al. (1988) reported that in situ moduli are larger than the triaxial counterpart by 50 to 75 percent.

Several explanations have been provided in the past to

Table 1 -- <u>Soil characteristics (Mississippi Soils)</u>

County/ Soil No.	Location Hwy	Passing 0.074 mm	Atterberg Limits		Proctor Test Data	
			LL	PI	Unit Weight kN/m^3	Optimum Moisture %
Forr./10[*]	US98	10	0	NP	18.8	9.5
Forr./2	US98	19	0	NP	19.2	10.4
Yalo./3	MS7	26	22	4	18.9	11.9
Sunfl./4	US49	70	32	13	18.4	15.1
Sunfl./5	US98	89	40	18	17.3	15.7

[*]subbase material

Table 2 -- <u>Comparison of backcalculated moduli</u>
<u>(Oregon Pavements)</u>

Road	Program	Base[a] M_r, kN/m^2 Mean/Std. Dev.	Subgrade M_r, kN/m^2 Mean/Std. Dev.
1[c].	BISDEF	131,050 / 31,640	79,920[b] / 7,750
	ELSDEF	26,600 / 380	79,920[b] / 7,750
	MODCOMP2	45,890 / 9,470	79,920[b] / 7,750
	MODULUS 4.0	31,010 / 1,690	103,210 / 5,240
	FPEDD1	76,070 / 12,000	87,140 / 7,420
2.	BISDEF	43,680 / 7,740	121,540[b] / 13,100
	ELSDEF	24,120 / 2,880	121,540[b] / 13,100
	MODCOMP2	25,630 / 5,730	121,540[b] / 13,100
	MODULUS 4.0	28,520 / 1,510	131,050 / 16,240
	FPEDD1	134,080 / 5,280	123,030 / 13,480
3.	BISDEF	50,620 / 21,770	59,250[b] / 4,020
	ELSDEF	34,590 / 5,000	59,250[b] / 4,020
	MODCOMP2	51,020 / 11,420	59,250[b] / 4,020
	MODULUS 4.0	46,850 / 16,850	72,350 / 8,460
	FPEDD1	84,610 / 10,610	61,110 / 4,800

[a]aggregate base, [b]input modulus fixed in the program
[c]1. King's Valley Hwy; 2. Willamina Hwy; 3. Lancaster Drive

explain this difference. At the outset it should be mentioned that a one-to-one correspondence between laboratory and backcalculated results cannot be expected. The laboratory sample provides a measure of the resilient modulus of a small, finite sample. Backcalculation, on the contrary, yields a gross, overall estimate for the entire depth of the subgrade. The ensuing section discusses the concepts/results proposed in soil dynamic response studies, and how those results can be employed to explain the aforementioned discrepancy.

SOIL MODULI CHARACTERIZED BY DYNAMIC RESPONSE STUDIES

The use of strain sensitive models for the evaluation of dynamic shear modulus, G, has been a popular approach in soil dynamics and earthquake engineering applications. Several important conclusions, relevant to the present study are extracted from previous research.

(1) Shear modulus, G, is a function of shear strain amplitude (Seed and Idriss 1970; Hardin and Drnevich 1972).

(2) At very low shear strain amplitude (below 10^{-3} percent), the dynamic shear modulus is strain independent and is typically referred to G_{max} (maximum dynamic shear modulus). Modulus associated with higher strain amplitude are strain sensitive (Seed and Idriss 1970).

(3) Dynamic shear modulus attenuation curves show identical trend in non-dimensional plots of G/G_{max} (normalized shear modulus) versus shear strain, as reported by Seed and Idriss (1970) for sands, Stokoe and Lodde (1978) for clays, and Seed et al. (1984) for gravelly soils. Figures 1 and 2 illustrate these unique relationships.

(4) As illustrated in Figures 1 and 2, for high strain amplitudes in the range of 10^{-3} to 10^{-1} percent, clays, sands and gravelly soils exhibit strain softening.

(5) If G_{max} is known, then G associated with any higher shear strain amplitude can be determined using the appropriate normalized shear modulus versus shear strain curve. G_{max} can be obtained in the field with seismic tests like the crosshole or downhole tests or by the surface waves techniques.

Recent laboratory dynamic test data (Kim and Stokoe 1992), using resonant column and torsional shear testing techniques further confirm the concept of modulus attenuation curves. Recall that the shear modulus, G, and Young's modulus, E, are related, therefore, the G/G_{max} data can be transformed to E/E_{max} values.

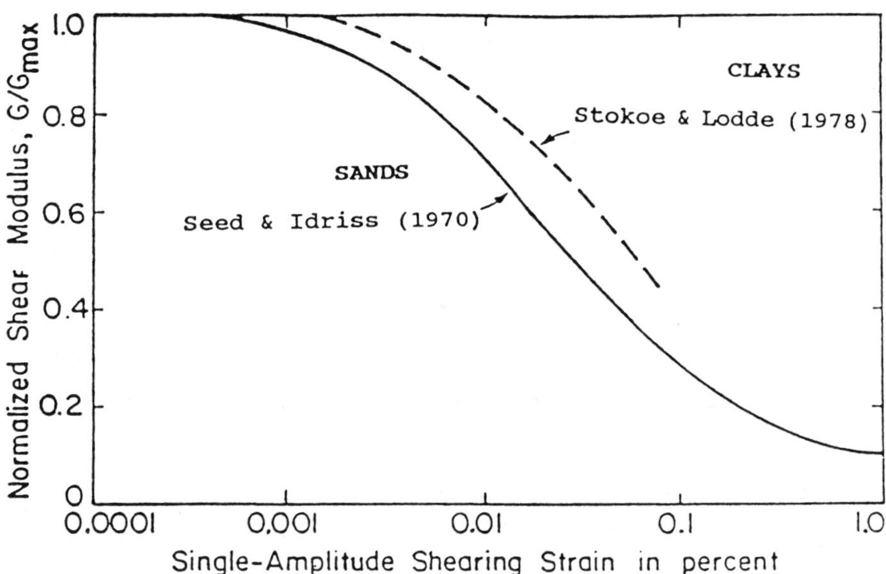

Figure 1: Normalized shear modulus attenuation curves for sands and clays (Stokoe and Lodde 1978; Seed and Idriss 1970)

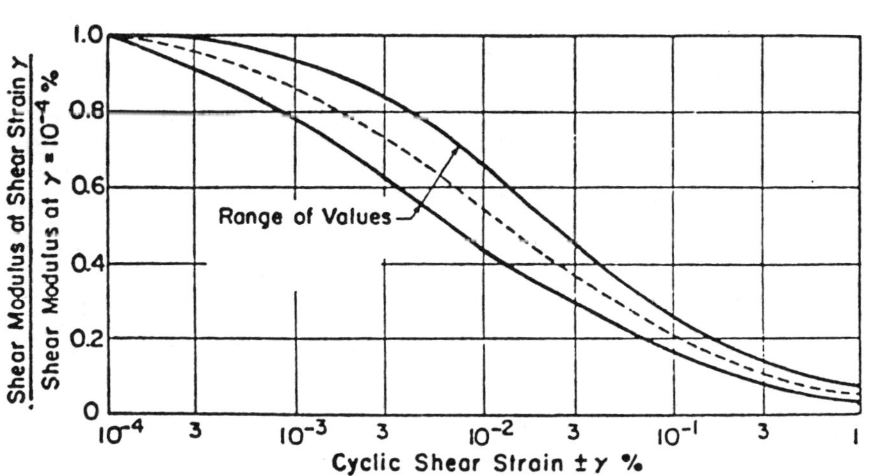

Figure 2: Normalized shear modulus attenuation curves for gravelly soils (Seed et al. 1984)

Kim and Stokoe (1992) have plotted the resonant column and torsional shear test data, as shown in Figure 3, which exhibit the strain-softening behavior for strains above the 10^{-3} percent threshold amplitude level. Curves with solid lines have been superimposed in this plot to show the range of the reported values. Kim and Stokoe (1992) also conducted the resilient modulus (M_r) test using the SHRP Protocol P46. These M_r results are plotted in Figure 4. Also graphed are the trend lines from figure 3. The authors show that the M_r results at low strain amplitude could not be obtained because of the inadequacy of the instrumentation. Nevertheless, Figure 4 illustrates two important points.

(a)   The limited laboratory M_r test data apparently agree with the seismic test data, though only for a small range of strain.

(b)   Owing primarily to large strain levels, the laboratory M_r values are expected to be lower than in situ modulus values determined in the field by seismic tests or backcalculated from light load NDT deflection data.

For example, Uddin et al. (1985) show shear strain amplitude in the range of 10^{-3} and 10^{-2} percent for 40 kN FWD loading force and 80 kN design axle load, and a strain amplitude of 10^{-4} percent for the light load Dynaflect device. Strain in typical triaxial laboratory sample is in the 10^{-1} percent level. Therefore, it can be surmised that laboratory resilient modulus values would be lower than the NDT-based field values for the reason that the laboratory tests are associated with relatively large shear strains.

To augment this observation that modulus decreases with increasing shear strain, FWD deflection data at four different load levels (27 to 57 kN range) were obtained from four sections of a road. This data afforded subgrade moduli values by backcalculation procedure, and the maximum shear strain in the subgrade by multilayer elastic analysis. Although, complete results are not included in this paper, it should be noted that the backcalculated moduli decreased by 8 percent when the strain increased from 0.035 to 0.049%. For the same strain variation in Figure 3, a reduction of moduli of 28 percentage points is noted. Suffice it to say that the static analysis performed for different load levels simply cannot duplicate the results in Figure 3, which were obtained from dynamic response studies. The results follow the general trend, however.

The on-going material characterization research program at the University of Mississippi has confirmed the above findings of strain amplitude variation under light NDT load and heavier loading forces. Figure 5 shows the result of GTM tests on compacted subgrade soils (at 0.1 degree

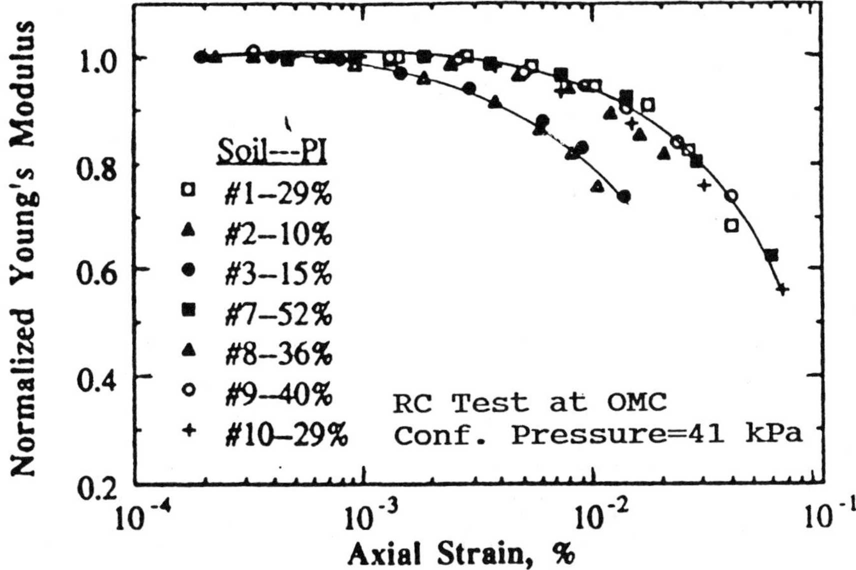

Figure 3: Normalized modulus attenuation curve for compacted subgrade soils (Kim and Stokoe 1992)

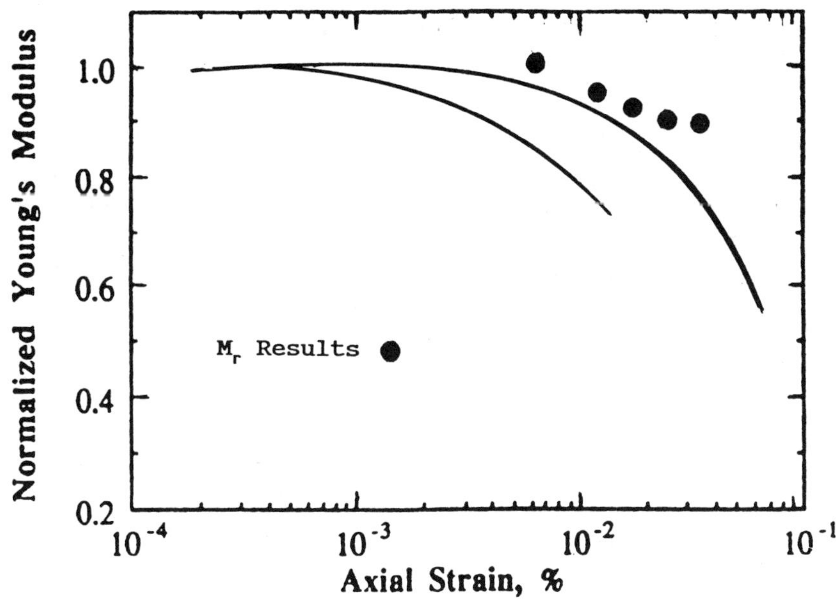

Figure 4: Normalized M_r versus strain data for compacted subgrade soils (Kim and Stokoe 1992)

gyratory angle) superimposed on the modulus attenuation
curves developed from Kim and Stokoe data (1992).
Obviously, there is more scatter in the authors' data than
reported in the M_r data of Figure 4, because in the GTM test
equipment the test specimen is subjected to rather complex
stress state. However, to a first degree of approximation,
it can be concluded that the repeated loads and GTM tests
tend to generate lower resilient modulus values than the
resilient moduli backcalculated from uncorrected NDT
deflection data because the results of both laboratory tests
are associated with higher strain amplitude.

<u>Correction of Backcalculated Moduli for Strain-Softening</u>

     Uddin et al. (1985) first recognized the importance of
the strain-softening behavior of unbound granular pavement
layers and roadbed soils and used the normalized shear
modulus versus shear strain attenuation curves shown in
Figure 1. The purpose was to correct the effective in situ
backcalculated moduli from the light load NDT deflection
data. The FPEDD1 and RPEDD1 programs (Uddin et al. 1985,
1986) incorporate self-iterative routines to correct the
Dynaflect-backcalculated moduli of all unbound layers and
subgrade soils for strain-softening nonlinear behavior. In
situ moduli from several sites are corrected using this
approach and compared with the laboratory values in the next
section.

BACKCALCULATED AND LABORATORY MODULI COMPARED

<u>Correction of Backcalculated Moduli</u>

     The Dynaflect-backcalculated moduli of the five asphalt
pavements were corrected for strain-softening nonlinear
behavior by the built in self-iterative correction procedure
in the FPEDD1 program. For implementing correction, the
program makes use of the normalized modulus versus shear
strain attenuation curves, as shown in Figure 1. Because
the strain amplitude calculated for 40 kN FWD matches that
resulting from 80 kN design axle load (Uddin et al. 1985),
it is surmised that FWD backcalculated moduli, derived from
FPEDD1 program, need not be corrected.

<u>Comparison of Laboratory and Backcalculated Moduli Values</u>

     As can be verified in Table 3, the corrected in situ
nonlinear values (backcalculated from Dynaflect) of four
soils are in good agreement with the laboratory GTM 0.1
degree kneading moduli, with deviations +10, -3, -13 and +9
percentage points, respectively in soils 10, 2, 3, and 4.
In the heavy clay soil 5, the corrected backcalculated value
is substantially larger than that from the GTM. That the
moduli obtained from the GTM are in satisfactory agreement
with the FPEDD1 corrected backcalculated Dynaflect moduli

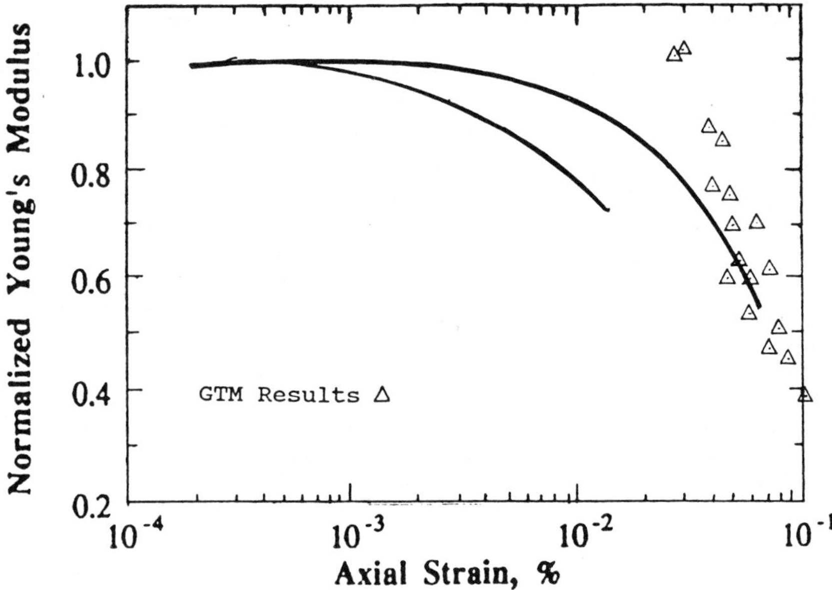

Figure 5: Illustration of GTM data on normalized modulus
attenuation curve

Table 3 -- GTM and backcalculated moduli compared

Soil Number (1)	Kneading Modulus, kN/m^2 (0.1 degree) (2)	Backcalculated Modulus, kN/m^2 (Dynaflect)		Backcalculated Modulus, kN/m^2 (FWD)	
		FPEDD1 (Uncorr.) (3)	FPEDD1 (Corr.) (4)	FPEDD1 (Uncorr.) (5)	MODULUS V.4 (6)
10[*]	108,790	198,750	121,270	235,020	229,570
2[*]	111,340	177,310	108,170	222,680	220,610
3[*]	88,520	178,000	78,390	221,300	199,240
4[**]	62,800	101,760	69,150	145,600	146,840
5[**]	43,640	114,230	76,870	148,840	149,600

[*]coarse-grained soil
[**]fine-grained soil

gives credence to the nonlinear correction methodology.

A comparison between the backcalculated values from two deflection devices --Dynaflect and FWD-- is pursued for proposing corrections, if any, for the FWD moduli. The backcalculated FWD and corrected Dynaflect moduli have been reported in fair agreement in past side-by-side deflection testing studies (Uddin et al. 1987; Hudson et al. 1987). Comparing the authors' results in columns 2, 4 and 5, it is noted that the FWD backcalculated moduli are higher than the laboratory GTM as well as the corrected Dynaflect moduli. Furthermore, that the FWD moduli are even higher than the uncorrected Dynaflect moduli (columns 5 and 3) may be attributed to inadequacies in testing methods as well as to using backcalculation procedures based on static load response in lieu of dynamic analysis. The results of two independent backcalculation algorithms (FPEDD1 and MODULUS 4.0) give rise to nearly identical FWD moduli values (compare columns 5 and 6). The higher FWD backcalculated moduli in Table 3 in part can be attributed to the following factors:

In situ Preconditioning

The FWD deflection data utilized in this investigation are obtained using SHRP FWD, with extensive preconditioning by specific load repetitions. Recall that only limited preconditioning was required in the previous FWD tests; for example, deflection data reported in Uddin et al. (1987) and Hudson et al. (1987). It is likely that, as a result of extensive preconditioning, the deflection was reduced with consequent increase in the backcalculation moduli.

Loading Mode and Duration

The static multilayered elastic analysis used in the backcalculation programs does not take into account the dynamic effects of loading mode and duration. Note that the FWD induces an impulse load whereas the Dynaflect and the Road Rater impart a steady state vibrating force. The load duration of each device is different as well: FWD 33 ms; Dynaflect 125 ms; Road Rater 40 ms; and GTM 1000 ms. Both of these effects can be evaluated by dynamic analysis of pavement systems, currently studied at the University of Mississippi utilizing a three-dimensional finite element dynamic analysis program.

Drainage of Test Sample

During deflection testing of in situ pavement, the subgrade material is less amenable to drainage than what can be expected in finite size samples, as the case in repeated triaxial testing. The impulse load of FWD tends to inhibit soil drainage process even further. The bulk compressibility of saturated, undrained soil, K_u, is effectively due to the bulk compressibility of the water phase Britto and Gunn (1987). For an unsaturated soil with adequate drainage, the bulk modulus approaches, K^1, which is equal to the bulk stiffness of the soil aggregate. Britto and Gunn (1987) have shown that K^1 is substantially smaller than K_u. The implication is that the bulk modulus, and, in turn, the resilient modulus, appropriate for drained behavior, a likely condition in a laboratory sample, tends to be lower than that can be expected for undrained behavior, a condition that may prevail during FWD testing.

<u>Seasonal Variation</u>

In addition to the three factors discussed above, another reason for the nonlinear corrected Dynaflect moduli not showing agreement with the FWD counterpart in this study is that the two tests were conducted at different times: Dynaflect soon after construction, and FWD after more than one year of trafficking. The additional densification during the one year period, and changes in the moisture regime could have contributed to a stiffer subgrade as measured by the FWD.

SUMMARY AND CONCLUSIONS

Backcalculation of in situ effective pavement layer moduli has been recognized as a powerful and cost effective nondestructive testing and evaluation tool. A number of backcalculation procedures exist in the literature, all of them employing the deflection matching technique for estimating in situ moduli of pavement materials. The reasonableness of five different backcalculation procedures is studied by employing deflection data (both Dynaflect and FWD) from the literature .

Based on the comparison of GTM laboratory moduli versus backcalculated moduli from light load NDT deflection data, it is determined that granular and soil material backcalculated moduli need correction for strain-softening behavior. Patterned after the results of two decades of soil dynamic studies, the authors propose correction curves as shown in Figure 1, where normalized shear modulus is related to shear strain, otherwise known as shear attenuation curve. Nonlinear correction procedure, especially applicable to light load NDT devices, assumes that due to the low strain level, the backcalculated moduli are the maximum values (Gmax or Emax). Now with the shear strain known under the design load (shear strain calculated by elastic layer analysis), design/field modulus can be estimated from the shear attenuation curve.

REFERENCES

American Association of State Highways and Transportation Officials, 1986, "AASHTO Guide for Design of Pavements Structure," Washington, D.C.

Britto, A.M. and Gunn, M.J., 1987, <u>Critical State Soil Mechanics via Finite Elements</u>, Ellis Horwood Limited, West Sussex, England.

George, K.P., 1992, "Resilient Testing of Soils Using Gyratory Testing Machine," Transportation Research Record 1369, Washington, D.C, pp. 63-72.

Hardin, B.O. and Drnevich, V.P., 1972, "Shear Modulus and Damping in Soils: Measurement and Parameter Effects," Journal of Soil Mechanics and Foundation Division, Vol 98, No. SM6, <u>Proceedings</u>, American Society of Civil Engineers, pp. 603-624.

Hudson, W.R., Elkins, G.E., Uddin, W. and Reilly, K.T., 1987, "Evaluation of Deflection Measurement Equipment," Report No. FHWA-TS-87-208, Pavement Condition Monitoring Methods and Equipment Phase I- Part I, ARE, Inc. for Federal Highway Administration.

Kim, D.S. and Stokoe, K.H. II, 1992, "Characterization of Resilient Modulus of Compacted Subgrade Soils Using Resonant Column and Torsional Shear Test," Transportation Research Record 1369, Washington, D.C., pp. 83-91.

Mamlouk, M.S., Houston, W.N., Houston, S.L. and Zaniewski, J.P., 1988, "Rational Characterization of Pavement Structures Using Deflection Analysis," Report No. FHWA-AZ88-254, I.CART, Arizona State University, Tempe,

May, R.W. and Witczak, M.W., 1981, "Effective Granular Modulus to Model Pavement Responses," Transportation Research Record 810, Washington, D.C., pp. 1-8.

Montalvo, J.R., Bell, C.A., Wilson, J.E., and Scofield, J.T., 1984, "Application of Resilient Modulus Test Equipment for Subgrade Soils," Report No. FHWA-OR-84-3, Oregon State University, Corvallis, Oregon.

Seed, H.B., Wong, R.T., Idriss, I.M. and Tokinatsu, K., 1984, "Moduli and Damping Factors for Dynamic Analysis of Cohesionless Soils," Report No. UBC/EERC-84/14, Earthquake Engineering Research Center, University of California, Berkley, California.

Seed, H.B. and Idriss, I.M., 1970, "Soil Moduli and Damping Factors for Dynamic Response Analysis." Report No. EERC-70-10, Earthquake Engineering Research Center, University of California, Berkley, California.

SHRP-National Research Council, 1991, "Resilient Modulus of Subgrade Soils and Untreated Base/Subbase Materials: SHRP Designation P46," Washington, D.C.

Stokoe, K.H. II and Lodde, P.F., 1978, "Dynamic Response of San Francisco Bay Mud," _Proceedings_, Earthquake Engineering and Soil Dynamics Conference, ASCE, Vol II, pp. 940-959.

Thompson, M.R. and Robnet, Q.C., 1976, "Resilient Properties of Subgrade Soils, Final Report," Transportation Engineering Series No. 14, University of Illinois, Urbana, Illinois.

Uddin, W., Meyer, A.H., Hudson, W.R. and Stokoe, K.H. II, 1985, Project-Level Structural Evaluation Based on Dynamic Deflections," Transportation Research Record 1007, Washington, D.C., pp. 37-45.

Uddin, W., Meyer, A.H. and Hudson, W.R., 1986, "Rigid Bottom Considerations for Nondestructive Evaluation of Pavements," Transportation Research Record 1070, Washington, D.C., pp. 21-29.

Uddin, W., Nixon, J.F., McCullough, B.F. and Kabir, J., 1987, "Diagnostic Evaluation of In-Service Pavement Performance Using Pavements Condition Data," _Proceedings_, Vol. I, 6th International Conference on Structural Design of Asphalt Pavements, Ann Arbor, Michigan, pp. 500-520.

U.S Army Engineer Waterways Experiment Station, Corps of Engineers, 1962, "Gyratory Compaction Method for Determining Density Requirements for Subgrade and Base of Flexible Pavements," Miscellaneous Paper No. 4-494, Vicksburg, Mississippi.

Zhou, H., Hicks, R.G and Huddleston, I.J., 1989, "Evaluation of the 1986 AASHTO Overlay Design Method," Transportation Research Record 1215, Washington, D.C., pp. 299-316.

Joao R. de Almeida[1] , Stephen F. Brown[2] , and Nicholas H. Thom[3]

A PAVEMENT EVALUATION PROCEDURE INCORPORATING MATERIAL NON-LINEARITY

REFERENCE: de Almeida, J. R., Brown, S. F., and Thom, N. H., "A Pavement Evaluation Procedure Incorporating Material Non-Linearity," Nondestructive Testing of Pavements and Back-calculation of Moduli (Second Volume), ASTM STP 1198, Harold L. Von Quintus, Albert J. Bush, III, and Gilbert Y. Baladi, Eds., American Society for Testing and Materials, Philadelphia, 1994.

ABSTRACT: Two new computer programs, LEAD and FEAD, have recently been developed at Nottingham for the back-calculation of elastic stiffnesses of pavement layers from deflections measured with the Falling Weight Deflectometer (FWD). These programs represent a significant improvement on the program PADAL, previously developed at Nottingham. A revised algorithm for the iterative back-calculation process, based on the Gauss-Newton method, has been successfully implemented. Appropriate stress-strain relationships for the granular layers and the subgrade were built into the programs, enabling the non-linear behaviour generally exhibited by these materials to be evaluated. The influence of overburden, pore pressures and "locked-in" horizontal stresses on the in situ stiffnesses is also considered. Additionally, a rigid bottom may be included in the analysis. The main difference between the two codes is in the sub-routines adopted for the computation of deflections. LEAD makes use of layered elastic analysis, while FEAD uses a finite element approach which, although more accurate for non-linear problems, is more time consuming. Data from FWD surveys have provided a means of validating the new software. Examples are presented, including comparisons with laboratory test data. The results suggest that the stress-dependent nature of the lower layers can be of significance for the overall behaviour of the pavement.

KEYWORDS: elastic stiffnesses, deflections, non-linear models, back-calculation algorithm, FWD data, validation

---

The computer program PADAL (Brown et al. 1987; Tam 1987) was developed at Nottingham for the back-calculation of elastic stiffnesses of pavement layers from deflection data measured using the Falling Weight Deflectometer (FWD). It has been extensively used to evaluate the

---

[1]Research assistant, Department of Civil Engineering, University of Nottingham, University Park, Nottingham, NG7 2RD, UK.

[2]Professor, Department of Civil Engineering, University of Nottingham, University Park, Nottingham, NG7 2RD, UK.

[3]Lecturer, Department of Civil Engineering, University of Nottingham, University Park, Nottingham, NG7 2RD, UK.

in situ condition of many different types of pavements. Meanwhile, research has continued aiming to improve the current procedures for interpreting the results of FWD surveys. To pursue these objectives, two numerical techniques have been considered for the analysis of pavement structures, layered elastic systems and finite elements.

In PADAL, the pavement is represented by a series of linear elastic layers. The non-linear properties of the subgrade are incorporated by dividing the subgrade into five sub-layers, each having a different stiffness based on a stress-dependent elastic model derived from the results of laboratory testing.

An assessment of the performance of PADAL revealed three important limitations:

1. Subgrade non-linearity is only modelled in the vertical direction, thus ignoring stiffness variation with radius.
2. PADAL does not consider the non-linear behaviour that the unbound granular layers are known to exhibit to a significant extent.
3. The iterative algorithm for the adjustment of layer stiffnesses during the back-calculation process does not take into account all deflections measured. This necessarily affects the final solution.

Two computer codes recently developed, called LEAD (Layered Elastic Analysis of Deflections) and FEAD (Finite Element Analysis of Deflections), have enabled the shortcomings listed above to be overcome. As their names indicate, LEAD makes use of layered elastic analysis for the computation of pavement deflections, whilst FEAD uses a finite element package. The essential characteristics of these programs are described in the following sections.

CONSTITUTIVE RELATIONSHIPS

For the subgrade, the model for fine grained soils proposed by Brown (1979) and already implemented in PADAL was selected, since experience has shown that it can realistically simulate subgrade behaviour (Tam 1987; Almeida et al. 1991). According to this model, developed from the results of laboratory repeated load triaxial tests, the subgrade resilient modulus, E_r, is given by:

$$E_r = A \left(\frac{p_o'}{q_r}\right)^B \qquad (1)$$

where: p_o' = effective mean normal stress due to overburden.
q_r = deviatoric stress due to wheel loading.
A, B = material constants.

For describing the non-linear response of unbound granular layers and coarse grained soils, the K-θ model (Hicks and Monismith 1971) was adopted. In this model, the resilient modulus, E_r, is given by:

$$E_r = k_1 \, \theta^{k_2} \qquad (2)$$

where: θ = sum of the peak values of the principal stresses (first stress invariant).
k_1, k_2 = material constants.

Although the K-θ model is often inaccurate when compared to more complex resilient models (Brown and Pappin 1985), it was chosen because of its simplicity. In fact, only two parameters, k_1 and k_2, are considered in the definition of the material behaviour, whilst other models generally involve a greater number of constants. This property is very important in back-calculation, where the number of material constants to be determined influences the speed and convergence of the iterative process. Thus, an excessive number of material constants will increase the amount of computation and make it more difficult to converge towards a unique solution. Furthermore, the K-θ model assumes a constant value for Poisson's ratio. This assumption, although may be regarded as a drawback (Brown and Pappin 1985), results beneficial from the point of view of a back-calculation procedure. In fact, the limited number of elastic parameters which can be back-calculated implies that only the most important ones, i.e., the elastic stiffnesses, can be taken as unknowns. Therefore, a model with a stress-dependent Poisson's ratio would possibly be inappropriate for back-calculation purposes.

MODELLING OF NON-LINEAR BEHAVIOUR

Program LEAD

In PADAL, the subgrade is divided into five sub-layers of thicknesses increasing with depth and stresses at the mid-depth of each sub-layer and directly beneath the load centre are considered for the calculation of subgrade elastic stiffnesses using Equation 1 (Tam 1987). However, since each sub-layer has a single stiffness, the stress-dependency of the material in the radial direction cannot be reproduced.

An approximate procedure, first outlined by Irwin and Speck (1986), has been implemented in LEAD to make stiffness variable with radius. Clearly, this can only be accurately achieved through finite element techniques and so simplifications have to be adopted for the layered analysis.

Essentially, the new approach computes the surface deflection at a certain radial position using a set of stiffnesses for the non-linear materials corresponding to the stresses at that same radial position. The subgrade and granular layer stiffnesses used for the calculation of surface deflection at a distance r from the load centre are obtained from Equations 1 and 2, considering the stresses existing at the mid-depth of each sub-layer at radius r.

It may be argued that this procedure is not entirely correct, since a single stiffness is still assumed for each layer in each deflection computation. Nevertheless, the adoption of stiffnesses taken from the stress state of points directly beneath the location where surface deflection (d) is to be computed can be justified by considering the following expression:

$$d = \int_{-\infty}^{0} \frac{1}{E} \{\sigma_z - \nu(\sigma_r + \sigma_\theta)\} \, dz \tag{3}$$

where: E, ν = Young's modulus and Poisson's ratio, respectively.
σ_z, σ_r, σ_θ = vertical, radial and tangential stresses, respectively.
z = depth.

From Equation 3, it can be concluded that the deflection depends mainly on the stiffnesses that exist below the point considered, for the integration variable is z (the radius is kept constant). However, as the stresses are calculated using constant stiffnesses in the radial direction, there will be some error because the evaluation of stresses does not take into account the variation of stiffness with radius. This drawback is inevitable in layered analysis, where a unique stiffness must be assigned for each layer to proceed with the computation. In spite of this, the new approach seems to be more appropriate than the one followed in PADAL.

As in PADAL, a division of the subgrade into five sub-layers is adopted in LEAD for modelling stress-dependent behaviour. With respect to the granular material, a study has been carried out to define the number of sub-layers necessary to model it accurately (Almeida 1993). The pavement analysed had a thin asphalt surfacing over a thick granular layer which was modelled according to the K-θ relationship and successively sub-divided in an increasing number of layers of equal thickness. For each case, the corresponding surface deflections were computed using a modified version of the layered elastic program BISTRO (Peutz et al. 1968) in which the stress-dependent properties of the granular material were taken into account by using the same approach as referred to above. It was concluded that, for the range of thicknesses of granular layers that can be found in pavement structures, two sub-layers are sufficient to model the behaviour of this material. A finer sub-division leads practically to the same results and it only serves to increase the computational effort, thereby lengthening the calculation.

Program FEAD

Some of the drawbacks inherent to layered elastic analysis may be removed by using a finite element formulation. This has led to the implementation of the back-calculation program FEAD. This code was derived from the program FENLAP (Finite Element Non-Linear Analysis of Pavements), also developed in Nottingham, which performs the finite element calculation of an axisymmetric solid using eight-node rectangular elements (Almeida et al. 1991).

In the finite element method, the structure is divided into a number of elements. Hence, this method is particularly adequate for non-linear problems since it can easily accommodate changes in material properties, allowing for variations in both the vertical and horizontal directions. Each point can then have an elastic stiffness consistent with its stress level.

Although finite element techniques can model non-linearity more correctly than layered elastic analysis, they are also much more time consuming. Considering the numerous calculations involved, the difference in computing time between the two approaches may become rather substantial. At the present time, these characteristics make FEAD more suitable for research purposes and for special cases, whereas LEAD, being faster and simpler, is preferable for routine pavement evaluation. Nevertheless, the continuous trend towards more powerful and cheaper personal computers may soon alter this situation.

ITERATIVE ALGORITHM

Gauss-Newton method

In PADAL, the adjustment of the layer stiffnesses from one iteration to the next is based on the deflections at selected points only, instead of considering the whole deflection bowl (Tam 1987). This requires a decision on which geophone location should be assigned to each layer stiffness and this may not always be an obvious choice. Furthermore, the remaining deflections are ignored and, consequently, more poorly matched.

It was, therefore, concluded that all measured deflections should be taken into account in the back-calculation procedure. Using the concept of the least squares method, an error function f requiring minimization may be written as follows (Uzan et al. 1989; Matsui et al. 1990):

$$f = f(X) = \frac{1}{2} \sum_{k=1}^{n} W_k \; \{dm_k - dc_k(X)\}^2 \qquad (4)$$

where: k = sensor index.
n = number of measured deflections (usually 7 for the FWD).
W = weighting coefficient or "weight".
dm = measured deflection.
dc = computed deflection.
X = unknown parameters (e.g., layer stiffnesses).

If all weights W_k are equal to 1, f represents half of the sum of the squares of the absolute deflection errors. If the weights W_k equal the inverse of the squares of measured deflections dm_k, then f represents half of the sum of the squares of the relative deflection errors.

There are several mathematical techniques for determining the minimum of a multi-variable non-linear function, such as f(X). Among them, the Gauss-Newton method is one of the simplest, being also steadily convergent (Inoue and Matsui 1990). Hence, this was adopted for back-calculation.

The derivation of the Gauss-Newton algorithm can be found in Matsui et al. (1990). Given an estimate for a set of unknown parameters {X}, the adjustment vector {ΔX} to be added to that set so as to minimize the error function defined by Equation 4 is obtained by solving the following simultaneous equations:

$$[S] \; \{\Delta X\} = \{R\} \qquad (5)$$

where: [S] = sensitivity matrix, with a generic term given by:

$$S_{ij} = \sum_{k=1}^{n} W_k \cdot \frac{\partial dc_k}{\partial X_i} \frac{\partial dc_k}{\partial X_j} \qquad i = 1,\ldots,m \quad j = 1,\ldots,m$$

{R} = vector of residuals, with a generic term given by:

$$R_j = \sum_{k=1}^{n} W_k \; (dm_k - dc_k) \frac{\partial dc_k}{\partial X_j} \qquad j = 1,\ldots,m$$

This procedure is iterative, i.e., at step p of the computation, after Equation 5 has been solved, the unknown parameters are updated by using:

$$\{X\}^p = \{X\}^{p-1} + \{\Delta X\} \qquad (6)$$

The process is repeated until the variations in the unknown parameters become very small (lower than a predefined limit). At this stage, computation stops and the system is said to have converged to an optimum point.

It can be observed that the evaluation of the coefficients of Equation 5 requires the knowledge of derivatives such as $\partial dc_i/\partial X_j$. This is the sensitivity of computed deflection dc_i with respect to an unknown parameter X_j. This sensitivity factor is obtained by increasing the value of X_j by a small increment (1%) while keeping all the remaining variables unchanged and determining the corresponding change in value of the deflection dc_i. In fact, for small increments, it is valid to assume that:

$$\frac{\partial dc_i}{\partial X_j} \approx \frac{\Delta dc_i}{\Delta X_j} \tag{7}$$

Back-calculation of non-linear parameters

Implementation of the Gauss-Newton method for back-calculation of non-linear model parameters, such as A and B in Brown's model (Equation 1) or k_1 and k_2 in the K-θ model (Equation 2), does not present any special difficulty, since these parameters can be treated as additional unknowns in the same way as the layer stiffnesses are considered for linear elastic materials. However, it was found that special provisions are needed to ensure convergence when non-linear relationships are adopted.

Non-linear materials, being stress-dependent, exhibit variable stiffness according to their level of stress. Hence, during back-calculation, their stiffnesses will be changing, not only due to successive adjustments of the elastic parameters but also due to variations in stress caused by those adjustments. Hence, the parameters are updated assuming a certain stress distribution but, after that parameter correction, the stresses will no longer be the same. This interdependency between stiffness and stress makes convergence extremely difficult unless appropriate assumptions are formulated.

For unbound granular materials, numerical instability has been overcome by use of the correlations proposed by Rada and Witczak (1981) between coefficients k_1 and k_2 of the K-θ model. Based on a comprehensive set of laboratory tests, the authors concluded that a relationship between the two parameters could be defined and regression analysis was performed to work out k_1-k_2 correlations for various types of granular materials. Using these expressions, the back-calculation of a non-linear granular layer is, therefore, reduced to the determination of just one coefficient (e.g., k_1), the other being given by the corresponding k_1-k_2 relationship. One unknown of the back-calculation procedure is then eliminated, which reduces the computing time and leads to a faster convergence.

For the subgrade, it was found that a preliminary evaluation of the equivalent elastic moduli would greatly simplify the task of back-calculating non-linear parameters. The equivalent or composite elastic modulus at a certain radius is defined as that of a linear elastic half-space which yields the same surface deflection as the one measured at that radius. It is known that, at increasing radial distances from the load, the surface deflections are increasingly influenced by the subgrade. For distances corresponding to the furthest geophones of the FWD, these deflections depend almost exclusively on the subgrade and, consequently, the equivalent modulus becomes practically equal to the subgrade modulus. Hence, a plot of composite modulus versus radius as shown in Figure 1, gives a clear indication of the degree of non-linearity of the subgrade (Rada et al. 1988). For linear elastic subgrades, the outer branch of the curve is approximately flat,

reflecting a constant subgrade stiffness. In the case of non-linear
behaviour of the subgrade, that branch is not horizontal, indicating
that the stiffness of the subgrade varies with radius.

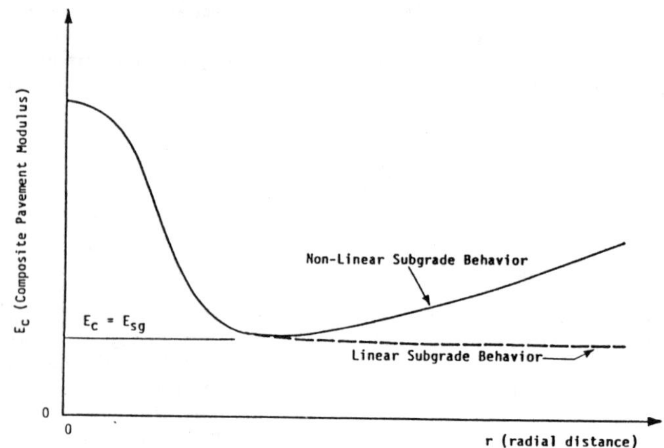

Fig. 1--Composite modulus versus radial distance (Rada et al. 1988)

It follows, therefore, that the first step in the back-calculation of
pavements with non-linear subgrades is to determine the equivalent
moduli. If these tend to a constant value for large radii, the
evaluation of stiffnesses is carried out assuming that the subgrade has
linear elastic behaviour. The non-linearity of the subgrade is taken
into account only if the rate of change of the equivalent modulus at the
locations of the outer geophones exceeds 10% per metre. If the variation
is smaller than this, it may even result from inaccuracies in the
measured deflections rather than from non-linear effects. In any case,
it is considered that material non-linearity is not significant for
changes in the equivalent modulus below 10% per metre.

Two non-linear models for the subgrade are provided in programs LEAD
and FEAD as options, the K-θ model, for coarse grained soils, and
Brown's model, for medium and fine grained soils. The geophone locations
selected are the ones corresponding to the outer branch of the
deflection bowl where the pavement response can be attributed solely to
the subgrade, according to a procedure described by Rada et al. (1988).
Then, for each position and assuming a one-layer system having an
elastic modulus equal to the equivalent modulus at the same location,
stresses are calculated at that radius and at a depth where the
deflection is 50% of the value recorded at the surface. Analysis of
typical pavement structures was performed to assess whether the stresses
at a depth where the deflection is 50% of the surface deflection at the
same radial offset are appropriate for characterizing the overall
resilient response of a stress-dependent subgrade (Almeida 1993). After
the surface deflections were computed, the equivalent moduli for the
outer part of the deflection bowl were determined and compared to the
stress-dependent subgrade moduli at depths where the deflection was half
the value obtained at the surface. The two sets of moduli were found to
be broadly similar and exhibit the same type of variation with radius.
Therefore, it was concluded that the degree of non-linearity, given by
the exponent parameter in the stress-dependent relationships used, can
be accurately estimated by considering that the equivalent modulus

varies in accordance to the stress state of points where the deflections are 50% of the surface deflections for the same radii.

A logarithmic regression of equivalent moduli on stresses yields estimates of the non-linear parameters. The estimates determined through this procedure will not differ much from the actual values (Almeida 1993). Therefore, to save computing time and avoid convergence problems, one subgrade parameter (k_2 in the K-θ model and B in Brown's) is set equal to the estimate obtained and remains unchanged throughout the iteration process. The initial value for the other parameter (k_1 or A) is also taken from the regression.

INPUT DATA

In addition to the input usually needed in pavement back-calculation (FWD platen radius and contact pressure, radial positions and corresponding measured deflections of each geophone, initial elastic stiffnesses, Poisson's ratios and layer thicknesses), programs LEAD and FEAD also require the values of unit weight, suction and coefficient of lateral pressure (K_O) for each pavement layer. These quantities are necessary for the evaluation of the initial stresses in the pavement, which must be taken into account when computing the stress-dependent elastic stiffnesses. In fact, it is known that the initial stress state due to overburden, pore pressures, compaction and other residual effects often plays an important role in the structural behaviour of paving materials (Brown 1979; Stewart et al. 1985).

Other parameters required as input by LEAD and FEAD are the weighting factors assigned to each deflection sensor. As noted above, the weights will depend on whether it is intended to minimize the sum of absolute or relative deflection errors. If the user is particularly interested in getting a very close match for some deflections, higher weights should be input for these. Conversely, if the user is not concerned with the goodness of fit for some geophones (e.g., if the accuracy of any measuring device is considered poor), the corresponding weights can be lowered or even set to zero (Uzan et al. 1989).

In order to prevent the system from yielding unrealistic solutions, upper and lower bound values are prescribed for the unknowns. If, at any step of the computation, a parameter tends to move beyond these limits, its value is set equal to the respective limit and the iteration is processed ignoring the contribution of that parameter.

Having in mind the importance that a shallow bed-rock can have on the pavement response, a procedure for predicting the depth to a rigid layer from the measured deflection bowl has been implemented. This procedure is based on the work presented by Rohde et al. (1990). Consequently, the programs offer three possibilities:

1. Infinite half-space beneath the pavement.
2. Rigid bottom at a depth assigned by the user.
3. Rigid bottom at a depth estimated by the program.

ANALYSIS OF FIELD DEFLECTION DATA

FWD survey near Wakefield

The experimental road near Wakefield, South Yorkshire, was designed as a haul road to carry waste material from a mine to a land reclamation area using 22 t lorries. It is 2.3 km long and includes 16 different sections with various materials and layer thicknesses, enabling direct comparisons to be made between their long term performance (Brunton and Akroyde 1990). The bituminous mixes used are Hot Rolled Asphalt (HRA) and Dense Bitumen Macadam (DBM). The sub-base is generally a Type 1 aggregate (crushed stone). These designations correspond to typical UK materials. Nottingham University is directly involved in 10 out of the 16 sections, the ones schematically shown in Figure 2.

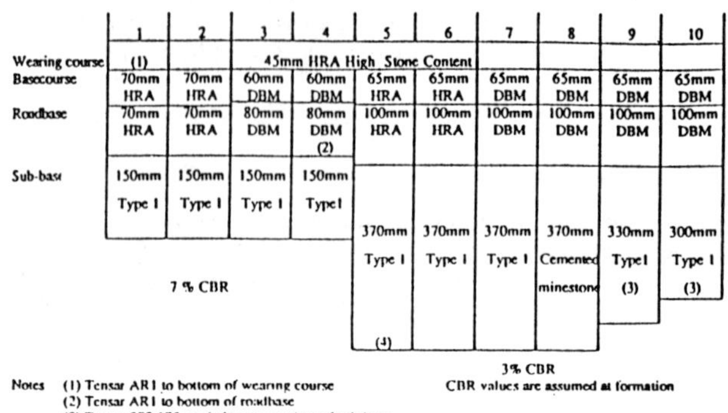

Fig. 2--Nottingham sections in Wakefield road (Brunton and Akroyde 1990)

In April 1991, a FWD survey was carried out on the Wakefield experimental road. FWD tests were made at 20 m intervals in the nearside wheel path. A 300 mm diameter platen and an applied pressure of approximately 700 kPa were adopted. Three drops were used in each test and the deflections recorded from all drops. Temperature readings taken from holes drilled 80 mm into the asphaltic layer at different locations and times of day indicated temperatures of 20 to 21°C, thus reflecting a practically uniform temperature throughout the whole test.

Deflection bowls representative of average pavement conditions, i.e., the ones closest to the theoretical 50th percentile bowl for each section, were selected for back-calculation and analysed using the program LEAD. The wearing course, basecourse and roadbase were combined into a single asphaltic layer for back-calculation purposes. The sub-base had a lower or similar thickness to that of the upper layers and, thus, its influence on the pavement response was not very significant. It was concluded that a non-linear model for the sub-base was unnecessary and linear elastic behaviour was adopted for this material. This option also has the advantage of enabling a direct comparison between the sub-base stiffnesses using programs LEAD and PADAL, since in PADAL linear elastic behaviour is assumed for granular layers.

The unit weights of the paving materials were obtained experimentally (Brunton and Akroyde 1990). Values of approximately 24, 22 and 19 kN/m^3 were used for the asphaltic layers, sub-base and subgrade, respectively. For Poisson's ratio, typical values of 0.40 for the asphaltic layers, 0.30 for the sub-base and 0.40 for the subgrade were adopted. Due to lack of information, the water table was assumed to be at formation level for all sections.

The subgrade was modelled as non-linear elastic, using Brown's relationship for fine grained soils (1979). The results of the back-calculation indicated a fairly stiff subgrade, which was expected, as a hard rock sandstone band had been detected at shallow depths (Brunton and Akroyde 1990). The stiffest subgrade was encountered over the first three sections, which is consistent with the higher CBR values indicated by the site investigation data.

With the exception of section 8, which had a cemented minestone sub-base, the subgrade exhibited strong stress-dependency. Table 1 shows the back-calculated non-linear parameters obtained from LEAD and the corresponding subgrade stiffnesses at two different depths and radii. The degree of non-linearity for the subgrade, which depends on the value of parameter B, is quite marked. The variation in stiffness, although more pronounced in the vertical direction, was also noticeable radially.

Table 1--Wakefield road: subgrade parameters and stiffnesses

| section | non-linear subgrade parameters | | subgrade stiffness (MPa) | | | |
| | A (MPa) | B | 0.3 m below formation | | 6.6 m below formation | |
			r=0.0 m	r=1.8 m	r=0.0 m	r=1.8 m
1	209	0.345	153	348	974	1029
2	285	0.348	185	482	1331	1411
3	273	0.375	182	486	1427	1521
4	114	0.281	114	219	523	547
5	112	0.333	111	192	517	541
6	129	0.248	125	193	408	421
7	112	0.151	109	142	228	233
8	linear elastic		294	294	294	294
9	76	0.312	74	126	327	342
10	100	0.248	90	147	310	321

The back-calculated elastic stiffnesses of the asphaltic layers and of the sub-base derived from LEAD were compared with the corresponding values obtained from PADAL on a previous survey (Brunton and Akroyde 1990), as a means of assessing the performance of both programs. The asphalt stiffnesses were also matched against elastic stiffnesses determined in the laboratory using the repeated load indirect tensile test (Cooper and Brown 1989) on cores taken from the road. Separate values were obtained in the laboratory for the basecourse and roadbase. The corresponding values are shown in Table 2, where all stiffnesses are referred to a temperature of 20°C.

Table 2--Wakefield road: laboratory, PADAL and LEAD stiffnesses

| section | elastic stiffness (MPa) | | | | | |
| | indirect tensile test | | PADAL | | LEAD | |
	basecourse	roadbase	asphalt	sub-base	asphalt	sub-base
1	2200	2500	4200	170	8300	140
2	2800	1800	5100	70	4400	380
3	3000	1500	3700	40	4200	310
4	1900	3200	5600	170	5500	250
5	2200	2600	7100	100	6700	130
6	2800	1600	6500	50	5100	100
7	...	...	5300	120	4700	110
8	2200	5000	...	...	6100	3100
9	4200	4400	5000	30	4500	90
10	3300	2600	4800	40	3300	80

It can be seen that for all sections the stiffnesses determined in the laboratory are lower than the back-calculated stiffnesses from both PADAL and LEAD. This may be explained by the fact that the laboratory stiffnesses were obtained using longer loading times than those of the FWD. Nevertheless, the LEAD results generally compare better with the laboratory values.

The value found for the cemented minestone sub-base of section 8 was remarkably high (approximately 3000 MPa). In the remaining sections, the unbound granular sub-base exhibited stiffnesses ranging from 30 to 170 MPa according to PADAL and 80 to 380 MPa according to LEAD. It is thought that the higher values determined by LEAD are more plausible since field testing showed the material to be well compacted and well drained, which was additionally confirmed by its high unit weight.

The scatter found in the values of sub-base stiffness can be explained by the fact that the contribution of this layer to the structural response of the pavement is small relative to that of the thick asphaltic layers and the hard subgrade. Therefore, it is difficult to obtain a reliable estimate of the granular layer stiffness, since even large differences in its value do not change the overall pavement load response significantly. This is particularly true for the first four sections, where the sub-base was just 150 mm thick. The relatively high stiffnesses of the sub-base back-calculated using LEAD for those sections should, therefore, be regarded with some caution.

FWD survey at Bothkennar

The experimental road at Bothkennar, Scotland, has been monitored since 1989, as part of a research project investigating geosynthetic reinforcement funded by the Science and Engineering Research Council (1991). It is an unsurfaced road, constructed on a soft clay site and composed of 16 sections, each one approximately 20 m long. Access to the testing site is provided by a thinly surfaced road having a length of 800 m. FWD surveys were carried out in May 1991, on both the experimental road and the access road. Only the analysis of the access road is reported here.

Measurements were taken at 20 m intervals. The seven geophones were placed at 300 mm spacings such that the outer geophone was 1.80 m away from the platen centre. In all tests, a 300 mm diameter platen was used and three drops per test were executed, with the deflections associated with each drop being recorded. A temperature of 21°C was measured in the asphaltic layer at the time of testing. The access road had two main sections, one pre-existing and the other newly constructed. Due to lack of reliable information on the pavement structure of the older section, it was decided to concentrate exclusively on the the newly constructed section, which is composed of 75 mm of Bituminous Macadam overlaying 150 mm of granular sub-base, with a high tensile polymer geogrid at the interface between the granular layer and the soft clayey subgrade.

The FWD tests carried out on the newer pavement showed a reasonable degree of uniformity. In view of that and of the reduced length of the section (200 m), no division into sub-sections was considered. In order to identify the influence of the load level on the pavement response, two drop heights were used, corresponding to applied pressures of approximately 370 and 680 kPa. For a given load level, the representative deflection bowl selected for analysis was taken as the measured one which most closely resembled the 50th percentile deflection bowl resulting from the average of deflections from all FWD drops executed in the section.

The representative bowls found for each drop height were located at different chainages. However, it was desirable to compare results at the same location. Consequently, since both the representative bowl for load level 1 and the second closest bowl to the 50th percentile values for load level 2 were recorded at the same chainage, it was decided to analyse the deflections measured at that chainage.

The Bothkennar clay has a crust 0.8 m thick above softer underlying soil (SERC 1991). The water table is located at approximately 1 m below the bottom of the crust. Therefore, for back-calculation purposes, the subgrade was divided into two parts (crust and deep subgrade) and the water table was placed at a depth of 2 m beneath the surface. Poisson's ratios of 0.35, 0.30 and 0.45 were assumed for the asphaltic layer, sub-base and subgrade, respectively. The unit weights adopted for the same materials were 24, 22 and 20 kN/m^3.

The non-linear behaviour of the granular layer and of the subgrade was considered by including the K-θ and Brown's model in the back-calculation procedure. The representative deflection bowls for each load level were analysed using LEAD. The solutions are shown in Table 3. The excellent match achieved seems to confirm the adequacy of the material relationships which were used. Furthermore, the non-linear parameters were almost identical in both cases, as should be the case. The non-linearity of the subgrade was not very pronounced, as illustrated by the low values of subgrade parameter B.

Table 3--Bothkennar access road: non-linear back-calculation

	load level 1		load level 2	
	measured	computed	measured	computed
d_1 (μm)	578.0	578.0	1046.0	1046.3
d_2 (μm)	335.0	335.1	642.0	642.6
d_3 (μm)	171.0	170.6	344.0	343.2
d_4 (μm)	107.0	107.3	212.0	212.2
d_5 (μm)	79.0	79.4	151.0	152.8
d_6 (μm)	63.0	62.5	120.0	118.2
d_7 (μm)	51.0	51.0	95.0	95.0
$E_{asphalt}$ (MPa)	...	2540	...	3259
k_1(σ in MPa)	...	423	...	441
k_2 (σ in MPa)	...	0.407	...	0.435
E_{crust} (MPa)	...	79	...	71
A (MPa)	...	55	...	47
B	...	0.060	...	0.100

To investigate the possible influence of compaction induced lateral stresses in the back-analysed stiffnesses, another LEAD calculation was performed for load level 1, this time assuming the earth pressure coefficient, Ko, equal to 3.0 for the granular layer (instead of 0.43, the value corresponding to linear elastic conditions). The overall stiffness profile did not vary significantly, as may be seen in Table 4. Nevertheless, the different stress distribution within the granular layer due to the residual lateral stresses necessarily produced changes in the sub-base stiffnesses.

Table 4--Back-calculated stiffnesses with different coefficients Ko

layer \ r(m)	0.00		0.30		0.60		0.90		1.20		1.50		1.80	
asphalt	2540	2542	linear elastic											
granular-top	177	177	122	119	67	75	50	66	49	65	49	65	49	65
granular-bot.	127	129	113	118	79	95	63	86	61	84	60	84	60	84
soil crust	79	79	linear elastic											
subgrade-top	60	61	61	61	61	61	62	62	63	64	65	65	66	66

Notes: Values are in MPa.
For each radius, the values on the left and right correspond to Ko=0.43 and 3.0, respectively.

The deflection bowl corresponding to the lower load level was also back-calculated using FEAD. The finite element mesh adopted was such that the vertical subdivision of the non-linear layers was the same as in LEAD, so the results of both programs could be directly compared. From the values presented in Table 5, it is evident that the solutions obtained with the two programs are similar.

Table 5--Bothkennar access road: LEAD and FEAD results

	measured	LEAD	FEAD
d_1 (μm)	578.0	578.0	578.0
d_2 (μm)	335.0	335.1	335.0
d_3 (μm)	171.0	170.6	170.8
d_4 (μm)	107.0	107.3	107.4
d_5 (μm)	79.0	79.4	78.7
d_6 (μm)	63.0	62.5	62.3
d_7 (μm)	51.0	51.0	51.8
$E_{asphalt}$ (MPa)	...	2540	2986
k_1 (σ in MPa)	...	423	431
k_2 (σ in MPa)	...	0.407	0.421
E_{crust} (MPa)	...	79	80
A (MPa)	...	55	51
B	...	0.060	0.060

CONCLUSIONS

A new approach for the back-calculation of layer stiffnesses, taking into account stress-dependent behaviour of the granular material and the subgrade, has been developed. It is believed that the procedure offers a more rational way to perform the back-calculation of pavements than the one adopted in the program PADAL. Two codes based on the new procedure, LEAD and FEAD, have been implemented. FWD testing carried out on full scale pavements has assisted with the validation of the developments introduced. The results indicate that the performance of the analytical models may be considered satisfactory. However, in view of the limited number of cases analysed, further validation is necessary.

ACKNOWLEDGEMENTS

The research described in this paper was made possible by the financial support of the Commission of the European Communities, the UK Science and Engineering Research Council and the Transport Research Laboratory, UK. The Authors are also grateful for the assistance rendered by SWK Pavement Engineering Ltd. UK and by their former colleague, Dr Janet Brunton.

REFERENCES

Almeida, J. R. de, Brunton, J. M., and Brown, S. F., 1991, "Structural Evaluation of Pavements" (Report Submitted to TRRL), University of Nottingham.

Almeida, J. R. de, 1993, "Analytical Techniques for the Structural Evaluation of Pavements", Ph.D. Thesis, University of Nottingham (in Press).

Brown, S. F., Tam, W. S., and Brunton, J. M., 1987, "Structural Evaluation and Overlay Design: Analysis and Implementation", Proc. 6th Int. Conf. on Struct. Design of Asphalt Pavements, Vol.I, Ann Arbor, pp.1013-1028.

Brown, S. F., 1979, "The Characterisation of Cohesive Soils for Flexible Pavement Design", Proc. 7th Euro. Conf. Soil Mech. and Found. Eng., Vol.2, pp.15-22.

Brown, S. F., and Pappin, J. W., 1985, "Modelling of Granular Materials in Pavements", Transportation Research Record 1022, Transportation Research Board, Washington DC, pp.45-51.

Brunton, J. M., and Akroyde, P. M., 1990, "Monitoring the Performance of a Full Scale Experimental Pavement", Proc. 3rd Int. Conf. on Bearing Capacity of Roads and Airfields, Vol.2, Trondheim, pp.585-594.

Cooper, K. E., and Brown, S. F., 1989, "Development of a Simple Apparatus for the Measurement of the Mechanical Properties of Asphalt Mixes", Proc. Eurobitume Symposium, Madrid, pp.494-498.

Hicks, R. G., and Monismith, C. L., 1971, "Factors Influencing the Resilient Response of Granular Materials", Highway Research Record 345, Highway Research Board, Washington DC, pp.15-31.

Inoue, T., and Matsui, K., 1990, "Structural Analysis of Asphalt Pavement by FWD and Backcalculation of Elastic Layered Model", Proc. 3rd Int. Conf. on Bearing Capacity of Roads and Airfields, Vol.1, Trondheim, pp.425-434.

Irwin, L. H., and Speck, D. P. T., 1986, "NELAPAV User's Guide", Cornell University Local Roads Program, Ithaca.

Matsui, K., Uchida, S., Tukano, H., and Inoue, T., 1990, "Pavement Rehabilitation Design Using Prognosis of Elastic Layered Properties", Proc. 6th Conf. Road Eng. Assoc. of Asia and Australasia, Vol.1.

Peutz, M. G. F., Van Kempen, H. P. M., and Jones, A., 1968, "Layered Systems under Normal Surface Loads - Computer Program BISTRO", Koninklijke/Shell-Laboratorium, Amsterdam.

Rada, G., and Witczak, M. W., 1981, "Comprehensive Evaluation of Laboratory Resilient Moduli Results for Granular Material", Transportation Research Record 810, Transportation Research Board, Washington DC, pp.23-33.

Rada, G. R., Witczak, M. W., and Rabinow, S. D., 1988, "Comparison of AASHTO Structural Evaluation Techniques Using Nondestructive Deflection Testing", Transportation Research Record 1207, Transportation Research Board, Washington DC, pp.134-144.

Rohde, G. T., Yang, W., and Smith, R. E., 1990, "Inclusion of Depth to a Rigid Layer in Determining Pavement Layer Properties", 3rd Int. Conf. on Bearing Capacity of Roads and Airfields, Trondheim, pp.475-486.

Science and Engineering Research Council, 1991, "The Science and Engineering Research Council Soft Clay Site, Bothkennar, Scotland".

Stewart, H. E., Selig, E. T., and Norman-Gregory, G. M., 1985, "Failure Criteria and Lateral Stresses in Track Foundations", Transportation

Research Record 1022, Transportation Research Board, Washington DC, pp.59-64.

Tam, W. S., "Pavement Evaluation and Overlay Design", 1987, Ph.D. Thesis, University of Nottingham.

Uzan, J., Lytton, R. L., and Germann, F. P., 1989, "General Procedure for Backcalculating Layer Moduli", Non-Destructive Testing of Pavements and Backcalculation of Moduli, ASTM STP 1026, Philadelphia, pp.217-228.

Richard N. Stubstad[1], Joe P. Mahoney[2], Nick F. Coetzee[1]

EFFECT OF MATERIAL STRESS SENSITIVITY ON BACKCALCULATED MODULI AND PAVEMENT EVALUATION

REFERENCE: Stubstad, R. N., Mahoney, J. P., and Coetzee, N. F., "Effect of Material Stress Sensitivity on Backcalculated Moduli and Pavement Evaluation," _Nondestructive Testing of Pavements and Backcalculation of Moduli (Second Volume), ASTM STP 1198_, Harold Von Quintus, Albert J. Bush, III, and Gilbert Y. Baladi, Eds., American Society for Testing and Materials, Philadelphia, 1994.

ABSTRACT: The Falling Weight Deflectometer (FWD) has become a popular device for evaluating both airfield and highway pavements. Typically, deflection tests with the FWD or HWD (Heavy Weight Deflectometer) are performed at predetermined load levels up to approximately 120 kN (27,000 lbf) for the FWD and 240 kN (54,000 lbf) for the HWD. This paper addresses the choice of load level for pavements exhibiting stress-sensitive behavior and the effect both on backcalculated moduli and the resulting rehabilitation designs for a number of pavement evaluation projects performed over the last few years.

It is well known that many unbound materials used in pavements exhibit non-linear behavior in laboratory tests, as discussed in the paper. The effect of such non-linear response in pavement structures suggests that deflection testing should be performed using load levels at or near the expected design loads, or in a manner which allows reasonable determination of material parameters defining the non-linear behavior.

Several pavement studies have been selected that exhibit non-linear response using various HWD test load levels, from approximately 60 kN (13,000 lbf) to the full 240 kN (54,000 lbf), with all tests on a given project performed on the same series of test points, at the same time. Results of backcalculation and evaluation analyses for these pavements are presented in the paper.

Analysis of deflection basins measured at the different load levels show significantly different backcalculated moduli which appear to be consistent with the stress-stiffening or stress-softening behavior exhibited by similar materials in laboratory tests. As a result, evaluation of structural capacity for a given (expected) traffic loading is

---

[1] President and Senior Engineer, respectively, Dynatest Consulting, Inc, P.O. Box 71, Ojai, California 93024, USA

[2] Professor, University of Washington, 121 More Hall, FX-10, Seattle, Washington 98195, USA

significantly affected by these moduli values, and required overlay thicknesses may vary by a factor of two or more for different deflection test load levels, at a given pavement section. The analyses presented in the paper clearly illustrate the necessity for adequately defining actual material response in evaluating the pavement system response. Asphalt surfaced pavements tend to be more prone to these non-linear response characteristics, due to the significant structural contribution of the unbound layers often found in these pavements.

KEYWORDS: backcalculation, Falling Weight Deflectometer, Heavy Weight Deflectometer, stress-sensitive materials, pavement evaluation, overlay design, airfield pavements, flexible pavements, asphalt pavements

---

INTRODUCTION

The process of pavement rehabilitation design often includes the following steps:

1) Test the candidate pavement with a nondestructive, load-deflection testing device.

2) Determine other physical constants, such as existing layer thicknesses, traffic volume and distribution (past and future), types of existing materials, and feasible rehabilitation options.

3) Backcalculate existing layered elastic properties of the pavement, using the NDT-determined load and deflections, if possible coupled with associated laboratory data.

4) Use various measures of expected future pavement performance to calculate remaining life, as a function of the theoretical stresses and strains imposed on the pavement structure, or layers, by the expected wheel load(s), adjusted for the seasonal or climatic variations likely in the region.

5) Based on these measures of future performance, design a new pavement structure (eg, an overlay) based on the allowable stresses and strains imposed by the expected traffic loads.

Step 1, above, is fairly straightforward — assuming that the device is accurate, quick, reproducible, etc. Step 2 should be easy if good records exist and if good predictions of future traffic can be made. Step 3 is becoming more and more common, both in the airfield and roadway pavement sectors. Several methods are now available to backcalculate *in situ* moduli from load-deflection data, while laboratory tests can be conducted if time and funds are available to do so.

Steps 4 and 5, meanwhile, are not only a function of the traffic assumed in Step 2, but the load and deflections measured in Step 1 and the backcalculation technique and laboratory data from Step 3.

How, then, do all these variables interact with one another, and what is the effect of a variation or "error" in one or more of the initial rehabilitation design steps?

CONSTITUTIVE MODELS OF UNBOUND MATERIAL BEHAVIOR

The moduli of unstabilized materials (such as granular base courses and subgrade soils) depend, as will be illustrated in the following, to a large extent on the state of stress in each layer. Other factors influencing layer moduli include dry density, moisture content, degree of saturation, gradation, load duration and frequency, to name a few.

Commonly, relationships between the modulus and stress state for these unstabilized base courses and subgrade soils are as follows:

$$E_{bs} = k_1 \times \theta^{k_2} \quad \text{for (granular) base courses}$$

$$M_R = k_3 \times \sigma_d^{k_4} \quad \text{for (cohesive) subgrade soils}$$

where:

E_{bs} = resilient modulus of granular materials or base courses,

M_R = resilient modulus of natural or embankment subgrade soils,

$\theta = \sigma_1 + \sigma_2 + \sigma_3$ = bulk stress,

$\sigma_d = \sigma_1 - \sigma_3$ = deviator stress,

$\sigma_1, \sigma_2, \sigma_3$ = principal stresses, and

k_1, k_2, k_3, k_4 = regression constants.

It should be noted that other relationships that account for the stress sensitive behavior of unstabilized materials have been developed; however, the above basic relationships have been widely reported in the literature and used in practice. Further, it is noted that if the above equations are utilized as presented, the units are inconsistent, so the derived k_1 and k_3 values will depend upon the units used (eg, SI or US Customary). This inconsistency can be avoided by dividing the stress invariant θ or σ_d by a reference stress, but for the sake of simplicity this has not been done here. The exponents k_2 and k_4, on the other hand, will not be affected by the choice of units or form of the equation.

For coarse-grained materials and soils, k_2 and k_4 are generally positive (but less than 1), thus indicating that the modulus increases with increasing stress state (and hence with heavier wheel loads). For fine-grained soils, these same regression constants are often negative, thus indicating that the material is "stress softening", ie, its modulus decreases with increasing stress state (hence with heavier loads).

To illustrate typical values for these constants, results from the
AASHO Road Test are shown (Table 1 - in units of MPa only) for the
granular crushed limestone base course [AASHTO 1986]. With θ expressed
in MPa, k_1 ranged from a high of 1,300 MPa (10,000 psi, with θ in psi)
in a "dry" condition to a low of 275 MPa (2,000 psi) in a "wet" condition. The regression constant k_2 ranged between 0.5 and 0.7, averaging
0.6, as shown in the table.

TABLE 1 — Crushed limestone modulus - stress state relationships from
AASHO Road Test [AASHTO 1986]

EQUATION (MPa Units Only)	MOISTURE STATE	STRESS STATE (MPa)			
		$\theta = 0.05$	$\theta = 0.10$	$\theta = 0.15$	$\theta = 0.20$
$E_{bs} \cong 1,100 \times \theta^{0.6}$	DRY	180	270	350	410
$E_{bs} \cong 550 \times \theta^{0.6}$	DAMP	90	135	175	205
$E_{bs} \cong 450 \times \theta^{0.6}$	WET	70	110	140	170

For different crushed stone base courses in Washington State, recent
laboratory triaxial testing of sampled base materials throughout the
state (14 test sites) was completed [Washington DOT 1992]. The triaxial tests were conducted in a manner similar to AASHTO Test Method T-
274. By averaging the collective k_1 and k_2 values, the following relationship evolved:

$$E_{bs} = 380 \times \theta^{0.375} \quad \text{(in MPa)}$$
$$[\text{or } E_{bs} = 8,500 \times \theta^{0.375} \quad \text{(in psi)}]$$

Thus, for comparison purposes and using $\theta = 0.17$ MPa (25 psi), the
above equation results in a modulus of about 195 MPa (28,500 psi).
The corresponding AASHO Road Test relationship, for damp conditions,
would be 190 MPa (27,500 psi) — a surprisingly close match. However,
the k_1 and k_2 values are quite different for the two separate relationships. Other similar relationships have been developed for both
subgrades and unbound granular bases, such as those reported in Alaska
[Hicks and McHattie 1982].

An important point is that both θ and σ_d are dominated by σ_1, the
vertical (principal) stress. Further, the dynamic portion of σ_1 is
directly proportional to the applied wheel load. If the NDT device

applied a "light" load such that θ was, say, 0.06 MPa (8.7 psi) in the base course (one third of that previously used), then the moduli would be 130 MPa (19,000 psi) for the Washington State relationship and 100 MPa (14,500 psi) for the AASHO Road Test relationship, corresponding to a 33 and 47 percent decrease, respectively. In such cases, the modulus values used in subsequent analyses may result in overly conservative or non-conservative rehabilitation decisions, as shown in the following sections.

NDT FIELD MEASUREMENTS OF LOADS AND DEFLECTIONS

Although several nondestructive testing (NDT) devices are presently in widespread use, this paper is limited to only one of those types. The Falling Weight Deflectometer (FWD) has, in recent years, become quite popular for measuring deflection basins used for backcalculating layered elastic properties of pavement systems. This device employs an impulse load derived from the kinetic energy of a free-falling mass to impose a load on a pavement surface which is much greater in magnitude than the dead weight of the mass. In addition, the imposed load is transient as opposed to vibratory or static, thus simulating the effect of a moving wheel load. Three common FWD-type devices used in North America are shown in *Figure 1*.

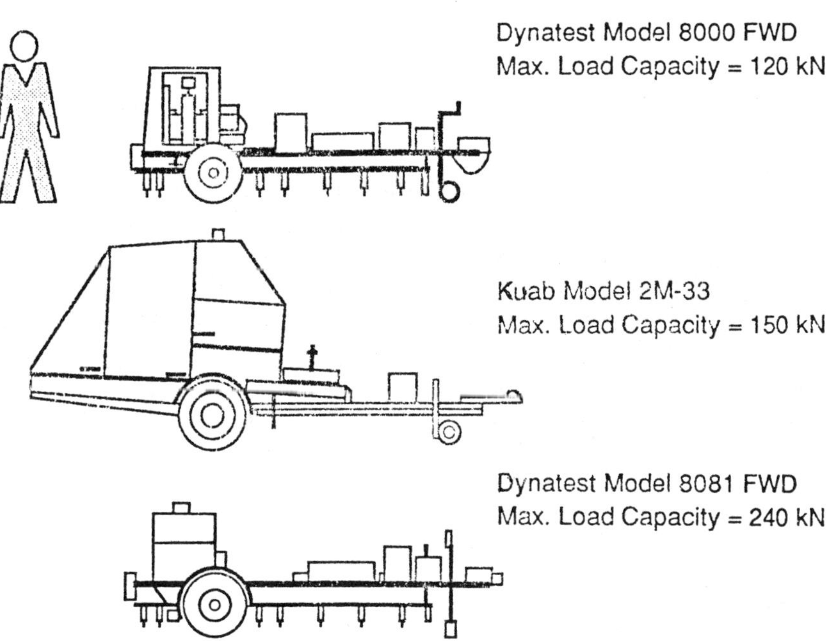

FIGURE 1 — Outline of Three Common FWD-type Devices

A problem that can arise when using an FWD with a maximum load capacity of 100 - 150 kN (22,000 - 33,000 lbf) will be illustrated for airfield pavements. This range of loads is more than adequate for roadway pavement evaluation, since the half-axle loads likely to pass over a roadway seldom exceed 70 kN (16,000 lbf). On the other hand, aircraft wheel loads often exceed 220 kN (50,000 lbf) _each_, so it was considered desirable to develop an FWD capable of simulating loads of this magnitude, and the Heavy Weight Deflectometer (HWD) sketched at the bottom of *Figure 1* (also shown in a photograph in *Figure 2*) was introduced in 1988. This device was used to generate the load-deflection data used in this study.

Generally, when an FWD is used on an airfield pavement, its maximum load capacity of 120 kN is used. In addition, if the standard 300 mm (12") diameter FWD loading plate is used in conjunction with this maximum load level, a reasonable pressure is achieved (ie, approx. 1.7 MPa, or 245 psi) but the footprint size of a heavy aircraft is not. To simulate the correct footprint, both the FWD and HWD are equipped with a larger loading plate 450 mm (18") in diameter. Thus the footprint size of a large aircraft is achieved, but with the FWD the maximum loading pressure becomes only 0.75 MPa (110 psi), which is far lower than heavy aircraft tire pressures. With the HWD's maximum loading capacity and the 450 mm loading plate, a plate pressure of some 1.5 MPa (220 psi) can be achieved, which fairly closely approximates the effect of a heavy aircraft wheel load.

FIGURE 2 — The Model 8081 HWD Used in the Study

In applying the FWD-generated data for backcalculation purposes, it is generally assumed that the pavement material response is linearly elastic, so that moduli are assumed to remain unchanged from those backcalculated for a 120 kN FWD test load when they are used with a much larger design load, eg, a 220 kN wheel load of the Boeing 747. Data has been gathered on a number of airfield pavements with the HWD to investigate the effect, if any, of HWD or FWD load level on moduli and the subsequent bearing capacity assessments based on NDT pavement testing.

BACKCALCULATION AND OVERLAY DESIGN RESULTS

Two fairly typical airfield pavements were selected to illustrate the effect of material non-linearity on pavement rehabilitation alternatives. One of these is the North Runway at Sky Harbor Airfield in Phoenix, Arizona. The other is the Chino Airport Runway south of Ontario, California. Based on both FWD and HWD data obtained to-date, it appears that only a relatively small percentage of asphalt surfaced pavements _do_ respond in a more-or-less linear elastic manner; most follow one of the two examples illustrated below. Most PCC pavements, on the other hand, generally exhibit a linear elastic response. In all examples illustrated in this section, the ELMOD program [Ullidtz and Stubstad 1985] was used.

At Sky Harbor, the asphalt surfaced Runway 8L/26R was tested in December, 1989. Three HWD test load levels were used, approximately 70, 160 and 235 kN (16, 36 and 53 kips). The 450 mm diameter loading plate was used for all load levels. One of the fairly uniform sections of runway, consisting of 10 HWD test points between Stations 25+00 and 45+00, was selected for this illustration.

In _Figures 3 and 4_, the backcalculated moduli are plotted as a function of HWD load [both average and design (average less one standard deviation) levels], for the subgrade and base, respectively. In these figures it can be seen that both the subgrade and base moduli are _increasing_ with increasing HWD test load. In particular, the granular base appears to be significantly "stress hardening", increasing from a value of some 275 MPa (40,000 psi) derived at the lowest test load up to over 410 MPa (60,000 psi) at the highest test load. Considering the Boeing 747, which imparts a single wheel load of about 220 kN (50,000 lbf), as being the representative design aircraft for this runway, the moduli backcalculated at the higher test load (235 kN) are appropriate for use in the rehabilitation analyses, since these moduli are more consistent with the 220 kN (50 kip) design load than the values derived at lower load levels.

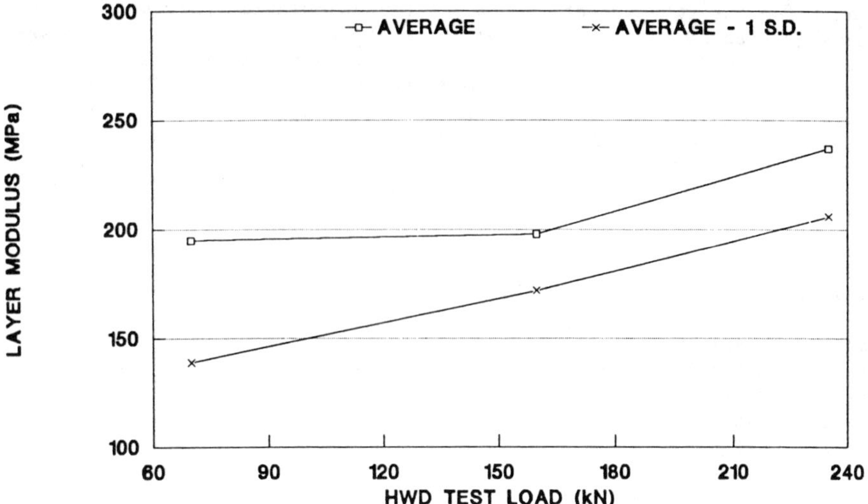

FIGURE 3 — Backcalculated subgrade modulus vs. HWD test load level –
Runway 8L/26R at Sky Harbor Airport

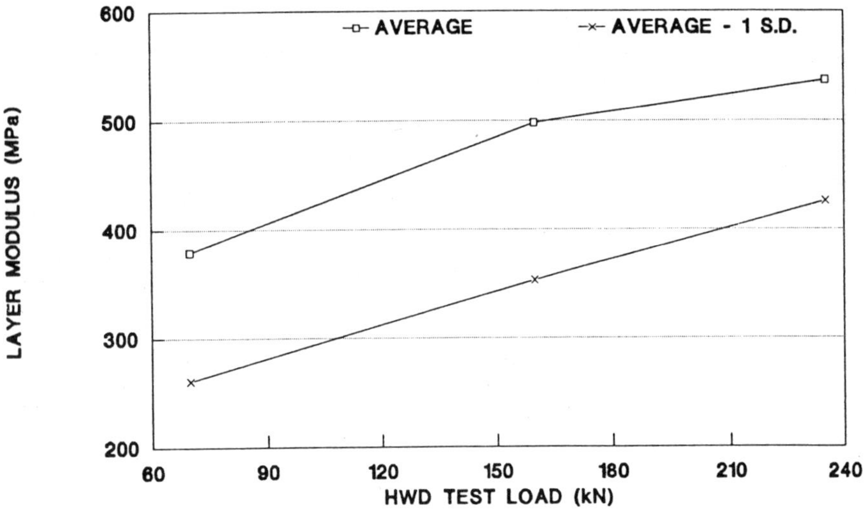

FIGURE 4 — Backcalculated base modulus vs. HWD test load level –
Runway 8L/26R at Sky Harbor Airport

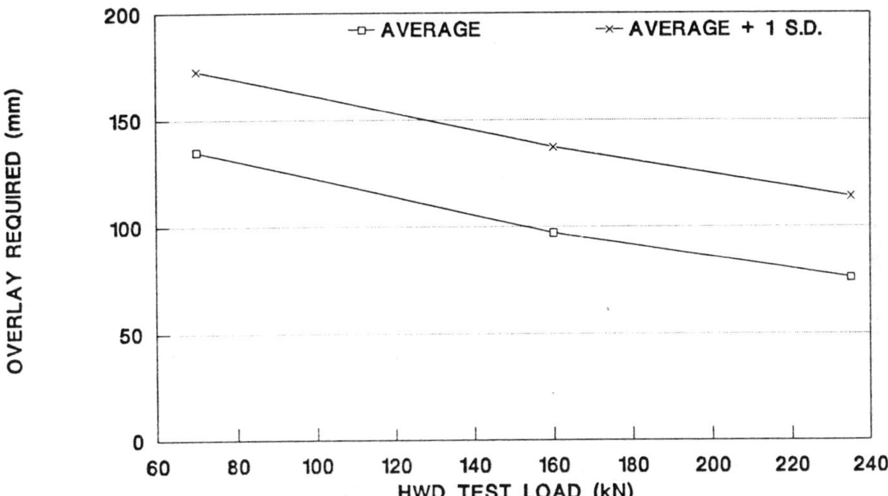

FIGURE 5 — Required AC overlay thicknesses for a 220 kN design load
vs. HWD test load level - Runway 8L/26R at Sky Harbor Airport

The corresponding AC overlay thickness requirements are plotted in *Figure 5* [both average and design (average less one standard deviation) levels]. Overlay design criteria were based on asphalt tensile strain and unbound layer (base and subgrade) stress levels. The asphalt strain criterion developed by Shell [Shell International 1978] was used, while an allowable maximum stress relationship [Kirk 1973] for unbound materials was applied to the base and subgrade. In the example shown, the required (design) overlay is 115 mm (4½") of new AC using a test load level that corresponds to the B-747 design wheel load. If a lower test load-derived overlay design had been chosen, the needed overlay is about 175 mm (7") at a test load of 70 kN, and a 150 mm (6") overlay is required at a test load of 120 kN. Thus it can be seen that the overlay would have been <u>over</u> designed by some 60 mm (2½") if the design for a B-747 had been based on tests carried out at a load level of 70 kN rather than the more correct 240 kN load capability of the HWD. As a result, use of the HWD for testing pavements that show significant stress stiffening will typically translate into a cost-savings in terms of reduced overlay material requirements when designing for heavy wheel loads.

A similar exercise was carried out for Runway 03/21 at Chino Airport in San Bernadino County (south of Ontario), California. Here, the design aircraft was a Boeing 737, consistent with expected air traffic at Chino. The facility was tested using test load levels of approximately 60, 95 and 140 kN (13, 21 and 31 kips), since the design B-737 aircraft has a wheel load of some 135 kN. To best simulate the footprint of the B-737, the 450 mm loading plate was used. Once again, a limited section of the runway was selected for this example in order to ensure a relatively uniform section of pavement, in this case between Stations 17+00 and 27+00.

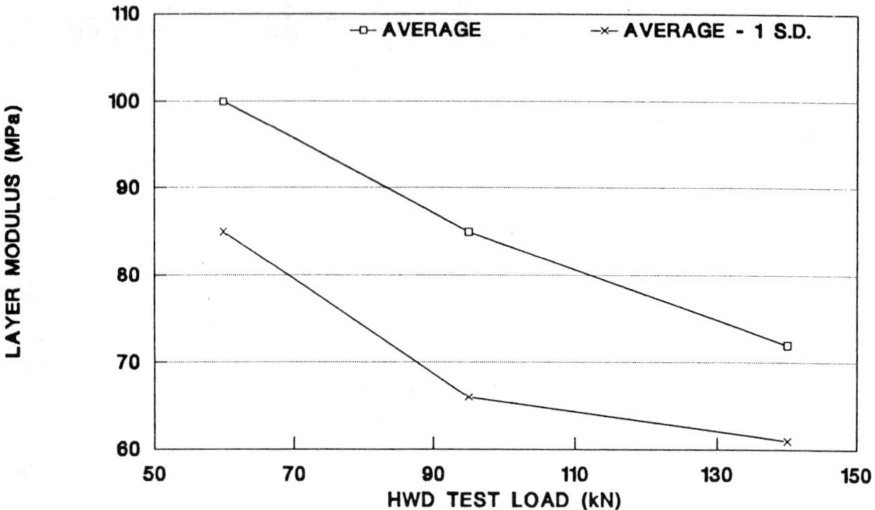

FIGURE 6 — Backcalculated subgrade modulus vs. HWD test load level – Runway 03/21 at Chino Airport

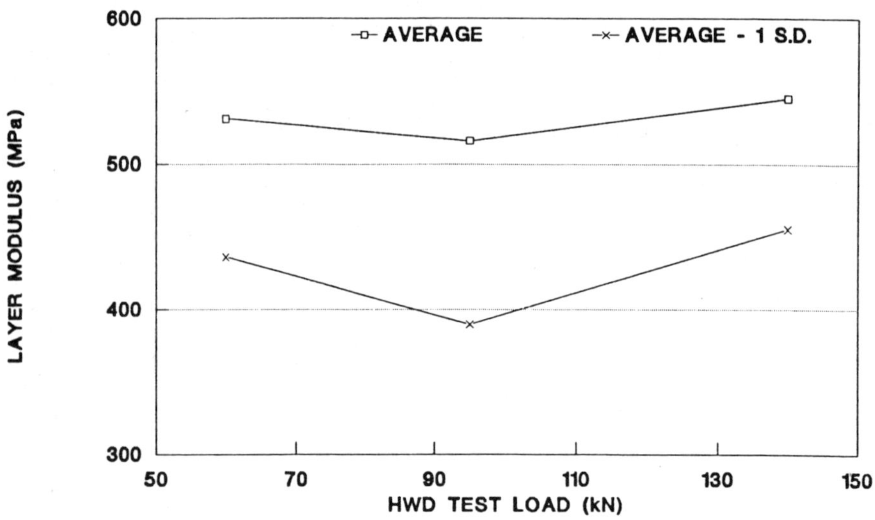

FIGURE 7 — Backcalculated base modulus vs. HWD test load level – Runway 03/21 at Chino Airport

In *Figures 6 and 7*, the backcalculated moduli are plotted as a function of load level, both average and design (average less one standard deviation), for the subgrade and base respectively. In these figures it can be seen that the base modulus is relatively constant, while the subgrade modulus *decreases* with increasing test load level. The effect of the subgrade turns out to be the critical factor, since it is stress softening, ie, decreasing from a value of some 85 MPa (12,500 psi) derived at the lowest test load down to about 60 MPa (9,000 psi) at the highest test load. In this case, the *lowest* derived subgrade modulus should be used for design purposes since it corresponds to the design wheel load of approximately 135 kN.

For this example, the corresponding AC overlay thickness requirements are plotted in *Figure 8*, and the overlay design criteria are identical to those used in the Sky Harbor analyses above. The calculated (design) overlay is about 190 mm (7½") of new AC using the test load level corresponding to the B-737 wheel load of some 135 kN (30 kips). If a lower load-derived overlay design had been chosen, the needed overlay would have been about 115 mm (4½") using an HWD or FWD test load of 60 kN (13,000 lbf), or about 175 mm (7") at the normal maximum FWD test load of 120 kN (27,000 lbf). In this case the overlay would have been under designed by 75 mm (3") if the 60 kN test results had been extrapolated to the 135 kN B-737 design load level.

In cases like this, inadequate overlay thickness would typically result in a significant reduction of pavement service life, along with additional maintenance or repair costs not normally included in an agency's planned budget.

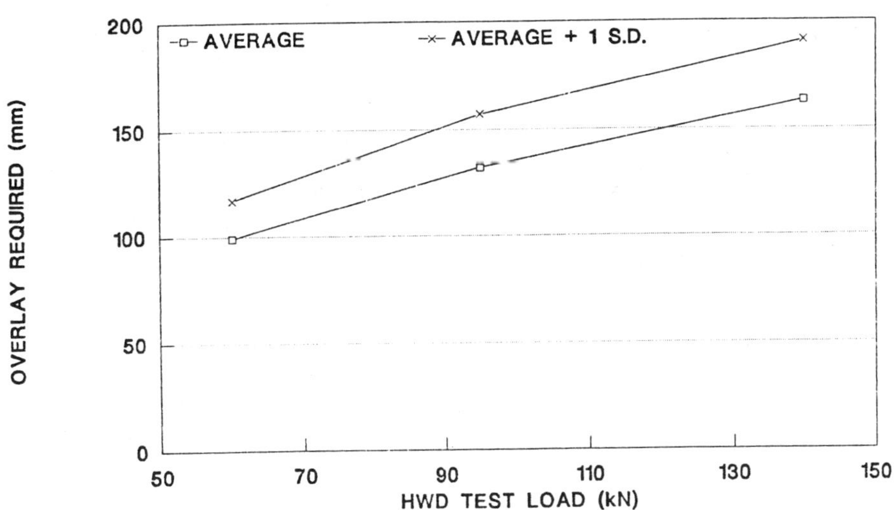

FIGURE 8 — Required AC overlay thicknesses for a 135 kN design load vs. HWD test load level - Runway 03/21 at Chino Airport

The two examples presented serve to illustrate the effect of non-linear
response of unbound materials on rehabilitation designs based on deflection analysis, and are fairly typical of AC pavement behavior. A
response similar to the Sky Harbor stress stiffening base has been
reported at Oakland Airport in California [Coetzee et al. 1989], for
instance.

ANALYSIS OF BACKCALCULATED MODULI

In examining the four unstabilized layers (base and subgrade at Chino
and Phoenix), two exhibited significant stress sensitive moduli at the
time the HWD measurements were conducted. These were the subgrade soil
at Chino and the base course at Phoenix. Although moisture content,
temperature and suction, etc, obviously have an effect on these relationships, the HWD tests were conducted nondestructively and the moisture content or temperature of the soils were not measured; however,
they may be assumed constant since the measurements at each site were
conducted within minutes of each other.

The associated equations (also referring to the discussion under
CONSTITUTIVE MODELS OF UNBOUND MATERIAL BEHAVIOR, above) are as
follows:

Phoenix:

$$E_{bs} = 1,450 \times \theta^{0.40} \qquad [\text{ie, } k_1 = 1,450 \text{ (in MPa), } k_2 = 0.40]$$
$$[\text{or } E_{bs} = 29,000 \times \theta^{0.40} \qquad \text{(in psi)}]$$

Chino:

$$M_R = 38 \times \sigma_d^{-0.37} \qquad [\text{ie, } k_3 = 38 \text{ (in MPa), } k_4 = -0.37]$$
$$[\text{or } M_R = 35,000 \times \sigma_d^{-0.37} \qquad \text{(in psi)}]$$

The base course relationship at Phoenix is quite different from those
found for Washington State crushed stone and the AASHO Road Test.
Recall that k_1 = 380 and k_2 = 0.375 for Washington State bases and k_1
ranged from 450 to 1100 and k_2 = 0.6 for the AASHO Road Test crushed
limestone base. The primary difference for the Phoenix base is in the
constant k_1, as k_2 is within a typical range. Generally, however,
major air carrier airfield pavements are better constructed and compacted, and therefore stiffer than similar highway pavements, so the
moduli found for the Phoenix base course are not all that surprising.
Unstabilized base used in such pavements is often thicker as well.
Figure 9 shows a comparison of various base course materials.

The subgrade soil relationship for Chino has a negative exponent which
is typical of a fine-grained soil such as a silt. In Washington State,
similar k_3 and k_4 values have been observed for silty soils, based on
laboratory triaxial tests. See *Figure 10* for a comparison of various
subgrade materials, including those reported in "Pavement Analysis"
[Ullidtz 1987] based on an earlier Ph.D. Dissertation [Fossberg 1970].
The two Fossberg curves presumably relate to two different moisture
contents (ie, 18% and 30%) for the same soil.

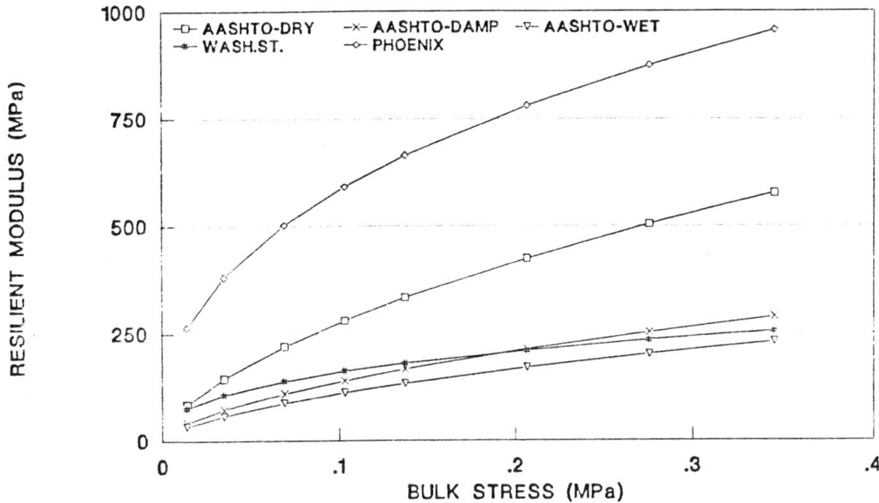

FIGURE 9 — Resilient modulus of various granular base courses vs. bulk stress level

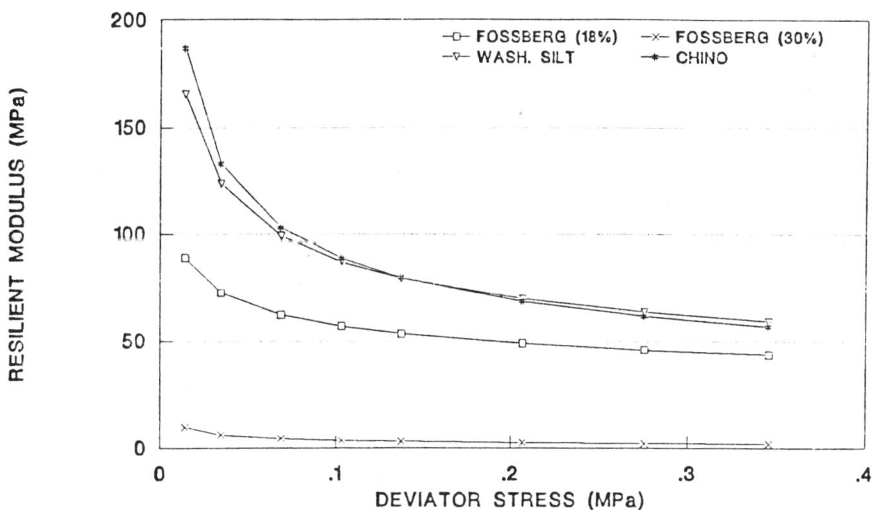

FIGURE 10 — Resilient modulus of various subgrades vs. deviator stress level

TABLE 2 — Measured (backcalculated) base and subgrade moduli at Chino and Phoenix airfields

Pavement Location	Unstabilized Layer	HWD Load Level (kN)	Average Elastic Modulus (MPa)	Stress State (MPa)
Chino	Base (Middle)	60 95 140	530 515 545	$\theta = 0.048$ $\theta = 0.033$ $\theta = 0.015$
	Subgrade (Top)	60 95 140	100 85 70	$\sigma_d = 0.083$ $\sigma_d = 0.124$ $\sigma_d = 0.170$
Phoenix	Base (Middle)	70 160 235	380 500 535	$\theta = 0.028$ $\theta = 0.060$ $\theta = 0.094$
	Subgrade (Top)	70 160 235	195 200 235	$\sigma_d = 0.014$ $\sigma_d = 0.032$ $\sigma_d = 0.048$

One of the unstabilized layers which showed no apparent stress sensitive-moduli relationship was the Chino base course. This base is quite stiff, with an average elastic modulus of about 530 MPa; however, the estimated bulk stresses were quite low as shown (Table 2). Please note, however, that the bulk stresses shown in this table were estimated using the ELSYM5 computer program at the middle of each base layer. Further, the bulk stress is the _sum_ of the principal stresses, discussed previously. For the Chino base course, the load-related vertical stress (σ_1) is compressive while the load-related lateral stresses (σ_2, σ_3) are tensile. Thus σ_2 and σ_3 for this specific case somewhat overwhelms σ_1, thus resulting in unrealistically low θ's. This helps to explain the lack of a strong stress sensitive-moduli relationship for this specific case. Also, elastic layer programs typically consider load-associated stresses only, ignoring both the fact that there are static overburden stresses present which, in effect, should make σ_2 and σ_3 compressive in reality, thus it may be misleading to draw any conclusions based on bulk stresses calculated using this type of analysis.

Similarly, the subgrade soil moduli at Phoenix indicate only a moderate stress sensitive relationship. The backcalculated moduli [average of about 200 MPa (30,000 psi)] as shown (Table 2) are fairly high for a subgrade soil, which suggests a coarse-grained subgrade. It is fairly common to not see a strong stress sensitive relationship for sandy subgrade soils.

CONCLUSIONS

Two examples of AC pavement deflection response have been cited which
are considered typical of a fairly general trend in flexible pavements
to exhibit significant stress-sensitivity in unbound layers. The
authors have attempted to show the importance in the design process
of either considering the non-linearity of such unbound layers using
adequate material response models, or by ensuring that the NDT load
imposed on the pavement during evaluation is as close as possible to
the actual, expected design wheel loads destined to use the pavement.

REFERENCES

American Association of State Highway and Transportation Officials,
    1986, <u>AASHTO Guide for Design of Pavement Structures</u>, AASHTO,
    Washington, DC.

Coetzee, N.F., Maree, J.H. and Van Wyk, A., 1989, "Impulse Deflection
    Measurement," <u>Proceedings</u>, Fifth Conference on Asphalt Pavements
    for Southern Africa, Swaziland.

Fossberg, P.E., 1970, "Load-Deformation Characteristics of Three-Layer
    Pavements Containing Cement-Stabilized Base," <u>Ph.D. Dissertation</u>,
    University of California, Berkeley.

Hicks, R.G. and McHattie, R.L., 1982, "Use of Layered Theory on the
    Design and Evaluation of Pavement Systems," <u>Report</u>, No. FHWA-AK-
    RD-83-8, State of Alaska, Fairbanks, Alaska, p B-11.

Kirk, J.M., 1973, "Revised Method of Design of Asphalt Pavements,"
    <u>Asphalt (Magazine)</u>, No. 42, Copenhagen, Denmark (in Danish).

Shell International Petroleum Company Limited, 1978, <u>Shell Pavement
    Design Manual</u>, London, England.

Ullidtz, P. and Stubstad, R.N., 1985, "Analytical-Empirical Pavement
    Evaluation Using the Falling Weight Deflectometer," <u>Transportation
    Research Record</u>, No. 1022, TRB, Washington DC., pp 36-44.

Ullidtz, P., 1987, <u>Pavement Analysis</u>, Elsevier, Amsterdam, Holland, pp
    62-63.

Washington State Department of Transportation, 1992, <u>WSDOT Pavement
    Guide for Design, Evaluation and Rehabilitation</u>, Washington DOT,
    Olympia, Washington.

Problems/Errors Associated with Backcalculation Methods and Design Parameters

Jim W. Hall, Jr.[1] and Patrick S. McCaffrey, Jr.[2]

MISLEADING RESULTS FROM NONDESTRUCTIVE TESTING—A CASE STUDY

REFERENCE: Hall, Jr., J. W., and McCaffrey, Jr., P. S., "Misleading Results from Nondestructive Testing-A Case, Study," Nondestructive Testing of Pavements and Backcalculation of Moduli (Second Volume), ASTM STP 1198, Harold L. Von Quintas, Albert J. Bush, III, and Gilbert Y. Baladi, Eds., American Society for Testing and Materials, Philadelphia, 1994.

ABSTRACT: A new apron and taxiway at the Cape Canaveral Air Force Station skid strip, Cape Canaveral, Florida, showed depressions as a result of parking various aircraft overnight. The U. S. Army Engineer Waterways Experiment Station performed a pavement investigation which included nondestructive falling weight deflectometer (FWD) tests and test pits for California Bearing Ratio (CBR), moisture, density, and sampling. A study of the field and laboratory results showed serious inconsistencies between the test methods. FWD data indicated the new pavement to be of higher strength than the older portion. The test pit data (CBR and density) indicated the recycled base layer to be weak (due to low density), which was in direct disagreement with the FWD. A possible explanation for the failure of the FWD to correctly characterize the pavement materials is that the faster rate of loading of the FWD as compared to the static load of the parked aircraft caused the difference.

KEYWORDS: nondestructive testing, pavement investigation, recycled base

In early 1989, a taxiway and apron extension project was completed at the Cape Canaveral Air Force Station (CCAFS) skid strip. The project included construction of a new apron and taxiway and overlay of the existing adjacent apron and taxiway. As soon as the pavement was returned to service, depressions began to appear in the pavement surface where heavy aircraft (C-5A) were parked overnight. Fig. 1 is an overall view of the depressions and Fig. 2 is a close-up of one depression filled with water. By June 1989, surface damage was extensive, and the U. S. Army Engineer Waterways Experiment Station (WES) was asked to perform a pavement investigation (Hall and Potter 1991) to determine the cause of distress and recommend a repair strategy.

The CCAFS skid strip airfield consists of a runway, parking apron, and two taxiways. Fig. 3 shows the apron and taxiways that were investigated. Area 1 is the older pavement while areas 2 and 3 indicate the new pavement. The old apron and taxiway (area 1) were milled approximately 13 mm to 19 mm (1/2 to 3/4 in.) and overlaid with 51 mm (2 in.) of asphalt concrete (AC). Area 1 consists of approximately 102 to 152 mm (4 to 6 in.) AC surface and 254 mm (10 in.) limerock base over the native sand subgrade. The new apron pavement (area 2) shown as hatched in Fig. 3 was comprised of 102 to 152 mm (4 to 6 in.) AC and 305 mm (12 in.) recycled AC base over the sand subgrade. Both heater-

---

[1]Chief, Systems Analysis Branch, U. S. Army Engineer Waterways Experiment Station, 3909 Halls Ferry Road, Vicksburg, MS 39180.

[2]Civil Engineering Technician, U. S. Army Engineer Waterways Experiment Station, 3909 Halls Ferry Road, Vicksburg, MS 39180.

planed and cold-milled asphalt mixtures were placed for this base course. Area 3 is a new section of taxiway constructed with 114 mm (4.5 in.) AC and 254 mm (10 in.) limerock base (new construction).

FIG 1.--Depressions resulting from parked aircraft.

FIG. 2--Close-up of depression filled with water.

TEST PROGRAM

Field Tests

Field tests consisted of nondestructive deflection measurements using the falling weight deflectometer (FWD) and test pits for in-place measurements of moisture, density, and CBR (California Bearing Ratio). The locations of the four test pits placed in the taxiway and apron are indicated in Fig. 3. Fig. 4 shows the FWD test locations.

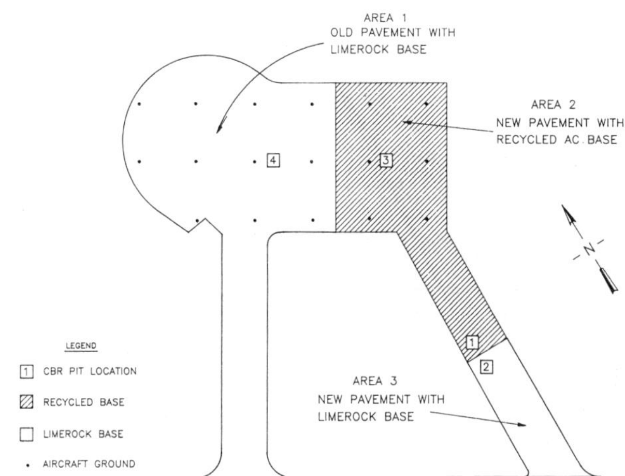

FIG. 3--CCAFS pavement layout with test pit locations.

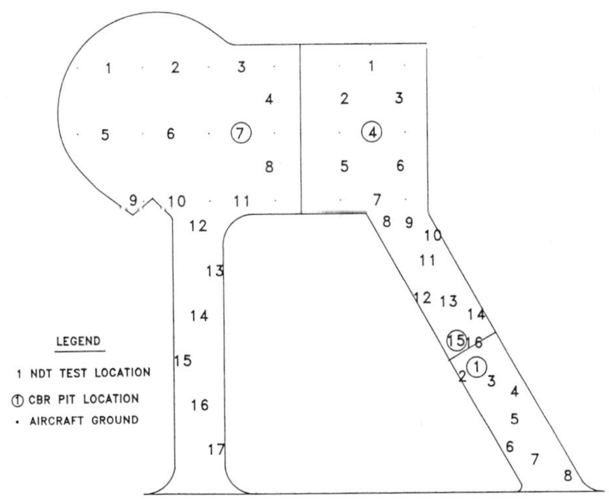

FIG.4--NDT locations at CCAFS.

Tests conducted in the test pits were in-place CBR, moisture, and density in accordance with MIL-STD-621A (Department of Defense 1964). Results from the test pits are indicated in Fig. 5. Test pits 1 and 3, constructed with the recycled asphalt base, were placed in the new pavement where depressions from parked aircraft were evident. Test pit 2 was in the new taxiway pavement where limerock base was used. Test pit 4 was placed in the older apron where the base course was limerock. No depressions existed in the areas of test pits 2 and 4.

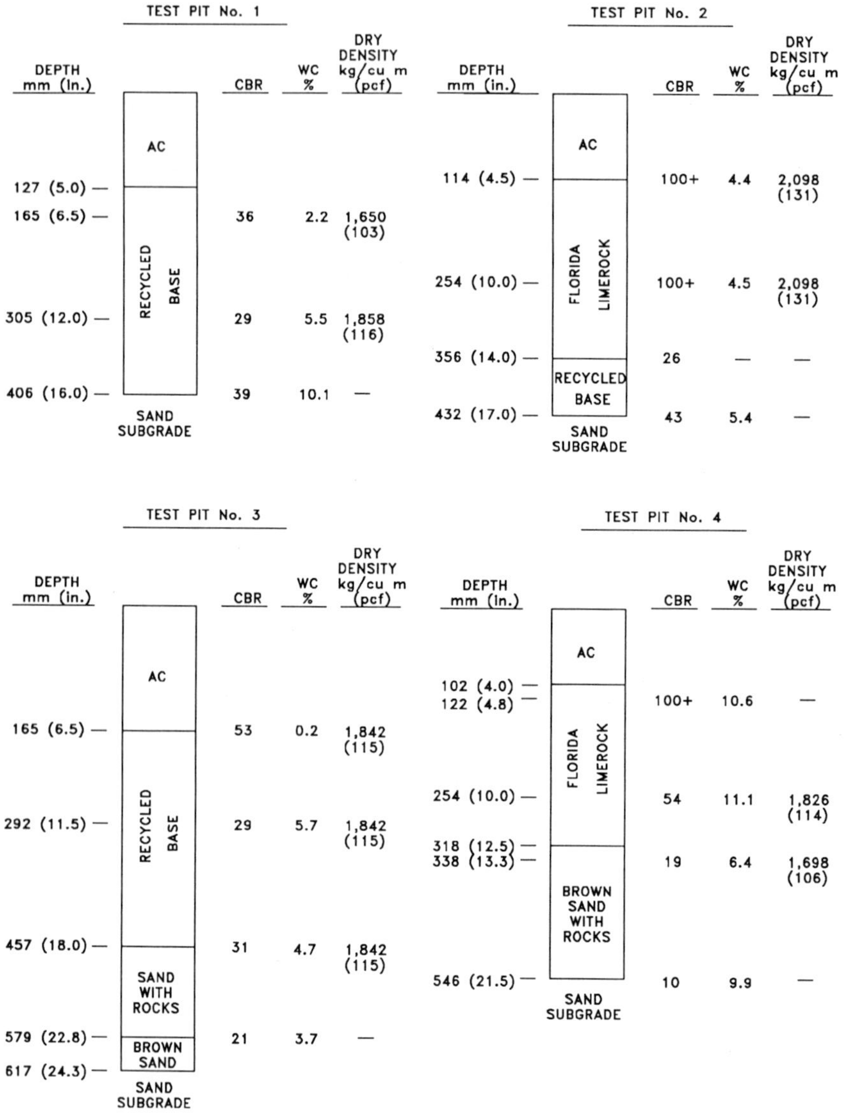

FIG. 5--Test pit results.

A study of the test pit data in Fig. 5 gives an indication of potential problems with the recycled base material. From test pits 1 and 3, the density of the recycled layer ranges from 1,650 to 1,858 kg/cu m (103 to 116 pcf) which is relatively low for this type material. Also, the average CBR is only 36, which is quite low for base course (design requirements are normally 80 or 100 CBR). In contrast, the limerock base course in test pits 2 and 4 has a density of that ranges from 1,826 to 2,098 kg/cu m (114 to 131 pcf) and an average CBR of 100 plus.

Observations made during the field investigation by cutting through the pavement (test pit 1) in a depression showed that the depression (consolidation) did not occur in the AC surface layer. The thickness of the AC had not changed in the depression indicating that the movement was in the base. The consolidation had occurred in the recycled base layer.

Nondestructive testing (NDT) was performed in a grid pattern with test numbers as indicated in Fig. 4. Several drops of the load were made at each test, typically, 33.4, 66.7, and 111.2 kN (7.5, 15, and 25 kips (or maximum attainable). Plots of the deflections measured at the different load levels for tests in the three pavement areas (NDT numbers 1, 4 and 7) are shown in Fig. 6. Note that the new pavement, area 2, appears to be the strongest section. This same trend is true for the deflection basins as shown in Fig. 7. In order to make relative comparisons of the NDT data, the ISM (impact stiffness modulus) is defined as the applied dynamic load (maximum value) divided by the corresponding deflection. ISM data for the three sections are plotted in Figs. 8 and 9. Note that the data in Fig. 8 were collected on 29 October 1990, and the data in Fig. 9 were taken on 16 March 1991. A study of this data indicates area 2 to be much stiffer (higher ISM values) than areas 1 and 3. Note that the transition between areas 2 and 3 was not well defined, and test numbers 13 through 16 may have actually been part of area 3. Areas 1 and 3 (both with limerock base) show approximately the same ISM values.

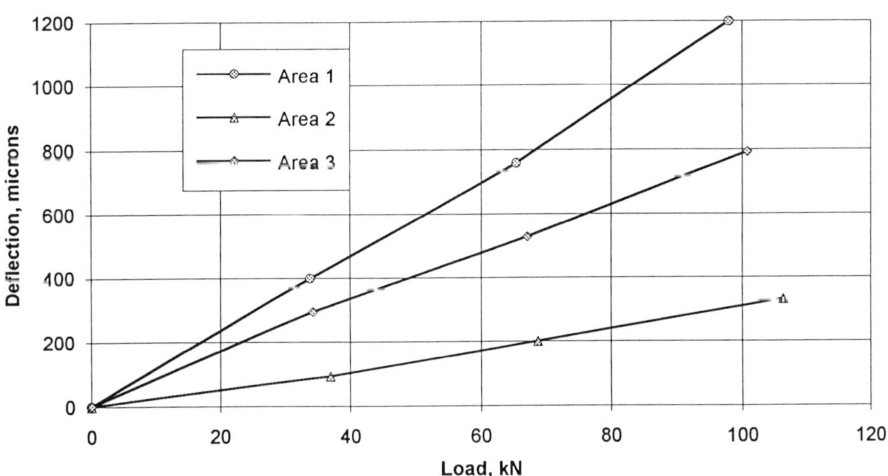

FIG. 6--Deflection versus load from FWD tests.

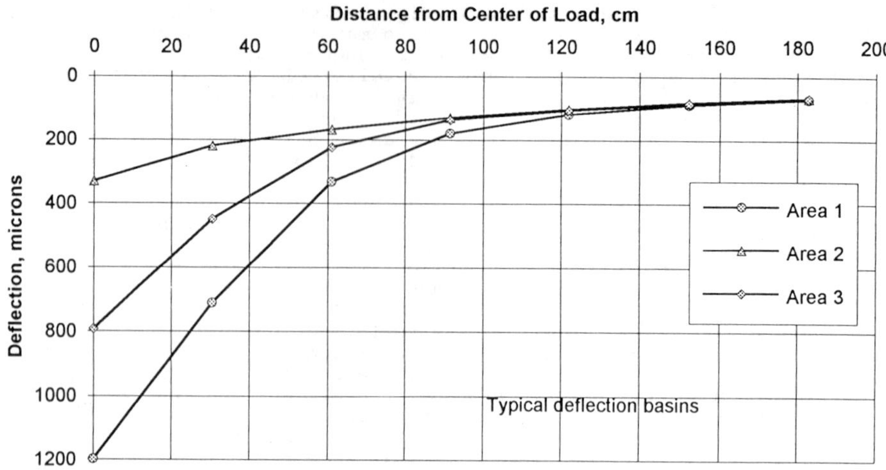

FIG. 7--Deflection basins for areas 1, 2, and 3.

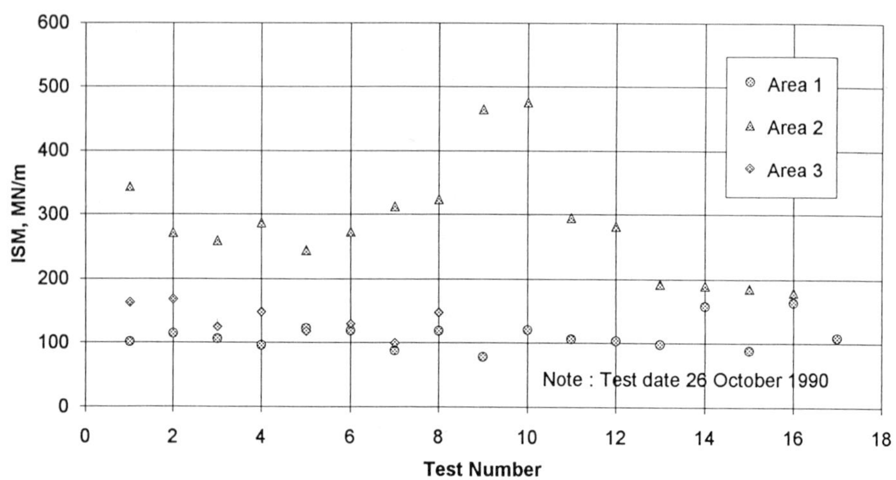

FIG. 8--ISM data from CCAFS, 29 October 1990.

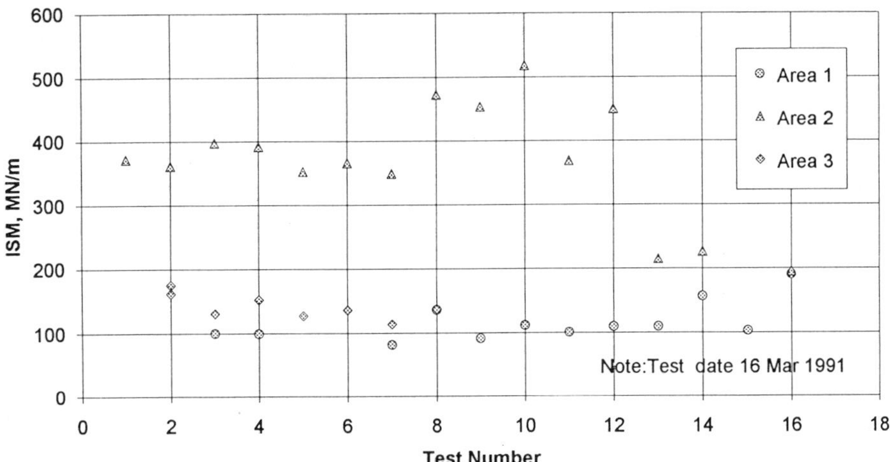

FIG. 9--ISM data from CCAFS, 16 March 1991.

Laboratory Data

Results of laboratory tests on the AC surface and recycled base materials are shown in Tables 1 and 2. Extraction tests were performed to measure asphalt content and gradation. Samples of the surface layer were recompacted in order to compare mix properties to project specifications.

Table 1 gives test results from the AC surface material. Comparisons are shown with Corps of Engineers (CE) requirements (Departments of the Army and the Air Force 1987) for a 75-blow Marshall mix. From Table 1, it can be noted that gradations of the surface mix are within specification, the penetration of the asphalt is low, voids in the mix are slightly above specification but are not excessively high, stability and flow meet specification, and field density is slightly below the required 98 percent. Overall, the surface mix appears to be a satisfactory material.

Test results from the recycled material used as base course are shown in Table 2. Comparisons were made with the project specification and with CE specification (Department of the Army 1989) for cold-mix recycling. The base material at CCAFS was placed in a cold state. It can be noted from Table 2 that the recycled material did not meet specifications. The material contained excessive fines. The in-place density was only 83.2 percent of the 75-blow Marshall compaction at 123°C (250°F). Attempts were made to compact the material in the laboratory at 31°C (77°F), but the specimens would not hold together after compaction. For the laboratory-compacted specimens, the voids in the mix were 9.7 percent. The voids for the field-compacted material would then be approximately 19 percent. The low field density was most likely due to improper compaction during construction. Consolidation under aircraft loads would likely be expected with such high voids.

DISCUSSION

The pavement failures at CCAFS are attributed to the low density of the recycled AC base course. This conclusion was determined from both test pit measurements and laboratory test results. Overnight parking of C-5A aircraft has caused significant depressions in the pavement constructed with the recycled material, but the pavement constructed with limerock base course is performing well. Aircraft taxiing across the new pavement caused no apparent distress.

The NDT results contradict the test pit and laboratory results and imply that the new pavement with recycled base material is the stronger (higher stiffness). The test pit results are obviously true since they confirm the failure under the aircraft loads. The pavements with limerock base, whether old construction (area 1) or new construction (area 3) have about the same stiffness which is considerably lower than area 2.

The only plausible explanation for the FWD tests to measure lower deflections on the new pavement with the recycled base is due to the loading rate. The FWD load pulse has been reported (Bentsen et al. 1989) to be approximately 20 Hz. An aircraft (C-5A) traveling at a speed of approximately 80 kmph (50 mph) would have a loading frequency of about 20 Hz at the center of the base layer. The viscoelastic response of the recycled material allows it to provide high stiffness response at the high rate of load but to yield and deform under static loads.

A check was made on the adequacy of the pavement thickness, and the design thickness (Departments of the Army and Air Force 1978) for a medium load pavement (includes the C-5A) is 127 mm (5 in.) AC over 279 mm (11 in.) base (80 CBR) over the subgrade (20 CBR). Pavement thicknesses shown in Fig. 5 for all the pavement areas are within this design range.

TABLE 1--<u>Test results from asphalt concrete surface.</u>

| Sieve Size | Gradation, Percent Passing | | CE Specification |
	Test Pit No. 1	Test Pit No. 2	
19 mm (3/4 in.)	100.0	100.0	100
13 mm (1/2 in.)	98.6	97.3	82-96
10 mm (3/8 in.)	89.1	86.8	75-89
No. 4	69.3	68.2	59-73
No. 8	50.7	49.5	46-60
No. 16	43.3	42.5	34-49
No. 30	39.4	38.9	24-39
No. 50	34.5	34.2	15-27
No. 100	16.3	16.2	8-18
No. 200	4.9	4.1	3-6
	Mix Properties		
Percent Asphalt	5.8	5.6	
Penetration	26	30	
Viscosity at 60°C (140°F), poises	25,519	18,438	
Viscosity at 137°C (275°F), centistokes	1,125	1,075	
Voids Total Mix, %	5.9	6.1	3-5
Voids Filled, %	67.4	66.1	70-80
Density, kg/cu m (pcf)	2,209 (137.9)	2,209 (137.9)	
Stability, kg (lb)	1,527 (3,367)	1,345 (2,965)	816 min. (1,800)
Flow, mm (0.01 in.)	1.8 (7)	2.0 (8)	4.1 max. (16)
Field Density, kg/cu m (pcf)	2,130 (133)	2,146 (134)	
Percent Density, %	96.4	97.1	98

TABLE 2--<u>Test results from recycled material, test pit 1.</u>

| | Gradation, Percent Passing | | | |
Sieve Size	Sample No. 1	Sample No. 2	Project Specification	CE Specification
19 mm (3/4 in.)	100.0	100.0	100	100
13 mm (1/2 in.)	97.2	96.2	82-100	73-92
10 mm (3/8 in.)	93.5	91.5	68-90	63-81
No. 4	79.7	77.2	50-79	45-63
No. 8	66.2	63.1		32-50
No. 10			36-67	
No. 16	53.4	50.8		23-41
No. 30	43.4	41.3		15-33
No. 40			17-44	
No. 50	32.1	30.7		10-24
No. 80			9-29	
No. 100	19.1	17.9		7-17
No. 200	12.1	11.0	3-8	3-7

<u>Mix Properties</u>

Percent Asphalt	6.0	5.5		
Penetration	10	9	85-100	
Viscosity at 60°C (140°F), poises	1,564,601			
Viscosity at 137°C (275°F), centistokes	4,384			
Voids Total Mix, %	9.7*		3-8	4-6
Voids Filled, %	54.9*		60-70	65-75
Density, kg/cu m (pcf)	2,154* (134.5)			
Stability, kg (lb)	2,077* (4,579)		544 min. (1,200)	816 min. (1,800)
Flow, mm (0.01 in.)	2.3* (9)		2-4.1 (8-16)	4.1 (16) max.
Field Density, pcf	1,794** (112)			
Percent Density, %	83.2		98	86 theoretical max. density

  * Recompacted samples using 75-blow Marshall at 122°C (250°F).
 ** Average field density from test pits 1 and 3.

SUMMARY

The pavement investigation at CCAFS indicated failures due to consolidation of a recycled asphalt base which was deficient in density. NDT tests were misleading in that they indicated the failed pavement to be stronger than the adjoining pavement (with no distress) that contained a limerock base.

The explanation for the discrepancy in the NDT results is due to the rate of loading. Failures were caused by static loads of the C-5A aircraft. The FWD loading of approximately 20 Hz equates to a loading rate of approximately 80 kmph (50 mph) of the aircraft. Aircraft at taxi speeds did not cause any damage to the pavement.

Users of NDT to evaluate the structural capacity of flexible pavements need to use caution and judgement to avoid a potential discrepancy as described in this paper.

ACKNOWLEDGEMENTS

This research was sponsored by the U. S. Air Force. The support of the U. S. Army Engineer Waterways Experiment Station in preparing this paper is gratefully acknowledged. This paper is published with the permission of the Chief of Engineers.

DISCLAIMER

The views expressed in this paper are those of the authors who are responsible for the facts and accuracy of the data. The contents do not necessarily reflect the official views or policies of the Waterways Experiment Station, the Department of the Army, or the Department of Defense. This paper does not constitute a standard, specification, or regulation.

REFERENCES

Bentsen, R. A., Bush, A. J., and Harrison, J. A., 1989, "Evaluation of Nondestructive Test Equipment for Airfield Pavements, Phase I, Calibration Test Results and Field Data Collection," Technical Report GL-89-3, Waterways Experiment Station, Vicksburg, Mississippi.

Department of the Army, Office, Chief of Engineers, 1989, "Cold-Mix Recycling," Guide Specification CEGS-02591, Washington, DC.

Departments of the Army and the Air Force, 1978, "Flexible Pavement Design for Airfields," Technical Manual TM 5-825-2/AFM 88-6, Chapter 2, Washington, DC.

Departments of the Army and the Air Force, 1987, "Bituminous Pavements Standard Practice," Technical Manual TM 5-822-8/AFM 88-6, Chapter 9, Washington, DC.

Department of Defense, 1964, "Test Method for Pavement Subgrade, Subbase, and Base-Course Materials," Military Standard MIL-STD-621A, Washington, DC.

Hall, J. W., Jr. and Potter, J. C., 1991, "Cape Canaveral Air Force Station Skid Strip Pavement Failure Investigation," Miscellaneous Paper GL-91-19, Waterways Experiment Station, Vicksburg, Mississippi.

K.M. Vennalaganti[1], C. Ferregut[2] and S. Nazarian[3]

STOCHASTIC ANALYSIS OF ERRORS IN REMAINING LIFE DUE TO MISESTIMATION OF PAVEMENT PARAMETERS IN NDT

REFERENCE: Vennalaganti, K. M., Ferregut, C., and Nazarian, S., "Stochastic Analysis of Errors in Remaining Life Due to Misestimation of Pavement Parameters in NDT," _Nondestructive Testing of Pavements and Backcalculation of Moduli (Second Volume), ASTM STP 1198_, Harold L. Von Quintas, Albert J. Bush, III, and Gilbert Y. Baladi, Eds., American Society for Testing and Materials, Philadelphia, 1994.

ABSTRACT: The recent mechanistic-empirical approaches for predicting the remaining life of flexible pavements are mainly based upon the predicted strains at the interfaces of different layers. To determine these strains, nondestructive testing techniques are utilized. Unfortunately, uncertainties in determining the strains may result in significant errors in the predicted remaining life. A number of major factors that contribute to these inaccuracies include the imprecise knowledge of thickness and Poisson's ratio of each pavement layer and the errors in measuring the loads and the deflections using a NDT device.

A methodology which accounts for these uncertainties in the assumed pavement parameters and measured responses is suggested herein. With this methodology the influence of these parameters on the calculation of the remaining life of the pavement can be quantified. The methodology is based on Monte Carlo simulation techniques and has been used to analyze four pavement sections, representing a wide range of highways from secondary to interstate. The results of the probabilistic analysis show that the variability in pavement parameters increases the probability of failure of the pavement.

KEY WORDS: nondestructive testing, remaining life, flexible pavement, statistical simulation

---

INTRODUCTION

The recent mechanistic approaches for predicting the remaining life (number of repeated 18-kip loads that can be further applied to the pavement) of a flexible pavement are mainly based upon predicting the

---

[1]   Graduate Research Assistant, The University of Texas at El Paso, TX 79968-0516

[2]   Assistant Professor, The University of Texas at El Paso, TX 79968-0516

[3]   Associate Professor, The University of Texas at El Paso, TX 79968-0516

strains at the interfaces of different pavement layers. To predict the remaining life, a nondestructive testing (NDT) device that impacts a load on the flexible pavement and measures deflections on the surface of the pavement is generally employed. The deflections along with the pavement parameters are then entered in backcalculation programs to obtain the stiffness profiles of the existing pavement. These backcalculated moduli are then used to compute the strains at the interfaces of the pavement layers. The remaining life of the pavement is finally determined by using a semi-empirical relationship between the number of loads applied on the pavement and the critical strains at the interfaces of the different paving layers.

From rehabilitation/maintenance point of view, it is important to understand and quantify the errors that may affect the accurate prediction of remaining life of the existing pavements, since the final design of the pavement system depends on that value. Moreover, if the uncertainties associated with the pavement remaining life are not accounted for, the remaining life may be over- or under-estimated. This may lead to excessive initial construction cost, excessive maintenance, or premature rehabilitation of the pavement.

Uncertainty in the prediction of the remaining life of the pavement structure is a function of the uncertainties that are associated with the flexural strains at the bottom of the asphalt concrete layer and the compressive strain at the top of the subgrade layer. In general, these strains are functions of pavement parameters, which themselves can only be determined with a degree of uncertainty. The uncertainties in the pavement parameters consist of random deviations from those values which are assumed or specified during the design of the pavement. For example, the random deviations on the thickness of each pavement layer are due to the inconsistency in the construction of the pavement structures, and the variability in the Poisson's ratio may be due to nonuniform compaction. The variation in the load and in the deflections may be due to the imprecise measurements during nondestructive testing. The variation in the backcalculated moduli is due to the errors that are associated with the thicknesses, Poisson's ratios, and measured load and deflections, since the backcalculated modulus is a function of these parameters. The critical strains can also vary with environmental conditions (moisture, temperature) and age. These parameters are not considered in this study.

Researchers have attempted to understand the sources and quantify the level of variability in some pavement parameters. For example, the inaccuracies in deflection measurements can be due to a multitude of pavement characteristics, each resulting in minor effects to create the whole effect on the variability of deflections. The variables that have been found to contribute most significantly to the variability in the measured deflections are: 1) pavement structural characteristics, such as layer thickness, stiffness and plasticity of clay of the subgrade layer; and 2) environmental characteristics, such as air temperature during testing and annual precipitation (Rahut and Jordahl 1993) . Hudson et al. (1986) and Bentsen et al. (1989) show that deflections and loads are known within an accuracy of 2 to 5 percent.

The impact of variabilities of the pavement parameters on the backcalculated moduli were recently studied by Rodriguez-Gomez et al. (1991). The authors found that the variability in the Poisson's ratio of the subgrade layer is the most influencing parameter on the variability of the backcalculated modulus for that layer. For the paving layers, the variability in the backcalculated moduli is mostly influenced by the variability of the thickness of the layers. The paper also showed that the backcalculated moduli may be underestimated between 40% and 60% of the time. Most of the methodology followed in this paper is based on the methodology proposed by Rodriguez-Gomez et al.

In this study a stochastic analysis was conducted to assess the influence of the errors or the random deviations of layer thickness, Poisson's ratio and the load and deflections of the NDT device on the predicted remaining life of a pavement system. In addition, probabilistic models for the main variables included in the computation of the pavement's remaining life were obtained. To the authors' knowledge this is the first time such an exercise has been conducted.

METHODOLOGY

The methodology used to carry out the study mentioned in the foregoing can be summarized in the following steps: 1) the pavement parameters were taken as random variables. Since no statistical data for the parameters were available, the variability and statistical distribution of each parameter were assumed using the authors' experiences and judgements. A range of values for these variabilities was considered. 2) Using the assigned distributions, sample values for each pavement parameter were generated using Monte Carlo simulation techniques (Ang and Tang 1984a, 1984b). 3) For each simulated sample set, the pavement layer moduli were backcalculated. 4) Using the simulated set of pavement parameters and the set of backcalculated moduli, critical stresses and strains within the pavement were computed. 5) Finally, a sample of values representing the predicted remaining life of the pavement was compiled. At each step, the statistics of each of the computed variables were determined and probabilistic models were fitted to each of them. The flow diagram shown in Figure. 1 illustrates the different steps adopted in this procedure. Each step is discussed in detail in the following paragraphs.

Pavement Models

Four pavement sections representing a wide range of highways (from secondary to interstate) were considered for this study. Each pavement section had 3 layers, the top most being the asphaltic concrete (AC) layer with a granular base beneath it. The thinnest pavement section (Pavement 3-6) had an asphalt layer thickness of 75 mm and base thickness of 150 mm. The second pavement section (Pavement 3-12) was a moderate pavement with a thin AC layer 75 mm thick and a base layer of 300 mm. A moderate pavement section (Pavement 5-6) with a relatively thick AC layer of 125 mm, and a base layer thickness of 150 mm, was also considered. The thickest pavement section (Pavement 5-12) had an asphalt layer of 125 mm and a base layer of 300 mm. The depth to a rigid layer for all these pavement sections was assumed to be 6000 mm as recommended by Bush and Alexander (1985). Poisson's ratios were assumed to be 0.35, 0.40 and 0.45 for the AC, base and subgrade, respectively. These are the typical values used by pavement engineers. Even though different layer thicknesses were used, the modulus of each paving layer was assumed to be the same. The moduli of the AC, base and subgrade were taken as 2800 MPa, 210 MPa, and 70 MPa, respectively. The pavement design parameters are shown in Table 1.

Deflections at seven locations (corresponding to those that would be measured with a Falling Weight Deflectometer, FWD) were theoretically determined using Program BISAR (Dejong et al. 1973). These deflections were assumed to represent the field measurements of the deflection bowls. The deflections were determined for each pavement section using their corresponding pavement parameters. A 4080 kg (9000 lb) load was used while developing the deflection bowls. Program BISDEF (Bush and Alexander 1985) was used to backcalculate the modulus of each layer.

Stochastic Model

To estimate the uncertainties associated with the prediction of

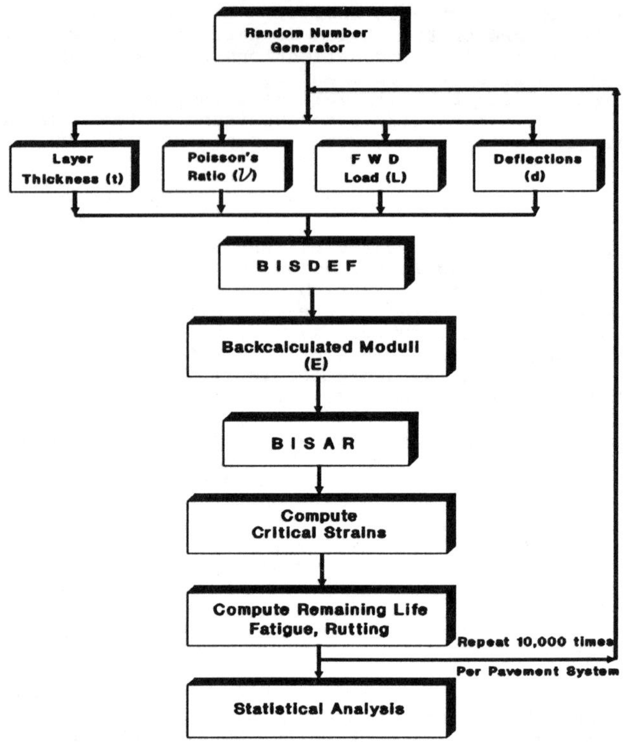

Figure 1.   Flow Diagram Showing Statistical Simulation of Remaining Life of Flexible Pavement

Table 1 - Mean (Design) Values of the Pavement Parameters

Pavement Parameters		Pavements			
		3-6	3-12	5-6	5-12
Thickness (mm)	Layer 1 (T_1)	75	75	125	125
	Layer 2 (T_2)	150	300	150	300
	Layer 3 (T_3)	6000	6000	6000	6000
Modulus (MPa)	Layer 1 (E_1)	2800	2800	2800	2800
	Layer 2 (E_2)	210	210	210	210
	Layer 3 (E_3)	70	70	70	70
Poisson's Ratio	Layer 1 (ν_1)	0.35	0.35	0.35	0.35
	Layer 2 (ν_2)	0.40	0.40	0.40	0.40
	Layer 3 (ν_3)	0.45	0.45	0.45	0.45

the remaining life of each flexible pavement, the following procedure
was adopted. The pavement parameters shown in Table 1 (except for T_3)
were modelled as random variables and statistical distributions were
assigned to each of them. The applied load and deflection basins were
also taken as random variables. All these random variables were assumed
to be statistically independent and normally distributed. Monte Carlo
simulation methods were used to numerically draw several sets of
pavement parameters from the assumed normal probability distributions.
The mean values used to generate these normally distributed variables
were the specified design values given in Table 1. Coefficients of
variation of 0.2, 0.1, 0.05 and 0.02 were assumed for the thickness of
the layers, Poisson's ratio, load on the pavement and the deflection
values, respectively. The distribution of the Poisson's ratios were
truncated at an upper bound of 0.50 and at a lower bound of 0.15
considering the practical impossibility of having paving materials with
Poisson's ratio out of those bounds. Since the accuracy of the Monte
Carlo simulation is heavily influenced by the size of the sample, it is
very much necessary to have a large sample. In this project 10,000 sets
of pavement parameters were randomly generated for each pavement
section.

<u>Backcalculation and Computation of Critical Strains</u>

Each of the generated sets of pavement parameters was entered
into program BISDEF. This process generated a random sample of
backcalculated moduli for each of the pavement layers. Each set of
pavement layer moduli along with other pavement parameters were then
used to compute the strains at the interfaces of the pavement layers.
Tensile strain at the bottom of the asphalt concrete layer and the
compressive strain at the top of the subgrade layer were computed using
program BISAR . Dual wheels with a load intensity of 2040 kg on each
wheel were considered in computing the strains. The contact area of each
wheel was assumed to be circular with a radius of 113 mm. Probabilistic
models were fitted to the generated sets of backacalculated moduli and
to the computed critical strains, to analytically describe the
variability observed in the data.

<u>Computation of the Remaining Life of Pavement</u>

Fatigue and rutting are the two major factors that contribute to
the structural loss of life in a pavement structure. The number of
repeated ESALs (remaining life) which cause the fatigue cracking damage
to the pavement is a function of the tensile strain at the bottom of the
asphaltic layer, ε_t and the modulus of the asphalt layer, E_R. The
remaining life of the pavement due to fatigue cracking, N_F, is assumed
as (Finn et al. 1977):

$$Log\ N_F = 15.947 - 3.291\ \log\left(\frac{\varepsilon_t}{10^{-6}}\right) - 0.854\ \log\left(\frac{E_R}{10^3}\right) \quad \cdot \cdot \cdot \quad (1)$$

The number of ESALs which cause the rutting failure, N_R, is a
function of the compressive strain at the top of the subgrade, ε_{vs}. For
computing the remaining life due to rutting, the equation developed by
Shook et al. (1982) was used:

$$N_R = 1.077\ X\ 10^{18} \left(\frac{10^{-6}}{\varepsilon_{vs}}\right)^{4.4843} \quad \cdot \cdot \cdot \cdot \cdot \cdot \cdot \cdot \cdot \quad (2)$$

Equations (1) and (2) are mechanistic-empirical equations that are
accepted by several organizations such as the Texas Department of
Transportation (TxDOT). Using equations (1) and (2) and the sample of

simulated critical strains, and moduli, a sample of (random) remaining
lives was generated for each failure mode, again probabilistic models
were fitted to describe the variability observed in N_f and N_R. These
models were later used in the computation of the probabilities of
failure of the pavement designs.

<u>Traffic Data</u>

Statistical traffic data for four types of highways: Interstate
Highway (IH), US roads (US), State Highway (SH) and Farm to Market road
(FM) were obtained from the TxDOT. The data comprises the number of
equivalent single axle loads (ESALs) that are applied to these pavements
in one year. The mean values and the coefficients of variation of the
traffic data are summarized in Table 2. The best fit probability
distribution to the traffic data for each of these four pavements was
found to be a lognormal distribution.

Table 2 - Mean and Coefficient of Variation of Traffic Data Per Year
          (Thousands of ESALs).

Statistical Estimates	Pavement IH	Pavement US	Pavement SH	Pavement FM
Mean	711	126	55	10
COV	0.69	1.42	1.79	0.60

RESULTS

<u>Statistical Analysis</u>

Statistical analyses were conducted on the data resulting from the
Monte Carlo simulation (backcalculated moduli, critical strains and the
remaining life of the pavement). Table 3 shows the mean values of the
pavement response variables. The values in the parentheses are the
deterministic values of these variables. The computed mean values of the
moduli of layer 1, and the mean values of the remaining lives are larger
than their respective deterministic values.

Table 4 illustrates the computed coefficients of variation of
pavement response variables. The COV of the AC and base moduli were
larger than those of the subgrade layer for all four pavement sections.
The computed coefficients of variation of the remaining life due to
fatigue criteria was higher for pavement sections with thinner base
layers. This in fact is a reflection of the larger variabilities in
strains and moduli associated with pavements with thin base layers.
However, the difference in the computed COV of the remaining life due
to rutting criteria is similar for the four pavement sections.

To arrive at the analytical equations to describe the variability
of the simulated data, several probability distributions were fitted to
the histograms of the simulated variables. The best fit probability
distributions to the simulated variables were similar for all of the
four pavement sections. As an example, the best fit distributions for
all variables of pavement 5-12 are shown in Figure 2. The best fit
distribution to the moduli of AC and base layers was lognormal. The
best fit for both the moduli of subgrade and the compressive strain was
normal. The tensile strain and the remaining life due to fatigue
cracking were best modeled using the extreme value distributions of type
I and type II, respectively. The best fit distribution to the remaining
life due to rutting was a lognormal distribution. As mentioned before,
similar results were found for the other pavement sections.

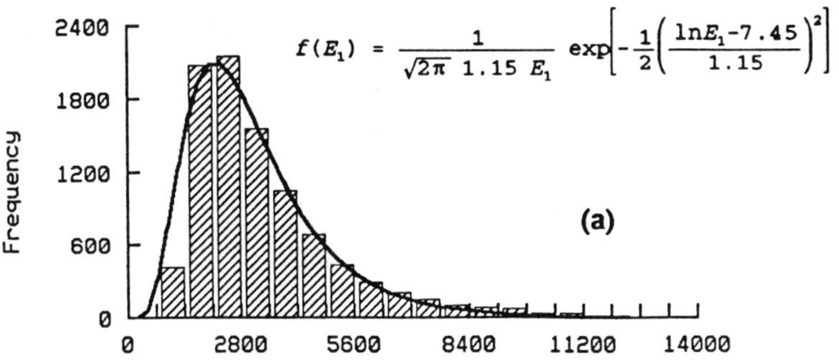

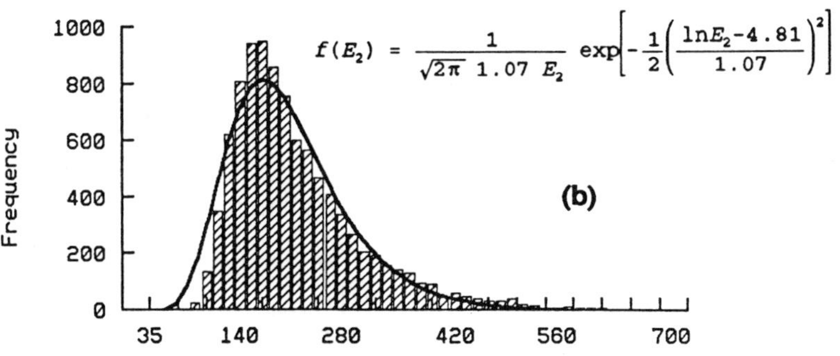

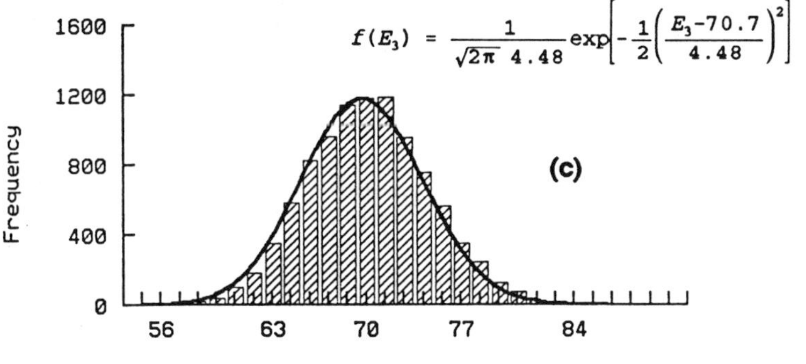

Figure 2. Distributions of Computed Variables

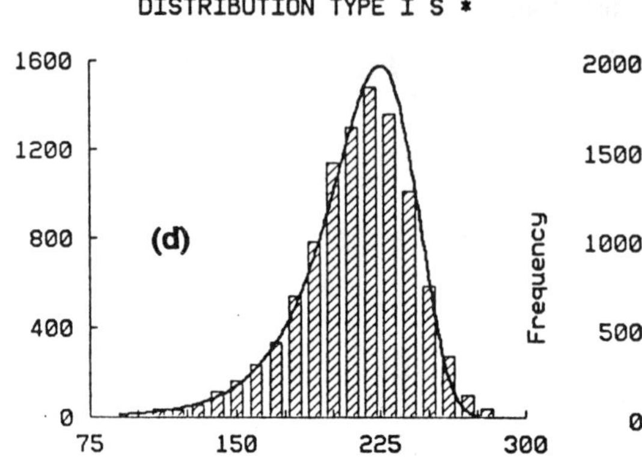

DISTRIBUTION TYPE II L

$$f(N_f) = \frac{4.04}{1082}\left(\frac{1082}{N_f}\right)^{5.04}\exp\left[-\left(\frac{1082}{N_f}\right)^{4.04}\right]$$

Tensile Strain at bottom of AC
micorstrain

Remaining Life, Fatigue Cracking
Number of loads, Thousands

* $f(\varepsilon_t) = 0.05\ \exp[(0.05(\varepsilon_t - 220)]\ \exp[-\exp(0.05(\varepsilon_t - 220))]$

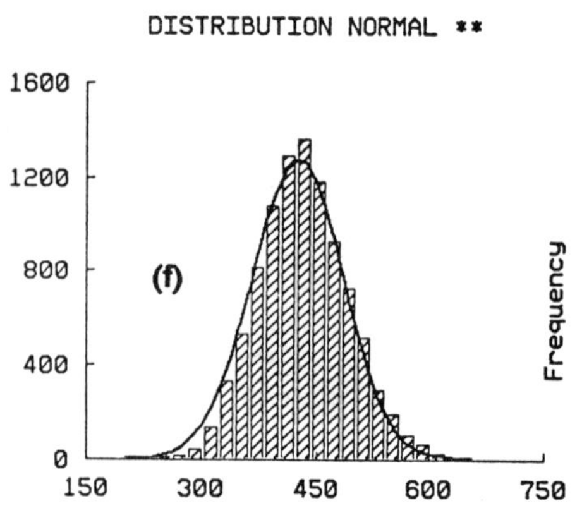

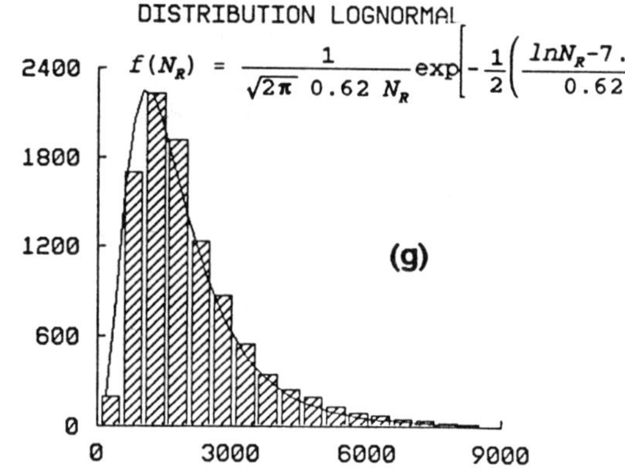

DISTRIBUTION LOGNORMAL

$$f(N_R) = \frac{1}{\sqrt{2\pi}\ 0.62\ N_R}\exp\left[-\frac{1}{2}\left(\frac{\ln N_R - 7.44}{0.62}\right)^2\right]$$

Comp. Strain at Top of SG
microstrain

Remaining Life, Rutting Criteria
Number of ESALs, Thousands

** $f(\varepsilon_v) = \dfrac{1}{\sqrt{2\pi}\ 58.4}\exp\left[-\frac{1}{2}\left(\frac{\varepsilon_v - 431}{58.4}\right)^2\right]$

Figure 2 (Cont'd). Distributions of Computed Variables

Table 3-Mean Values of Response Variables

Computed Variables	Pavement 5-12	Pavement 3-12	Pavement 5-6	Pavement 3-6
E_1, MPa	3346 (2800)**	3710 (2800)	3423 (2800)	3752 (2800)
E_2, MPa	217 (210)	209 (210)	214 (210)	200 (210)
E_3, MPa	70.7 (70)	70.7 (70)	71.4 (70)	70.7 (70)
ε_t, microstrain	209 (218)	285 (298)	230 (242)	320 (331)
ε_v, microstrain	432 (431)	598 (602)	579 (611)	873 (913)
N_f ESALs, thousands	1287 (1067)	477 (378)	988 (749)	309 (255)
N_R ESALs, thousands	2061 (1637)	470 (369)	514 (344)	79 (57)

**Numbers in the parentheses are deterministic "actual" values

Table 4-Coefficients of Variation of Response Variables

Computed Variables	Pavement 5-12	Pavement 3-12	Pavement 5-6	Pavement 3-6
E_1	0.559	0.626	0.611	0.651
E_2	0.387	0.668	0.747	0.520
E_3	0.063	0.062	0.072	0.088
ε_t	0.121	0.147	0.167	0.200
ε_v	0.135	0.129	0.108	0.109
N_f	0.580	0.800	0.690	0.830
N_R	0.680	0.668	0.578	0.579

Probabilistic Analysis

To quantify how the uncertainty in the pavement parameters influences the predicted performance of the pavement, a scalar quantity that measures the performance of a pavement under random loads is needed. The probability of failure of the pavement design was chosen as this scalar quantity. In general, the probability of failure of a pavement is proportional to the overlap area between the distribution of traffic and the distribution of the remaining life (see Figure 3a). A failure event of the pavement occurs when the remaining life of the pavement, N is less than the applied number of ESALs, L. Accordingly, the probability of failure is the probability that N-L<0. The probability of failure of each pavement was computed under the basic assumption that the distribution of the remaining life of the pavement estimated on the day the NDT is carried out is the same through out the life of the pavement, (even though the remaining life of the pavement changes with time). Conversely, the distribution of the traffic was considered to change with time. The mean value of the traffic was assumed to change every year by a certain percentage growth. Coefficients of variation of the yearly traffic distributions were assumed to remain constant over time. This assumption had to be made because no data were available to predict the changes in the COV for future years.

Probabilities of failure were computed using two different methods. The probability of failure due to rutting was calculated as (Ang and Tang 1984b):

$$P_f = \Phi\left(\frac{\ln \bar{N} - \ln \bar{L}}{\sqrt{v_N^2 + v_L^2}}\right) \quad \ldots \ldots \ldots \ldots \quad (3)$$

where $\bar{N}$ and $\bar{L}$ are the mean values and v_N and v_L are the COV of the remaining life and the traffic data, respectively and, $\Phi()$ is the standard normal cumulative distribution. This equation gives accurate results when both the distribution of the traffic and the distribution of remaining life are statistically independent lognormal variables (Ang and Tang 1984b).

A more general method proposed by Rackwitz and Fiessler (1978) was used to compute the probabilities of failure of the pavement due to fatigue. This method allows to handle non-normal distributions very efficiently.

Using the yearly traffic distributions derived from the data of Table 2 and the results of the stochastic analysis, the probabilities of failure for the four pavements considered were computed. The probabilities of failure were also computed with the deterministic values of the remaining life as shown in Table 3. A comparative study between these probabilities of failure was conducted, in order to assess the influence of the uncertainty in the pavement parameters on the estimation of remaining life. Figures 4 and 5 show how variability in the pavement parameters influences the probability of failure for rutting and fatigue, respectively. In the figures, probabilities of failure for the four pavement sections considered using all four types of traffic distributions are shown. The traffic distributions are derived for a 5% growth rate in the traffic. In the figures, the curves labeled without variability represent the probabilities of failure computed using a deterministic value of the remaining life. In the case of rutting, a significant difference between the curves labeled with variability and without variability is noticed. For example, in Figure 4a the difference in the predicted life of the pavement between the curves marked with variability and without variability for US traffic at the 20% probability of failure is approximately 4 years. For SH traffic, the same difference at the same level of probability is about 6 years. This shows how, accounting for the variability in the pavement parameters significantly influences the estimated remaining life of the pavement. However, in the case of IH traffic, the curves cross each other after 2 years, showing an inverse effect of the variability. This happens when the mean value of the traffic distribution exceeds the mean value of the remaining life distribution, an undesirable solution since the failure of the pavement is practically guaranteed. These contradictory results are explained graphically in Figure 3b, when the mean value of traffic distribution exceeds the mean value of the remaining life. The overlap area of the traffic and the remaining life distributions (cross hatched area) is less than the area to the right of the deterministic value, thus the probability computed with variability in the parameters is less than the probability computed without variability. Intuitively, when the mean value of the traffic exceeds the mean value of the remaining life, the pavement may be considered failed.

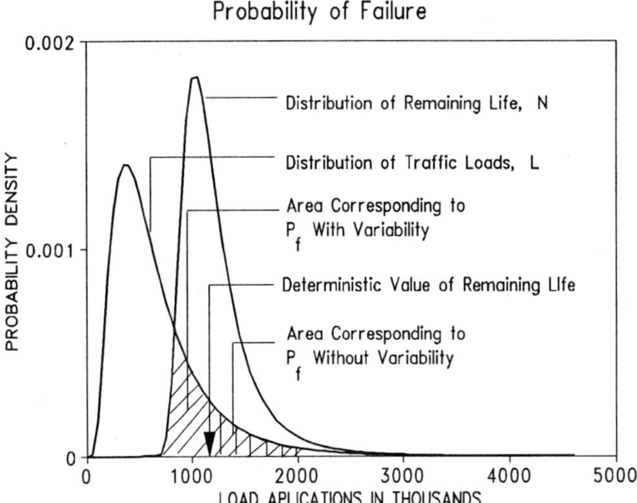

(a)   Probability of Failure When Mean Traffic Less Than Mean
Remaining Life

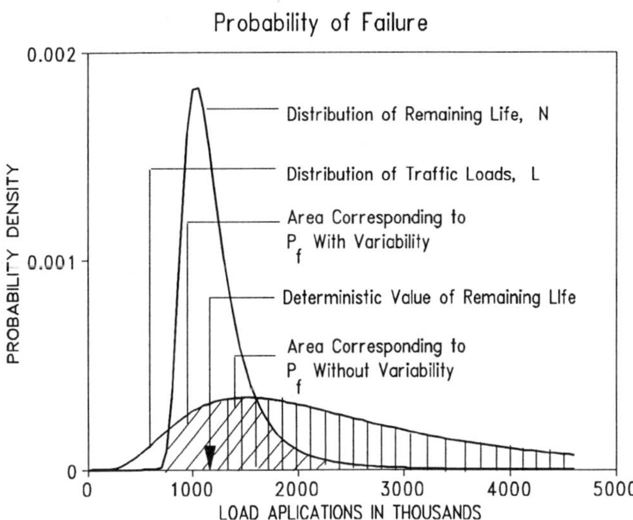

b)   Probability of Failure When Mean Traffic Exceeds Mean Remaining
Life

Figure 3.   Graphical Representation of Probability of Failure
Distributions.

□ without varibility x with variability

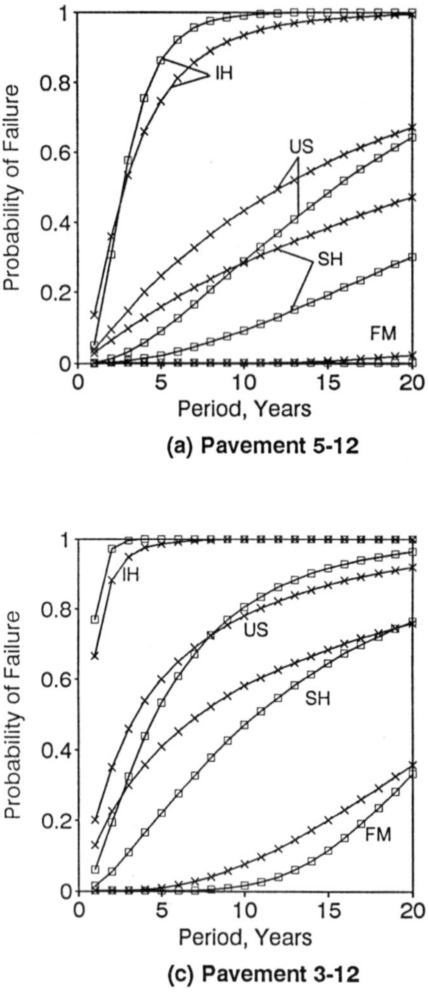

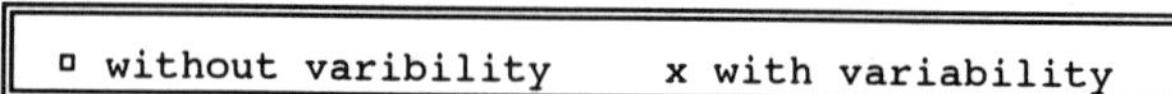

Figure 4. Influence of Variability in Remaining Life (Rutting)

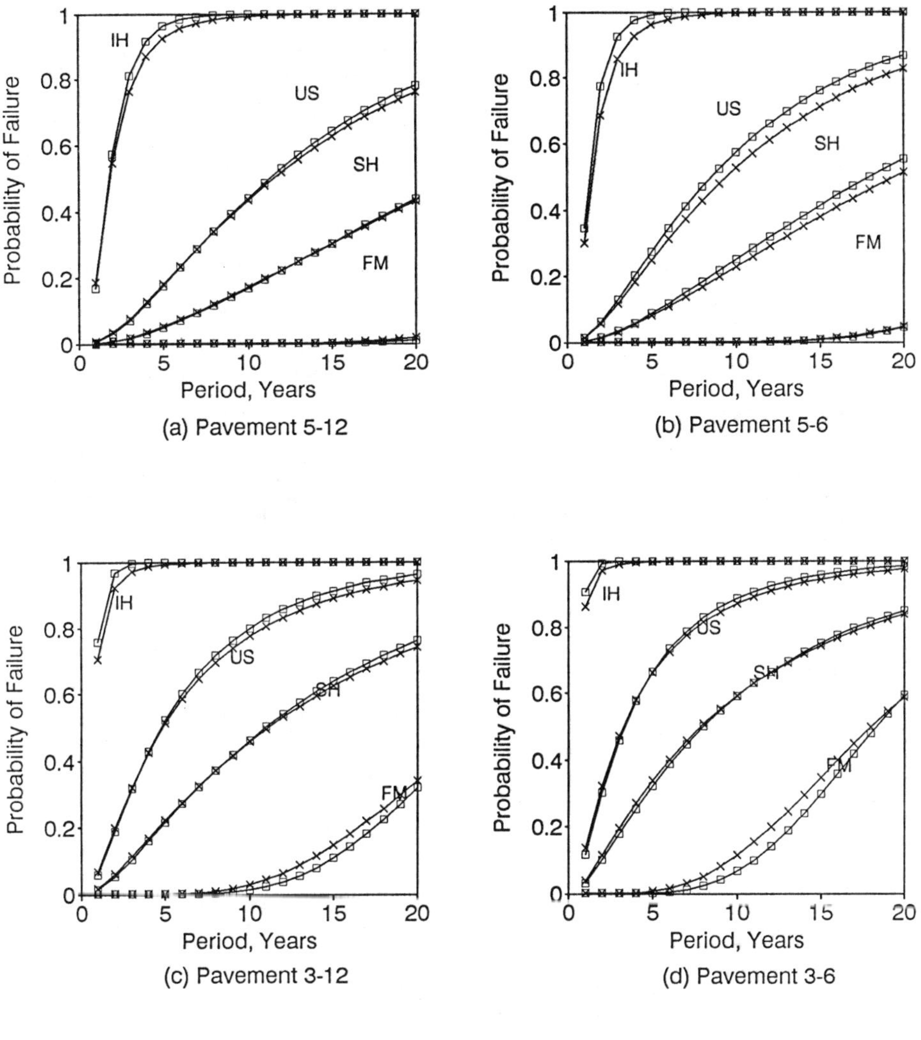

Figure 5. Influence of Variability in Remaining Life (Fatigue)

Beyond this point the results are of no practical value.

For the remaining life computed using the fatigue failure criteria, the difference in the probabilities of failure of the pavement is negligible (see Figure 5). This suggests that the probability of occurrence of this failure mode is mostly influenced by the distribution of traffic loads.

<u>Sensitivity Analysis</u>

To study the influence of each individual pavement parameter on the remaining life, a sensitivity analysis was conducted. For this analysis the pavement remaining life was simulated considering only one pavement parameter as a random variable at a time, while all the other pavement parameters remained constant at their mean values (see Table 1). The coefficients of variation and the mean values for each pavement parameter were described earlier. Cumulative frequency distributions of the remaining life were developed for both fatigue and rutting of pavement 5-12. The results are shown in Figure 6. Each curve in the figures represents the cumulative distribution of the remaining life when a particular parameter is considered a random variable. These curves are compared with the cumulative distribution of the remaining life when all the pavement parameters are considered random variables. Layer thicknesses of the flexible pavement are the most influencing parameters, since most of the variability is contributed by these parameters. For example, in Figure 6a the cumulative curve of thickness of layer two is closest to the cumulative distribution of the remaining life due to rutting. A possible explanation is that the thickness of layer two has a significant effect on the stiffness of that layer. Therefore, the strain at the top of subgrade layer is also effected. So, the properties of this layer dominate the critical strain and the remaining life. Similarly in Figure 6b, the thickness of layer one dominates the variability in the remaining life due to fatigue cracking. Variability on the thickness of layer one, load and deflections of FWD, and the Poisson's ratio of layer three also influence the variability of the remaining life in a moderate way. The influence of the Poisson's ratio of layer one and layer two may be negligible in both failure modes.

CONCLUSIONS

This paper shows that the variability in the pavement parameters has a significant effect on the predicted remaining life of a pavement. The variabilities in the pavement parameters should be taken into account if more realistic values of the remaining life of a pavement are desired. The proposed methodology is a step towards the achievement of this goal.

It has been shown that uncertainty on the remaining life due to fatigue cracking of the flexible pavement may be modelled using an extreme type II distribution and the uncertainty in the remaining life due rutting may be modelled with a lognormal distribution. These distributions of the remaining life have long been assumed normal by the paving engineers during the pavement rehabilitation/maintenance.

The study also shows that when the remaining life of the pavement due to rutting is estimated considering all the variabilities in pavement parameters, the probability of failure of the pavement is larger. However, this is not the case for fatigue failure criteria for which the variability of the pavement parameters did not influence the probability of failure.

The sensitivity analyses show that the thickness of layer two is

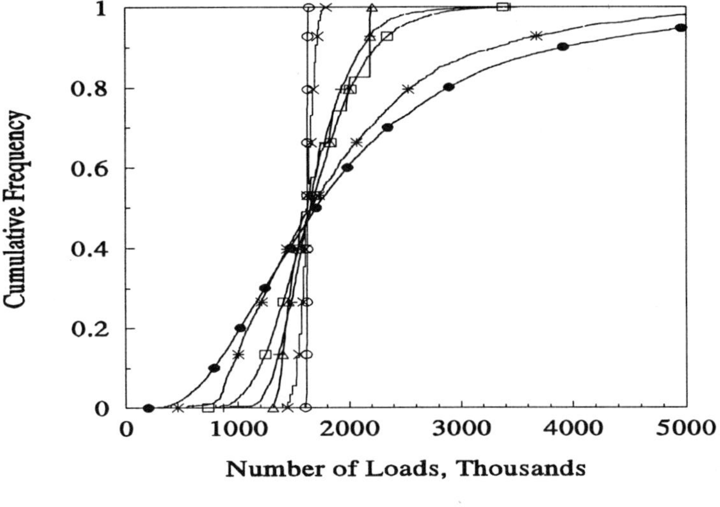

(a) Rutting

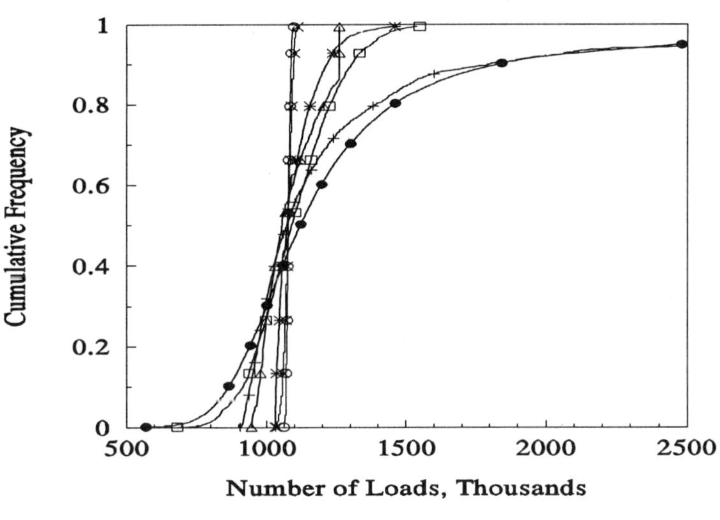

(b) Fatigue

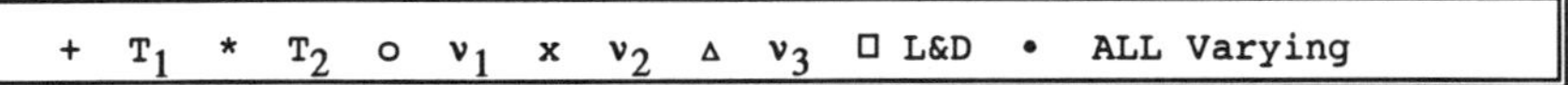

Figure 6. Influence of Pavement Parameters on Remaining Life

what influences the most on the predicted remaining life of the pavement. The thickness of layer one, the load and deflections of the FWD and Poisson's ratio of layer three also influence the variability of the remaining life of the pavement although in a moderate way.

As in probabilistic studies of this nature, the need for better statistical information for the input parameters was evident. The authors hope that this paper will encourage researchers to start data gathering studies.

REFERENCES

[1]   Ang, A.H-S., and Tang, W.H., 1984a, _Probability Concepts in Engineering Planning and Design - Basic Principles,_ Vol. I, John Wiley and Sons, New York.

[2]   Ang, A.H-S., and Tang, W.H., 1984b, _Probability Concepts in Engineering Planning and Design - Decision, Risk, and Reliability,_ Vol. II, John Wiley and Sons, New York.

[3]   Bentsen, R.A., Nazarian, S., and Harrison, J.A., 1989, "Reliability Testing of Seven Nondestructive Pavement Testing Devices," _STP 1026_, American Society for Testing and Materials, Philadelphia, pp 278-290.

[4]   Bush III, A.J. and Alexander, D.R., 1985, "Computer Program BISDEF", U.S. Army Corps of Engineers WES, Vicksburg, MS, November.

[5]   Dejong, D.L., Peutz, M.G.F. and Korswagen, A.R., 1973, "Computer Program BISAR," External Report, Koninklijkel/Shell-laboratorium, Amsterdam, Netherlands.

[6]   Finn, F., Saraf, C., Kulkarini, R., Nair, K., Smith, W. and Addullah, A., 1977, "The Use of Distress Prediction Subsystems for the Design of Pavement Structures", _Proceedings,_ Fourth International Conference on the Structural Design of Asphalt Pavements, pp 3-37.

[7]   Hudson, W.R., Elkins, G.E., Uddin, W., and Reilley, K.T., 1986,"Pavement Condition Monitoring Methods and Equipment Phase I -Part I Evaluation of Deflection Measuring Equipment, "_Report No FH67/1_, ARE Inc., Austin Tx., pp. 113.

[8]   Rackwitz, R., and Fiessler, B., 1978, "Structural Reliability Under Combined Random Load Sequence", Computer and Structures, Pergamom Press, Vol. 9, pp. 489-494.

[9]   Rahut J. B. and Jordahl, P. R., 1993, "Variability in Measured Deflections and Backcalculated Moduli for the SHRP Southern Region", Transportation Research Record No. 1337, TRB, pp 45-56.

[10]   Rodriguez-Gomez, J., Ferregut, C., Nazarian, S., 1991, "Impact of Variability in Pavement Parameters on Backcalculated Moduli", Road and Airport Pavement Response Monitoring Systems, Special Publications, American Society of Civil Engineers, Edited by Vincent C. Janoo and Robert A. Eaton, Mar., 1991, pp. 261-275.

[11]   Shook, J.F., Finn, F.N., Witczak, M.W. and Monismith, C.L., 1982, "Thickness Design of Asphalt Pavements - The Asphalt Institute Method", _Proceedings,_ Fifth International Conference on the Structural Design of Asphalt Pavements.

Philippe Lepert[1], Philippe Caprioli[2]

INTERPRETATION OF DYNAMIC SURVEY MEASUREMENT ON PAVEMENT WITH TREATED ROADBASE

REFERENCE: Lepert, P., and Caprioli, P., "Interpretation of Dynamic Survey Measurement on Pavement with Treated Roadbase," _Nondestructive Testing of Pavement and Backcalculation of Moduli, (Second Volume),ASTM STP 1198_, Harold L. Von Quintas, Albert J. Bush, III and Gilbert Y. Baladi, Eds., American Society for Testing and Materials, Philadelphia, 1994.

ABSTRACT : For over 50 years, the structural condition of a conventionnal pavements has been characterized by the bearing capacity, measured through the deflection produced by a rolling or impact load. However, modern pavements exhibit specific defects which have small or even no effect on deflection. To overcome this difficulty, two types of dynamic survey methods have been developed, the so-called "dispersive methods", and the "impedance method". The first group involves single-, two- and multi-station dispersive methods. Theoretical and experimental studies show that, although the multi-station dispersion method seems the most powerfull one, it requires considerable resources, and thus it is, today, inadequate for routine analysis of pavements. The two-station method provides, when pavement layers are correctly bonded, a simple but promising approach for determining the mechanical parameters of the pavements. The impedance method, which was recently developed by the french "Laboratoires des Ponts et Chaussées (LPCs)", provides a reliable means to assess the state of pavements interlayers, and thus complete the two-station dispersive method. Therefore, the LPCs developed a new piece of equipement, the COLIBRI system, which applies both methods on a routine basis.

KEYWORDS : Pavement survey, investigation, dynamic measurement, treated roadbase, waves propagation, layered media

INTRODUCTION

For over 50 years, the parameter considered to best characterize the structural condition of a pavement has been its bearing capacity, measured through the deflection produced by a rolling or impact load on its surface. Different apparatus have been developed for performing this measurement: Benkelman beam, Lacroix Deflectograph (Lacroix 1963;

---

[1] : Engineer, Laboratoire Central des Ponts et Chaussées, 44340 Bouguenais, France

[2] : Ph.D. Student, Laboratoire Central des Ponts et Chaussées, 44340 Bouguenais, France

Autret 1972), Falling Weight Deflectometer (Sorensen et al. 1982, Ali et al. 1987), etc. Various studies (Sauteret and Autret 1977; Kennedy 1987) have made it possible to link this bearing capacity to the structural condition of the pavement and then to its maintenance requirements or to its residual service life. These surveys were concerned essentially with conventional pavements, i.e. with an untreated base covered with a bituminous layer of greater or lesser thickness.

It has been over 25 years since industrial countries undertook to modernize their major highway networks, using thick courses of materials treated with hydrocarbon or hydraulic binders to build or overlay pavements. Structural analysis methods have attempted to adapt with varying degrees of success to these new conditions. Thus, to take into account the greater rigidity of modern pavements, deflection measurement apparatus have been refined (De Boissoudy et al. 1984). However, these pavements exhibit certain specific defects having no effect on deflection (unbonded wearing course, for example), so new investigation methods had to be developed to identify them. Original techniques were devised, such as Ovalisation (Kobisch and Peyronne 1979). Those which employed dynamic surveying (Jones 1968; Gramsammer et al. 1983) are certainly among the most promising. They are based upon two main theories: firstly, the theory of surface wave propagation in layered structures, involving dispersive methods, and the theory of stationary dynamic phenomena, involving the impedance method (Caprioli 1991; Bats-Villard 1991; Lepert et al. 1992).

Today we have more perspective with regard to all these developments. However, though yet to be finalized, research conducted by the french "Laboratoires des Ponts et Chaussées" (LPCs) (Lepert et al. 1992) shows that dynamic surveying methods reveal the most serious defects of structures with treated bases: loss of modulus in treated courses (Caprioli 1991), separation of wearing course (Bats-Villard 1991), deterioration of load transfer conditions at shrinkage cracks in hydraulic binder treated bases. This research has led to the development of a multi-function dynamic surveying apparatus: the COLIBRI system.

DISPERSIVE METHODS

Mechanical waves propagating on the surface of a stratified medium have the particular feature of being dispersed: the phase velocity of these waves varies with their frequency. More precisely, the propagation of these waves takes place according to natural modes, each mode having its own dispersion curve. These dispersion curves are distinct below a few kHz, but tend to intermingle beyond those frequencies. Knowing their characteristics theoretically makes it possible, based upon certain assumptions, to determine the characteristics of the medium through which the waves propagate (Caprioli 1991).

The surveying of pavements by the analysis of surface wave dispersion was introduced towards 1950. (Jones 1968). The proposed method consisted in measuring the mechanical wavelengths, frequency by frequency, using a Goodman vibrator. This entailed a long setup time and interpretation difficulty. Followed by other authors in France in the 1970s (Frémond 1972, Avramesco and Guillemin 1974), the development of this technique was then oriented towards investigation methods offering higher performance but also greater complexity, in particular by Nazarian (Nazarian 1984) in the United States. In 1991, Caprioli (Caprioli 1991) conducted a detailed theoretical and experimental survey of the different dispersive methods: one requiring a single receiving station, one using a complete network of receiving stations and, between these two extremes, one method requiring two stations. Let us note that all the methods analyzed by Caprioli are based upon the measurement and analysis of the response of a pavement to a shock.

"Single-Station" Method

The principle of the single-station method is simple: a shock applied to the surface of the pavement generates a wave train which moves away from the point of impact. The velocity of the different components of these waves differs according to their frequency (dispersion phenomenon). These components will consequently be gradually separated: an accelerometer placed at a sufficient distance from the point of impact will then record first the passage of the fastest components and then that of the slowest. A time-frequency analysis - consisting basically in calculating the energy contained in the acceleration signal at different frequencies, and at different instants after the shock - should theoretically allow the establishment of the "group velocity / frequency" relation characterizing the propagation of the energy carried by the waves along the surface of the pavement. Figure 1 illustrates this method in an academic case: the study of the group velocity of the Love wave from a synthetic seismogram calculated on a theoretical structure (in this case, a half-space under a "softer" layer).

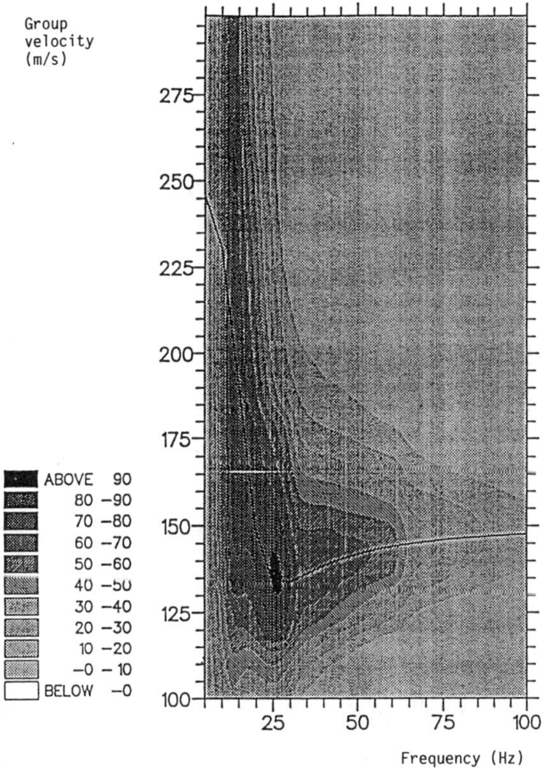

FIG. 1 -- "Single-station" method - group velocity of the Love's wave derived from a synthetic seismogram calculated on a theoretical structure (a half-space + a "softer" layer).

The main drawback of this method is that the receiving station (the accelerometer) must be located quite far from the point of impact so that the waves have time to separate sufficiently. But a pavement with a treated base is rarely homogeneous and continuous over distances greater than a few metres. Moreover, even if such were the case, many modes will have been attenuated, and will have even completely disappeared before the signal reaches the receiving station. Finally, it is much more difficult to identify the parameters of a pavement structure from the group speed curve than from the phase speed curve. For all these reasons, Caprioli did not go any further in the development of dispersive methods of this type.

"Multi-Station" Method

In this method, several tens of receiving stations are placed at the surface of the pavements along a line going through the point of impact. The shock generates a wave train whose passage is recorded by the different accelerometers. A double Fourier transform, in time and in space, makes it possible to obtain a diagram of the distribution of the energy transported by the waves in the "wavelength/frequency inverse" plane (cf. Fig. 2). This diagram clearly separates the first two natural modes (longitudinal and bending) of the waves in the structure.

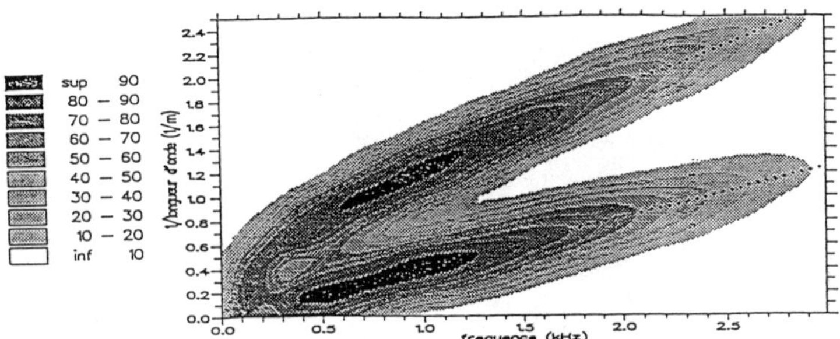

FIG. 2 -- "Multi-station" method - diagram of the distribution of the energy vehiculed by surface waves in a pavement, showing the separation of the longitudinal and bending natural modes of the pavement.

This method is of particular interest. It is in fact known that a wideband frequency analysis increases the amount of information on the traversed medium only provided that it is possible to differentiate the dispersion curves from the modes contained in the recorded signals. Unfortunately, Caprioli shows that this separation inevitably requires considerable resources: to prevent aliasing as a result of Fourier transformation in the space domain, the spacing of the receiving stations (space sampling) should be less than half of the smallest wavelength contained in the recorded signals; to obtain a resolution allowing the separation of the two propagation modes of the waves, the length of the station network must be sufficient; finally, Caprioli is led to propose, for the most current cases, a network of 20 receiving stations at 20 cm spacing. Even if we follow Cara (Cara 1978) who proposes an approach enabling the doubling of the space between sensors without any loss of information, the resources required are still significant. In addition, the analysis of the energy diagrams in the "frequency / wavelength inverse" plane requires powerful computation facilities. Finally, pavements with treated bases exhibit discontinuities (base edges, cracks) which are quite close to each other and are not taken into account by the theory.

Without totally abandoning this approach, the LCPC consequently preferred to orient its efforts towards the development of the "two-station" method.

"Two-Station" Method

This method is derived directly from the method applied by the "high rate vibrator" (Gramsammer and Guillemin 1972). Changing from a harmonic source to an impulse source allows significant progress in test rates and in the resolution of dispersion curves. A shock at the surface of the pavement generates a wave train which propagates along this surface. Two accelerometers record the passage of these waves at points aligned with the point of impact. The first point is located 0.30 m from the point of impact, to comply with the surface wave development condition, the second about 1.10 m away, so that the dimensions of the test device are compatible with routine use. The analysis of the test consists first of all in calculating the complex transfer function between the two accelerograms measured, $a_1(t)$ and $a_2(t)$, defined as :

$$G(\omega) = \rho(\omega) . e^{j\Phi(\omega)} = S_{12}(\omega) / S_{11}(\omega) \qquad [1]$$

in which $S_{12}(\omega)$ represents the cross spectrum of the signals $a_1(t)$ and $a_2(t)$, and $S_{11}(\omega)$ the power spectrum of the signal $a_1(t)$. In practice, the test is repeated 3 to 5 times and the function $G(\omega)$ is calculated from the average values of these two spectra. The coherence function between the two accelerograms is also calculated, from the averaged spectra, according to the formula:

$$H(\omega) = |S_{12}(\omega)|^2 / [S_{11}(\omega) . S_{22}(\omega)] \qquad [2]$$

where $|S_{12}(\omega)|$ represents the modulus of the cross spectrum $S_{12}(\omega)$. It will be recalled that the real coherence function is a transfer function validity indicator because it specifies the degree of causality existing between the two accelerograms. Its value varies between 0 (the two accelerograms appear independent; they are in fact very noisy) and 1 (the two accelerograms result from the same mechanical phenomenon which dominates the measurement noise). The phase velocity of the waves is deduced directly from the phase of the transfer function:

$$c(\omega) = d.\omega / (\Phi(\omega) + 2k\pi) \qquad [3]$$

where d represents the distance between the points P1 and P2. By plotting $c(\omega)$ as a function of the wavelength $l(\omega) = 2\pi.c(\omega) / \omega$, we obtain the sought dispersion curve (cf. example in Figure 3). The determination of the number k is an important problem and the finer the resolution with which the function $c(\omega)$ is calculated, the more reliable will the solution of the problem be (Gramsammer and Guillemin 1972). However, present spectral analysis facilities make it possible to carry out, in a few milliseconds, the calculation of $c(\omega)$ over a very wide frequency band and with a very fine resolution. In practice, the device developed by the LCPC allows analysis within the band from 300 to 5,000 Hz with a 10 Hz frequency resolution. Note that another solution for determining the number k consists in installing a third receiving station between the first two (Caprioli 1991).

This dispersion curve leads to a simple but interesting interpretation. The structure of the pavement is generally assimilated with a two-layer plate, which is an acceptable hypothesis in the case of a pavement with a treated base resting on an untreated subgrade. The first bending mode of this two-layers arrangement determines largely the dispersion curve measured by the "two-station" method, within the considered frequency range. The first longitudinal mode can introduce certain disturbances, in particular towards the higher frequencies; however, Guillemin (Guillemin 1974) has proposed a method for dissociating these two modes. The interpretation of the dispersion curve takes place using an identification technique based upon either the use of charts or upon numerical method of the "least squares" type (Gramsamer and Guillemin 1972). If the thicknesses of the layers are known, it provides a first estimate of the moduli of these layers. Presented in the form of a longitudinal profile, this information may facilitate the analysis of the structural condition of the pavement. A computer code is under development at the LCPC, which will apply this procedure. It should however be emphasized that this interpretation of the dispersion curves is valid only if the pavement layers are correctly bonded.

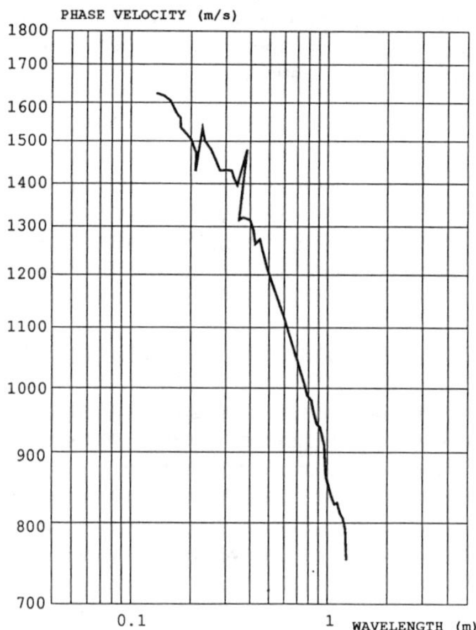

FIG. 3 -- "Two-station" method : dispersion curve measured on a single layer plate - this dispersion curve was obtained on section T2 of the experimental pavement of figure n° 5.

IMPEDANCE METHOD

Principle

The impedance method was proposed by Bats-Villard (Bats-Villard 1991). The theory of stationary mechanical vibrations defines the dynamic impedance $I(\omega)$ of a structure as the ratio between the amplitude of a sinusoidal force F(t) applied to this structure and the amplitude of the sinusoidal movement A(t) resulting therefrom, measured at the same point and in the same direction:

$$I(\omega) = A(\omega) / F(\omega) \qquad [4]$$

It is easily shown that $I(\omega)$ is the transfer function (as defined by [1]) between a point shock $F(t)$ and the response of the structure at the same point, $A(t)$. By analogy, the impedance of a pavement is measured by applying a shock $F(t)$ to it by means of a hammer, instrumented by a load cell. The response $A(t)$ is measured by a piezoelectric accelerometer located within 10 cm of the point of impact (cf. Fig. 4). The impedance is equal to the transfer function between the response $A(t)$ and the excitation $F(t)$, i.e.:

$$I(\omega) = S_{AF}(\omega) / S_{FF}(\omega) \qquad [5]$$

where $S_{AF}(\omega)$ represents the modulus of the cross spectrum of the signals $A(t)$ and $F(t)$, and $S_{FF}(\omega)$ the power spectrum of the signal $F(t)$. Here also, the test is repeated 3 to 5 times and the function $I(\omega)$ is calculated over the average values of these two spectra. The calculation of the coherence function (cf. § II.2) allows the validation of the impedance function.

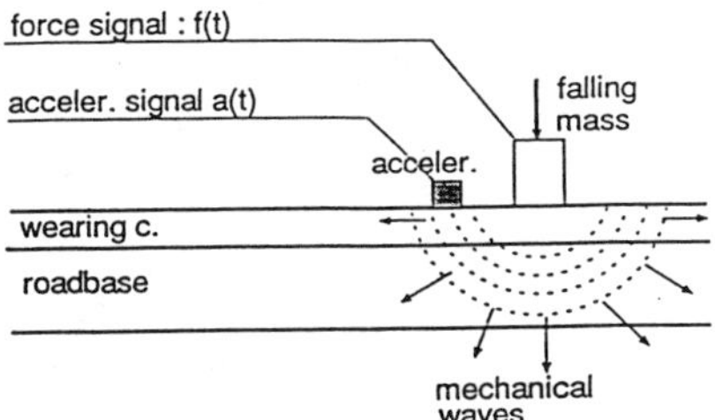

$$Imp(f) = \frac{Fourier\ Transf.\ [a(t)]}{Fourier\ Transf.\ [f(t)]}$$

FIG. 4 -- Principle of impedance method.

<u>Experimental Validation</u>

<u>Basic study and experiment</u> -- A numerical study (Bats-Villard 1972) showed that, below 300 Hz, the modulus of this function is sensitive only to global deteriorations of the roadbase. On the other side, over 2,500 Hz, it is influenced by all the variations, even small, of the geomechanical parameters of the structure. Finally, between 1,000 Hz and 2,000 Hz, the impedance is especially sensitive to the condition of the interface. These theoretical results received a first experimental confirmation on a semi-rigid pavement consisting of a bituminous wearing course over a cement treated base (6 AC / 25 CTB). This pavement involves four sections 100 metres long (Fig. 5), the wearing course of which is :

* bonded (sections T1, T4) onto the base course,
* simply separated (section T3) from the base course,
* or separated with the insertion of a polyane film (section T2).

Figure 6 enables us to compare a series of impedance functions measured on section T4 (dashed lines) with a series of functions measured on section T3 (solid lines). If, below 300 Hz, the two families of curves coincide, on the other hand, between 1,000 and 2,500 Hz, they are clearly differentiated owing to the difference in interface condition.

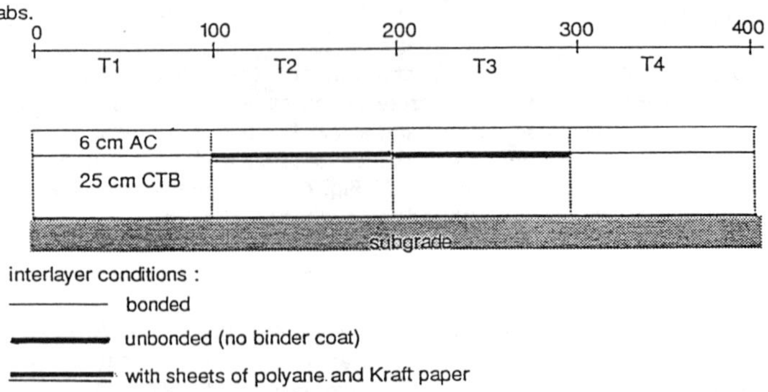

FIG. 5 -- Experimental semi-rigid section (along the A11 motorway, western France)

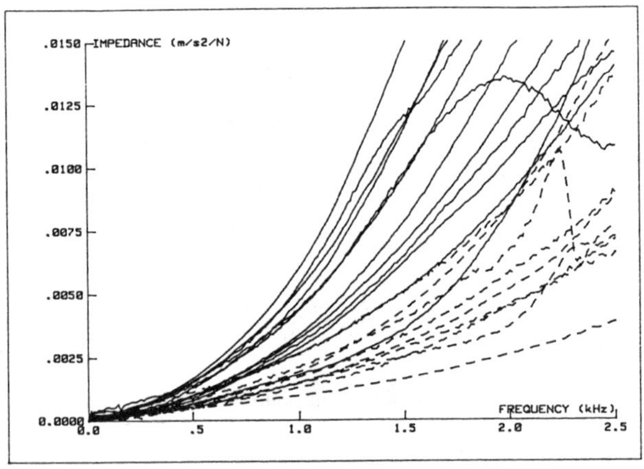

FIG. 6 -- Impedance curves measured on sections T3 (solid lines) and T4 (dashed lines) of the experimental pavement of figure n° 5.

Interpretation of the impedance curve around 3 00 Hz -- The statement that, belov or around 300 Hz, the modulus of this function is sensitive only to global deteriorations o the roadbase, was supported by several experiments. As an example, figure 7a shows th impedance values measured at 300 Hz on a new all bitumen pavement : the values ar almost constant. On the contrary, the value of the impedance function measured at th same frequency on an old, deeply and densely cracked pavement are given on figure 7b They are much more scattered. It should be emphasize on the fact that this last pavemen was just covered by a thin wearing course and no damage was visible. Also, th deflectograph brought no pertinent indication of damages on this section, as the slabs wer resting on a firm subgrade.

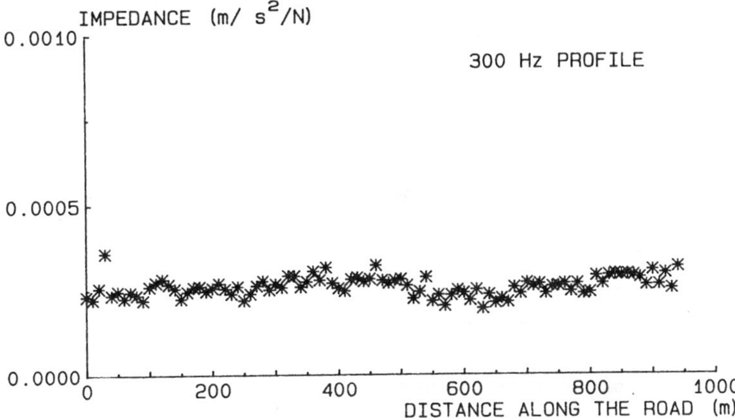

FIG. 7a -- amplitude at 300 Hz of the impedance function measured at different locations along a new all bitumen pavement (A83 motorway, western France)

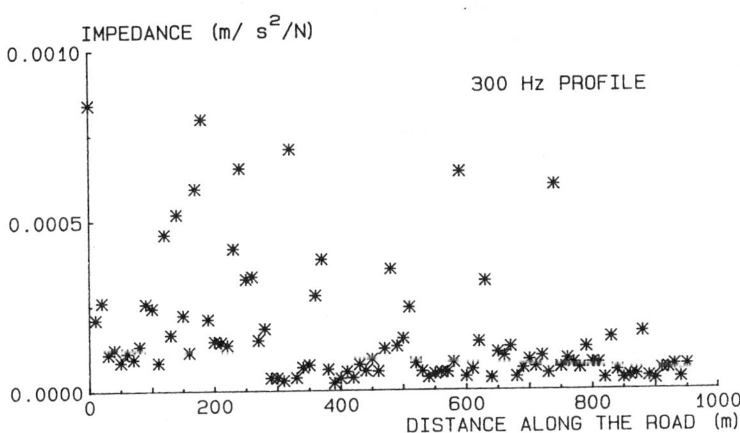

FIG. 7b -- amplitude at 300 Hz of the impedance function measured at different locations along an old densely cracked pavement (A62 motorway, southern France)

<u>Interpretation of the impedance curve between 1,000 and 2,000 Hz</u> -- It was also stated that, between 1,000 Hz and 2,000 Hz, the impedance is especially sensitive to the condition of the interface. A second experimental confirmation of this statement was obtained in April 1990 on a roadway specially constructed to study the detection of interface problems (Guillemin 1974). This pavement, on the emergency stopping shoulder of a motorway, includes six sections (Figure 8) representing three bituminous structures and, for each, two or three possible interface conditions:

* completed according to standard rules of practice,
* completed without a binder coat,
* completed without a binder coat and with an intermediate sand bed.

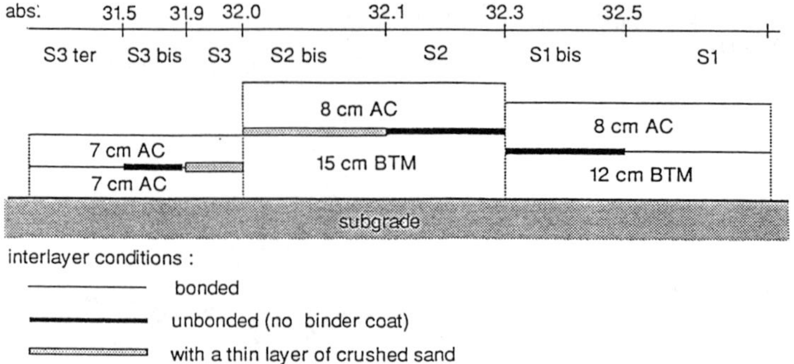

FIG. 8 -- Experimental all-bitumen section (shoulder of the A38 motorway, eastern France)

Let us point out that core samples taken during the tests demonstrated that, on the sections completed without a binder coat but without an intermediate sand bed, the interface was nevertheless bonded, which is understandable considering that the pavement did not experience any traffic. Figure 9 gives the impedance values measured at 1,500 Hz on different sections of this pavement. Sections S1 and S1 bis give the same values: this is understandable if we consider that, on these two sections, the interface was bonded.

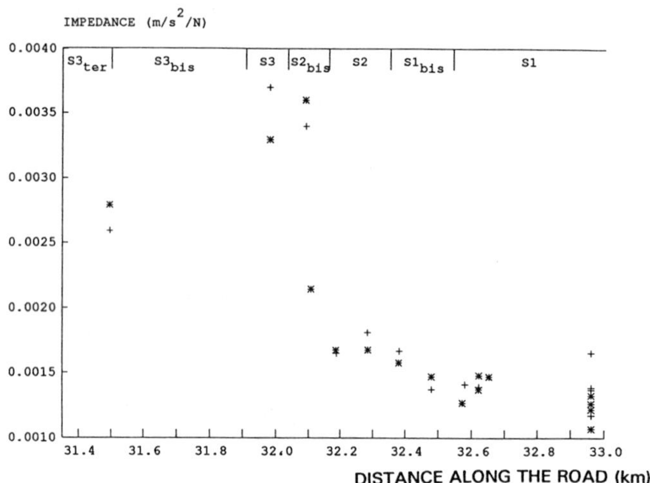

FIG. 9 -- Amplitude at 1,500 Hz of the impedance functions measured at different locations along the experimental sections of figure n° 8.

On the other hand, section S2, without a binder coat, but nevertheless bonded, exhibits a response very different from that of layer S2 bis in which the separation was obtained by a thin sand layer. This observation is also valid for sections S3 and S3 bis, one separated by a sand layer, the other without a binder coat but bonded as shown by the core samples.

<u>Practical Application</u>

From this research, the LCPC derived a new surveying method for pavements with treated bases: from impedance measurements, two longitudinal profiles of the pavement are made - one at 300 Hz, the other at 1,500 Hz. Non-uniformity in the 300 Hz profile indicates serious degradations concerning the base of the pavement (for example, partial or total break-up of the base). If the 300 Hz profile is relatively uniform, the variations of the 1,500 Hz profile indicate specific interface problems. As an example, Figure 10 shows the longitudinal profile of the impedance at 1,500 Hz measured on the semi-rigid pavement of Figure 5. Each section is characterized by a certain scattering of the measurements around an average value characteristic of the section. These values are in perfect agreement with the condition of the interface: on T1 and T4, these values are identical and low; on T3, the average is higher, showing an interface problem more or less serious depending on the position in the section; same observation on T2 where the problem is much more significant. Finally, it was possible to verify, by reproducing the measurements exactly at the same points within intervals of several months, that the scattering of the values was repetitive and hence representative of a heterogeneity of the interface condition.

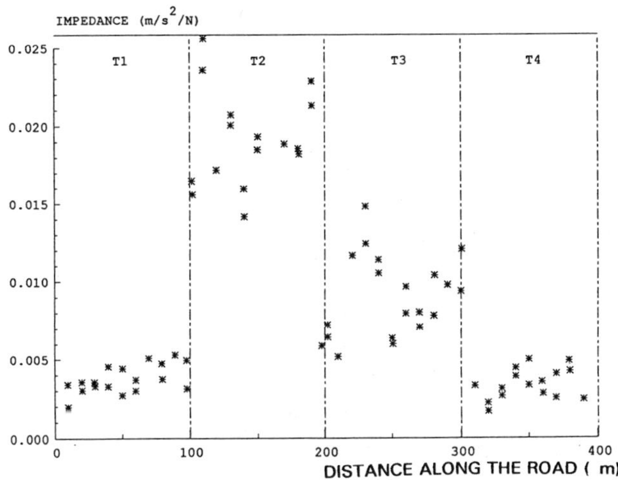

FIG. 10 -- Amplitude at 1,500 Hz of the impedance functions measured at different locations (every ten metres) along the experimental sections of figure n° 5

THE COLIBRI DYNAMIC SURVEYING SYSTEM

The analysis of dispersion curves makes it possible to produce a longitudinal profile of the estimated moduli of the wearing course and of the base of a pavement, under the assumption that this pavement is assimilable with a two-layer structure whose layers are correctly bonded. The analysis of the impedance curves at 300 and 1,500 Hz allows the characterization of the condition of the interface. It was thus tempting to combine these two measurements, especially as they can be carried out simultaneously by the same apparatus. This combination gave birth to the COLIBRI system, which is in the form of a lightweight trailer (cf. Fig. 11) supporting:

- a lightweight mechanical impactor capable of delivering 3 blows in 10 seconds,
- three accelerometers suitably coupled to the pavement.

The accelerometers measure the response of the pavement 10 (P1), 30 (P2) and 110 cm (P3) from the point of impact (P0). The characteristics of the impactor (1 kg) enable the system to cover a frequency band ranging from 300 to 4,000 Hz. Thanks to these arrangements, COLIBRI measures simultaneously the impedance of the pavement between P0 and P1 and its dispersion curve between P2 and P3. The measurements are repeated every 10 metres so as to provide a fairly dense longitudinal profile of the measured parameters : presently, impedance at 300 Hz, I_3, impedance at 1,500 Hz, $I15$; in the future, estimated moduli of the wearing course, G_r, and of the road base, G_a. As previously said, the measurement rate, presently 1 to 2 points/minute, should reach 6 points/minute thanks to ongoing technological developments.

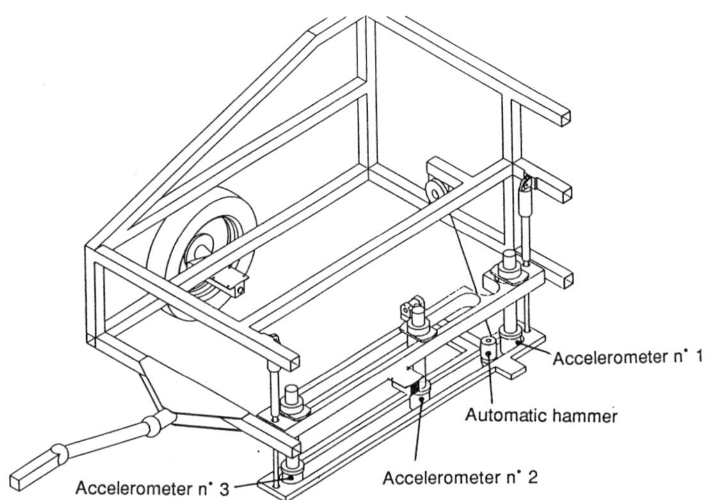

FIG. 11 -- Sketch of the COLIBRI testing equipment showing the measurement beams.

The operating method is currently being finalized. Its principle is however already well defined: it begins with the observation of the longitudinal profile of the 300-Hz impedance of the pavement, I_3. This profile shows the zones in which the base exhibits total or partial break-up, zones in which the analysis of the dispersion curves is of no great interest (which is not troublesome because the problem is identified by the simple observation of the impedance curve). The profile at 300 Hz can also reflect the presence of insufficiently rigid zones (incorrectly designed or poorly compacted base materials), this information then being supported or detailed by the profile of the estimated modulus of the road base, G_a. In other zones, the observation of the longitudinal profile of the impedance at 1,500 Hz allows the detection and location of zones in which the interface is defective. In these zones, the processing of the dispersion curves is generally impossible. On the other hand, in sound zones, the analysis of the dispersion curves is possible and permits the determination of the actual characteristics of the pavement.

CONCLUSIONS

On pavements with treated bases, evaluating the structural condition from the bearing capacity of a pavement alone is very difficult, in particular because the deflections measured on these pavements become very small and the deflection basin very long. Dynamic surveying offers a possible alternative for overcoming this difficulty. By combining the measurement of the dispersion curve and that of the impedance curve, the COLIBRI system represents a precious aid for the diagnosis of the structural condition of pavements with treated bases: the identification of interface problems, problems of rigidity and cohesion of treated layers, falls within its field of application.

REFERENCES

Ali, N.A., Khosla, N.P., January 1987, "Determination of layer-moduli using Falling Wheight Deflectometer, Report presented at the Transportation Board Record Meeting

Autret, P., July 1972, "Evolution du Déflectographe Lacroix : pourquoi ?", Bulletin de Liaison des Laboratoires des Ponts et Chaussées, n° 60, pp 11-17

Bats-Villard, M., January 199, "Influence des défauts de liaison sur le dimensionnement et le comportement des chaussées", Thèse de doctorat, Université de Nantes

Cara M., 1978, Thèse d'état, Université Pierre et Marie Curie, Paris VI.

Caprioli, Ph., June 1991, "L'auscultation structurale des sols et des chaussées routières à partir de la propagation d'ondes mécaniques totalement ou partiellement guidées", Thèse de doctorat, Université de Strasbourg

De Boissoudy, A., Gramsammer, J.C., Keryell, P., Paillard, M., January 1984, "Le Deflectographe 04", Bulletin de Liaison des Laboratoires des Ponts et Chaussées, n° 129, pp 81-98

Frémond, M., 1972, " Etudes des chaussées soumises à des charges vibrantes", Bulletin de Liaison des Laboratoires Routiers, Spécial S, pp 74-84

Gramsammer, J.C., Kerzrého, J.P., Petitgrand J.C., July 1983, "Le collographe", Bulletin de Liaison des Laboratoires des Ponts et Chaussées, n° 126, pp 77-90

Gramsammer, J.C., Guillemin, R., September 1972, "Dynamic non destructive testing of pavement in France", Third International Conference on the Structural Design of Asphalt Pavements, London

Guillemin, R., July 1974, "Auscultation au vibreure léger : programme d'exploitation automatique des courbes de phase", Bulletin de Liaison des Laboratoires des Ponts et Chaussées, n° 72, pp 31-35

Jones, R., July 1968, "Historique de l'auscultation par ondes superficielles", Bulletin de Liaison des Laboratoires des Routiers, Spécial J, pp 24-35

Kennedy, 1987, "Structural investigation of road for the design of strengthening", TRRL research report, n° 189

Kobich, R., Peyronne, C., July 1979, "L'ovalisation : une nouvelle méthode de mesure des déformations élastiques des chaussées", <u>Bulletin de Liaison des Laboratoires des Ponts et Chaussées</u>, n° 102, pp 59-71

Lacroix, J., Septembre 1963, "Déflectographe pour l'auscultation rapide des chaussées", <u>Bulletin de Liaison des Laboratoires des Ponts et Chaussées</u>, n° 3, pp 191.1-191.6, Septembre 1963

Lepert, Ph., Poilane, J.P., Bats-Villard, M., August 1992, " Evaluation of various field measurement techniques for the assessment of pavement interface condition", <u>Seventh International Conference on Asphalt Pavement</u>, pp 224-233, Nottingham

Nazarian, S., 1984, "Non destructive testing of pavements, using surface waves", <u>Transportation Research Record</u> 993, TR3, NRC, Washington DC, pp 67-79

Sorensen, A., Hayven, M., June1982, "The Dynatest Falling Wheight Deflectometer Test System", <u>International Symposium on Bearing Capacity of Roads and Airfields</u>, Tronheim, Norway, pp 464-470

Sauteret, R., Autret, P., 1977, "Guide pour l'auscultation des chaussées souples", <u>Eyrolles</u>, Paris

Peter E. Sebaaly[1] and Srikanth Holikatti[2]

PHASE LAG EFFECTS ON ANALYSIS OF FWD DATA

REFERENCE: Sebaaly, P. E., and Holikatti, S., **"Phase LAG Effects on Analysis of FWD Data,"** _Nondestructive Testing of Pavements and Backcalculation of Moduli (Second Volume)_, ASTM STP 1198, Harold Van Quintas, Albert J. Bush, III, and Gilbert Y. Baladi, Eds., American Society for Testing and Materials, Philadelphia, 1994.

ABSTRACT: The research documented in this paper examines the effect the phase lag has on the analysis of FWD data. The research examined the use of a deflection basin generated from the peak sensors' deflections, a deflection basin generated from instantaneous deflections based on peak load, and a deflection basin generated from instantaneous deflections based on the peak deflection at sensor #1. The influence of the various analyses is evaluated by comparing the material properties produced by each individual analysis and their relative impact on the recommended overlay thickness.

The analysis of the data indicated that almost all of FWD field data have instantaneous deflection basins which greatly differ from the deflection basin generated by the peak response. The backcalculation data indicated that the type of analysis may have a significant effect on the magnitudes of the moduli of the layers. Finally, the overlay analysis showed that there are significant differences among the recommended overlay thicknesses based on the different analyses. The differences were as high as 1.5 inches in some cases.

KEYWORDS: backcalculation, falling weight deflectometer, phase lag, deflection basin, overlay thickness.

[1]Associate Professor, Civil Engineering Department, University of Nevada, Reno, NV 89557

[2]Research Assistant, Civil Engineering Department, University of Nevada, Reno, NV 89557

291

Introduction

Evaluating the in situ properties of pavement requires two major steps: a) field testing of the pavement section, and b) analysis of the test data. The first step, field testing, has seen great progress in the past several years. Measuring the load deflection response of the pavement system has progressed from static, to vibratory, and finally to impulse loading.

Studies have been conducted by Bensten et al. (1989) at U.S. Army Waterways Experiment Station (WES) to evaluate the reliability of seven NDT devices. The findings showed that all the seven NDT devices were capable of acquiring accurate, consistent and reliable NDT data. Currently, the falling weight deflectometer (FWD) is widely used because of its ability to simulate traffic loading. Field measurements have shown close correlations between both the shape and the maximum surface deflection under truck and FWD loading.

On the other hand, step (b), the analysis of the test data, is an area still evolving. Recently several techniques have been developed to analyze the FWD test data. Each technique has its own assumptions and limitations.

One obvious limitation of any technique of data analysis is the inconsistency in simulating the loading source and the resulting response. This is the area of interest to the research presented in this paper. It examines the effect of the assumptions made in the data analysis phase on the final product.

FWD data are used in three different analyses based on both the summary of peak deflections and the full history of the deflection function. The in situ material properties evaluated from each analysis have been compared and the evaluated properties are used in a mechanistic overlay design procedure in order to evaluate their impact on the recommended overlay thickness.

Today numerous microcomputer methods (programs) developed by various agencies are available to pavement engineers. All these programs are capable of evaluating in situ pavement layers moduli from field deflection measurements. In addition, they all have common aspects in their approach to the solution. Most of them use the multilayer elastic solution to generate the theoretical deflection basin. Generally, most of these programs have been validated through extensive use by different agencies. A discussion of the limitations of different programs is beyond the scope of this paper. However, it is sufficient to say that all of them are capable of backcalculating pavement layer moduli from field deflections to an acceptable degree of accuracy (Uzan et al. 1988; Lee 1988; Irwin at al. 1989).

Methods of Analysis

FWD applies an impulse load to the pavement surface,

which generates a response by the pavement system. Research studies have shown that both the inertia of the pavement system and the damping characteristics of the pavement materials would have an impact on the shape of the response function (Sebaaly 1987). Figure 1 shows the deflection responses at five FWD sensors as a result of the impulse load imparted by the FWD device. The deflection curves (Figure 1) clearly show the existence of phase lags among the peaks of the various responses. The magnitude of these phase lags is a function of both pavement inertia and the damping characteristics of the various materials. Therefore, each pavement section would have its own set of five phase lags (that is, one phase lag between each two consecutive sensors, if seven sensors are used there will be seven phase lags including a phase lag between the load and the first deflection). Field measurements indicated that any individual phase lag could range between 0 and 15 msec (Sebaaly et al. 1992).

The backcalculation process uses a theoretical analysis to define the set of layer moduli which reproduce the measured surface deflection basin with the least amount of error. Therefore, the reliability of the backcalculation analysis is anticipated to be highly dependent on the analysis procedure and the deflection basin. The most commonly used backcalculation procedures utilize the multilayer elastic solution. This solution assumes that the material exhibits a perfectly linear elastic behavior. In other words, the deflection basin generated by the multilayer elastic solution assumes that the peaks at all sensors occur at the same time. This assumption contradicts field data which shows the existence of phase lags among the various sensors (Sebaaly et al. 1992).

Using the deflection basin generated from peak sensors deflections violates the primary assumption of the multilayer elastic solution. To avoid this violation, the instantaneous deflection basins taking into consideration the phase lags between consecutive sensors should be used in the backcalculation analysis. In this research the following three deflection basins were analyzed.

Analysis 1: Deflection basin generated from the peak sensors deflections
Analysis 2: Deflection basin generated from instantaneous deflections based on peak load (Figure 2).
Analysis 3: Deflection basin generated from instantaneous deflections based on peak deflection at sensor #1 (Figure 3).

The influence of the various analyses is evaluated by comparing the material properties produced by each individual analysis and their relative impact on the recommended overlay thickness.

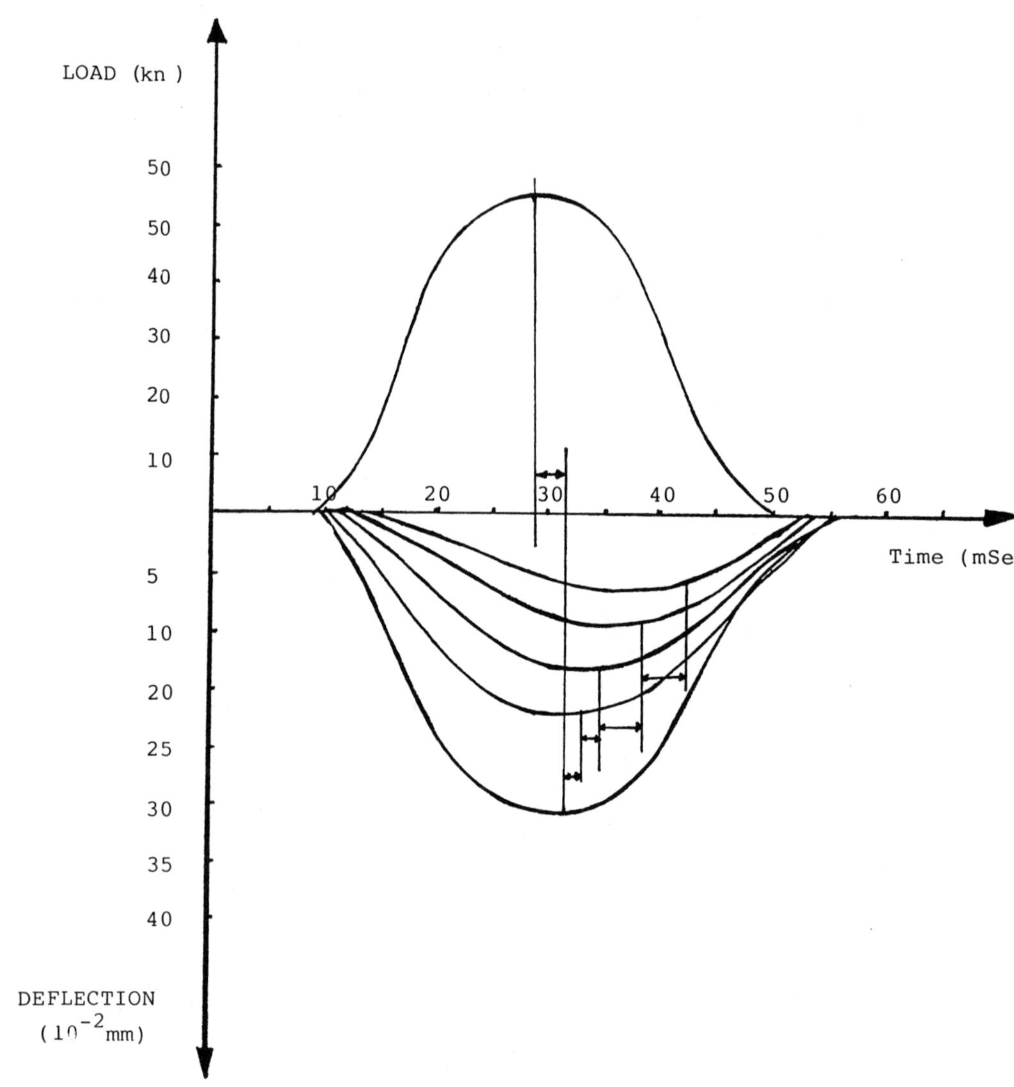

Figure 1.   Typical load and deflection response under FWD
loading

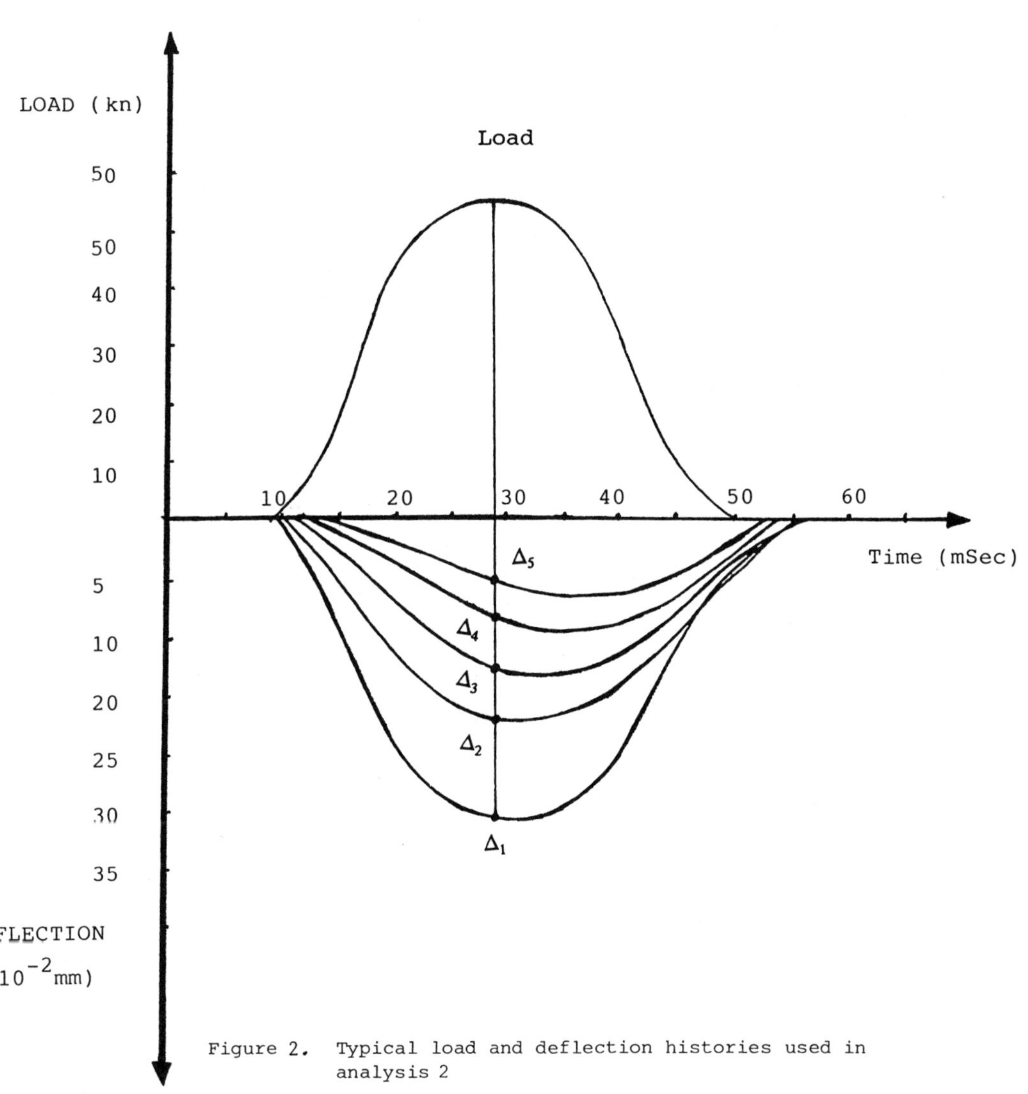

Figure 2.   Typical load and deflection histories used in analysis 2

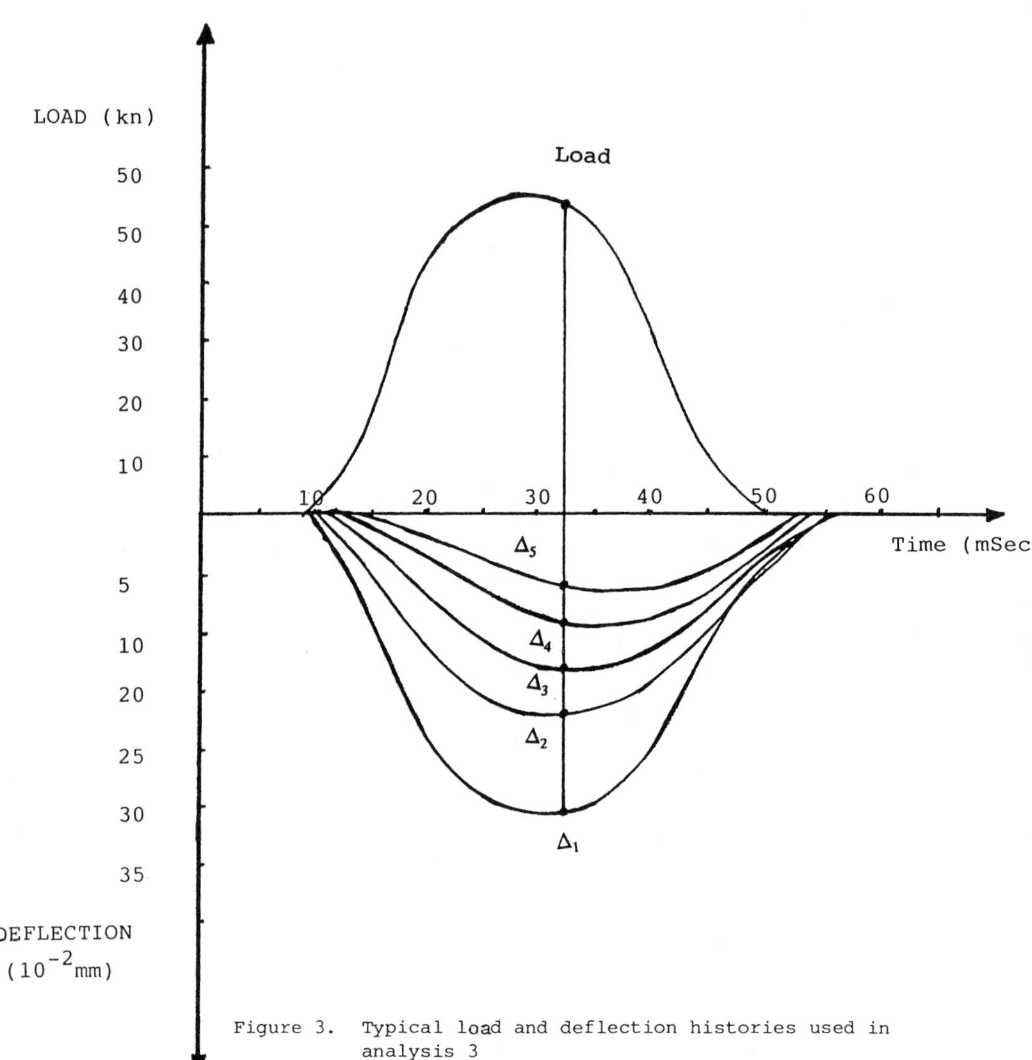

Figure 3.   Typical load and deflection histories used in analysis 3

Pavement Sections

As discussed earlier, the magnitude of the phase lag is a function of both pavement inertia and material damping. Therefore, in order to obtain a wide range of phase lags and deflection basins, several pavement sections were tested and are analyzed in this paper. Table 1 summarizes the structures of the pavement sections tested in this study. The range of surface and base course thicknesses is expected to provide a wide range of deflections basins for the three types of analysis. The Dynatest FWD device was used in the testing of the pavement sections.

Table 1. Structures of the Pavement Sections

Pavement Section	AC Surface (cm)	Base Course (cm)
1	33	15
2	17	28
3	23	25
4	26	18
5	23	28
6	22	18
7	28	17
8	13	22
9	10	23

Backcalculation Method

A growing body of knowledge exists with regard to the backcalculation of layer moduli under FWD loadings. The aim each method is to match measured deflections with those calculated using assumed layer moduli. The layer moduli are appropriately changed in order to minimize the error between measured and calculated deflection basins.

In this paper, the MODULUS backcalculation model has been used (Uzan et al. 1988). It is a generalized backcalculation procedure which uses the 3-point Lagrange interpolation technique along with deflection basins data base generated ahead of the backcalculation analysis. The deflection computation routine used in MODULUS is based on the multilayer elastic solution. Therefore, the backcalculation process is founded on the basic assumption that the deflection basin is an instantaneous one.

As discussed earlier, using the complete history of the

FWD load deflection curves, three different basins are generated. Analysis 1 represents the classical approach which is based on the peak load and peak deflections. Analyses 2 and 3 on the other hand present a new approach which tries to identify the instantaneous deflections basin based on either peak load or peak deflection at the first sensor. Table 2 presents typical deflection basins as generated from the different analyses.

For each analysis (that is, 1, 2, or 3) a deflection basin is generated from the FWD data on the pavement sections described in Table 1. The deflection basins along with the corresponding FWD load and pavement structure properties are provided as input to the MODULUS program. In addition, the following moduli ranges are also provided.

	Maximum (Mpa)	Minimum (Mpa)
Asphalt concrete:	21000	350
Base course:	1400	70
Subgrade:	Most probable value	35

Analysis of Backcalculation Data

As a result of the backcalculation analysis, three sets of layer moduli were generated for each pavement section. The objective of this part of the research was to evaluate the influence of analysis type on both the magnitudes of the in situ properties and their use in overlay design. These objectives were accomplished through two separate evaluations as follows.

Effect of Analysis on In Situ Properties

The FWD testing was conducted at several stations within the pavement section. Therefore each pavement section has a mean, a standard deviation, and a coefficient of variation for each modulus. Figure 4 shows the distribution of coefficients of variation for the three analyses and all pavement sections. In summary, the data indicate that analysis 2 has the lowest coefficients of variation for all pavement sections. Analysis 1 shows the highest coefficients of variation. This indicates that the use of analysis 1 violates the basic assumption of the multilayer elastic solution regarding the instantaneous deflection basin and therefore introduces a large variability into the results.

The second part of this evaluation deals with comparing the layer moduli generated from the various analyses. Figures 5, 6, and 7 compare the layer moduli as evaluated from the different analyses for AC, base and subgrade layers, respectively. It should be noted that pavement sections 1 through 7 represent thick structures while sections 8 and 9 represent thin ones (Table 1). By looking at the AC layer data

Table 2. Typical Deflection Basins Generated from All Three Analyses

Pav. Sec.	Ana. Type	Load (KN)	Radial Distance from Center (cm)						
			0	20	31	46	61	91	152
			Deflections (microns)						
1	1	52	105	94	89	74	66	48	30
	2	53	102	92	87	70	63	44	24
	3	47	106	96	90	75	67	47	27
2	1	40	272	231	207	139	111	59	32
	2	42	262	225	200	131	102	46	10
	3	37	277	233	209	138	109	50	15
6	1	52	388	290	239	127	92	46	29
	2	52.5	382	285	235	123	85	35	9
	3	48	397	294	242	127	88	38	15
8	1	52	419	338	302	173	131	64	36
	2	53.4	438	349	305	178	130	37	4
		48	440	350	307	180	131	42	8

in figure 5, it can be seen that analysis 1 predicted higher
moduli for the thin sections while for the thick sections all
three analyses were relatively close. The base course data,
figure 6, indicate that there is no clear trend except that
analysis 1 predicted higher values for sections 1, 7,and 8. In
the subgrade case, in general all the three analyses predicted
close values except for section 8, where analysis 1 predicted
a higher value, and section 7, where analysis 2 predicted a
higher value.

In general, the evaluation indicates that in the case of
AC and base course layers, analysis 1 has the tendency to
generate relatively higher values than the other two analyses.
In the case of subgrade the three analyses traded places for
the various sections.

Effect of Analysis on Overlay Thickness

The objective of this part of the research is to evaluate
the effect of different analyses on the recommended overlay
thickness.   The overlay analysis is based on a mechanistic
based overlay design procedure recently developed for the
State of Nevada (Sebaaly et al. 1992). Using this procedure,
the required overlay thickness will be evaluated based on in
situ material properties generated from three different
analyses.   Each pavement section will have three different
recommended overlay thicknesses.   The significance of the
analysis type will be evaluated by comparing the recommended
overlay thicknesses for each pavement section.   It is
anticipated that the product of this evaluation will provide
an indication as to what effect the analysis type will have on
the final decision. Since the majority of pavement management
systems use the results of FWD testing to evaluate the
required overlay thickness, this type of evaluation was
considered important.

The Nevada's mechanistic based overlay design procedure is
based on the multilayer elastic solution. It uses the in situ
material properties as evaluated from the backcalculation
analysis of FWD data. Pavement performance is defined as a
function of both fatigue and permanent deformation failures.
The performance analysis is conducted on a seasonal basis and
further combined using the cumulative damage concept. Axle
load, tire spacing, tire pressure, and the location of
critical pavement responses are all variables that can be
changed throughout the analysis (Sebaaly et al. 1992).

The overlay material is assumed to have a modulus of 575
kpa at 21°C.   The backcalculated moduli for the asphalt
concrete layer are adjusted to 21°C using the ASSHTO design
guide (1986) method along with temperature measurements taken
during the FWD testing. The loading vehicle is assumed to
have dual tires spaced at 37 cm, a 689 Kpa tire inflation
pressure, and an axle load of 98 KN.

For each analysis the moduli values for asphalt concrete,

base course, and subgrade layers are imput along with the
depth of subgrade as calculated by MODULUS. Based on each
analysis, an overlay thickness is recommended for each
pavement section. A total of three overlay thicknesses are
recommended for each pavement section.

It is important to note that, the overlay analysis combines
the effect of analysis type on the backcalculated moduli of
all three layers. As mentioned earlier, for each analysis the
corresponding moduli values for each of the layers are used.
Therefore, the final effect of the analysis type on the
overlay thickness is the combined effect of that analysis on
the in situ properties of all the pavement layers. Using this
analogy, two analysis may show the same recommended overlay
thickness eventhough they do not generate the same in situ
material properties. This can only occur if the effect of one
type of analysis on the surface moduli cancels out the effect
of another analysis on the base moduli or vice-versa.   It
should be noted that the recommendations from the overlay
design procedure used in this study may differ from
recommendations by other procedures. However, since the same
procedure is used with the data generated from all three
analyses, the comparisons are valid. In order to perform the
overlay design analysis, design periods of 5 and 10 years were
assumed. Different traffic levels were used for thick and
thin sections in order to obtain realistic overlay thickness
for all the sections. The traffic levels used in the analysis
were 1.5 million and 0.35 million equivalent single axle loads
(ESAL) per year for thick and thin sections, respectively.

Table 3 summarizes the recommended overlay thicknesses for
5 and 10 years design periods. It can be seen from this data
that by varying the type of analysis, the recommended overlay
thickness will vary. It is clearly shown that the effect of
analysis varies from one pavement section to another. This is
expected since the phase lags among the various sensors are
highly dependent on the thickness of layers and their relative
stiffness.

The data indicated that analysis 3 always recommended the
highest overlay thickness for both 5 and 10 year design
periods. In the majority of cases, analysis 2 recommended the
lowest overlay thickness. The differences in the recommended
overlay thicknesses among all the three analyses vary between
0 and 1.5 inches. Considering the construction cost of
overlays, any difference larger than 0.25 inch is considered
significant. The data presented in table 2 indicate that
analysis 2 always provided the lowest deflection basin along
with the highest load level. This is always true unless the
pavement section will have zero inertia and zero damping. In
other words, as long as there are phase lags between the load
and deflections and between the various sensors analysis 2
will represent the most realistic instantaneous basin.

The above evaluation has shown that the type of analysis
has a significant effect on the overlay thickness for 5 and 10
years design periods. Figures 8, 9 and 10, show the
recommended overlay thickness as a function of number of years

Table 3. Summary of the Recommended Overlay Thicknesses

Pavement Section	Analysis Type	Analysis Period	
		5 Years	10 Years
		Overlay Thickness (cm)	
1	1	0	6
	2	0	6
	3	0	6
2	1	15	22
	2	15	22
	3	18	23
3	1	15	22
	2	14	20
	3	15	22
4	1	8	14
	2	6	13
	3	9	15
5	1	19	23
	3	22	23
6	1	23	23
	2	23	23
	3	23	23
7	1	13	18
	2	17	22
	3	17	23
8	1	17	22
	2	15	21
	3	17	22
9	1	10	15
	3	10	15

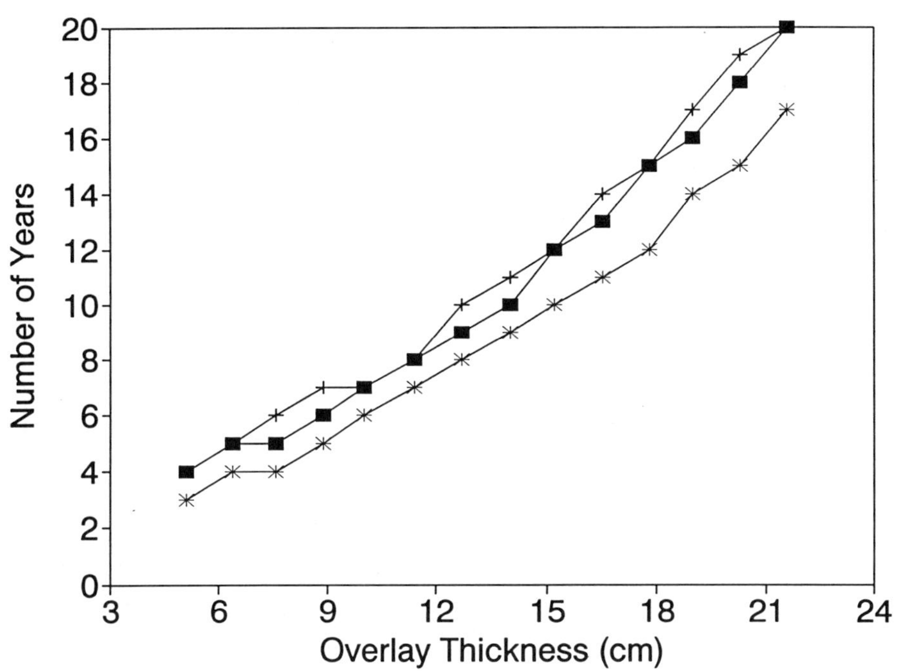

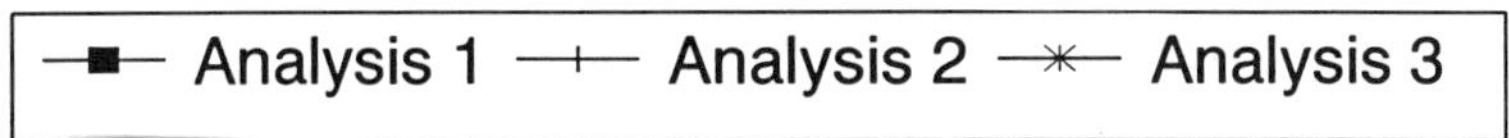

Figure 8.   Overlay thickness as a function of service year for
section 4, 1.5 million ESAL/year

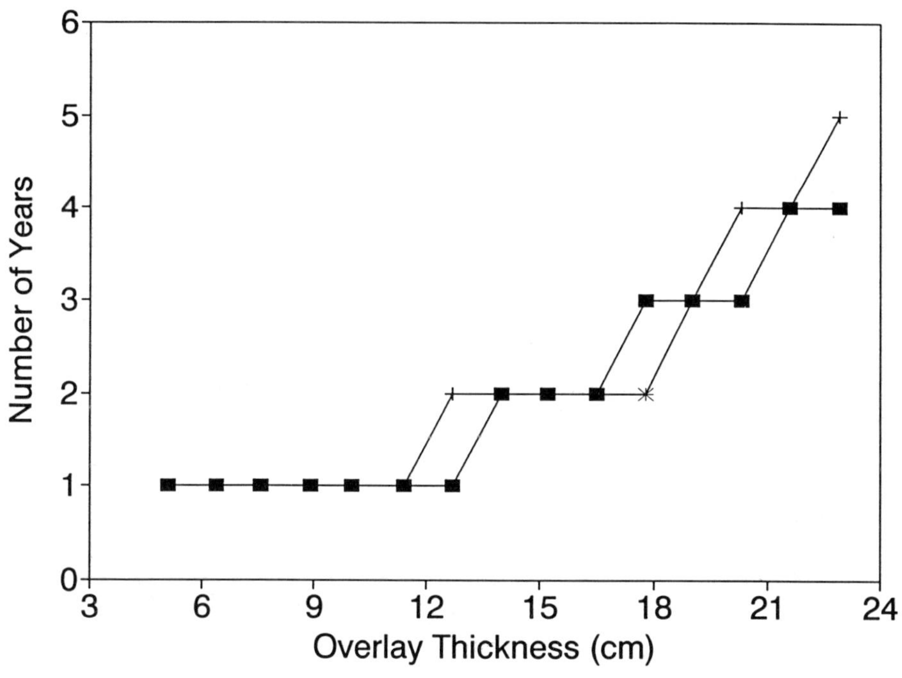

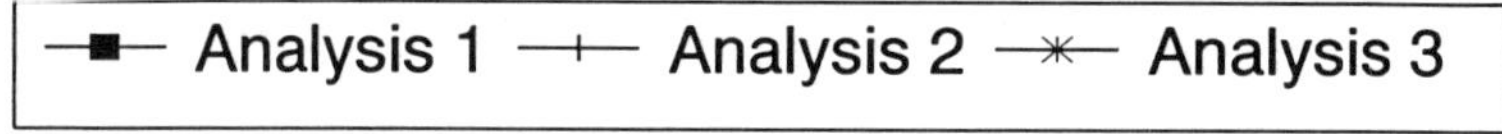

Figure 9. Overlay thickness as a function of service year for section 6, 1.5 million ESAL/year

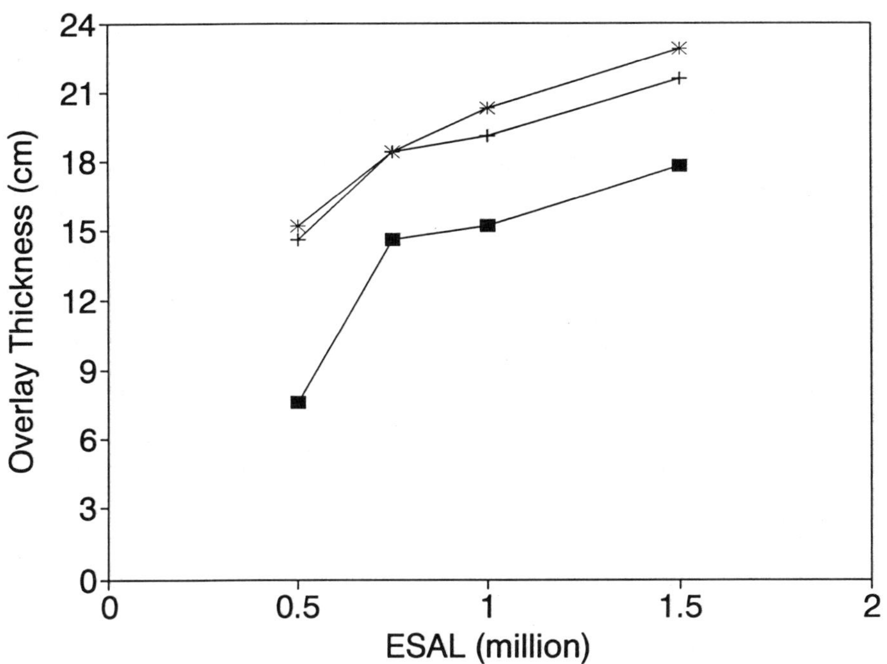

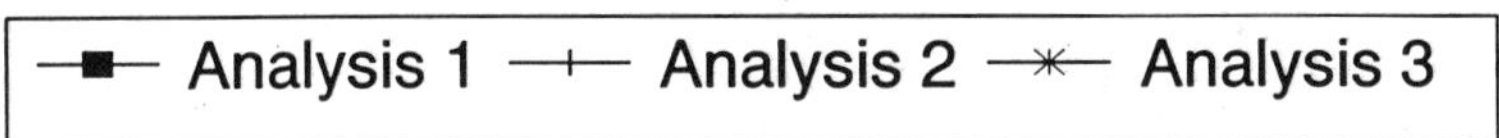

Figure 10.   Effect of analysis type on the thickness of overlay
as a function of ESAL, section 7

for sections 4, 6, and 7, respectively. The data in these figures show that the three different analysis may generate very close recommendations in one case such as section 6 (figure 9), yet in other cases they may generate highly different recommendations. This indicates that it would be unreasonable to assume that one type of analysis is more conservative than the others and should always be used.

Since section 7 showed the highest variability among the various analyses, it was selected for further evaluation. This evaluation consists of investigating the effect of analysis type at various traffic levels. The overlay analysis was conducted for section 7 using all three analyses and traffic levels of 0.5, 0.75, 1.0, and 1.5 million ESALs per year. The objective of the investigation was to see if the effect of analysis type would become more or less significant at different traffic levels. Figure 10 shows the distribution of the recommended overlay thickness as a function of traffic level for all three analyses. The data indicate that the effect of analysis type becomes more significant as the traffic level drops below 1.0 million ESAL per year. The recommended overlay thickness by analysis three is double the thickness recommended by analysis 1. Analyses 2 and 3 generated close recommendations at all traffic levels.

One positive finding of this evaluation is that the recommendations of the various analysis did not switch rankings as a function of traffic level. In other words, analysis 1 always recommended the lowest overlay thickness, while analysis 3 recommended the highest followed closely by analysis 2.

Conclusions

Based on the analysis of the FWD data presented in this paper, the following conclusions may be drawn:

• All FWD data analyzed in this paper indicate the existence of phase lags between the load and deflections and among various deflection sensors. As a result of these phase lags, the instantaneous deflection basins greatly differ from the deflection basin generated by the peak responses. Most significantly, the deflection basins generated from analysis 2, which considers the instantaneous response along with the peak applied load, are consistently lower than the basins generated from analysis 1 and 3.

• The backcalculation data indicate that the type of analysis may have a significant effect on the magnitudes of layers moduli. However, there was no constant trend in which any specific analysis provided moduli values that are always higher or lower than the others for all pavement sections. An important finding of this investigation is that analysis 2 generated the lowest coefficients of variations for all pavement layers. This indicates that analysis 2 is the most stable one, because it represents the actual instantaneous response of the pavement system. In addition, analysis 1 may be introducing more variability into the data by assuming that the peak deflections basin represent the instantaneous response of the system. To evaluate the influence of analysis type by comparing the layer moduli values was a little difficult, since the effect of the analysis is present on all layers and on the evaluated depth of subgrade. Therefore, the investigation was extended to include the determination of the overlay thickness.

• The overlay analysis showed that there are significant differences among the recommended overlay thicknesses based on the different analyses. The differences among the recommended overlay thicknesses were as high as 3.8 cm in some cases. Analysis 3 always recommended the highest overlay thickness. Analysis 2 recommended the lowest overlay thicknesses for the majority of the cases.

• The evaluation of overlay thickness versus number of years was conducted for three representative sections. This investigation indicated that the relative ranking of various analyses varies from one section to another. This is highly expected since the basins generated from analysis 2 and 3 are totally dependent on the pavement structure and the characteristics of various layers. Therefore, it is not appropriate to assume that one analysis may be more conservative than the other in all the cases.

• The effect of traffic level on the difference among the various analyses was shown to be significant when the number of ESAL drops below 1.0 million per year. The U.S. highway system has a great number of roads that are loaded below 1.0 million annual ESALs. Therefore, the difference among the various analyses should be considered very seriously.

• Finally, the research presented in this paper has shown that the effect of analysis type on the final product (that is, overlay thickness) is significant and can not be ignored. The deflection basin generated by analysis 2 represents the actual instanteneous basin which does not violate the main assumption of the multilayer elastic solution. It is recommended that such analysis should be used in the backcalculation process. Especially, since most of the current FWD devices are capable of providing the required data.

Acknowledgement

The authors would like to thank Steve Seeds from Nichols Consulting Engineers for offering technical advise to this research.

References

1.   American Association of State Highway and Transportation Officials, "ASSHTO Guide for Design of Pavement Structures," AASHTO, Washington, D.C. 1986.

2.   Bentsen, R.A., Nazarian, S., and Harrison, J.A., "Reliability Testing of Seven Nondestructive Pavement Testing Devices," ASTM STP 1026, <u>Nondestructive Testing of Pavements and Backcalculation of Moduli</u>, November 1989.

3.   Irwin, L.H., Yang, W.S., Stubstad, R.N., "Deflection Reading Accuracy and Layer Thickness Accuracy in Backcalculation of Pavement Layer Moduli," ASTM STP 1026, <u>Nondestructive Testing of Pavements and Backcalculation of Moduli</u>, November 1989.

4.   Lee, S.W., "Backcalculation of Pavement Moduli by Use of Pavement Surface Deflections," Ph.D. Dissertation, University of Washington, 1988.

5.   Sebaaly, P.E., "Dynamic Models for Pavement Analysis," Ph.D. Dissertation, Arizona State University, May 1987.

6.   Sebaaly, P.E., Siddharthan, R., and Javaregowda, M., "Evalaution of FWD Data for NDOT Overlay Design Procedure," Research Report No. 410-3, Nevada Department of Transportation, Carson City, NV, June 1992.

7.   Sebaaly, P.E., Siddharthan, R., Javaregowda, M., and Srikantiah, S., "Mechanistic Overlay Design Procedure for the State of Nevada," Research Report No. 410-4, Nevada Department of Transportation, Carson City, NV, June 1992.

8.   Uzan, J., Scullion, T., Michalek, C.H., Parades, M., and Lytton, R.L., "A Microcomputer Based Procedure for Backcalculating Layer Moduli from FWD Data," Texas Transportation Institute Research Report No. 1123-1, July 1988.

James A. Crovetti[1] and Mohamed Y. Shahin[2]

THE EFFECT OF ANNULAR LOAD DISTRIBUTIONS ON THE BACKCALCULATED MODULI OF ASPHALT PAVEMENT LAYERS

REFERENCE: Crovetti, J. A., and Shahin, M. Y., "The Effect of Annular Load Distributions on the Backcalculated Moduli of Asphalt Pavement Layers," _Nondestructive Testing of Pavements and Backcalculation of Moduli (Second Volume), ASTM STP 1198_, Harold L. Von Quintas, Albert J. Bush, III, and Gilbert Y. Baladi, Eds., American Society of Testing and Materials, Philadelphia, 1994.

ABSTRACT: A linear-elastic computer program is used to model uniform and annular loading distributions beneath the area of loading. Load distributions were selected to approximate those obtained using conventional impulse loading equipment on a variety of in-service asphalt pavement surfaces. Maximum surface deflections are computed for a factorial of pavement structures using the ELSYM5 computer program. The maximum deflections calculated under uniform load distributions are used to develop regression equations for the prediction of the asphalt layer modulus for twenty four separate pavement families. These equations are used in conjunction with deflections calculated under annular load distributions to backcalculate asphalt moduli. The errors which are introduced in asphalt layer moduli prediction due to various annular loading conditions are presented.

KEYWORDS: falling weight deflectometer, load distributions, annular loading, backcalculation, asphalt layer moduli.

The proliferation of impulse deflection testing devices in use throughout the world, commonly known as falling weight deflectometers (FWDs), has led to a dramatic increase in the use of collected deflection data as inputs in mechanistic analyses of asphalt pavement systems. Backcalculation of the elastic moduli of the component pavement layers, based on measured surface deflections, is one such usage which has been exploited in the recent past. Backcalculation programs based on

---

[1] Vice President, ERES International, Inc., Savoy, IL; current affiliation: Department of Civil & Environmental Engineering, Marquette University, Milwaukee, WI 53233
[2] Research Engineer, USA-CERL, Champaign, IL 61826

linear elastic theory, such as ELSDEF, BISDEF, WESDEF, MODULUS etc., are commonly used for this purpose. It is important to note that during layer moduli backcalculation, the applied load is commonly assumed to be uniformly distributed over the entire area of loading, representing flexible plate loading conditions.

FWD loading plates are circular in shape, typically have a radius of 15 cm, and fall into two basic categories; segmented and non-segmented. Both categories of loading plates are constructed of a semi-rigid upper portion with one or more ribbed rubber pad(s) attached to the underside which rests on the pavement surface during testing. The semi-rigid upper portion of these load plates may be classified as: 1) non- segmented, solid plate; 2) 2-part segmented, utilizing two semi-circular portions with the axis of segmentation running in the longitudinal direction; and 3) 4-part segmented, utilizing four quarter-circle portions with axes of segmentation running in both the transverse and longitudinal directions. The circular ribbed rubber pad(s) bridge any existing plate segmentation gaps.

It has been suggested that the ribbed rubber pad(s) act to transform the semi-rigid loading plate into a flexible-type loading plate, thus producing uniform pressure distributions under the full area of the load plate. This paper presents the results of a study conducted to:

1.   investigate pressure distributions under two types of FWD load plates,
2.   investigate the effects of variable load pressure distributions on maximum surface deflection, and
3.   determine the magnitude of error which may be introduced into the backcalculation of asphalt layer moduli if annular loading conditions exist.

PAVEMENT LOADING STUDY

Two types of FWD loading plates, a 4-part segmented plate and a solid, non-segmented plate were used to perform routine-type testing on three asphalt pavements located in Central Illinois. Each load plate was mounted on an FWD following manufacturers guidelines. (Note: The 2-part segmented plate was not available at the time of testing.) Pressure sensitive prescale film manufactured by Fuji Film I & I, sensitive in the range from 482 to 2,413 kPa, was placed between the load plates and the pavement surfaces to obtain footprints of the applied pressure distributions. The selected prescale film was available in widths up to 27 cm on continuous rolls, 5 meters long. This maximum width did not allow for complete coverage beneath the 30 cm diameter load plates; therefore, the paper was positioned as shown in Figure 1 to cover the load plate as completely as possible.

At each test location, one impulse load of approximately 62 kN was generated with each FWD to produce applied pressures of approximately 877 kPa, assuming uniform load distribution. Under this applied pressure, microcapsules within the prescale film were broken up and absorbed by a color-developing material yielding a color record of applied pressures with intensity of color proportional to applied pressure.

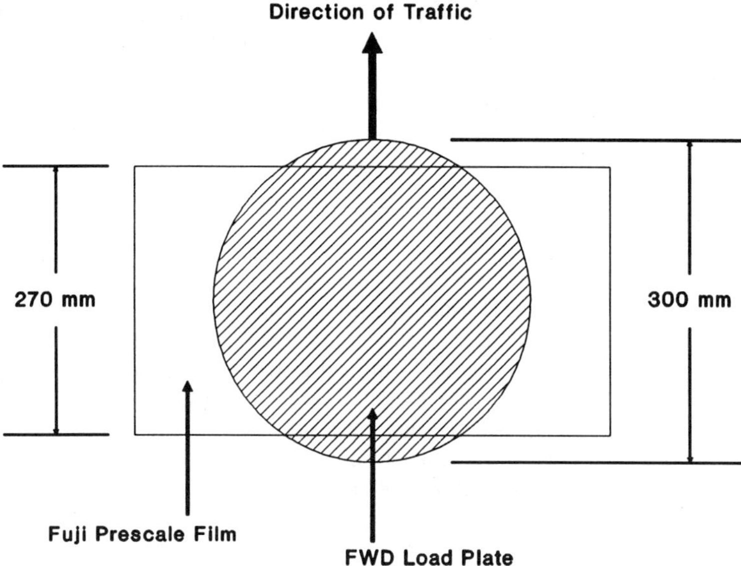

FIGURE 1-- Load plate positioning on prescale film.

Tests were conducted on three different pavements as follows:

1. a relatively strong, recently overlaid asphalt pavement which had received very few traffic loads subsequent to the overlay placement,

2. a relatively strong, moderately trafficked asphalt pavement which had a rut depth of 3 mm, as measured across the load plate, and

3. a relatively weak, lightly trafficked chip-seal pavement which had a flat profile under the loading plate.

Prior to pavement loading, the FWD mounted with the 4-part segmented plate was positioned on the pavement and the load plate was lowered to just above the pavement surface. A spray painted outline of the load plate location was then made. The load plate was raised and a strip of prescale film taped onto the pavement surface as shown in Figure 1. The FWD was then switched to computer operation to produce a single load of approximately 62 kN (877 kPa). The prescale film was then removed and the FWD driven off the pavement surface.

The FWD equipped with the non-segmented loading plate was then positioned on the pavement such that the loading plate would rest as closely as possible within the previously painted load plate outline. A second strip of prescale film was taped to the pavement surface and a single impulse load of approximately 62 kN was produced under computer operation. The prescale film was removed and the second FWD driven off the pavement surface.

Digitized copies of the pressure distributions obtained during the load tests, previously published by Touma et al. [1991], indicated annular load distribution under the non-segmented plate on Pavement 3 (essentially unloaded under central portion with an approximate unloaded radius of 10 cm), non-uniform load distribution under the non-segmented plate on Pavement 2 (essentially unloaded along a central longitudinal strip approximately 85 mm wide), and relatively uniform load distribution under all other load plate/pavement combinations.

COMPUTER ANALYSIS

The pressure distributions recorded during the pavement loading study were modeled by Touma et al. [1991] using multiple circular loadings to cover as much as possible the recorded contact areas. Touma indicated that surface deflections calculated using either type of non-uniform load distribution produced essentially similar deflections and that only maximum deflection was significantly different from the full contact case. This paper presents the results of a second computer analysis, which investigated the effects of annular loadings using the linear elastic computer program ELSYM5 [Kopperman, et al., 1986].

Load Modeling

Annular loadings were modeled using the principles of superposition, as shown in Figure 2. Inner radii, a_1, were varied from 0 cm to 12 cm in 3 cm increments.

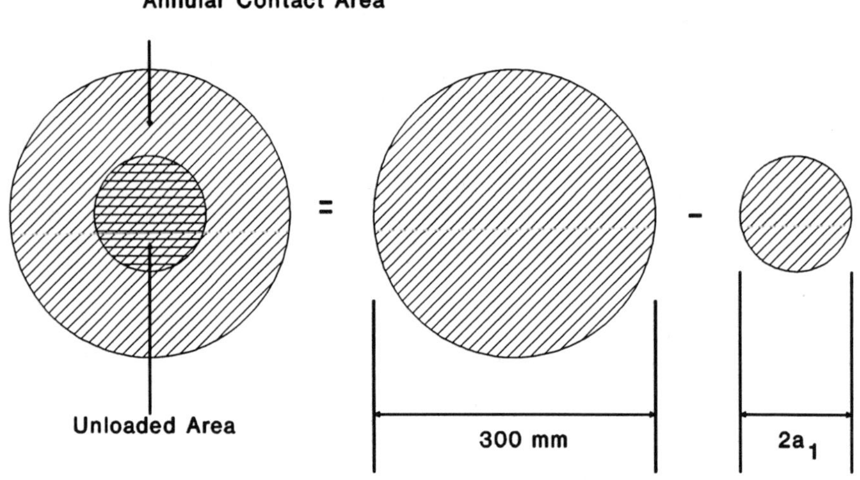

FIGURE 2-- Superposition used for annular load modeling.

The contact area of each annular loading is given by:

$$AREA = \pi\,(15^2 - a_1^2)\tag{1}$$

where: $AREA$ = Contact area of annular loading, cm^2
              a_1 = Inner radius of loading, cm

A total applied load of 40 kN was used for each annular loading. The contact pressure appropriate for each annular loading is given by:

$$\text{Contact Pressure (kPa)} = 400{,}000\,/\,[\pi\,(15^2 - a_1^2)]\tag{2}$$

ELSYM5 input variables used for each annular loading are provided in Table 1.

TABLE 1-- ELSYM5 input values for annular loadings.

Inner Radius, a_1 cm	Annular Contact Area cm^2	Contact Pressure kPa
0	706.9	566
3	678.6	589
6	593.8	674
9	452.4	884
12	254.5	1572

Superposition was used to calculate the maximum surface deflection, D_{max}, resulting from each annular loading by 1) selecting contact pressure appropriate for annular load (Table 1); 2) calculating the maximum deflection, D_1, using selected contact pressure with an applied load radius of 15 cm; 3) calculating the maximum deflection, D_2, using selected contact pressure with an applied load radius = a_1; and 4) by equating $D_{max} = D_1 - D_2$.

Pavement Factorial

All pavements were modeled as three-layer systems with a rigid layer located beneath layer three. Layer descriptions and input variables are as follows:

Layer 1: Asphalt Surface Layer
    Thickness = 5.08, 10.16, 20.32, and 30.48 cm
    Elastic Modulus = 862, 1724, 3448, 5172 and 6895 MPa
    Poisson's Ratio = 0.35

Layer 2: Aggregate Base Layer
   Thickness = 20.32 cm
   Elastic Modulus = 172.4, 344.8 and 517.2 MPa
   Poisson's Ratio = 0.40

Layer 3: Subgrade
   Thickness = 609.6 cm
   Elastic Modulus = 34.5 and 103.4 MPa
   Poisson's Ratio = 0.45

The above factorial represents 120 separate pavement systems which may be grouped into 24 pavement "families", with each family having equal layer thicknesses and base/subgrade moduli but varying asphalt layer moduli.

RESULTS

Maximum surface deflections calculated for each pavement system and annular loading condition are provided in Table 2. The limiting case of $a_1 = 0$ cm represents the control condition where full contact and uniform pressure is obtained under the load plate. Increasing a_1 values represent situations which may arise during routine deflection testing due to:

1.   a weak surface layer (low moduli) which results in "rigid-type" plate loading conditions, and
2.   localized surface depressions in the area of loading.

As shown in Table 2, maximum surface deflections continually decline as the contact area decreases (increased a_1) for all pavement systems investigated.

Using deflections obtained under the control loading condition ($a_1 = 0$ cm), family regression equations were developed for backcalculating asphalt surface layer moduli based on calculated maximum deflection. Family specific equations were selected to improve the goodness of fit. Equations of this type represent the most basic form of backcalculation tool available, in contrast to iterative deflection basin matching techniques or data base search techniques. However, when only one layer modulus (e.g., the asphalt layer) is to be backcalculated, these simple equations provide efficient and reliable results. Each equation is of the general form:

$$\text{Log (Eac)} = A + B\,(D_{max}) + C\,(D_{max})^2 \tag{3}$$

where: Eac = Young's Modulus of the Asphalt Layer, MPa
                  D_{max} = Maximum Surface Deflection, microns
                  A, B, C = Regression Constants (See Table 3)

Table 3 presents the results of this regression analysis. As shown, each equation has an R-squared value greater than 0.999 which indicates excellent correlation.

TABLE 2--Maximum surface deflections calculated with ELSYM5.

Run No.	Pavt Family	Layer Moduli AC MPa	Base MPa	SG MPa	Thickness AC cm	Maximum Deflection, microns Inner Radius a1, cm 0	3	6	9	12
1	1	862	172	34	5.08	1572	1520	1450	1372	1295
2	1	1724	172	34	5.08	1486	1450	1397	1334	1245
3	1	3448	172	34	5.08	1389	1366	1326	1278	1219
4	1	5171	172	34	5.08	1326	1308	1275	1234	1176
5	1	6895	172	34	5.08	1278	1261	1234	1199	1151
6	2	862	345	34	5.08	1224	1180	1133	1087	1036
7	2	1724	345	34	5.08	1156	1124	1085	1044	991
8	2	3448	345	34	5.08	1090	1070	1036	1001	955
9	2	5171	345	34	5.08	1052	1035	1005	975	937
10	2	6895	345	34	5.08	1024	1011	982	955	922
11	3	862	517	34	5.08	1072	1033	993	963	914
12	3	1724	517	34	5.08	1006	979	942	914	881
13	3	3448	517	34	5.08	945	929	894	869	828
14	3	5171	517	34	5.08	912	899	868	846	813
15	3	6895	517	34	5.08	892	879	850	828	800
16	4	862	172	103	5.08	902	855	790	719	650
17	4	1724	172	103	5.08	848	819	768	711	650
18	4	3448	172	103	5.08	782	764	730	686	635
19	4	5171	172	103	5.08	739	724	699	660	617
20	4	6895	172	103	5.08	706	695	671	638	602
21	5	862	345	103	5.08	668	629	587	541	503
22	5	1724	345	103	5.08	632	609	575	538	498
23	5	3448	345	103	5.08	599	581	556	526	490
24	5	5171	345	103	5.08	574	563	543	516	483
25	5	6895	345	103	5.08	559	548	528	505	478
26	6	862	517	103	5.08	572	535	505	472	439
27	6	1724	517	103	5.08	541	519	491	462	434
28	6	3448	517	103	5.08	513	497	477	452	424
29	6	5171	517	103	5.08	495	483	466	445	417
30	6	6895	517	103	5.08	483	473	457	437	411

Note: For all program runs: $\mu_{ac} = 0.35$; $\mu_{base} = 0.40$, base thickness = 20.32 cm; $\mu_{subgrade} = 0.45$, subgrade thickness = 690.6 cm.

TABLE 2--Maximum surface deflections calculated with ELSYM5.

Run	Pavt	Layer Moduli			Thickness	Maximum Deflection, microns				
		AC	Base	SG	AC	Inner Radius a1, cm				
No.	Family	MPa	MPa	MPa	cm	0	3	6	9	12
31	7	862	172	34	10.16	1252	1212	1176	1135	1087
32	7	1724	172	34	10.16	1116	1093	1066	1039	999
33	7	3448	172	34	10.16	983	969	948	930	901
34	7	5171	172	34	10.16	905	894	875	862	838
35	7	6895	172	34	10.16	849	841	821	811	791
36	8	862	345	34	10.16	1019	984	945	917	867
37	8	1724	345	34	10.16	919	900	868	847	820
38	8	3448	345	34	10.16	830	819	792	775	754
39	8	5171	345	34	10.16	777	770	744	729	711
40	8	6895	345	34	10.16	740	734	709	695	679
41	9	862	517	34	10.16	909	876	835	810	782
42	9	1724	517	34	10.16	818	800	766	745	723
43	9	3448	517	34	10.16	742	733	703	685	666
44	9	5171	517	34	10.16	701	695	666	649	632
45	9	6895	517	34	10.16	672	668	640	622	607
46	10	862	172	103	10.16	714	680	645	610	569
47	10	1724	172	103	10.16	619	600	580	554	525
48	10	3448	172	103	10.16	530	519	507	490	469
49	10	5171	172	103	10.16	481	472	464	451	433
50	10	6895	172	103	10.16	447	440	434	422	406
51	11	862	345	103	10.16	566	535	507	478	450
52	11	1724	345	103	10.16	503	485	467	447	424
53	11	3448	345	103	10.16	445	434	423	410	392
54	11	5171	345	103	10.16	412	404	395	385	370
55	11	6895	345	103	10.16	389	382	375	366	353
56	12	862	517	103	10.16	500	468	442	419	394
57	12	1724	517	103	10.16	444	427	410	394	374
58	12	3448	517	103	10.16	398	388	376	365	350
59	12	5171	517	103	10.16	372	364	355	346	333
60	12	6895	517	103	10.16	354	347	339	332	320

Note: For all program runs: $\mu_{ac} = 0.35$; $\mu_{base} = 0.40$, base thickness = 20.32 cm; $\mu_{subgrade} = 0.45$, subgrade thickness = 690.6 cm.

TABLE 2--Maximum surface deflections calculated with ELSYM5.

Run No.	Pavt Family	AC MPa	Base MPa	SG MPa	AC cm	0	3	6	9	12
		Layer Moduli			Thickness	Maximum Deflection, microns				
						Inner Radius a1, cm				
61	13	862	172	34	20.32	874	842	802	777	752
62	13	1724	172	34	20.32	722	707	675	657	640
63	13	3448	172	34	20.32	594	589	559	539	525
64	13	5171	172	34	20.32	529	528	499	474	459
65	13	6895	172	34	20.32	487	488	461	433	414
66	14	862	345	34	20.32	762	732	690	660	638
67	14	1724	345	34	20.32	642	629	595	569	551
68	14	3448	345	34	20.32	545	541	512	485	467
69	14	5171	345	34	20.32	494	494	467	438	417
70	14	6895	345	34	20.32	461	463	438	407	384
71	15	862	517	34	20.32	701	673	627	597	571
72	15	1724	517	34	20.32	594	583	548	518	497
73	15	3448	517	34	20.32	512	510	481	450	428
74	15	5171	517	34	20.32	470	471	445	413	389
75	15	6895	517	34	20.32	441	445	422	388	362
76	16	862	172	103	20.32	503	471	449	427	406
77	16	1724	172	103	20.32	398	381	368	356	342
78	16	3448	172	103	20.32	316	307	298	293	284
79	16	5171	172	103	20.32	276	270	261	257	251
80	16	6895	172	103	20.32	250	246	237	233	229
81	17	862	345	103	20.32	437	406	382	363	345
82	17	1724	345	103	20.32	352	336	322	311	298
83	17	3448	345	103	20.32	287	279	268	262	254
84	17	5171	345	103	20.32	255	249	240	234	228
85	17	6895	345	103	20.32	234	230	220	215	210
86	18	862	517	103	20.32	404	372	348	330	312
87	18	1724	517	103	20.32	326	310	294	284	273
88	18	3448	517	103	20.32	269	261	249	242	235
89	18	5171	517	103	20.32	241	236	225	219	213
90	18	6895	517	103	20.32	222	219	209	202	197

Note: For all program runs: μ_{ac} = 0.35; μ_{base} = 0.40, base thickness = 20.32 cm; $\mu_{subgrade}$ = 0.45, subgrade thickness = 690.6 cm.

TABLE 2--Maximum surface deflections calculated with ELSYM5.

| Run | Pavt | Layer Moduli | | | Thickness | Maximum Deflection, microns | | | | |
| | | AC | Base | SG | AC | Inner Radius a1, cm | | | | |
No.	Family	MPa	MPa	MPa	cm	0	3	6	9	12
91	19	862	172	34	30.48	683	656	613	579	556
92	19	1724	172	34	30.48	542	532	499	467	443
93	19	3448	172	34	30.48	438	438	414	379	350
94	19	5171	172	34	30.48	391	394	378	342	308
95	19	6895	172	34	30.48	362	367	357	322	285
96	20	862	345	34	30.48	622	600	555	516	488
97	20	1724	345	34	30.48	505	497	466	430	401
98	20	3448	345	34	30.48	419	419	399	363	330
99	20	5171	345	34	30.48	378	382	369	334	298
100	20	6895	345	34	30.48	353	358	350	317	279
101	21	862	517	34	30.48	589	564	523	480	450
102	21	1724	517	34	30.48	481	473	445	406	374
103	21	3448	517	34	30.48	404	405	388	352	316
104	21	5171	517	34	30.48	368	372	361	327	290
105	21	6895	517	34	30.48	345	351	344	313	275
106	22	862	517	103	30.48	404	374	351	333	317
107	22	1724	172	103	30.48	301	285	269	260	250
108	22	3448	172	103	30.48	228	221	208	200	194
109	22	5171	172	103	30.48	195	191	179	170	165
110	22	6895	172	103	30.48	176	173	163	153	146
111	23	862	345	103	30.48	368	339	314	295	279
112	23	1724	345	103	30.48	278	263	246	235	225
113	23	3448	345	103	30.48	215	209	196	186	179
114	23	5171	345	103	30.48	186	183	172	162	155
115	23	6895	345	103	30.48	169	167	157	147	139
116	24	862	517	103	30.48	348	320	292	272	257
117	24	1724	517	103	30.48	263	249	232	219	210
118	24	3448	517	103	30.48	206	200	187	177	169
119	24	5171	517	103	30.48	181	177	166	155	148
120	24	6895	517	103	30.48	165	163	153	142	134

Note: For all program runs: $\mu_{ac} = 0.35$; $\mu_{base} = 0.40$, base thickness = 20.32 cm; $\mu_{subgrade} = 0.45$, subgrade thickness = 690.6 cm.

TABLE 3-- Regression analysis of pavement families - full contact loading.

Pavement Family	A	B (x 10^{-3})	C (x 10^{-6})	R^2
1	3.694	2.693	-2.02	0.999998
2	9.257	-5.941	0.635	0.999973
3	12.417	-13.50	4.340	0.999875
4	3.186	5.231	-6.106	0.999825
5	5.487	1.504	-7.977	0.999191
6	8.771	-10.20	-0.020	0.999743
7	5.848	-2.446	0.095	0.999986
8	6.944	-4.871	0.917	0.999924
9	8.387	-8.944	3.241	0.999887
10	5.742	-4.808	1.227	0.999999
11	6.391	-7.569	2.591	0.999964
12	7.408	-12.84	7.795	0.999877
13	5.627	-4.432	1.549	0.999953
14	6.168	-6.303	2.704	0.999998
15	6.722	-8.449	4.347	0.999993
16	5.360	-7.336	5.005	0.999968
17	5.647	-9.517	7.577	0.999988
18	5.928	-11.84	10.97	0.999997
19	5.842	-7.016	4.047	0.999638
20	6.198	-8.611	5.415	0.999861
21	6.530	-10.22	6.989	0.999936
22	5.133	-8.940	8.678	0.999583
23	5.326	-10.81	11.75	0.999663
24	5.468	-12.30	14.46	0.999715

Using the calculated maximum deflections (Table 2), Equation 3 was used to backcalculate asphalt layer moduli for the control and all annular loadings. These backcalculated moduli were then compared to the moduli input during the computer modeling to determine the asphalt layer moduli prediction error associated with each annular loading condition.

The backcalculated asphalt moduli and prediction errors are presented in Table 4. As shown, significant errors (i.e., prediction errors greater than 25%) result as the annular loading progressively deviates from the assumed full-contact, uniform loading condition (increased a_1). The magnitude of the prediction error is typically increased as:

1.   the annular contact area is decreased (increased a_1),
2.   the asphalt layer weakens (decreased elastic moduli), and/or
3.   the asphalt layer thickness decreases.

As summary of the prediction error occurrence rate is provided for each annular loading condition in Table 5.

CONCLUSIONS

This paper has presented an analysis of the variation in maximum surface deflection and backcalculated asphalt moduli for a variety of asphalt pavement systems subjected to annular loading distributions. A limited pavement loading study has indicated that, on weak pavements, annular pressure distributions with inner radii approaching 10 cm may be obtained under non-segmented load plates.

Maximum surface deflections were calculated for a factorial of pavement systems with the linear elastic program ELSYM5, using uniform and annular load distributions. Maximum surface deflections calculated under the uniform load distribution were used to develop regression equations for backcalculating asphalt surface moduli. The calculated maximum deflections under all modeled load distributions were used to backcalculate asphalt layer moduli using the developed regression equations. It has been demonstrated that significant errors can be introduced into the backcalculated moduli for asphalt layers if measured deflections are obtained under annular loadings but analyzed assuming uniform pressure distributions.

The moduli prediction errors provided herein serve to highlight a potential source of error which may arise during routine deflection-based asphalt layer moduli backcalculation. The authors suggest that direct measurements of applied pressure distributions under FWD load plates be made, wherever possible, to verify uniform load distribution and/or to identify local pavement conditions where non-uniform distributions exist. Identification of non-uniform load distributions would then facilitate appropriate modifications to the basic premise of uniform load distribution during routine pavement layer moduli backcalculation.

TABLE 4--Backcalculated asphalt layer moduli and prediction errors.

Run No.	Pavt Family	Backcalculated AC Modulus, MPa Inner Radius a1, cm					Eac Prediction Error, % Inner Radius a1, cm				
		0	3	6	9	12	0	3	6	9	12
1	1	860	1320	2242	3868	6206	-0.2	53.1	160.1	348.8	620.0
2	1	1720	2251	3265	4933	8254	-0.2	30.6	89.4	186.2	378.8
3	1	3437	4005	5170	6876	9434	-0.3	16.2	50.0	99.5	173.6
4	1	5170	5775	6976	8713	11677	-0.0	11.7	34.9	68.5	125.8
5	1	6876	7533	8713	10452	13131	-0.3	9.2	26.4	51.6	90.4
6	2	862	1354	2196	3539	6056	-0.0	57.1	154.8	310.6	602.6
7	2	1734	2402	3635	5584	9885	0.6	39.4	110.9	224.0	473.4
8	2	3446	4222	6056	8860	14531	-0.0	22.5	75.7	157.0	321.5
9	2	5151	6122	8459	11654	17642	-0.4	18.4	63.6	125.4	241.2
10	2	6935	7945	10851	14531	20850	0.6	15.2	57.4	110.7	202.4
11	3	857	1263	1951	2774	5028	-0.6	46.5	126.3	221.8	483.3
12	3	1694	2287	3543	5028	7759	-1.7	32.7	105.5	191.7	350.1
13	3	3435	4159	6528	9222	16375	-0.4	20.6	89.3	167.5	375.0
14	3	5195	6146	9319	12686	20483	0.5	18.9	80.2	145.3	296.1
15	3	6774	8002	11892	16375	24771	-1.8	16.1	72.5	137.5	259.3
16	4	867	1572	3221	6182	10132	0.6	82.3	273.7	617.2	1075.5
17	4	1695	2378	3986	6574	10132	-1.7	38.0	131.2	281.4	487.8
18	4	3478	4162	5645	7973	11107	0.9	20.7	63.8	131.3	222.2
19	4	5207	5904	7242	9496	12261	0.7	14.2	40.1	83.6	137.1
20	4	6843	7425	8828	10943	13252	-0.8	7.7	28.0	58.7	92.2
21	5	855	1896	4181	9243	16818	-0.8	119.9	385.0	972.3	1851.1
22	5	1768	2791	5153	9635	18142	2.5	61.9	198.9	459.0	952.5
23	5	3328	4665	7229	11819	20291	-3.5	35.3	109.7	242.8	488.6
24	5	5269	6430	8977	13858	22645	1.9	24.3	73.6	168.0	337.9
25	5	6864	8287	11397	16187	24335	-0.5	20.2	65.3	134.8	252.9
26	6	861	2025	4125	8864	19278	-0.1	134.9	378.5	928.3	2136.4
27	6	1765	2987	5696	11258	21725	2.4	73.3	230.4	553.1	1160.3
28	6	3407	4964	8008	14298	27591	-1.2	44.0	132.3	314.7	700.3
29	6	5176	6855	10355	17106	33008	0.1	32.6	100.2	230.8	538.3
30	6	6979	8759	12840	20465	37198	1.2	27.0	86.2	196.8	439.5

Note: Asphalt moduli backcalculated using $\text{Log}(E_{ac}) = A + B*D_{max} + C*D_{max}^2$

TABLE 4--Backcalculated asphalt layer moduli and prediction errors.

Run No.	Pavt Family	Backcalculated AC Modulus, MPa Inner Radius a1, cm					Eac Prediction Error, % Inner Radius a1, cm				
		0	3	6	9	12	0	3	6	9	12
31	7	859	1053	1267	1561	2001	-0.3	22.1	47.0	81.1	132.1
32	7	1728	1941	2237	2571	3153	0.2	12.6	29.7	49.1	82.9
33	7	3431	3685	4113	4513	5251	-0.5	6.9	19.3	30.9	52.3
34	7	5154	5451	6031	6473	7320	-0.3	5.4	16.6	25.2	41.6
35	7	6902	7223	7996	8470	9405	0.1	4.8	16.0	22.8	36.4
36	8	859	1096	1446	1772	2573	-0.3	27.1	67.8	105.6	198.5
37	8	1740	2010	2544	2997	3672	0.9	16.6	47.6	73.9	113.1
38	8	3422	3716	4586	5247	6197	-0.7	7.8	33.0	52.2	79.7
39	8	5142	5463	6697	7562	8763	-0.6	5.6	29.5	46.2	69.5
40	8	6935	7271	8929	10041	11446	0.6	5.5	29.5	45.6	66.0
41	9	859	1092	1503	1853	2363	-0.4	26.7	74.4	114.9	174.2
42	9	1741	2021	2737	3315	4102	1.0	17.3	58.8	92.3	138.0
43	9	3421	3725	5008	6071	7338	-0.8	8.1	45.3	76.1	112.8
44	9	5125	5425	7338	8888	10664	-0.9	4.9	41.9	71.9	106.2
45	9	6953	7220	9799	11954	14269	0.8	4.7	42.1	73.4	106.9
46	10	862	1095	1420	1849	2533	-0.0	27.0	64.8	114.5	193.9
47	10	1722	1983	2330	2846	3597	-0.1	15.1	35.2	65.1	108.7
48	10	3445	3764	4149	4785	5707	-0.1	9.2	20.4	38.8	65.6
49	10	5175	5551	5944	6678	7778	0.1	7.3	15.0	29.1	50.4
50	10	6884	7316	7727	8512	9788	-0.2	6.1	12.1	23.4	42.0
51	11	861	1209	1663	2331	3250	-0.1	40.2	92.9	170.4	277.0
52	11	1737	2145	2643	3341	4446	0.8	24.4	53.3	93.8	157.9
53	11	3434	3911	4488	5307	6645	-0.4	13.4	30.2	53.9	92.8
54	11	5158	5694	6351	7279	8864	-0.3	10.1	22.8	40.8	71.4
55	11	6932	7545	8270	9253	11044	0.5	9.4	19.9	34.2	60.2
56	12	861	1279	1809	2488	3640	-0.2	48.4	109.8	188.7	322.3
57	12	1747	2230	2835	3626	4957	1.3	29.4	64.5	110.3	187.5
58	12	3420	3986	4777	5753	7425	-0.8	15.6	38.6	66.9	115.4
59	12	5143	5826	6784	7889	9880	-0.5	12.7	31.2	52.6	91.1
60	12	6961	7720	8881	10149	12448	1.0	12.0	28.8	47.2	80.5

Note: Asphalt moduli backcalculated using $\mathrm{Log(Eac)} = A + B*D_{max} + C*D_{max}^2$

TABLE 4--Backcalculated asphalt layer moduli and prediction errors.

| Run No. | Pavt Family | Backcalculated AC Modulus, MPa | | | | | Eac Prediction Error, % | | | | |
| | | Inner Radius a1, cm | | | | | Inner Radius a1, cm | | | | |
		0	3	6	9	12	0	3	6	9	12
61	13	865	987	1171	1312	1481	0.4	14.5	35.9	52.2	71.8
62	13	1715	1852	2189	2425	2667	-0.5	7.4	27.0	40.7	54.7
63	13	3467	3574	4299	4878	5343	0.6	3.7	24.7	41.5	55.0
64	13	5205	5239	6332	7462	8315	0.7	1.3	22.4	44.3	60.8
65	13	6859	6800	8199	9957	11397	-0.5	-1.4	18.9	44.4	65.3
66	14	861	1007	1280	1530	1773	-0.1	16.8	48.5	77.5	105.7
67	14	1723	1874	2378	2855	3289	-0.0	8.7	37.9	65.6	90.8
68	14	3440	3536	4480	5591	6542	-0.2	2.6	29.9	62.2	89.8
69	14	5177	5166	6528	8423	10199	0.1	-0.1	26.2	62.9	97.2
70	14	6885	6748	8442	11245	14036	-0.1	-2.1	22.4	63.1	103.6
71	15	862	1012	1356	1689	2057	-0.1	17.4	57.3	95.9	138.6
72	15	1728	1879	2499	3252	3947	0.2	9.0	45.0	88.7	129.0
73	15	3439	3512	4630	6290	7983	-0.3	1.9	34.3	82.5	131.6
74	15	5161	5109	6652	9419	12410	-0.2	-1.2	28.6	82.1	140.0
75	15	6907	6688	8558	12484	17168	0.2	-3.0	24.1	81.1	149.0
76	16	864	1036	1188	1383	1604	0.2	20.2	37.9	60.5	86.1
77	16	1717	1961	2178	2406	2741	-0.4	13.7	26.3	39.6	59.0
78	16	3464	3785	4145	4385	4803	0.5	9.8	20.2	27.2	39.3
79	16	5201	5548	6105	6365	6808	0.6	7.3	18.1	23.1	31.6
80	16	6865	7225	8014	8348	8801	-0.4	4.8	16.2	21.1	27.6
81	17	862	1080	1307	1548	1835	-0.0	25.2	51.6	79.6	112.9
82	17	1716	2011	2345	2634	3031	-0.4	16.7	36.0	52.8	75.8
83	17	3454	3809	4346	4709	5213	0.2	10.5	26.1	36.6	51.2
84	17	5179	5554	6341	6796	7395	0.1	7.4	22.6	31.4	43.0
85	17	6868	7265	8322	8950	9599	-0.4	5.4	20.7	29.8	39.2
86	18	862	1099	1372	1639	1994	0.0	27.6	59.2	90.1	131.3
87	18	1721	2047	2480	2830	3281	-0.2	18.8	43.9	64.2	90.3
88	18	3455	3840	4542	5048	5624	0.2	11.4	31.8	46.4	63.1
89	18	5165	5580	6593	7270	7999	-0.1	7.9	27.5	40.6	54.7
90	18	6893	7270	8630	9565	10437	-0.0	5.4	25.2	38.7	51.4

Note: Asphalt moduli backcalculated using $\text{Log}(Eac) = A + B*D_{max} + C*D_{max}^2$

TABLE 4--Backcalculated asphalt layer moduli and prediction errors.

Run No.	Pavt Family	Backcalculated AC Modulus, MPa Inner Radius a1, cm					Eac Prediction Error, % Inner Radius a1, cm				
		0	3	6	9	12	0	3	6	9	12
91	19	866	957	1155	1368	1554	0.5	11.0	34.0	58.7	80.3
92	19	1696	1800	2238	2815	3368	-1.6	4.4	29.8	63.3	95.4
93	19	3514	3521	4266	5826	7641	1.9	2.1	23.7	69.0	121.6
94	19	5234	5083	5867	8228	11571	1.2	-1.7	13.4	59.1	123.8
95	19	6787	6472	7139	10024	14751	-1.6	-6.1	3.5	45.4	113.9
96	20	864	958	1220	1576	1934	0.2	11.2	41.5	82.9	124.4
97	20	1704	1808	2300	3151	4136	-1.2	4.9	33.4	82.8	139.9
98	20	3486	3478	4198	6108	8812	1.1	0.9	21.8	77.2	155.6
99	20	5198	5023	5738	8482	13016	0.5	-2.9	11.0	64.0	151.7
100	20	6823	6454	7038	10292	16399	-1.0	-6.4	2.1	49.3	137.8
101	21	860	974	1245	1716	2228	-0.2	13.1	44.4	99.0	158.5
102	21	1705	1815	2323	3395	4837	-1.1	5.3	34.8	97.0	180.6
103	21	3469	3441	4133	6304	9898	0.6	-0.2	19.9	82.9	187.1
104	21	5176	4952	5629	8601	14293	0.1	-4.2	8.8	66.3	176.4
105	21	6839	6403	6904	10382	17814	-0.8	-7.1	0.1	50.6	158.4
106	22	867	1005	1161	1316	1477	0.6	16.6	34.7	52.6	71.3
107	22	1695	1941	2259	2488	2751	-1.6	12.6	31.1	44.4	59.6
108	22	3521	3812	4454	4928	5308	2.1	10.6	29.2	42.9	54.0
109	22	5239	5503	6440	7268	7882	1.3	6.4	24.5	40.5	52.4
110	22	6780	6970	8110	9343	10329	-1.7	1.1	17.6	35.5	49.8
111	23	868	1030	1234	1449	1671	0.7	19.5	43.1	68.1	93.8
112	23	1700	1967	2383	2721	3079	-1.4	14.1	38.3	57.9	78.6
113	23	3513	3826	4584	5256	5848	1.9	11.0	33.0	52.5	69.6
114	23	5236	5500	6551	7697	8610	1.3	6.4	26.7	48.8	66.5
115	23	6790	7011	8257	9822	11243	-1.5	1.7	19.8	42.5	63.1
116	24	869	1031	1288	1560	1837	0.8	19.6	49.4	81.0	113.2
117	24	1704	1994	2481	2918	3353	-1.2	15.7	43.9	69.3	94.5
118	24	3518	3850	4682	5565	6353	2.0	11.7	35.8	61.4	84.3
119	24	5228	5519	6665	8043	9241	1.1	6.7	28.9	55.5	78.7
120	24	6813	7027	8343	10257	12040	-1.2	1.9	21.0	48.8	74.6

Note: Asphalt moduli backcalculated using $Log(Eac) = A + B*D_{max} + C*D_{max}^2$

TABLE 5-- Occurrence of moduli prediction errors for all loading conditions.

Inner Annular Radius a_1 (cm)	Percent of Occurrences Where Prediction Error Falls Within Specified Limits						
	0 - 5% Error	5 - 10% Error	10 - 15% Error	15 - 25% Error	25 - 50% Error	50 - 100% Error	> 100% Error
0.0	100.0	0.0	0.0	0.0	0.0	0.0	0.0
3.0	20.0	23.3	15.8	20.8	14.2	4.2	1.7
6.0	2.5	0.8	3.3	19.2	44.2	17.5	12.5
9.0	0.0	0.0	0.0	3.3	25.8	45.0	25.8
12.0	0.0	0.0	0.0	0.0	7.5	37.5	55.0

REFERENCES

Touma, B.E., J.A. Crovetti and M.Y. Shahin, "Effects of Various Load Distributions on Backcalculated Moduli Values in Flexible Pavements," TRR No. 1293, Transportation Research Board, National Research Council, Washington, D.C., 1991.

Kopperman, S., et.al., "ELSYM5; Interactive Microcomputer Version," Users Manual: IBM-PC and Compatible Version, Report No. FHWA-T-87-206, Federal Highway Administration, 1986.

Krishna M. Boddapati[1] and Soheil Nazarian[2]

EFFECTS OF PAVEMENT-FALLING WEIGHT DEFLECTOMETER INTERACTION ON MEASURED PAVEMENT RESPONSE

REFERENCE: Boddapati, K. M., and Nazarian, S., "Effects of Pavement-Falling Weight Deflectometer Interaction on Measured Pavement Response," _Nondestructive Testing of Pavements and Backcalculation of Moduli (Second Volume), ASTM STP 1198_, Harold L. Von Quintas, Albert J. Bush, III, and Gilbert Y. Baladi, Eds., American Society of Testing and Materials Philadelphia, 1994.

ABSTRACT: In any linear elastic program used in backcalculation, a uniform pressure distribution is assumed for the applied load. As such, the loading system of any FWD should be designed so that the load transferred to the pavement is uniform. However, the pressure distribution is also affected by the pavement profile being impacted on.

A finite element study has been carried out to determine the cases where the imparted load would or would not yield a uniform pressure distribution and to assess its significance on the measured and backcalculated parameters. The FWD-pavement interaction has little significance when rigid pavements are tested. For flexible pavements, the deflections from the first sensor may be affected by this interaction. The softer the pavement system, the more significant this interaction will be.

KEY WORDS: Falling Weight Deflectometer, flexible pavement, rigid pavement, pressure distribution, nondestructive testing, remaining life, backcalculation.

INTRODUCTION

Nondestructive testing techniques are widely used to obtain effective pavement moduli of different layers. Moduli obtained in this manner are subsequently used with structural response and pavement performance models to estimate the existing load carrying capacity and overlay thickness needed for future traffic.

For applying a wide range of load pulses, from light to heavy, Falling Weight Deflectometers (FWD) are versatile NDT devices. In the backcalculation process using FWD deflections, a uniform distribution of load is assumed, so the loading system of any FWD should be designed so that the load transferred to the pavement is uniform. Manufacturers typically address this problem utilizing a composite loading plate. The composite loading plate consists of a steel plate, a PVC plate, and a rubber pad placed on the lower face of the plate. Some FWD manufacturers utilize segmented load plates to distribute the load more evenly. The

---

[1]Grad. Research Assist., The Univ. of Texas, El Paso, TX 79968

[2]Assoc. Prof., The Univ. of Texas at El Paso, El Paso, TX 79968-0516

effects of segmenting the plate were not studied. However, as in this study an axi-symmetrical solution has been adapted, the results should be valid for segmented plates as well. The rubber pad should transform the semi-rigid loading plate to a "flexible" loading plate; thus producing uniform pressure distribution over the pavement. Unfortunately, the pressure distribution under plate is also affected by the pavement profile being impacted on. If the uniform stress distribution assumption is violated, the deflection of the sensors may be in error. As such, the backcalculated moduli and critical stresses and strains, and naturally the predicted pavement life may be in error.

A finite element study was carried out to determine the impact of FWD plate and pavement structure on these parameters. Two models were used. In one model, the typical composite FWD loading plate on top of pavement system was discretized in the finite element mesh. In the other, a uniform load distribution was assumed on top of the pavement. Through sensitivity analyses the effects of different components of the FWD composite loading plate, as well as the stiffness and thickness of each pavement layer were studied. The variation in the measured deflection basins, pressure distributions under the load, critical stresses and strains were discussed and quantified.

The results reported herein are limited to the linear elastic cases. The load-induced nonlinear behaviors of the base and subgrade were not considered in this study. Boddapati (1992) has conducted such as study and the reader is referred to that documentation for more information.

FINITE ELEMENT MODEL DEVELOPMENT

The computer program ABAQUS, developed by Habbit, Karlsson and Sorensen, Inc., was used throughout this study. Four-node axi-symmetrical elements were selected.

A semi-infinite subgrade is an inherent assumption in the use of elastic layered theory for evaluating the in-situ moduli of the pavement materials. The presence of a stiff layer at a finite depth can significantly affect the deflection basins measured in situ (Briggs et al 1989 and Chang et al 1992). In practice, unless the thickness of the subgrade is known through some geological information, a subgrade thickness is usually assumed.

The characteristics of the finite element mesh (i.e., location of the lateral and bottom boundaries, the element size and shape, and the distribution of the elements) affect the accuracy of the calculated stresses , strains, and deflections. Previous research established that fixing the bottom boundary at 50 radii and the lateral boundary at 12 radii would simulate a layered-half-space. In our preliminary studies the lateral boundary was fixed at 6 m and the bottom boundary was fixed at 7.5 m. The results from this system were then compared to those obtained from computer program BISAR. Large differences were observed. A model with the lateral boundary fixed at 12 m was found to be the most appropriate mesh and selected for this analysis. This mesh was well refined along the interface between plate and pavement sections and contained 7500 elements. The results from this mesh were typically less than 1 percent different from the results from BISAR (refer to Boddapati 1992 for the detailed study).

ASSUMED PLATE AND PAVEMENT SECTIONS

The composite loading plate of the FWD, was assumed to consist of a steel plate with an elastic stiffness of 70 GPa, and a Poisson's ratio of 0.3 over a PVC plate with an elastic stiffness of 7 GPa, and a

Poisson's ratio of 0.3. The diameter of the FWD loading plate was
assumed to be 300 mm with a 25-mm-diameter hole at the center. The steel
and PVC plates rested over a rubber pad with an elastic stiffness of 35
MPa, and a Poisson's ratio of 0.49. The steel and PVC plates were
assumed to be 25 mm thick and the rubber pad 6.2 mm thick.

The standard flexible pavement section used in this study was
assumed to have three layers, an asphalt-concrete (AC) layer over a
granular base over a subgrade (see Fig. 1). The thicknesses of the AC
and base layers were assumed to be 75 mm and 300 mm, respectively. The
moduli of AC, base and subgrade were assumed to be 3500 MPa, 350 MPa and
70 MPa, respectively. The Poisson's ratio of the AC and base layers was
assumed to be 0.35. A Poisson's ratio of 0.45 was assigned to the
subgrade.

A rigid pavement consisting of a 225-mm-thick concrete layer with
an elastic stiffness of 28 GPa and a Poisson's ratio of 0.15 over a
subgrade with an elastic stiffness of 70 MPa and a Poisson's ratio of
0.45 was also studied.

The above mentioned hypothetical cross sections for the FWD
loading plate and the pavement sections are referred to as the standard
plate and standard pavement sections, hereafter.

APPROACH TO THE PROBLEM

A falling load of 67 KN is assumed to impact on the composite
loading plate of the FWD. As the problem is assumed to be elasto-static
in nature, the magnitude of the loading does not affect the generality
of the conclusions. As indicated before, two input data files were
generated. In one model (called STATFWD) the typical composite FWD
loading plate on top of the pavement system was discretized in the
finite element mesh. In the second model (named STATPAV) a uniform load
distribution was assumed on top of the pavement. These data files were
input to ABAQUS. In the STATPAV, a falling mass of 67 KN resulted in a
uniform stress of about 930 KPa over the pavement. After executing
ABAQUS program with the two data files (STATFWD and STATPAV), the stress
distribution over the pavement surface and the deflection bowls along
the pavement surface at radial distances of 0 mm to 1800 mm (300 mm
interval) from the center of the loading plate were compared.

The pressure distributions under the loading plate at the pavement
surface are graphically presented in Fig. 2, for the standard flexible
and rigid pavements. The horizontal line represents the uniform pressure
distribution of 930 KPa on the pavement as normally assumed. For the
flexible pavement, some concentration of stress at the edges is
apparent. Little stress is imparted to the pavement near the center of
the plate (see Fig. 2). For the rigid pavement, the pressure is more or
less constant except for the inner and outer 25 mm of the plate.

To determine the energy concentration in the loaded area due to
the FWD/pavement interaction, a parameter was introduced. The stress
recovery ratio, S_r was defined as:

$$S_r(\%) \quad = \quad \frac{p_i}{p_u} * 100 \tag{1}$$

where,

    p_u = stress assuming uniform stress distribution ($\approx$ 931 KPa here)
    p_i = average stress applied to the pavement

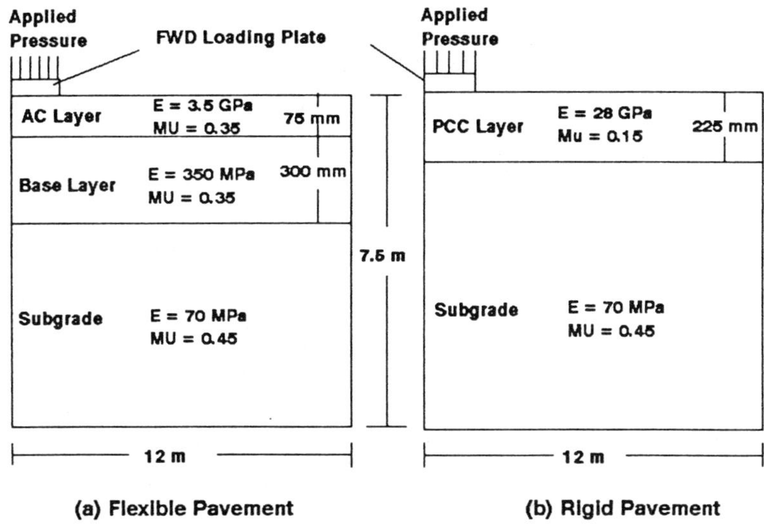

FIG. 1--Standard pavement profiles used in this study

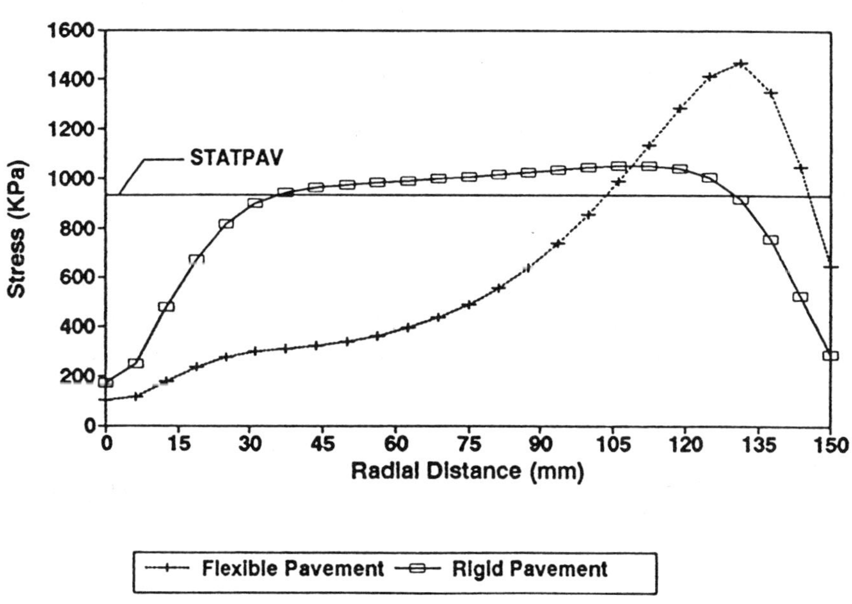

FIG. 2--Pressure distribution under composite loading plate

To determine p_u, the applied load (67 KN) is divided by the area of the plate, whereas, p_i is determined by calculating average stress obtained considering the FWD/pavement interaction. For the standard flexible pavement, the stress recovery ratio (S_r) was 70.3 percent. To the contrary, the standard rigid pavement recorded an S_r of 94 percent. Practically speaking, the closer the value of S_r is to unity, the more uniform the stress distribution will be.

The deflections calculated with STATPAV (uniform pressure distribution) and STATFWD (FWD/pavement interaction considered) on flexible and rigid pavements are compared in Table 1. For the standard flexible pavement, the difference is 6.3 percent for the first sensor and practically zero for the other sensors. For the standard rigid pavement, the differences are less than 1 percent. For the rigid pavements, the FWD/pavement interaction results in small variation in the outcome. As such, the interaction should not be of any concern. No further results for the rigid pavements are reported in this paper.

The above study shows that the nonuniformity in the pressure distribution over a flexible pavement influences the measured response, and naturally calculated remaining lives. Sensitivity analyses have been done on different pavement layers and composite loading plate components. The results are discussed in the next sections.

TABLE 1--Impact of plate/pavement interaction on surface deflections for standard flexible and rigid pavements

Type of Pavement	Model	Deflection (in micron) Measured at (mm)						
		0	300	600	900	1200	1500	1800
(1)	(2)	(3)	(4)	(5)	(6)	(7)	(8)	(9)
Flexible Pavement	STATPAV	870	543	345	235	165	118	88
	STATFWD	815	543	345	235	165	118	88
		(6.3)	(0.1)	(0)	(0)	(0)	(0)	(0)
Rigid Pavement	STATPAV	265	240	208	170	138	108	83
	STATFWD	263	240	208	170	138	108	83
		(0.4)	(0)	(0)	(0)	(0)	(0)	(0)

Numbers in parentheses denote percent differences in deflections obtained with STATPAV and STATFWD.

SENSITIVITY STUDY

The stiffness and thickness of each component of the FWD plate were varied to determine their influences on the deflection basins. In addition, moduli and thicknesses of the pavement layers were varied to produce a realistic range of in-service pavement systems. In each case, the influence of assuming a uniform pressure in STATPAV on the pavement performance is estimated by comparing deflections calculated by STATPAV to those by STATFWD.

Plate Components

To conduct the sensitivity analysis on each parameter the following approach was followed. The standard plate and pavement was

considered. The parameter of interest was than varied symmetrically with
respect to the standard value. For example, to investigate the
contribution of the steel plate modulus, its stiffness was doubled (to
140 GPa) and halved (to 35 GPa). The differences between the results of
the three cases (i.e. standard, stiffer plate and softer plate) were
compared. These results are shown in Table 2. For the case of the
standard plate, the difference between the central deflections obtained
with STATPAV and STATFWD is 6.3 percent. By doubling and halving the
stiffness of the plate the differences in the central deflections vary
between 6.1 and 6.9 percent, not much different than 6.3 percent
obtained for the standard plate. Therefore, one can conclude that the
impact of the stiffness of the plate on the overall FWD/pavement
interaction is small. In practical terms, the user and manufacturer
should not be much concerned with the actual stiffness of the steel
plate. One can replace the steel plate with hardened steel or aluminum
and still obtain similar results. However, it should not be forgotten
that irrespective of the type of metal used, the central deflection on
the standard pavement would be different by about 6.5 percent relative
to the case when a uniform stress distribution is considered.

The stiffness of the steel plate is the major source of
contribution to the rigidity of the semi-rigid FWD loading plate. The
pressure distribution along the interface and central deflections are
slightly affected by the variation in the stiffness and thickness of the
steel plate. The differences in central deflections obtained from
STATPAV and STATFWD are about 6.5 percent for steel plate stiffnesses of
35 GPa, 70 GPa and 140 GPa.

The second parameter studied was the thickness of the steel plate.
Contrary to the plate stiffness variations, plate thickness may
influence the central deflection measured. As the thickness decreases
(to 12.5 mm) the plate becomes more flexible and it can better conform
to the shape of pavement. In this case, the difference in deflection is
about 4.5 percent which is 2 percent less than that of the standard
pavement. Thickening of the steel plate would naturally result in a more
rigid plate system, which in turn results in a larger deviation in the
values of central deflection between the cases when the interaction is
and is not considered.

The PVC plate is placed to distribute the imparted load to the FWD
loading plate more evenly. Once again the variation in the PVC plate
stiffness or thickness has a small influence on pressure distribution
along the interface and central deflections. The variation in the PVC
plate stiffnesses from 3500 MPa to 14000 MPa resulted in differences of
central deflections from 6 to 7 percent. The variation of PVC plate
thickness from 12.5 mm to 37.5 mm causes a difference in deflections
ranging from 5.5 percent to 7 percent.

The rubber pad facilitates the distribution of the load to the
pavement. To find the influence of the stiffness of the rubber pad on
the overall distribution of the loads, two cases were studied. In one
case, the stiffness of the pad varied from 7 MPa to 175 MPa. In the
other case, the thickness of the pad varied from 0 (no-pad) to 9.4 mm.
As illustrated in Fig. 3, with the decrease in the modulus of the pad
the stress distribution becomes more uniform. The stress recovery ratio
increases from 65 percent to about 82 percent. The differences in
deflections calculated by two approaches (STATPAV and STATFWD) decrease
from 10 percent to about 3 percent (see Table 2). This implies that the
stiffness of the rubber pad is of significant importance and should not
be ignored.

Unlike pad stiffness, the variation in rubber pad thickness from
9.4 mm to about 3.1 mm has small influence on the pressure distribution
and central deflections. The exclusion of the pad results in significant

TABLE 2--Influences of plate components on deflections, backcalculated parameters and remaining lives

Parameter	Plate Moduli (MPa)			Plate Thickness (mm)			SRR (percent)	Central Deflections (μm)	Backcalculated Moduli (MPa)			Basin Fitting Mismatch[c] (percent)	Remaining Life (million ESAL)	
	Steel	PVC	Pad	Steel	PVC	Pad			AC	Base	Subgrade		Rutting	Fatigue
Standard Plate	70000	7000	35	25.0	25.0	6.2	71	32.8 (6.3)[a]	6664 (-90)[b]	329 (6)	72.1 (-3)	0.7	1.4 (-33)	1.3 (3)
Steel Plate Modulus	35000						74	32.8 (6.1)	6286 (-80)	322 (6)	72.1 (-3)	0.9	1.3 (-16)	1.4 (-7)
Steel Plate Modulus	140000	7000	35	25.0	25.0	6.2	68	32.4 (6.9)	7049 (-101)	322 (7)	71.4 (-2)	0.9	1.3 (-19)	1.5 (-14)
Steel Plate Thickness				12.5			80	33.2 (4.5)	5537 (-58)	329 (5)	72.1 (-3)	0.9	1.2 (-13)	1.3 (-1)
Steel Plate Thickness	70000	7000	35	37.5	25.0	6.2	65	32.2 (7.5)	7910 (-126)	322 (8)	71.4 (-2)	0.9	1.3 (-23)	1.6 (-22)
PVC Plate Modulus		3500					74	32.8 (6.0)	6237 (-78)	329 (6)	72.1 (-3)	0.9	1.2 (-15)	1.4 (-7)
PVC Plate Modulus	70000	14000	35	25.0	25.0	6.2	68	32.4 (6.9)	7112 (-103)	322 (7)	71.4 (-2)	0.9	1.3 (-19)	1.5 (-15)
PVC Plate Thickness					12.5		75	32.9 (5.5)	6146 (-76)	329 (6)	72.1 (-3)	0.9	1.2 (-15)	1.4 (-6)
PVC Plate Thickness	70000	7000	35	25.0	37.5	6.2	68	32.4 (6.9)	7049 (-101)	329 (6)	71.4 (-2)	0.9	1.3 (-19)	1.5 (-14)
Pad Modulus			7				82	33.7 (3.1)	4809 (-37)	336 (5)	72.1 (-3)	0.8	1.2 (-11)	1.2 (5)
Pad Modulus	70000	7000	175	25.0	25.0	6.2	65	31.4 (9.8)	9660 (-176)	315 (9)	72.1 (-2)	1.0	1.5 (-35)	1.8 (-39)
Pad Thickness						0	57	29.5 (15.2)	24619 (-603)	231 (34)	72.1 (-3)	1.3	2.2 (-106)	3.7 (-183)
Pad Thickness	70000	7000	35	25.0	25.0	3.1	69	32.1 (7.7)	7749 (-121)	322 (8)	71.4 (-2)	0.9	1.3 (-22)	1.6 (-20)
Pad Thickness						9.4	74	33.0 (5.1)	5908 (-69)	329 (6)	72.1 (-3)	0.9	1.2 (-14)	1.4 (-4)

Number in parantheses corresponding to percent difference between actual and backcalculated values

a = [(Deflections from STATPAV - Deflections from STATFWD) / Deflections from STATPAV] * 100

b = $\sum$ (Actual Value - Backcalculated Value) / Actual Value] * 100

c = $\sum$ [(Actual Deflection - Backcalculated Deflection)/(Actual Deflection)]/7 * 100

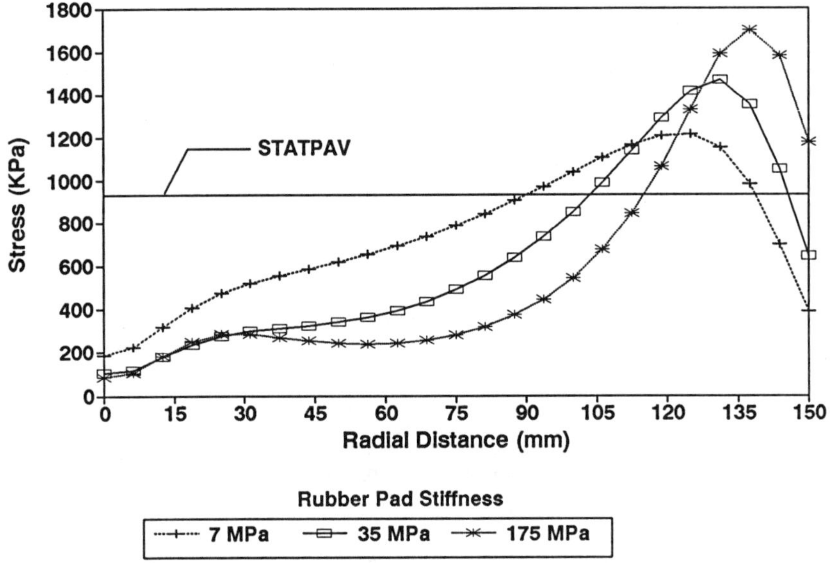

FIG. 3--Influence of pad stiffness on pressure distribution under plate

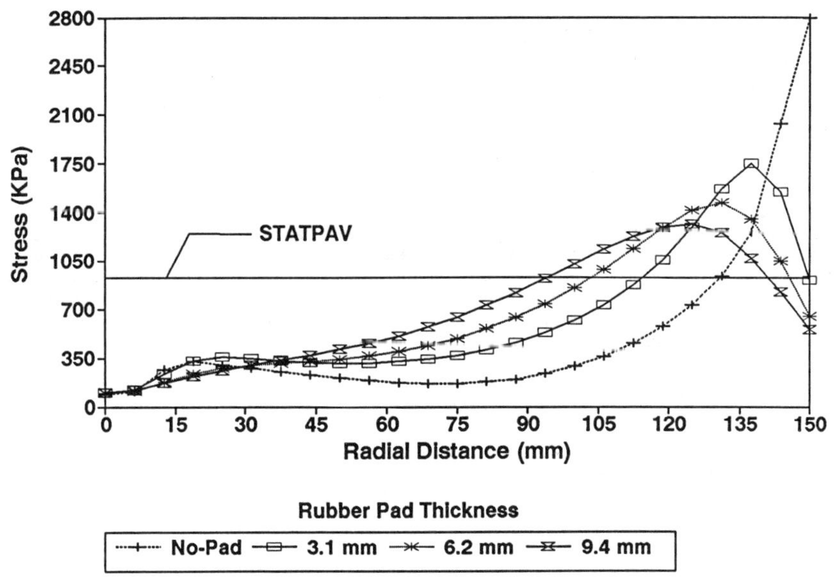

FIG. 4-Influence of pad thickness on pressure distribution under plate

concentration of stress at the outer edge of the plate (see Fig 4). In addition, a low pressure zone is developed along the central part of the plate. This reduces the stress recovery to 57 percent. In this case, the central deflections calculated with and without plate differ by 15 percent (see Table 2). When the pad was used, the variation in deflections is about 5 to 8 percent. Practically speaking, the pad should be periodically checked to ensure that its properties have not changed. In addition, to ensure uniform central deflection, the pad should be acquired from the same source.

<u>Pavement Layers</u>

So far the effects of different components of the FWD plate on the central deflection were investigated. The other parameters that play a role in the FWD/pavement interaction are the thickness and modulus of each layer. The impact of each of these parameters on the response of the pavement is presented in the next section, utilizing a procedure similar to that followed for different plate components. The results of these sensitivity analyses are summarized in Table 3.

The variation in the AC layer stiffness results in large variations in the deflection of central sensors. The variation in deflection measured with the STATFWD and STATPAV ranged from 9 percent to 4 percent, as the stiffness of the asphalt layer increased from 1750 MPa to 7000 MPa.

The thin asphalt layer (thickness of 25 mm) results in a low pressure zone along the central parts of the loading plate. As the thickness increases from 25 mm to 125 mm, the stress distribution under the plate becomes more uniform as shown in Fig. 5. The stress recovery decreased from 74 percent to about 71 percent, as the thickness of the asphalt layer increased from 25 mm to 75 mm. An S_r of about 79 percent is recorded with further increase in the asphalt layer thickness to 125 mm. As shown in Table 3, the differences in deflections calculated by STATFWD and STATPAV decreased from 12 percent to about 4 percent, with the change in thickness from 25 mm to 125 mm. Therefore, the thickness of the AC layer significantly affects the deflection measured at the center of the plate. The thicker the AC layer, the less important is the interaction.

The change in the base layer stiffness influences the pressure distribution under the plate. As shown in Fig. 6, base layer with a stiffness of 87.5 MPa develops a non-contact zone along the central parts of the plate. This results in the concentration of the entire load along the outer edge. As the base layer stiffness increases from 87.5 MPa to 1400 MPa, the stress distribution under the plate becomes more uniform. The S_r increases from 55 percent to about 84 percent with the increase in the base layer moduli from 87.5 MPa to 350 MPa. Such large increase in the stress recovery with the increase in the base layer stiffness results in a difference in deflections obtained from STATFWD and STATPAV and ranging from 10 to about 3 percent (see Table 3).

The change in the base layer thickness, on the other hand, exerts a small influence on the pressure distribution under the plate. The difference between deflections obtained from the two approaches is about 6.5 percent with the change in thickness from 150 to 300 mm.

To determine the influence of the stiffness of the subgrade on the overall response of the pavement system, the stiffness of the subgrade varied from 17.5 MPa to 280 MPa. The variations in the subgrade stiffness have some influence on the pressure distribution along the interface. The differences in the deflections obtained from the two approaches vary from 4 to 10 percent, as the stiffness of the subgrade increases from 17.5 MPa to 280 MPa.

TABLE 3--<u>Influences of different pavement profiles on central deflections, backcalculated parameters and remaining lives</u>

Parameter	Moduli (MPa)			Thickness (mm)			SRR (percent)	Central Deflections (μm)	Backcalculated Moduli (MPa)			Basin Fitting Mismatch (percent)	Remaining Life (million ESAL)	
	AC	Base	Subgrade	AC	Base	Subgrade			AC	Base	Subgrade		Rutting	Fatigue
Standard Pavement	3500	350	70	75	300	7125	71	32.6 (6.3)[a]	6664 (-90)[b]	329 (6)	72.1 (-3)	0.7	1.4 (-33)	1.3 (3)
AC Modulus	1750	350	70	75	300	7125	72	34.7 (9.3)	5488 (-214)	301 (14)	71.4 (-2)	1.1	1.1 (-35)	1 (38)
	7000						72	30.3 (4.1)	13524 (-93)	280 (20)	72.1 (-3)	0.9	2.1 (-33)	1.5 (-5)
AC Thickness	3500	350	70	25	300	7175	74	39.9 (11.7)	1400 (60)	441 (-26)	72.1 (-3)	1.2	0.5 (-40)	698.8 (-3318)
				125		7075	79	26.5 (3.5)	3500 (0)	350 (0)	71.4 (-2)	1.0	4.3 (-5)	2.8 (0)
Base Modulus	3500	87.5	70	75	300	7125	55	53.8 (10.5)	5936 (-70)	81.2 (7)	72.1 (-3)	0.8	0.4 (-93)	0.2 (-55)
		1400					84	19.8 (3.2)	3668 (-5)	1386 (1)	72.1 (-3)	1.0	27.4 (-4)	139.1 (14)
Base Thickness	3500	350	70	75	150	7275	66	41.5 (6.5)	8624 (-146)	224 (36)	72.1 (-3)	0.6	0.2 (-96)	0.5 (26)
					450	6975	72	28.0 (6.6)	7959 (-127)	308 (12)	72.1 (-3)	1.0	0.9 (-17)	1.5 (1)
Subgrade Modulus	3500	350	17.5	75	300	7125	69	66.9 (3.8)	13027 (-272)	238 (32)	18.2 (-4)	0.9	0.1 (-14)	0.8 (34)
			280				73	17.8 (9.5)	6713 (-92)	336 (4)	294 (-5)	1.1	70.4 (-59)	1.7 (-18)

Numbers in parantheses corresponding to percent differences between actual and backcalculated values

[a]   [(Deflections from STATPAV - Deflections from STATFWD) / Deflections from STATPAV] * 100
[b]   [(Actual Value - Backcalculated Value) / Actual Value] * 100

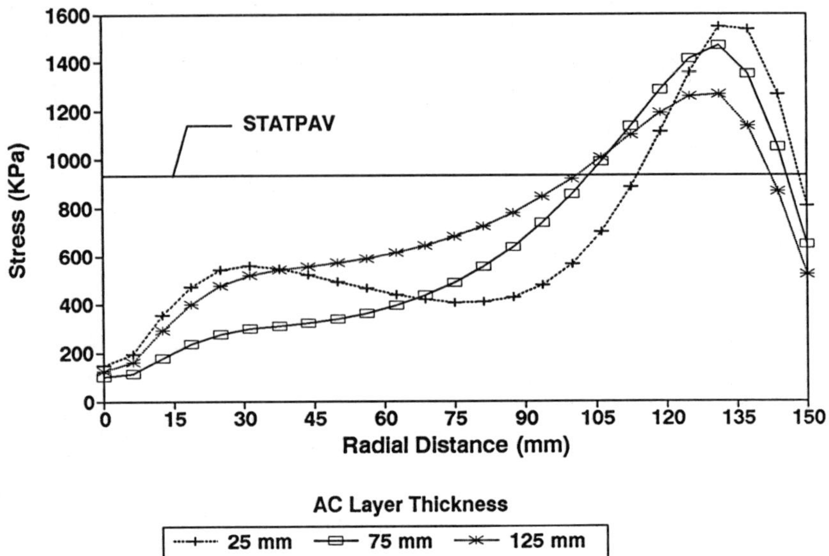

FIG. 5--Influence of AC layer thickness on pressure distribution under plate

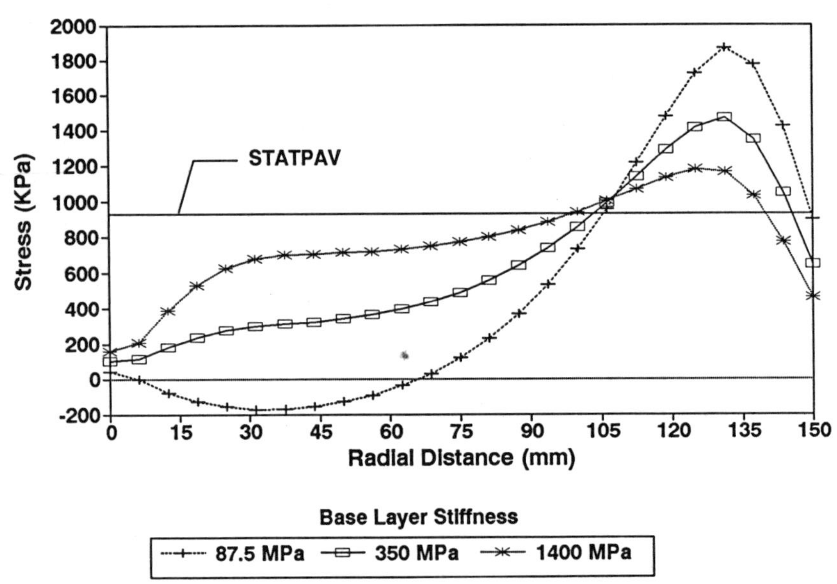

FIG. 6-Influence of base layer stiffness on pressure distribution under plate

PRACTICAL SIGNIFICANCE

As indicated, deflection basins are utilized to determine the backcalculated moduli and remaining lives. Therefore, it would be desirable to determine how significantly the remaining lives and backcalculated moduli are affected by the variations in the deflection basins due to the FWD/pavement interaction. A procedure typically employed by a design engineer to obtain these parameters was followed herein.

The surface deflections calculated by executing STATFWD and STATPAV were used as input to BISDEF to backcalculate the pavement layer properties. The critical strains were then calculated by inputting the backcalculated moduli to BISAR.

Load-induced (fatigue) cracking is directly related to the strain that develops at the bottom of the asphaltic layers. The form of the fatigue law used to predict fatigue cracking life in this study is (Finn et al 1977):

$$Log\ N_F\ =\ 15.947 - 3.291\ \log(\varepsilon_t) - 0.854\ \log(E_R) \tag{2}$$

where,

N_F = number of load application,
ε_t = maximum tensile strain at the bottom of the asphaltic layer, (in micro strain)
E_R = stiffness of the asphalt layer (in ksi)

Rutting is related to the compressive strain developed at the top of subgrade. The remaining life due to rutting can be predicted from (Shook 1982):

$$N_R\ =\ 1.077 * 10^{18} \left(\frac{1}{e_v}\right)^{4.4843} \tag{3}$$

where,

ε_v = compressive strain at the top of subgrade, (in micro strain)

Tables 2 and 3 illustrate the influences of the deflection variation (caused by FWD/pavement interaction and different plate and pavement profiles) on the backcalculated moduli and the remaining lives. In Tables 2 and 3, the differences between the actual and backcalculated values are shown in parentheses. Actual values were determined by using the known properties assumed for each case. The remaining lives obtained in this way are reported in Table 4.

Plate Components

The backcalculated moduli from different layers utilizing the deflection basins obtained from varying the properties of different plate components are reported in Table 2. The average basin fitting mismatch is also defined and reported in Table 2. The average mismatch is about 1 percent indicating the success of the basin fitting process.

The subgrade moduli are backcalculated quite accurately with about a 3 percent error in all cases. The moduli of the base layer are also relatively accurately determined. The maximum differences between the actual and backcalculated moduli are about 9 percent and typically around 5 percent. The case where no rubber pad was utilized under the loading plate is ignored because it is not a practical case. Unfortunately, the moduli of the AC layer are overestimated by a factor of 1.5 to 2 times. Therefore, it may be concluded that the plate components affect the moduli of the base slightly and the modulus of the AC significantly.

The remaining lives obtained from the backcalculated moduli are also presented in Table 2. The remaining lives due to rutting are typically overestimated. This is expected because the stiffnesses of the AC layer are always overestimated by 20 to 40 percent.

The remaining lives due to fatigue are also influenced by the variation in the properties of the plate components but not as significantly as for the rutting cases. Ignoring the case when the rubber pad was totally removed, the predicted fatigue remaining lives are typically within 10 percent and in extreme case not more than 20 percent of actual ones.

TABLE 4--Pavement parameters (actual values) calculated with BISAR

Parameter Studied	Modulus (MPa)			Thickness (mm)			Remaining Life (million ESAL)	
	AC	Base	Subgrade	AC	Base	Subgrade	Rutting	Fatigue
(1)	(2)	(3)	(4)	(5)	(6)	(7)	(8)	(9)
Standard Pavement	3500	350	70	75	300	7125	1.1	1.3
Modulus of AC	1750	350	70	75	300	7125	0.8	1.6
	7000						1.6	1.5
Thickness of AC	3500	350	70	25	300	7175	0.3	20.4
				125		7075	4.1	2.8
Modulus of Base	3500	87.5	70	75	300	7125	0.2	0.1
		1400					26.4	162.4
Thickness of Base	3500	350	70	75	150	7275	0.1	0.7
					450	6975	7.9	1.5
Modulus of Subgrade	3500	350	17.5	75	300	7125	0.1	1.3
			280				44.4	1.4

<u>Pavement Layers</u>

Once again, the basin-fitting process could be carried out successfully as the average mismatches between the measured and theoretical deflections were typically around 1 percent. The FWD/pavement interaction did not significantly affect the moduli of the subgrade. The variations in the properties of the pavement layers did not produce appreciably different subgrade moduli. As indicated in Table 3, the backcalculated moduli are typically within 5 percent of the actual values.

On the other hand, the moduli of the base are somewhat affected. For the overall stiffer pavements (i.e. thicker or stiffer base or AC layers), the base moduli are predicted with an accuracy of about 10 percent. However, for the less stiff pavement sections, the backcalculated base moduli deviate by as much as 35 percent from the actual values. Interestingly, in almost all cases, the base moduli are underestimated. The backcalculated moduli of the AC layer are significantly different from the actual values. Variations by factor of 2 to 3 are not uncommon. These variations can be attributed to two factors, (1) the pavement/FWD interaction and (2) to a less extent to the problems with backcalculating the moduli of thin AC layers. Contrary to the base case, the moduli of the AC layers are always overestimated.

The impacts of the FWD/pavement interaction on remaining lives are also summarized in Table 3. The remaining lives due to rutting are always overestimated resulting in an unsafe design. Once again for pavement with higher overall stiffnesses, the differences, between the actual and predicted remaining lives are small. However, for overall weaker sections, the rut lives are overestimated by a much as a factor of 2.

The predicted and actual fatigue lives are typically different. The differences are less pronounced as compared to the rut lives. The deviations between the actual and predicted fatigue lives exhibit a random pattern, that is, in some cases the fatigue lives are overestimated and in some underestimated. In the model used herein, the fatigue life is functions of radial strains at the bottom of the AC layer and the stiffness of the AC layer. As indicated before, the moduli of AC layers are typically overestimated. Even though not shown here, the radial strains are typically overestimated as well. From Eq. 2, depending on which parameter is more overestimated, the remaining lives can be larger or smaller.

CONCLUSIONS

In this paper the effects of the pavement/FWD interaction on the response of the pavement are addressed. The impacts of the pavement/FWD interaction on the backcalculated moduli and remaining lives of pavements are also investigated.

The nature of the stress distribution under the FWD plate was studied. It was found that the stress distribution under the FWD plate is reasonably uniform for rigid pavements, and should not be of concern. However, the stress distribution under the FWD plate for flexible pavements are influenced by the plate/pavement interaction.

Typically a composite FWD plate is utilized to approximate a uniform load distribution. Variations in the moduli of the steel and PVC plates marginally affect the stress distribution under the loaded area. The thicknesses of the steel plate, PVC plate and rubber pad have some effects on the stress distribution under the plate. The most significant parameter is the stiffness of the rubber pad.

Pavement components affect the stress distribution as well. The pavements with overall stiffer structure (i.e. pavements with thicker or stiffer AC or base Layers) are less affected. As a rule of thumb, the smaller the central deflection, the more uniform the stress distribution will be.

The deviation of the stress distribution from uniformity affects the deflection measured at the center of the loading pad. The deflections measured at the other six locations are typically not affected.

The backcalculated modulus of the subgrade is not affected by the pavement/FWD interaction. The base layer is typically affected to some degree. The modulus of the AC layer is significantly overestimated.

Unfortunately, the remaining life due to rutting is typically overestimated leading to an unconservative design. The remaining lives due to fatigue are also overestimated in many instances. However, the fatigue life is less affected by the FWD/pavement interaction as compared to the rut life.

REFERENCES

Boddapati, K. M., 1992, "Effects of Pavement-Falling Weight Deflectometer Interaction on Measured Pavement Response", <u>M.S. Thesis</u>, University of Texas at El Paso, El Paso, Texas.

Briggs, R.C., and Nazarian, S., 1989, "Effects of Unknown Rigid Subgrade Layers on Backcalculation of Pavement Moduli and Predictions of Pavement Performance", <u>Transportation Research Record</u>, No. 1227, Washington, D.C.

Chang, D.W., Kang, Y.V., Roesset, J.M., and Stokoe, K.H., II, 1992, "Effects of the Bedrock on Profile Backcalculation Using the Deflection Measurements of Pavements", <u>Transportation Research Board</u>, Washington, D.C.

Finn, F., Saraf, C., Kulkarni, R., Nair, K., Smith, W., and Abdullah, A., 1977, "The Use of Distress Prediction Systems for the Design of Pavement Structures", <u>Proceedings,</u> Fourth International Conference on the Structural Design of Asphalt Pavements.

Shook, J.F., 1982, "Thickness Design of Asphalt Pavement - The Asphalt Institute Method", <u>Proceedings,</u> Fifth International Conference on the Structural Design of Asphalt Pavements.

ACKNOWLEDGEMENT

This work has been supported by the Center for High Performance Computing (CHPC) of the University of Texas System. The financial support and technical advice obtained from CHPC are highly appreciated.

NDT For Other Pavement Uses

Kenneth R. Maser,[1] Tom Scullion,[2] W. M. Kim Roddis,[3] and Emmanuel
Fernando[4]

RADAR FOR PAVEMENT THICKNESS EVALUATION

REFERENCE: Maser, K. R., Scullion, T., Roddis, W. M. K., and
Fernando, E., "Radar for Pavement Thickness Evaluation," _Non-
destructive Testing of Pavements and Backcalculation of Moduli
(Second Volume), ASTM STP 1198_, Harold L. Von Quintas, Albert J.
Bush, III, and Gilbert Y. Baladi, Eds., American Society of
Testing and Materials, Philadelphia, 1994.

ABSTRACT: Pavement layer thickness data required for network and
project level pavement management is often not available due to the
cost, time, and traffic disruption involved in taking cores. A nondestructive, non-contact method for thickness measurement is available
and can be implemented from a survey vehicle moving at highway speed.
The technology incorporates horn antenna radar equipment coupled with
customized processing software. The reported work represents evaluations
of the method on 46 different pavement sections located in 12 different
states. The processed radar data has been correlated with results from
over 300 cores, showing an expected accuracy of +/-7.5% (typically +/-
0.5 inches or 12.7 mm) for asphalt thickness. Results from borings and
test pits show an expected accuracy of +/-12% (+/- 1 inch or 25.4 mm)
for base thickness. The radar data has also been successfully utilized
to identify changes in pavement section.

KEYWORDS: Pavement thickness, radar, ground penetrating radar, pavement
management, nondestructive testing, pavement evaluation

---

[1] President, INFRASENSE, Inc., 765 Concord Ave., Cambridge, MA
02138-1044.

[2] Research Engineer, Texas Transportation Institute, Texas A&M
University, College Station, TX 7874.

[3] Assistant Professor, Department of Civil Engineering, University
of Kansas, Lawrence, KS 66045.

[4] Assistant Research Engineer, Texas Transportation Institute,
Texas A&M University, College Station, TX 7874.

INTRODUCTION

Background

Pavement layer thickness data is important in many aspects of pavement engineering and management. Mechanistic models for pavement performance, and structural tests which employ these models for backcalculation, require pavement layer thicknesses as input. Pavement thickness measurements are required for quality control of new construction or overlays, and for determining the milling depth for mill and recycle projects. The layer thicknesses represent an important element of a Pavement Management System (PMS) database, and are needed for load rating and overlay design.

Many state highway agencies have layer thickness records which are inaccurate and/or difficult to access and use. In a recent FHWA survey (Botelho, 1992), it was found that one half of the States do not have pavement layer thickness in their pavement management database. Of the half that have it, the layer thickness data exists for only part of the network, usually the Interstate System.

Traditionally, core samples have provided the only means for accurate pavement layer thickness evaluation. However, these are time consuming and intrusive to traffic. Depending on their spacing, there is always uncertainty regarding thickness variations between cores. For network level pavement inventories, cores are impractical and inadequate as a means for pavement thickness characterization.

A non-destructive, non-contact method for thickness measurement is available and can be implemented from a survey vehicle moving at highway speed. The technology incorporates horn antenna radar equipment coupled with customized processing software. The studies reported in this paper have been carried out to evaluate the accuracy, reliability, and practicality of using this radar technology for continuous measurement of pavement layer properties.

This paper presents the results of five independent studies, each carried out for the purpose of evaluating radar for measuring pavement thickness. The studies have focused on accuracy and repeatability, and on the influence of survey speed, pavement type, and layer construction. The paper describes the individual studies that have been carried out, and the results which have been obtained.

Radar Operational Principles

Ground penetrating radar operates by transmitting short pulses of electromagnetic energy into the pavement using an antenna attached to a survey vehicle (Fig. 1). These pulses are reflected back to the antenna with an arrival time and amplitude that is related to the location and nature of dielectric discontinuities in the material (air/asphalt or asphalt/base, etc). The reflected energy is captured and digitized as a series of pulses that are referred to as the radar waveform. The waveform contains information related to the properties and thicknesses of the layers within the pavement. Figure 2 shows a typical set of pavement waveforms for a two-layer asphalt pavement.

FIG. 1--Photograph of radar van supplied by Pulse Radar, Inc.

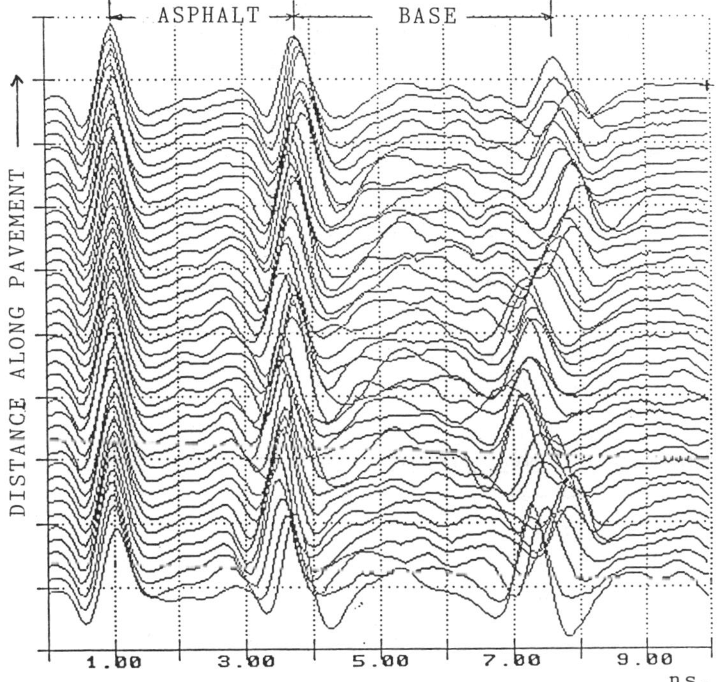

FIG. 2--Typical radar data.

The pavement layer thicknesses and dielectric properties may be calculated by measuring the amplitude and arrival times of the waveform peaks corresponding to reflections from the interfaces between the layers (Fig. 2). The equations for layer thickness and moisture content are presented elsewhere (Maser and Scullion, 1992). Typical results of these calculations are shown in later sections.

The radar equipment required to produce the data shown in Figure 2 is a wide-band, short pulse (1 nanosecond) air-coupled "horn antenna" operating at a center frequency of approximately 1 GHz (Fig. 1). More traditional "ground-coupled" radar equipment used by previous investigators (Berg, 1986; Rosetta, 1980; Eckrose, 1989) is less suitable to pavement layer evaluation due to poor resolution and slow operating speeds.

Radar waveform analysis procedures for pavement layer evaluation have been implemented into a software package called "PAVLAYER" (PAVment LAYer Evaluation using Radar). [5] This software uses automated signal processing algorithms to track the amplitudes and arrival times of the significant reflectors in the radar data. The layer properties and thicknesses are subsequently computed from this information, independently of core data. Such core data was previously required for ground-coupled systems and for manual analysis procedures as reflected in the current ASTM specification (1987). The remainder of this paper describes the results of studies carried out to evaluate the accuracy of the thickness measurement technology described above. The studies and their results are presented in chronological order.

TEXAS STUDIES

Initial Study (Maser and Scullion, 1992; Maser, 1990)

In the initial study, four asphalt pavement sites in the vicinity of College Station, Texas, were selected for evaluation, as described in Table 1. These sites had been designated by the Strategic Highway Research Program (SHRP) as General Pavement Study (GPS) sites. Each test section was 1500 feet (457.5 m) long, including 500 ft. (152.5 m) preceding the GPS site, the 500 ft. GPS site itself, and 500 ft. beyond the GPS site. It was understood that verification sampling could only take place in the first and last 500 ft. sections, since the 500 ft. GPS site could not be disturbed.

TABLE 1--Pavement properties for the Texas study.

Site	Asphalt Thickness (in.)		Type	Base Thickness (in.)
	top course	bottom course		
SH 30	1.0	7.0	Bituminous treated soil	6.0
SH 19	1.0	6.0	Lime-treated soil	6.0
SH 105	1.0	1.0	crushed stone	10.0
SH 21	2.0	6.0	crushed stone	10.0

Note: 1 inch = 25.4 mm

---

[5] PAVLAYER is copyrighted by INFRASENSE, Inc. Cambridge, MA.

Radar data was collected by INFRASENSE using a van-mounted horn antenna system provided and operated by Pulse Radar, Inc. of Houston, Texas. Data was collected in June and July of 1990 and analyzed by INFRASENSE. Based on the analysis, areas within each site were identified for direct sampling. Extraction of direct samples was carried out jointly by the Texas Transportation Institute (TTI) and the Texas Department of Transportation (TexDOT).

Four surveys were carried out at each site: one each at 5, 15, and 40 mph, and one repeat survey one month later. In each survey, data was collected continuously over the 1500 ft. (457.5 m), resulting in typical processed data as shown in Figure 3. Locations for ground truth were determined to represent significant variations in thickness and dielectric constant, as revealed by the radar data. Asphalt thickness was determined using wet 4" (101.6 mm) and dry 6" (152.4 mm) diameter cores, and base thickness was determined using a penetrometer and by visual observation in the dry coreholes. Ground truth data was also available from field data collected previously as part of the SHRP Program.

The ground truth data was subsequently correlated with the radar data at each sample location, as shown in Figures 4 and 5. The accuracy of the radar predictions for asphalt thickness was within +/- 0.32 inches (8.13 mm) using the radar data alone, and within +/- 0.11 inches (2.79 mm) when one calibration core was used per site. The accuracy of the radar predictions for base thickness was within +/- 0.99 inches (25.15 mm). The actual asphalt layer thickness was shown to vary by over 20% from values assumed from prior records and earlier cores. These variations have been shown to lead to errors of up to 95% in base moduli backcalculated from FWD data (Maser and Scullion, 1992).

The radar results were shown to be repeatable over time and independent of survey speed at speeds up to 40 mph. The radar data was analyzed automatically using software which operated directly on the raw radar waveforms and produced numerical layer thickness profiles. The results of this initial study showed that highly accurate measurements of pavement layer properties using the appropriate radar equipment and data processing software.

Followup Study (Maser, 1992a)

A followup study was carried out by TexDOT to (a) evaluate the influence of thin overlays and surface treatments on the accuracy of radar thickness predictions; (b) develop approaches which deal with these conditions in a manner which maximizes the accuracy; and (c) assess the capability of radar to accurately measure overlay thickness.

This effort was carried out by collecting and analyzing data collected on SHRP Specific Pavement Study (SPS) sites adjacent to the GPS sites surveyed in the initial study. This approach provided the means for investigating the influence of thin overlays, chip seals, and slurry seals under conditions where the pre-existing pavement had been extensively evaluated.

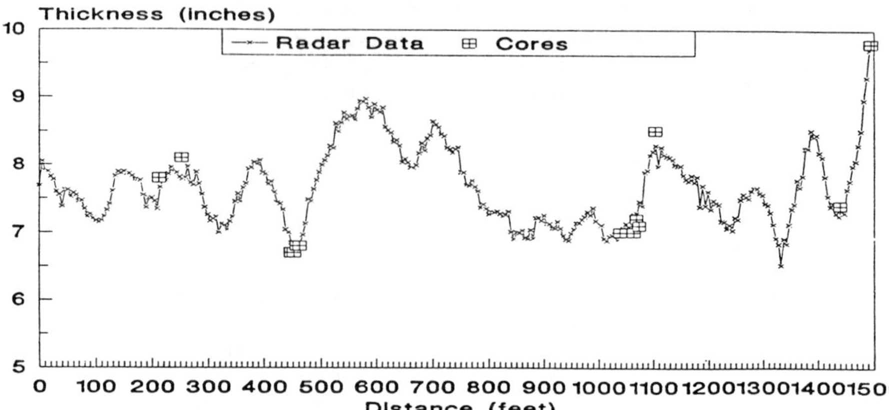

FIG. 3--Texas study: Asphalt thickness vs. distance.
Note: 100 Feet = 30.5 meters

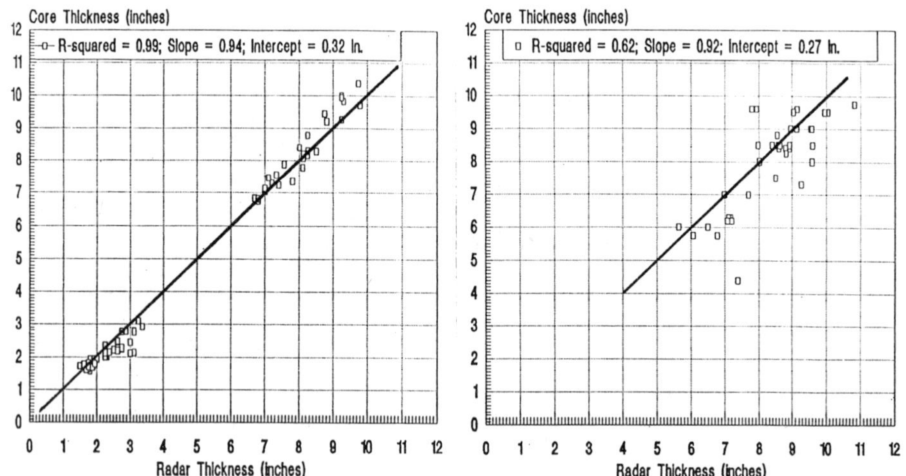

FIG. 4--Texas Study: Summary of
radar vs. cores; asphalt thickness
(based on 68 cores).
Note: 1 inch = 25.4 mm

FIG. 5--Texas study: Summary of
radar vs. cores; base thickness
(based on 40 cores).

Radar data for this followup study was collected by INFRASENSE and
TTI using TTI's Penetradar PS-24 horn antenna radar system in June and
July of 1991. The characteristics of the radar data produced by this
system are very similar to those of the Pulse Radar system used in the
initial study. The radar data was analyzed using PAVLAYER to compute
layer thicknesses, and thickness predictions are compared to core
samples. Test sections listed in Table 2 below were associated with the
GPS sections used in the initial study, plus one new GPS site also in
the vicinity of College Station.

TABLE 2--<u>Test sections for Followup Texas Study.</u>

Section	Highway	Description
SH30-1	SH30	GPS section
SH30-2	SH30	thin overlay
SH30-3	SH30	slurry seal
SH30-4	SH30	chip seal
SH105-1	SH105	thin overlay
SH105-2	SH105	slurry seal
SH105-3	SH105	GPS & chip seal
SH105-4	SH105	chip seal
US190-R	US190	GPS right wheelpath
US190-C	US190	GPS centerline
US190-I	US190	inside lane (not GPS)

The cross sectional properties of these sections for the SH30 and SH105 sites are those listed in Table 1. The US190 site was 3-4" (88.9 mm) of asphalt over a 9 inch (228.6 mm) cement treated base. The asphalt layer included a 1 inch asphalt overlay made from lightweight aggregate.

The data collected at these sites was analyzed for pavement layer thickness and dielectric constant. Ground truth asphalt thickness was obtained using 6 inch (152.4 mm) diameter wet cores. Coring was also used for the thickness evaluation of the cemented base on US190. Granular and treated base thickness evaluation was carried out using a grooved cylindrical sleeve. The sleeve was placed in the corehole, and driven through the base into the subgrade. The base and subgrade material remained embedded in the groove, and the boundary could be observed when the cylinder was removed. The location of this boundary provided a measure of the base thickness.

Table 3 summarizes the deviations between radar and ground truth measurements by pavement section, by site, and for the entire study.

Overall accuracies for asphalt thickness are comparable to those of the earlier study. Locally, however, higher accuracies were achieved at most sites. The presence of the expanded aggregate overlay at US190 and the chip seal at SH30 produced lower accuracies which adversely affected the overall average.

The accuracy of the base thickness calculations improved over those presented in the initial study. This improvement is attributed to the proper accounting of the surface layers, and to an algorithm which qualified the acceptability of the base thickness data. Note that for the cement treated base at US 190, there was inadequate contrast between base and subgrade to allow for detection at base layer thickness.

TABLE 3--<u>Followup Texas study,
deviation of radar data from ground truth.</u>

Site	RMS Deviation (inches)	
	Asphalt	Base
SH30 GPS	0.19	0.77
SH105 GPS	0.17	1.11
US190 GPS	0.56	...
SH30 Chip Seal	0.64	0.03
SH30 Slurry "	0.27	0.61
SH105 Chip Seal	0.11	0.34
SH105 Slurry "	0.15	0.76
SH105 Overlay	0.20	0.33
SH30 All Sites	0.35	0.77
SH105 All Sites	0.17	0.81
All Sites	0.33	0.77

Note: 1 inch = 25.4 mm

The results of this followup study illustrated the influence of
overlays and surface treatments on the accuracy of pavement thickness
calculations using radar. Where there was a significant contrast between
the overlay and the existing pavment, it was (a) possible to accurately
measure the overlay thickness for thickness as low as 1 inch; and (b)
necessary to account for the overlay in order to obtain accurate
thickness values for the deeper layers. On the other hand, it appears
that slurry and chip seals can be ignored in the thickness computation.
For chip seals, the results may be slightly less accurate due to the
distortion which is created on the surface reflection.

KANSAS STUDY (Roddis et al, 1992)

This study sought to evaluate radar thickness measurements on a
population of pavement types present in the Kansas, with a particular
emphasis on variations in base type and road history. Fourteen sites in
the Lawrence/Topeka area were selected by the University of Kansas (KU)
with assistance from the Kansas Department of Transportation (KDOT). The
characteristics of each of the sites are described in Table 4. The Table
shows sites which are older and more significantly layered than those
studied in Texas. The partially designed bituminous sites (KDOT road
categories 15, 21, and 22) are particularly unique to the radar
evaluation. These represent roads of previously unknown construction
which were acquired by the state and subsequently overlaid to meet
current service requirements.

The radar surveys at each site were carried out in July, 1991. The
test sections were 1000 feet (305 m) long, and the surveys were conducted at speeds ranging from 5 to 20 mph using Pulse Radar's R-II

equipment. The radar data was collected and analyzed by INFRASENSE using PAVLAYER. The resulting thickness calculations were correlated with 4" (101.6 mm) diameter wet core samples obtained by KDOT in coordination with the KU. Because of unforseen scheduling problems, ground truth data was collected on only 10 of the 14 sites which were surveyed. Core and test pit data from an 11th site (the SHRP site) were also used since these were available from the SHRP database.

TABLE 4--<u>Pavement layers from KDOT data base.</u>

KDOT Road Cat.	Asphalt Layer 1 (in.) Thick	Year	Layer 2 (in.) Thick	Year	Layer 3 [6] (in.) Thick	Year	Base Layer (in.) Thick	Matl.	Year
3	3.0	79	1.0	79	2.0	50-70	9.0	PCC	50
4	1.0	80	3.0	80			9.0	PCC	56
5	0.75	81	1.0	73	1.0	73	16.0	ACB	73
11	1.0	84	2.0	84			9.0	PCC	68
12	1.5	79					2.0	BIT	79
15	1.5	90	1.5	83	2.3	47-70	2.0	BIT	47
17	4.0	75					9.0	BM4	75
18	2.0	88	0.75	88	2.3	47-70	6.0	BM2	47
21	2.0	90	1.5	81	1.8	52-70	2.0	BIT	52
23	1.5	82	1.0	50	2.0	50-70	6.0	AB	50
SHRP [7]	4.0	72					7.5	ACB	72

Legend: AB = Aggregate binder; ACB = Asphalt concrete base mix; BIT = Bituminous cover on old pavement; BM2 = Bituminous mixture 2A; PCC = Portland cement concrete; 1 inch = 25.4 mm.

Two different types of radar data analyses were carried out on these tests sections. The first was where the various bituminous layers could be treated as a single monolithic layer. The second was where the thicknesses of each significant bituminous layer had to be calculated separately. The decision to use one of the two approaches depended on how clearly the individual bituminous layers appeared in the raw data. Figure 6 shows an example result for a two layer bituminous system, in which the original asphalt construction can be distinguished from the more recent overlays.

Figure 7 shows the correlation between the radar data at the 73 core locations, and the core data. The figure shows a very strong correlation, with an R-squared of 0.96, and an rms deviation of 10%.

---

[6] SHRP LTPP GPS 201005 Site, Road Category 17

[7] Includes chip seals applied approximately every 3 years.

This correlation was achieved using radar data alone, and included data from damaged cores and from locations where the asphalt bottom was unclear from the core (due to an asphalt treated base). Summary statistics for the data from the 11 sites is presented in Table 5.

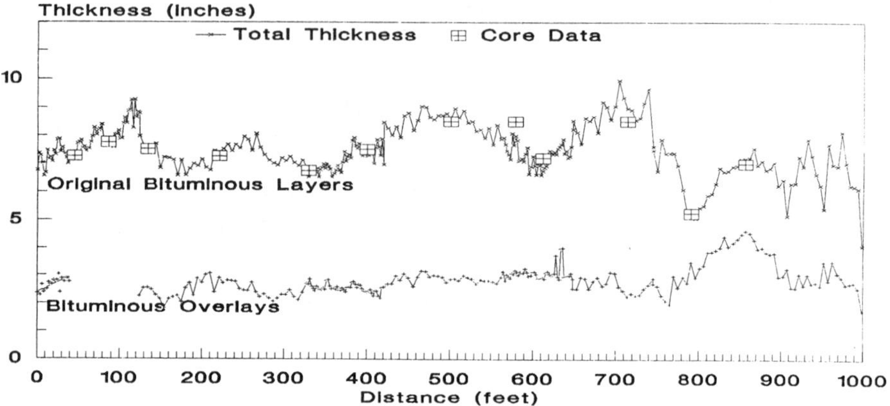

FIG. 6--Kansas study: Asphalt thickness vs. distance.
Note: 100 feet = 30.5 meters

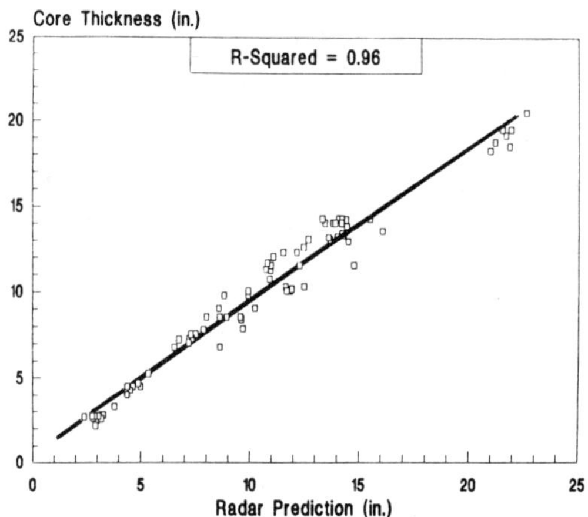

FIG. 7--Kansas study: radar vs. cores
Results for 73 core samples.
Note: 1 inch = 25.4 mm

The largest deviations between radar predictions and core data occur in sites 5, 15, and 18. The deviations for sites 15 and 18 can be attributed to the poor quality of the core data, as discussed earlier. In site 5, the core data revealed a total of 5 asphalt layers adding up to 18 (457 mm) to 20 (508 mm) inches in thickness. The radar data did not show any significant contrast between layers, and therefore the dielectric constant for the top layer was used for the entire thickness computation. Typical pavement conditions would suggest, however, that there is a gradient of moisture content with depth. This gradient would yield an increased dielectric constant and a reduced velocity with depth, which, if accounted for, would produce more accurate thickness computations.

TABLE 5--<u>Kansas study: comparison at radar to core data.</u>

KDOT Road Cat.	AVERAGES (in.) at Core Sites		DIFFERENCES Between Radar and Cores	
	Radar	Core	inches	%
3	4.64	4.41	0.23	5.30
4	3.30	2.80	0.50	17.86
5	21.66	19.17	2.49	13.00
11	2.82	2.63	0.19	7.31
12	7.37	7.41	0.05	0.62
15	9.60	8.36[8]	1.23	14.76
17	14.32	14.03	0.30	2.11
18	11.92	10.12[9]	1.79	17.70
21	10.91	10.71	0.20	1.87
23	12.46	12.55	0.09	0.70
17(SHRP)	14.23	13.35	0.89	6.64

Note: 1 inch = 25.4 mm

In summary, the Kansas project extended the results of the original Texas projects to a wider variety of pavement ages and types, and to thicker and more layered asphalt constructions.

FLORIDA PHASE I STUDY (Fernando and Maser, 1992)

The overall objective of this multi-phase project is to test and implement a ground penetrating radar system to supply accurate pavement layer data for Florida's PMS. The specific objectives of Phase I were to (1) evaluate the accuracy of radar-based layer thickness predictions; and (2) assess the ability of radar to identify (a) base layer material types and (b) changes in pavement structure. Five test sections were used for evaluation of layer thickness and base material type. Four of these were .3 miles in length, and a fifth was 420 feet (128.1 m) long. A sixth section 1.7 miles in length was used for identifying changes

---

[8]questionable data due to poorly defined interface between asphalt layers and asphalt/soil base.

[9]questionable data due to core damage during drilling.

in pavement structure. All six sites were located in the Tallahassee area.

The pavement layer structure at the five thickness evaluation sites is shown in Table 6. As can be seen in the table, these test pavements were more complex in their layer construction than those tested in the previous Kansas and Texas studies. Radar data was collected in August of 1991 by INFRASENSE and TTI using Pulse Radar's R-II equipment, at speeds ranging from 5 to 30 mph. The radar data was analyzed by INFRASENSE using PAVLAYER. A core and test pit sampling plan was prepared by TTI, and cores were taken and logged by the Florida Department of Transportation (FDOT).

TABLE 6--<u>Layer structure for Florida test sites</u>
<u>(determined from cores and test pits)</u>.

Layer	Site 1	2	3	4	6
Asphalt					
1	1" FCS	0.9" FCS	1.2" FCS	1.1" FCS	1.3" FC1
2	0.5" TYII	0.65" TYII	0.5" TYII	0.6" TYII	1.7" TYII
3	0.6" CRL	0.66" CRL	0.6" CRL	0.6" CRL	
4	0.36" LEV	0.31" LEV	0.5" LEV	0.5" LEV	
5	1" TYI	3.4" TYII	1.9" TYS	4.5" SAHM	
6	2" BND	1" BND			
Base	9.1" LR	8.5" LR	9.6" LR	8.5" CON (LWP) 6.7" SCM (RWP)	11.7" LR

Legend: FCS = Friction course & type S; TYII = Type II; CRL = Crack Relief Layer; LEV = Levelling; TYI = Type I; BND = Binder; LR = Limerock; SAHM = Sand-Asphalt Hot Mix; TYS= Type S; FC1 = Friction Course; CON = Concrete; SCM = Soil Cement; 1 inch = 25.4 mm.

The initial data analysis was carried out blind - i.e., without the benefit of core data or inventory data from the FDOT database. The computed radar data was then correlated with results from 73 core samples and 10 test pits. Site averages of the blind radar predictions for asphalt thickness were within 0.5 inches (12.7 mm) of core values for 4 of the 5 test sections, and within 1.2 inches (30.5 mm) for a fifth section. The discrepancy on this last section (#4) was due to an unusual response of a sand-asphalt hot mix layer which had not been encountered in previous work. Once this response was understood, the resulting radar computations for section #4 were revised and fell to within 0.6 inches (15.2 mm) of the core values.

Blind radar predictions for the base layer thickness were within 0.7 inches (17.8 mm) of core values for 3 of the 5 test sections, and within 2.1 inches (53.3 mm) for a fourth section (#1). Base thickness data for a fifth section (Section 4) could not be computed. The 2.1 inch discrepancy for Section 1 was reduced to 1.0 inch when the details of the asphalt layering were properly taken into account in the analysis. The average accuracy of radar-based thickness calculations for all sites was within 0.2 inches (5.1 mm) for the asphalt and 0.2 inches (5.1 mm) for the base when one calibration core was used for each section.

The blind radar analysis correctly identified the base layer material for 4 of the 5 sites listed in Table 6. Since this was the first use of radar for making this type of determination, the results were considered to be very promising. The radar analysis also correctly identified seven structural changes on the 1.7 mile section test section.

In summary, the results of this study showed that the PAVLAYER analysis procedures applied in Texas and Kansas could handle the more complex asphalt layering found in Florida with no significant effect on accuracy. The study also showed that the accuracy of the radar calculations can be enhanced by the availability of a calibration core.

SHRP STUDY (Maser, 1992b)

The objective of this study was to evaluate the accuracy of ground penetrating radar for measuring pavement thickness in LTPP test sections. Radar has the capability to detect thickness variations which would not otherwise be revealed. These variations can produce errors in the LTPP data analysis and in the performance models being developed under SHRP. Field surveys were carried out at 10 LTPP General Pavement Study (GPS) asphalt pavement sites representing a range of environmental conditions and pavement structures.

The radar data for this study was collected in February-March, 1992 using Pulse Radar's R-II equipment at the sites listed in Table 7 below. The data was collected by INFRASENSE on 1500 foot (475.5 m) sections as done in the original Texas study, and was analyzed by INFRASENSE using PAVLAYER. The accuracy of the radar-based calculations was evaluated using core data obtained from the SHRP LTPP database at 67 locations in 3 steps: a "blind" evaluation, an update using pavement structure information from inventory data, and an update using core data from one location for calibration. Figure 8 shows a typical radar thickness result for one of the 10 sites, and Figure 9 shows the overall correlation between radar and core data.

TABLE 7--<u>Sites surveyed in the SHRP study.</u>

GPS Site #	Location	Plan Data Layer Thickness(in.)		Base Material
		Asphalt	Base	
223056	Bunkie, LA	10.0	8.0	CTB
124108	Ft. Walton Beach, FLA	10.6	12.0	Soil/Agg
134112	Brunswick, GA	16.5		none
371645	Whiteville, NC	4.5	7.0	CTB
512004	Danville, VA	7.4	6.5	Cem/Agg
242401	Edgewood, MD	6.7	4.0	Cem/Agg
341033	Trenton, NJ	7.0	6.0	Cr.Stone
479024	Murfreesboro, TN	9.3	5.0	" "
053071	Rogers, AR	17.0		none
482108	Texas City, TX	3.0	14.0	Cem/Agg

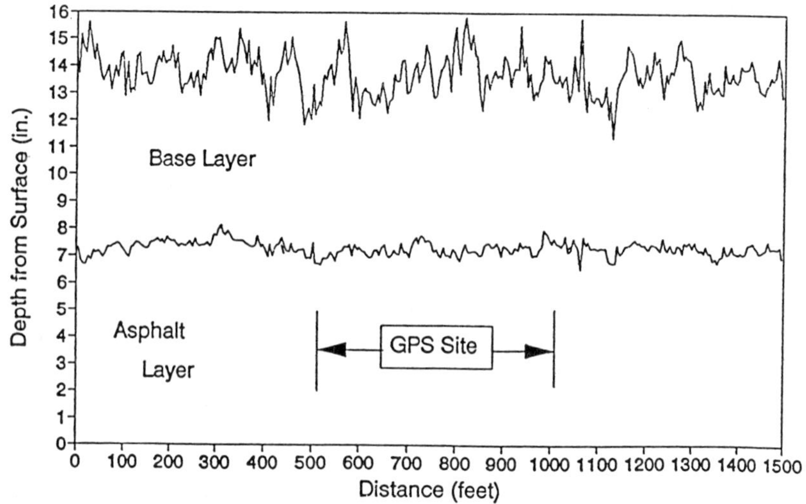

FIG. 8--SHRP study: sample analysis results.
Note: 100 feet = 30.5 meters

The results show that blind radar asphalt thickness data correlated with the core data with an R-squared of 0.98 and an RMS deviation of +/- 0.78 inches (19.8 mm), or +/- 7.1%. Plan data had little influence on revising the "blind" radar calculations for asphalt thickness. The availability of the approach end core data helped to identify an error in detecting the asphalt bottom at one site. Once corrected, the RMS deviation was reduced to +/- 0.68 inches (17.3 mm). Finally, the full set of radar data was calibrated with the approach end core information of each GPS site. The resulting RMS deviation was reduced to +/-0.51 inches (12.9 mm), or +/-5.1%.

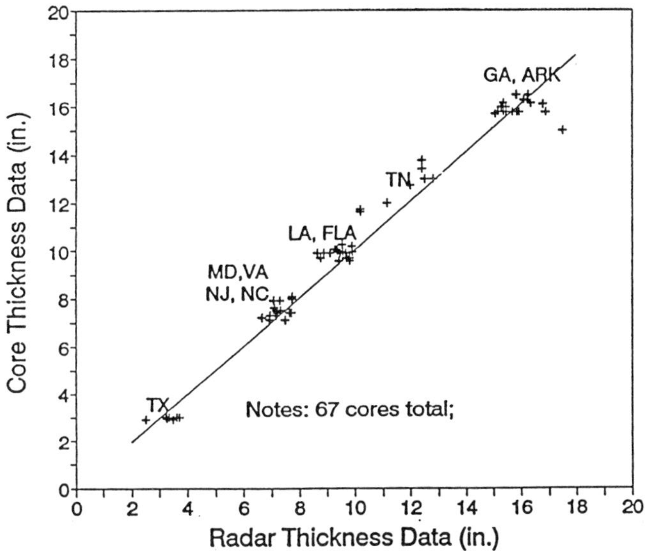

FIG. 9--SHRP study: Radar vs. core values for asphalt.
Note 1 inch = 25.4 mm

The radar data revealed deviations from the LTPP cores within the
GPS sections which exceeded 10% in 5 of the 10 sites surveyed. The
maximum deviation between LTPP core data and radar data within the GPS
site ranged from 6 to 21%. The potential errors in LTPP data analysis
and modelling associated with these deviations could be reduced if radar
thickness data were available for each site.

FHWA Study (Maser, 1992c)

The purpose of this project was to demonstrate and evaluate the
pavement layer thickness measurement technology described above for use
in State PMS systems. Since most of the previous studies focused on
asphalt pavement, this study confined its effort to concrete and
composite pavement.

The FHWA study was carried out in two states, Arkansas and
Louisiana. A 1 mile section of concrete pavement and a 1 mile section of
composite (concrete with an asphalt overlay) pavement was evaluated in
each state. The route in Louisiana selected for the study is in the
Alexandria area and the route selected in Arkansas is Highway 65 in the
Little Rock area. The participating States were to provide a
calibration core and conduct verification tests (coring) approximately
5 per mile.

The survey was conducted with Pulse Radar's R-II radar equipment,
and the data was analyzed by INFRASENSE using INFRASENSE's PAVLAYER
software. Layer thickness was analyzed every 10 (3.1 m) to 20 (6.1 m)

feet to represent "network-level" resolution needs, and results included statistics showing the mean and standard deviation of the thickness of each layer detected. Mean thicknesses and standard deviations for the four pavement sections are summarized below, along with inventory data available for each site.

Pavement Type	Location	Layer Thicknesses (in.)					
		Asphalt			Concrete		
		Plan	Radar		Plan	Radar	
			Mean	St.Dev		Mean	St.Dev
Concrete	I49 NB, LA		...	...	10.0	10.7	0.83
Composite	US71 NB, LA	11.0	9.8	0.7	7.5	8.1	0.7
Concrete	US65 NB, ARK		...	...	10.0	10.2	0.71
Composite	US65 NB, ARK	4.5	5.1	0.46	10.0	11.0	0.7

Note: 1 inch = 25.4 mm

The radar data and plan data are in general agreement, with a maximum difference between the two of 13%. The accuracy of the radar data at the Arkansas sites was evaluated by comparing radar data from specific locations to cores at those location. The radar concrete thickness data was accurate to within 0.08 inches (2.0 mm) and 0.88 inches (22.4 mm) for the concrete and composite pavements respectively. These results shows that the accuracies achieved for asphalt pavements can also be achieved for concrete. They also show that data analysis can be carried out at selectable distance intervals to suit project as well as network applications.

CONCLUSIONS

The studies described above have demonstrated the capability of a non-destructive test technology for pavement layer thickness evaluation. The technology incorporates short-pulse air-coupled horn-antenna radar equipment coupled with analysis software which applies electromagnetic models to the raw radar data. This technology can provide accurate pavement layer thickness data for project and network applications. Accuracy of thickness calculations of +/- 7.5% for asphalt and concrete, and +/-12% for unbound base layers can be expected. Data collection can be carried out at speeds up to 40 mph without any effect on accuracy and calibrating cores are not required. This technology is distinct from earlier radar-based systems using "ground-coupled" antennas.

Asphalt layer thicknesses can be calculated for multilayer asphalt systems, and the thickness of the individual layers can be calculated down to a minimum of one inch (25.4 mm). The radar data can also be effectively used to identify the locations of changes in pavement structure which are not apparent from surface conditions. All of these above capabilities appear to be independent of geographic location, pavement construction, and pavement age.

ACKNOWLEDGEMENTS

The authors would like to acknowledge the following individuals and organizations for their assistance and support in carrying out the projects described in this paper: Mr. Robert Briggs, of TexDOT; Mr. Andy Gisi of KDOT; Mssr. Bruce Dietrich and Bill Lofroos of FDOT; Ms. Cheryl Richter of SHRP; Ms. Sonya Hill and Mr. Frank Botelho of FHWA; and Mssrs. Dave Turner and Dave Gahr of Pulse Radar, Inc.

REFERENCES

American Society for Testing and Materials, 1987, <u>Standard Test Method for Determining the Thickness of Bound Pavement Layers using Short Pulse Radar.</u> ASTM Standard D4748-87.

Berg, F, Jansen, J.M., and Larsen, H.J.E., 1986, <u>Structural Pavement Analysis based on FWD, Georadar, and/or Geosonar Data.</u> Proc. 2nd International Conference on the Bearing Capacity of Roads and Airfields, Plymouth, U.K.

Botelho, Frank, 1992, FHWA Pavements Division, Personal Correspondence, February.

Eckrose, R.A.,1989, <u>Ground Penetrating Radar Supplements Deflection Testing to Improve Airport Pavement Evaluations.</u> Non-Destructive Testing of Pavements and Back Calculation of Moduli, ASTM STP 1026, A.J. Bush III and G.Y. Baladi, Eds., ASTM.

Fernando, E., and Maser, K., 1992, "Development of a Procedure for the Automated Collection of Flexible Pavement Layer Thicknesses and Materials: Phase 1: Demonstration of Existing Ground Penetrating Radar Technology," Florida DOT State Project 99700-7550.

Maser, K.R., 1990, <u>Automated Detection of Pavement Layer Thickness and Subsurface Moisture using Ground Penetrating Radar.</u> Final Report Prepared for the Texas Transportation Institute, College Station, Texas.

Maser, K.R.,1992a, <u>Influence of Asphalt Layering and Surface Treatments on Asphalt and Base Layer Thickness Computations Using Radar.</u> Report prepared for the Texas Transportation Institute.

Maser, K.R.,1992b, <u>Ground Penetrating Radar Surveys to Characterize Pavement Layer Thickness Variations at GPS Sites.</u> SHRP Contract 90-ID034.

Maser, K.R.,1992c, "Highway Speed Pavement Thickness Surveys using Radar" Draft Final Report, FHWA Contract DTFH 78-72-P-F198.

Maser, K.R. and Scullion, T., 1992, "Automated Pavement Subsurface Profiling Using Radar- Case Studies of Four Experimental Field Sites" <u>Transportation Research Record</u> No. 1344, Transportation Research Board, Washington D.C.

Roddis, W.M. Kim, K.R. Maser and A.J. Gisi, 1992, "Radar Pavement Thickness Evaluations for Varying Base Conditions," <u>Transportation Research Record</u> No. 1355.

Rosetta, Jr., J.V., 1980, <u>Feasibility Study of the Measurement of Bridge Deck Overlay Thickness using Pulse Radar.</u> Report R-35-80, Prepared for the Massachusetts Department of Public Works by Geophysical Survey Systems, Inc. Salem, New Hampshire.

David A. Van Deusen[1], Carl A. Lenngren[2], and David E. Newcomb[3]

A COMPARISON OF LABORATORY AND FIELD SUBGRADE MODULI AT THE MINNESOTA ROAD RESEARCH PROJECT

REFERENCE: Van Deusen, D. A., Lenngren, C. A., and Newcomb, D. E., "A Comparison of Laboratory and Field Subgrade Moduli at the Minnesota Road Research Project," Nondestructive Testing of Pavements and Backcalculation of Moduli (Second Volume), ASTM STP 1198, Harold L. Von Quintas, Albert J. Bush, III, and Gilbert Y. Baladi, Eds., American Society of Testing and Materials, Philadelphia, 1994.

ABSTRACT: A soil sampling and testing program was conducted on the pavement embankment at the Minnesota Road Research Project. Both destructive and nondestructive approaches were taken. Nondestructive load tests were conducted both before and after base construction using a falling-weight deflectometer; both disturbed and undisturbed soil samples were retrieved. Falling-weight deflectometer tests were conducted at regular intervals along three different offsets directly on the soil embankment. A large diameter plate was used with modest load levels. Other data obtained during the testing program included jar, bag, and thin-wall soil samples, and dynamic cone penetrometer soundings. The falling-weight deflectometer load and deflection data were used to backcalculate the elastic moduli for a multi-layered system. Due to the variability in the surface conditions special backcalculation techniques had to be employed. The variability of the deflections and estimated moduli were addressed using both general statistics and geostatistical analyses. The backcalculated moduli for the near-surface layers were softer and more variable than the deeper layer. Laboratory resilient modulus tests conducted on the thin-wall samples yielded values that compared well with the backcalculated values from the deeper layer.

KEYWORDS: non-destructive testing, falling-weight deflectometer, moduli backcalculation, subgrade testing, subgrade variability.

[1]Research Fellow, [2]Post-Doctoral Research Associate, and [3]Assistant Professor, University of Minnesota, Department of Civil and Mineral Engineering, 500 Pillsbury Drive S.E., Minneapolis, MN 55455.

361

INTRODUCTION

Background

The Minnesota Department of Transportation is currently constructing the Minnesota Road Research Project (Mn/ROAD) on Interstate 94 in central Minnesota. A total of 40 different pavement test cells will be instrumented with sensors designed to measure responses and parameters which in turn will be used to rationally explain the pavement performance. Many factors affect the structural design of a pavement, including the physical and mechanical characteristics of the subgrade soils. The study presented in the present paper was undertaken to assess the properties of the subgrade soils at the site.

Objectives

The primary objectives of this study were to arrive at an initial assessment of the variability of the subgrade properties at the facility and to compare two different methods of resilient modulus determination. Since the performance of the individual test cells may be influenced by the properties of the subgrade soils it is important to have a sound understanding of how these quantities vary along the site. The properties were determined by nondestructive test methods as well as disturbed and undisturbed soil samples. Secondary objectives were to assess the changes in measured deflections from tests conducted before and after base construction.

Scope

Mn/ROAD is located in east-central Minnesota on I-94 (see Fig. 1). The facility is divided into a high and low traffic volume experiment. The high volume mainline experiment consists of 23 different test cells and will be open to traffic. These test cells are divided between 5-year and 10-year design lives. The low volume experiment is a closed test track constructed of 17 different cells with two turnaround loops. Native soils at the site are primarily silty clay and the existing topography had no more than 3 to 4 m of relief prior to construction. The embankments for both the mainline and low volume experiment range in height from 1 to 3 m and are constructed on cuts ranging from 0.2 to 0.6 m. A detailed discussion of the experimental design of the facility is given by Newcomb et al. (1991).

The subgrade testing procedures discussed in this paper mark the beginning of the proposed material sampling and testing program at the facility. Falling-weight deflectometer (FWD) tests, soil borings for bag and jar samples, and thin-wall samples were taken simultaneously in the field. About half of the 40 test cells were tested during fall 1991 while the remaining cells were finished in early summer of 1992. The layout of the test cells are also shown in Fig. 1.

Disturbed and undisturbed soil samples were retrieved from selected locations within each cell on the same day as the FWD testing so that comparable material samples were obtained. Testing was also

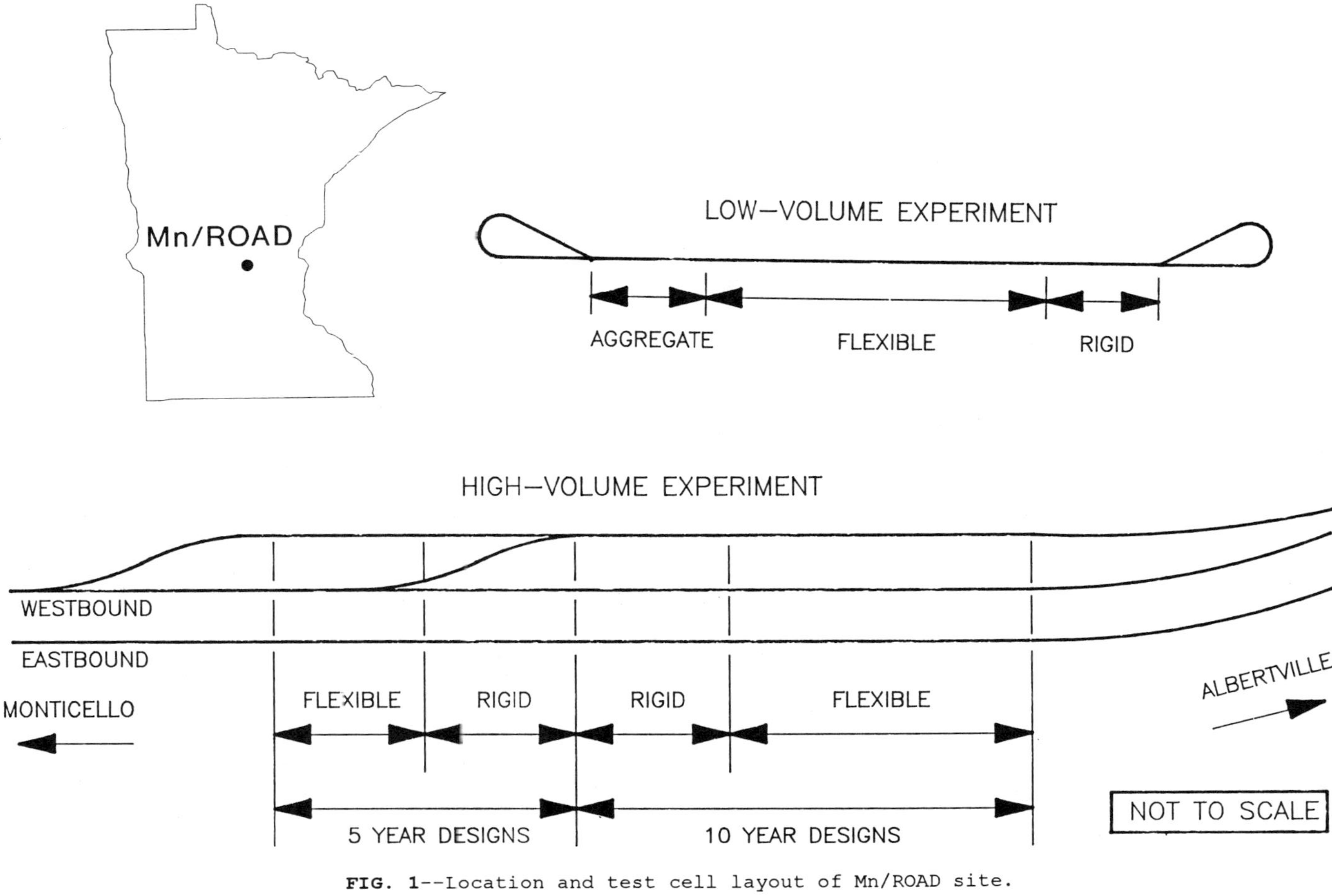

FIG. 1--Location and test cell layout of Mn/ROAD site.

conducted on the base layer surface of the 5-year test cells during
summer 1992 prior to paving. These tests included deflection testing,
and base layer soil sampling.

The current data are important to future data analysis at Mn/ROAD
in that the data (e.g., resilient modulus, moisture content, density)
obtained in this study will be used with future data. In essence, this
study is an attempt to characterize the pavement subgrade soils before
construction of the pavement test cells.

PREVIOUS WORK

Although the focus in the present paper is on the analysis of FWD data a
brief review of other site surveys is included here. Previous studies
of laboratory and in-situ property correlation have focused on
comparison of field (backcalculated) and laboratory (resilient modulus)
determined moduli and also site variability. The problem with arriving
at a comparison between field and laboratory subgrade moduli lies in the
differences between the two approaches. In particular, there are major
contrasts in the volume of material being tested by the FWD relative to
a laboratory resilient modulus test on thin-wall specimen. There is
also the additional question of the extent to which the laboratory
conditions can represent in-situ stress conditions.

Houston et al. (1992) discussed the problem of volume size
contrast and suggested that it is only exacerbated by the usual sample
retrieval procedures of taking material from the upper 30 to 60 cm of
the subgrade. They also suggested a remedy to the sampling volume
problem that involves retrieving between six and 10 samples from the
soil in the vicinity of an FWD station. While this approach may help to
address the volume-variability relationship it seems impractical to
employ in a large-scale field test.

Newcomb (1987) and Lee et al. (1988) performed FWD tests on
existing pavements in Washington. Each 300 m long test cell was tested
using the FWD at 15 m intervals. Disturbed soil samples were retrieved
from the shoulders of these pavements and recompacted in the laboratory.
The data from this study indicated that the values did not necessarily
compare but the means of the field and laboratory moduli were in good
agreement. Janoo and Berg (1990) conducted tests directly on a clay
layer prepared in a concrete test pit. This study indicated that the D_0
sensor reading exhibited a higher degree of variability relative to the
other sensors and oftentimes gave values which exceeded the 2 mm range
of the sensors. Houston and Perera (1991) conducted a series of tests
at 20 different pavement sites. At each site, FWD tests were sampled at
3 m intervals over a 28 m cell of the pavement. They concluded that
most of the variability in the deflections within a cell were due to
variability in the subgrade materials and that these variations occurred
over distances less than 28 m for the structures tested. They also
state that the variability is mostly contained within the native
(basement) foundation soils and not in the engineered (compacted) soils.

SUBGRADE TESTING PROCEDURES

FWD Testing

The particular FWDs used for the testing were Dynatest Model 8000 machines. The testing pattern for each cell was designed for a series of 10 test stations located at 15 m intervals along three different longitudinal lines. These lines were situated along the roadway centerline and at 3 m offsets to either side, i.e., the anticipated outer wheel paths of the two lanes. A large diameter (45 cm) plate was used for all of the tests and seven velocity sensors were located at the center of the plate and offsets of 30, 45, 60, 90, 135, and 180 cm away from the center.

The loading patterns for the 1991 and 1992 tests were slightly different. In 1991 three seating loads plus three load drops each at three different drop heights were used. The maximum loads ranged from 12-25 kN; the drop height was increased for each successive set. The loading pattern was changed slightly for 1992 to include four seating drops plus three load drops from three heights in progressive order. In addition, data from the seating drops were stored in the data file. This was done in order to address the effects of the seating load drops and load repetitions on the sensor deflections. Similar load levels were used for these series of tests.

Each test site was inspected prior to loading and the visual appearance of the surface was noted. In cases where the station was unsuitable for testing due to loose surficial material, wheel ruts, etc. attempts were made to remedy the cause. In most cases the preparation work amounted to clearing loose material with a shovel or filling in divots. In some cases the surface appeared to be uniform but out-of-range deflections resulted after a few load drops. This was most likely due to a soft layer of material that was perhaps saturated. In these cases the test station was shifted forward or backward 1 or 2 m and the test was repeated. If the test was not successful then it was simply abandoned.

Soil Sampling

The primary objective of the accompanying soil sampling program was to obtain the physical parameters that are useful in interpreting the properties of the soil. Since the silty-clay subgrade is quite moisture-sensitive it is important to have an estimate of the soil moisture at selected stations along the embankment.

Small jar samples were retrieved from three different locations within each cell. Two of the samples were taken at two different locations from the upper 30 cm of the subgrade while at the third location three samples were taken from depths down to 1 m. These samples were used to determine the in-situ moisture content of the soil.

Disturbed bag samples obtained from auger borings were also retrieved from selected locations within each cell. Several different tests were conducted on these samples including Atterberg limits (AASHTO T 99), sieve and hydrometer analyses (AASHTO T 88), and moisture-density tests (AASHTO T 99 and T 100).

Undisturbed samples were retrieved using thin-walled tubes from depths ranging from 30 cm to over 200 cm. Resilient modulus tests (SHRP Protocol P46) and unconfined compression tests (AASHTO T 208) were performed on these samples.

Results from the laboratory resilient moduli tests on the undisturbed (thin-walled) specimens are shown in Table 1. The resilient moduli listed in the table are given for a deviator stress of 30 kPa. In general, the results from all thin-walled specimens indicate a decrease in the resilient modulus with increasing deviator stress. A typical resilient modulus-deviator stress relationship from one of these tests is shown in Fig. 2. Only the undisturbed samples from the 10-year cells have been tested to date. Data from tests run on the disturbed samples are shown in Table 2. The variation of soil moisture as determined from the small jar samples varied from 10 to 18 percent.

Other researchers have investigated the correlation between resilient moduli and various soil parameters such as degree of saturation (Thompson and Robnett, 1979). The effect of the degree of saturation on the laboratory resilient modulus was investigated but over the range observed (75-85 percent) no significant effects were seen.

TABLE 1--Soil data from undisturbed (thin-walled) specimens.

Station	Comp. Strength, kPa	Dry Density, kg/m^3	Wet Density, kg/m^3	Moist. Cont.,%	Depth, cm	Resilient Modulus, MPa
1175.84	176	1858	2141	15.3	30	87
1175.84	242	1936	2226	15.0	86	145
1175.84	95	1749	2063	17.9	213	62
1175.91	357	1942	2210	13.8	61	185
1175.91	363	1898	2168	14.2	152	196
1191.70	250	1763	2096	18.9	30	75
1191.70	210	1832	2144	17.0	91	96
1197.20	431	1832	2131	16.3	30	173
1197.20	182	1816	2127	17.1	91	77
1197.20	129	1753	2077	18.4	183	62
1203.15	196	1752	2083	18.9	30	51
1208.65	381	1915	2163	13.0	30	124
1208.65	236	1915	2163	13.0	30	109
1214.25	351	2010	2258	12.3	30	194
1220.05	159	1810	2128	17.6	30	62
1225.55	233	1867	2168	16.2	30	73
1231.45	143	1779	2109	18.5	30	40
1237.25	167	1835	2143	16.8	30	62
1243.05	148	1819	2128	17.0	61	53

ANALYSIS

Data Validation

During the course of this study several techniques were used to analyze and interpret the data. The first part of the analysis approach

RESILIENT MODULUS VS. DEVIATOR STRESS

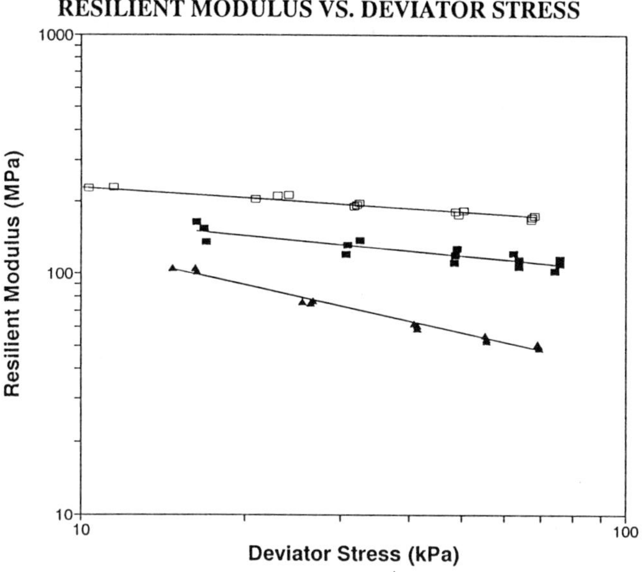

FIG. 2--Typical resilient modulus-deviator stress relationships for the silty clay.

TABLE 2--<u>Physical characteristics of subgrade soil from disturbed samples.</u>

Station	Depth, cm	Sp. Gr	LL, %	PI, %	Opt. Moist., %	Max. Dens., kg/m^3
1170.21	30	2.657	29.8	10.2	14.4	1808
1170.21	91	...	40.7	20.9	17.4	1673
1175.76	30	...	34.3	14.1	15.3	1738
1186.10	30	...	36.9	19.2	16.0	1762
1186.10	152	...	37.8	16.8	19.7	1646
1191.65	30	...	34.8	14.3	17.3	1725
1191.65	91	...	37.0	15.7	16.7	1749
1197.15	91	2.650	33.8	14.3	17.0	1758
1197.15	152	...	33.6	13.5	15.8	1754
1203.10	15	...	36.6	17.0	17.4	1752
1203.10	46	...	35.2	14.5	16.7	1726
1208.60	15	...	35.5	17.3	17.3	1726
1214.30	91	...	32.7	14.0	15.2	1805
1214.30	122	...	39.0	20.4	17.0	1689
1220.00	15	...	36.1	18.5	17.0	1749
1225.50	15	...	36.8	17.6	17.2	1742
1231.40	30	...	35.7	15.7	17.3	1749
1237.20	15	...	34.8	14.6	18.4	1712
1237.20	61	...	37.7	18.4	17.1	1714
1243.00	30	2.666	37.2	19.8	18.0	1725
1243.00	91	...	38.8	19.6	17.9	1681

involves the validation of the deflection data. In essence this procedure involved investigating the basins and verifying that they were sensible. A statistical analysis of the deflection data from each cell is also useful in establishing spatial correlation and trends of the data as well as detecting outlying values that may be suspect due to considerations discussed shortly.

Practical aspects of the analysis that were dealt with focused on detecting erroneous data. By inspecting the deflection basins and identifying sets with either (1) extremely large deflection values, (2) negative slopes, and (3) null deflections it is possible to sort out basins unsuitable for analysis. With respect to item (1) excessively large deflections were observed under the plate (D_0) sensor. This is presumably due to the variable nature of the soil surface with conditions of either a soft layer or brittle crust present at the surface that would lead to inelastic deformation (e.g., compaction or punching). Item (2) refers to basins in which an outer sensor yielded a deflection value higher than its interior neighbor. As for item (3) this may have occurred when the sensor was either resting on a piece of debris or not properly coupled with the soil surface.

Several special tests were conducted in which the repeatability of the sensor responses were investigated. In the first set of tests the outer wheelpath of two different cells were retested following a period of about 48 hours after the original test. Two cells (6 and 7) were retested at the same locations as in the original tests and were selected for the fact that the surface appearance was very smooth and uniform relative to the other cells. The deflection readings from each test were normalized to a uniform applied stress intensity level of 100 kPa and plotted on a scatter chart. Results from cell 6 showing the comparison between the two sets of D_0 sensor readings is given in Fig. 3 while the comparison for D_6 is shown in Fig. 4. As can be seen there is scatter owing to slight differences in FWD placement, soil moisture, and load level.

The coefficient of determination (r^2) for the sensor reading correlations were 0.85 and 0.67 for D_0 and D_6, respectively. The effects of load level (i.e., stress-sensitivity) are presumably small within the range of loads applied. The scatter in the D_0 sensor comparison is thought due to placement of the FWD which would obviously effect the plate contact and sensor location. Also, during the period between the two tests the weather conditions were sunny and windy which would allow for drying, especially near the surface, thus changing the soil properties. However, the effects of soil drying should be less for the D_6 sensor than for the D_0 sensor since at greater offsets from the loading plate responses from deeper down in the structure are being detected. Thus, scatter observed in the D_6 readings for these tests is probably due to sensor-soil contact. Comparable results were obtained from the test conducted in cell 7 with r^2 values ranging from 0.13 (D_0) to 0.97 (D_3) for the sensor correlations.

Another series of tests were conducted in which cells 3, 4, 24, and 27 were retested using 15 m spacing but with additional tests conducted within the cells on a close-set grid. In this additional test the FWD plate was positioned at nine different stations within a 3.7 m square forming a grid size of 1.85 m. Summary statistics from these data sets are shown in Table 3. For the most part the variability in the sensor deflections is consistently smaller for the cluster test, as

DEFLECTION CORRELATIONS

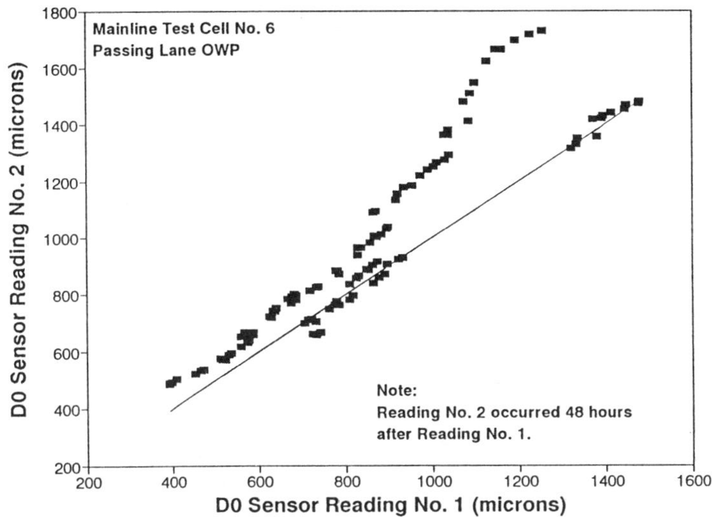

FIG. 3--Deflection correlations from repeatability study: mainline cell 6, D_0 sensor.

DEFLECTION CORRELATIONS

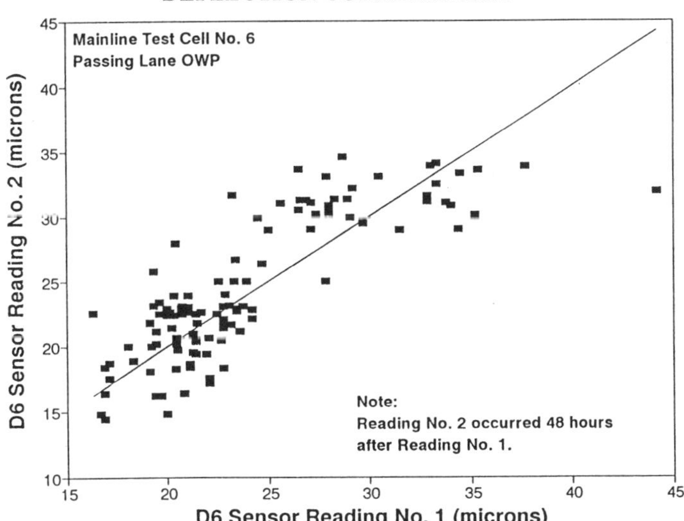

FIG. 4--Deflection correlations from repeatability study: mainline cell 6, D_6 sensor.

TABLE 3--Statistics for sensor deflection variability study.

		Entire Cell					9-pt. Grid (3.7 m by 3.7 m)				
Cell	Sensor	Mean	SD	CV	Min	Max	Mean	SD	CV	Min	Max
3	D_0	1133	455	0.402	347	2635	906	267	0.295	466	1469
	D_1	468	224	0.479	148	1382	399	132	0.330	199	759
	D_2	211	123	0.583	61	722	167	52	0.312	87	332
	D_3	110	72	0.649	35	402	81	20	0.253	52	146
	D_4	56	32	0.560	22	271	43	9	0.205	30	63
	D_5	35	12	0.348	6	99	32	8	0.238	20	47
	D_6	31	45	1.427	10	501	23	6	0.246	15	34
4	D_0	665	345	0.520	188	3276	652	202	0.310	318	1090
	D_1	381	182	0.478	87	1167	375	133	0.354	202	753
	D_2	198	90	0.454	55	680	202	60	0.296	115	354
	D_3	115	57	0.495	37	606	117	30	0.257	70	173
	D_4	44	16	0.359	17	144	50	11	0.225	34	76
	D_5	25	9	0.346	0	76	25	5	0.218	12	37
	D_6	21	7	0.339	2	65	20	5	0.265	6	33
24	D_0	516	273	0.528	227	3120	481	86	0.179	306	648
	D_1	197	43	0.219	103	393	193	31	0.163	149	277
	D_2	116	21	0.177	70	184	115	18	0.155	90	155
	D_3	86	17	0.196	43	182	85	13	0.157	65	110
	D_4	58	10	0.177	33	97	57	10	0.170	44	78
	D_5	38	7	0.179	17	57	38	7	0.175	27	53
	D_6	28	6	0.209	14	45	28	6	0.216	16	40
27	D_0	1494	812	0.543	257	3690	1767	601	0.340	666	3416
	D_1	612	372	0.607	79	3218	723	223	0.308	238	1228
	D_2	247	123	0.497	43	727	264	72	0.274	100	455
	D_3	103	41	0.399	26	299	98	29	0.299	45	184
	D_4	37	14	0.381	0	88	32	10	0.318	6	49
	D_5	28	9	0.346	0	132	27	7	0.257	10	41
	D_6	21	7	0.307	0	47	20	5	0.232	10	32

Note: all sensor deflections in microns.

would be expected. This is due to the fact that at small separation distances the subgrade properties are not varying as much relative to larger distances. However, even at the short separation distance the coefficient of variation (CV) of the deflections ranges from 18 to 35 percent. Again, these variations are attributable to both spatial changes in the soil parameters as well as test variability due to surface effects.

The effects of load repetition were addressed by investigating the change in deflections, possibly due to localized compaction of material, with successive load applications at each test station. In this analysis the sensor deflections were normalized to a standard load. Significant variations in the two innermost sensors occurred. In some cases an increase in normalized deflection for these sensors was observed while in others a decrease was noted. No trend could be established.

Static Multi-Layered Backcalculation Analysis

The load and deflection data were used in a backcalculation analysis to arrive at estimates of the moduli for an assumed layered structure. The program CLEVERCALC 3.5 was used for the backcalculation. CLEVERCALC is a derivative version of EVERCALC developed in the state of Washington (Washington State Transportation Center, 1987), but adapted to metric units and with certain features for research studies. The program is based on the CHEVRON linear elastic layer analysis program and the iteration process stops if one of the following occurs:
1. The mean root-mean-square of the relative difference between measured and backcalculated reading is less than a given value.
2. The combined change of modulus for all layers from one iteration to the next is less than a given value.
3. The maximum number of given iterations has been reached.
The first criterion is obvious, if the match is near perfect, the search is stopped. The difference between the two basins is referred to as the *deflection error.* It is expressed as the root-mean-square (RMS) deviation.

The second criterion requires some explanation though. The program will base the adjusted moduli on the effect a change of the logarithm of the modulus has on the deflection error. This could be expressed as a slope, and a slope is determined for each unknown layer. A steep slope will lead to a quick solution, but a shallow slope may take longer or even lead to non-convergence. Further, the resolution of the sensors may not be adequate for reaching the stipulated RMS-criterion. In addition, the shortcomings of the linear elastic model, faulty assumption of layer thicknesses, or both, may deter a perfect match of the basins. Thus, when the change of modulus is small for each layer, the program is unable to arrive at a better match anyway and the iteration procedure stops. Such basins should always be checked critically, but may be accepted if the deviation from the measured basin is not too large.

The last criterion is as obvious as the first and is needed for practical reasons. The criteria used in the present study was an RMS tolerance of one percent, a change of modulus of one percent, and a maximum of 10 iterations.

Initial attempts with a two-layer system were unsuccessful due to the previously mentioned variability in the inner sensor deflections. A

three-layer system was assumed for the subgrade structure to account for
variation of stiffness with depth. An important assumption regards the
selection of layer thicknesses. The approach taken in this analysis was
to choose an appropriate set of thicknesses and perform backcalculations
from deflection basins at representative test stations. These
preliminary calculations were performed by assuming two different
thicknesses for layer one, i.e. the top layer (15 and 25 cm), and
allowing the thickness of layer two to vary from 5 to 60 cm. These
thicknesses are believed to be reasonable as they are on the order of
the thicknesses of the lifts used in constructing the embankment. It
was found that the backcalculated moduli of the second layer were highly
sensitive to the ratio of the thicknesses of layer one and two,
regardless of the thickness of layer one, for thickness ratios greater
than about 0.3. Using this figure as a guide, the backcalculations were
performed for data from all tests using thicknesses of 15 cm and 45 cm
for layers one and two, respectively. Poisson's ratio was assumed to be
0.4 for all three layers.

It is important to address the differences between the test
conditions and the assumed model. Key deviations from the implicit
assumptions in the theory of elasticity are as follows:

1. Imperfect sensor contact giving rise to erroneous deflection
   measurements.
2. Imperfect plate contact giving rise to nonuniform surface pressure
   distributions.
3. The effects of repeated FWD load drops on the subgrade soil thus
   changing the properties of the soil.
4. FWD tests are quasi-dynamic by nature while the backcalculation
   analysis is based upon static, linear elasticity.

Item (1) above is a consequence of the surface condition, e.g., ruts,
desiccation cracks, loose material, etc. These conditions can
presumably result in zero deflections or negative slopes as discussed
above and were excluded from the analysis. Items (2) and (3) are also
due to surface conditions but it was assumed that the effects of
nonuniform plate contact were local, i.e., they were negligible at
distances away from the plate. Also, the observed change in normalized
deflections of the outer sensors with load repetition were small
compared with the innermost sensors. Item (4) cannot be accounted for
in the analysis and will induce a systematic error in the backcalculated
moduli which cannot be determined. Any stress-dependency of the
material will be reflected in an apparent change in the backcalculated
moduli as a function of load.

The deflections measured by the innermost sensors (e.g., D_0 and
D_1) were influenced by items 1-3 above to a greater degree than were the
outer sensors. In a static multi-layered system the modulus
backcalculated for the upper layer is largely dependent on the inner
sensor deflections. The technique adopted in this study was to exclude
the two innermost sensors (D_0 and D_1) from the backcalculation. Thus,
the moduli backcalculated for layer one, while essential to the
calculation, are not used in any subsequent analyses of moduli. In this
way, the local effects of surface variability are presumably excluded
from the analysis. Only final drops from each different height (load
level) were backcalculated.

The variation of backcalculated moduli for layers two and three
for the mainline test cells are shown in Fig. 5. This figure represents

the average of the backcalculated moduli from the three wheelpaths at each test station. Note that the direction of increasing station number opposes the direction of traffic. Traffic traveling on the Interstate would first encounter the 10-year design cells in cell 23 at the east end of the facility (station 1245) and leave the 5-year cells in cell 1 at the west end (station 1104). One increment in the station number represents about 30 m (100 feet).

There is a gap shown in Fig. 5 between the data for the 5 and 10-year cells from stations 1152 to 1168 to allow traffic to be diverted off of the 5-year cells. Similar plots are shown for the low volume test cells in Fig. 6 and 7. Gaps other than the transition zone in these figures indicate that either the stations were abandoned or the data were erroneous for reasons discussed earlier.

Another observation that was made concerns the RMS value from the calculations. It was noted that a large percentage of the basins failing the validation criterion discussed above yielded RMS values greater than 15 percent and corresponding moduli that were exceedingly large. As such these stations have been excluded from the figures. Note also that cells 24, 25, 36, and 37 in the low volume test loop are constructed of approximately 2 m of granular fill. These cells are indicated in Fig. 6 and 7. It is apparent that the moduli for these cells are less variable which suggests that the sand fill is more homogeneous than the native soil.

Comparison of Field and Laboratory Data

The laboratory and field data will be compared on a qualitative basis by comparing the distributions of the data and variability. Frequency distributions of the backcalculated moduli are shown in Fig. 8. The mean and coefficient of variation for the layer two moduli are 42 MPa and 1.09, respectively and the values for layer three are 97 MPa and 0.57. The distribution for layer two is quite variable and skewed relative to the layer three moduli. Note that the histograms of the data suggest a lognormal distribution, especially for layer two. A total of twenty different specimens were tested in the laboratory and the mean and coefficient of variation of these moduli were 92 MPa and 0.56 at a deviator stress of 30 MPa.

The spatial correlation of the data were investigated using geostatistics and expressed in the form of the variogram. A variogram represents the variance of the data as a function of separation distance: it is a measure of the spatial variability. The experimental variograms for the log-transformed moduli of layer two and three from the mainline test cells are shown in Fig. 9. In Fig. 9 the variogram value is the variance of the transformed backcalculated resilient moduli values. The transfomed values are obtained by simply taking the natural logarithm of the data. It is evident from these curves that the data for layer two are more variable than layer three. This effect may be due to greater variation in moisture content near the surface or to the effects of construction (e.g., placement, compaction, and traffic). With respect to the assumed multi-layered system model, layer three is largely comprised of the undisturbed foundation soils. The native soils at the facility are part of a fairly extensive, uniform glacial deposit.

A comparison of the field and laboratory moduli for the 10-year mainline cells are shown in Fig. 10. The field moduli represent the

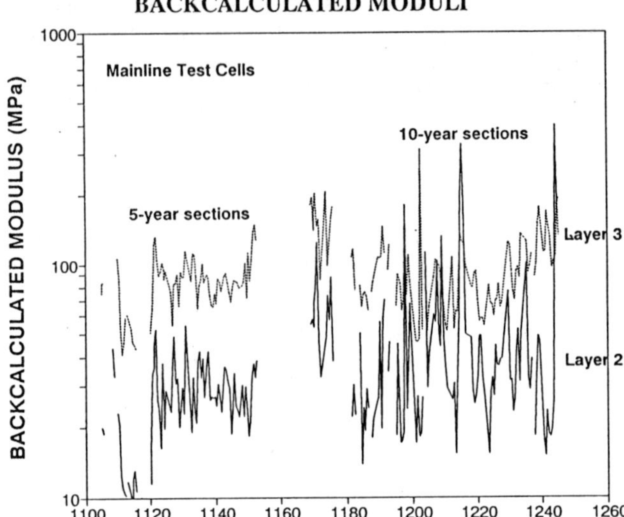

FIG. 5--Variation of backcalculated moduli with station: mainline test
cells.

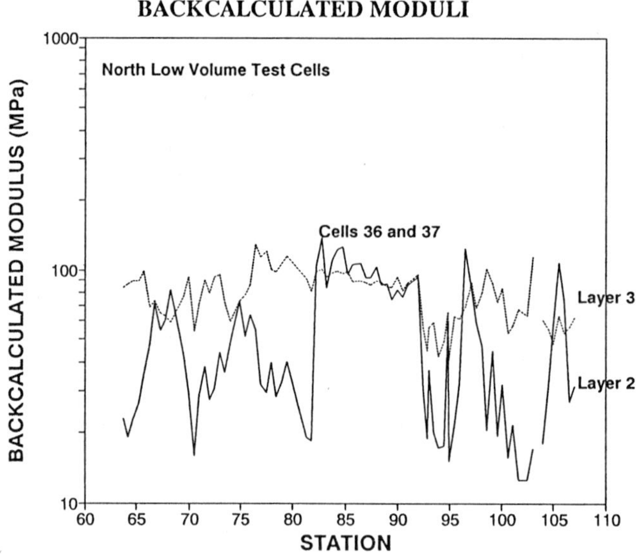

FIG. 6--Variation of backcalculated moduli with station: north low
volume test cells.

BACKCALCULATED MODULI

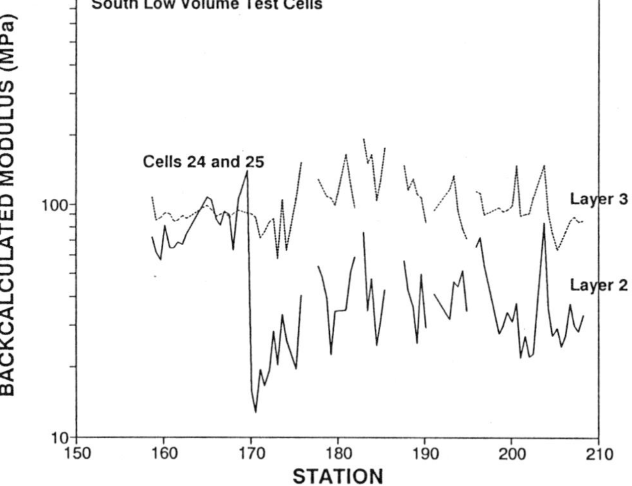

FIG. 7--Variation of backcalculated moduli with station: south low volume test cells.

DISTRIBUTION OF BACKCALCULATED MODULI

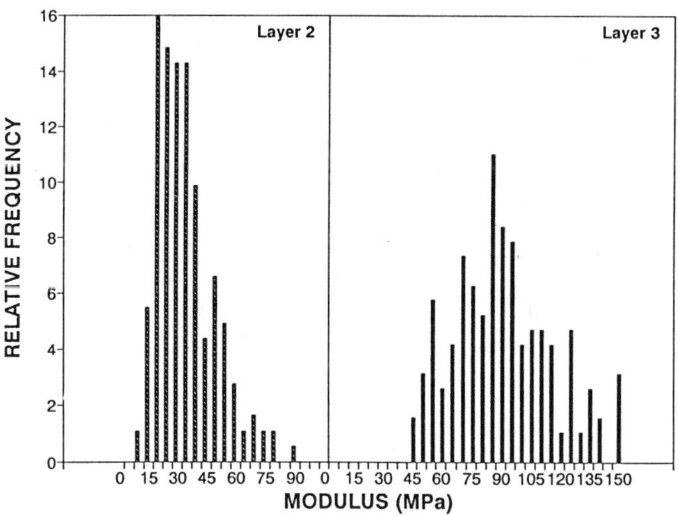

FIG. 8--Frequency distributions of backcalculated moduli: mainline 10-year test cells.

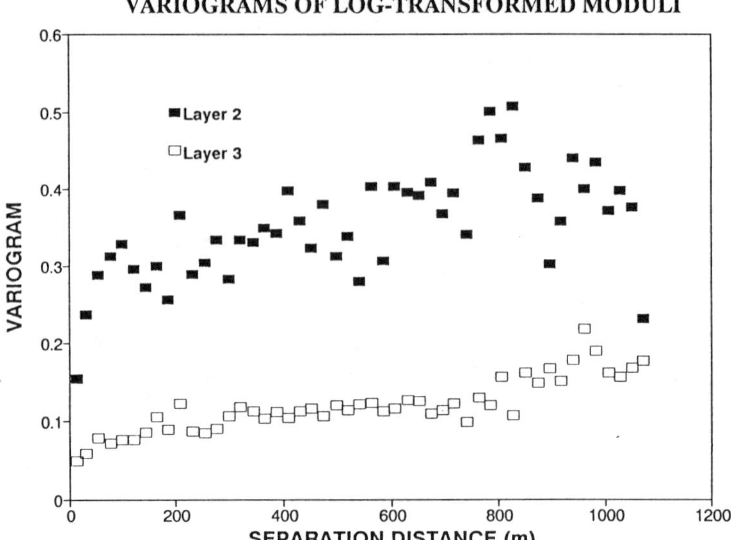

FIG. 9--Spatial correlation (variogram) of backcalculated moduli for 10-year test cells.

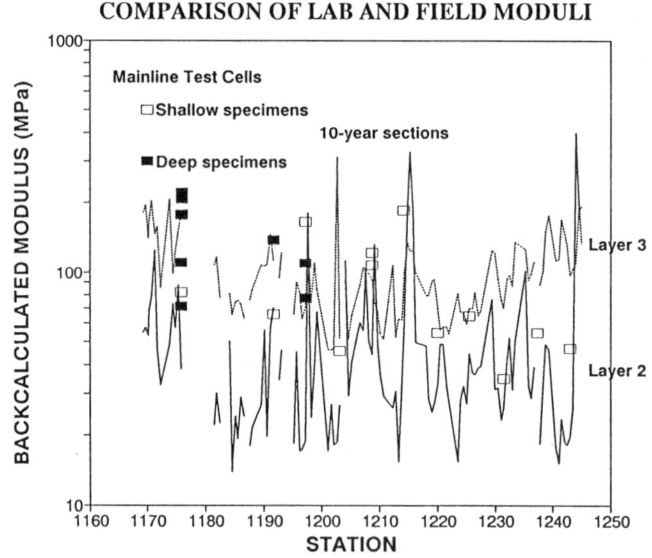

FIG. 10--Comparison of laboratory and backcalculated moduli.

average backcalculated value at each station and the laboratory
resilient moduli are referenced to a deviator stress of about 30 kPa for
the shallow specimens and 15 kPa for the deeper specimens. These
deviator stresses reflect average values given the load levels, layer
thicknesses, and moduli. Due to the large differences in the number of
backcalculated data relative to the laboratory data a correlation will
not be sought. However, a look at Fig. 10 suggests a fair but far from
strong correlation.

Effects of Base Layer

After base construction in cells 1-4 the FWD testing procedure was
repeated. The thickness of the granular layers in cells 1, 2, and 3
were 85, 70, and 85 cm, respectively. test cell 4 is actually a full-
depth design (i.e., no base layer) and the surface was simply fine-
graded prior to the testing. For these backcalculations a three-layer
system was modeled and the granular base layers were divided into two
layers, the top layer of which was assumed to be 20 cm. A layered
system identical to the one used in the subgrade calculations was used
for cell 4. The distributions of backcalculated moduli for layer 3 from
both the base layer and subgrade tests are shown in Fig. 11. The
surface quality of the base layer in these cells during the tests was
far better than during the subgrade testing. In general, it was much
smoother and consistent owing to finer construction tolerances. This is
reflected in lower variability in the deflections and subsequently
moduli (see Fig. 11). Also note that there is an increase in the post-
base moduli due to stress-sensitivity effects and that the moduli in
cell 4 remain essentially unchanged.

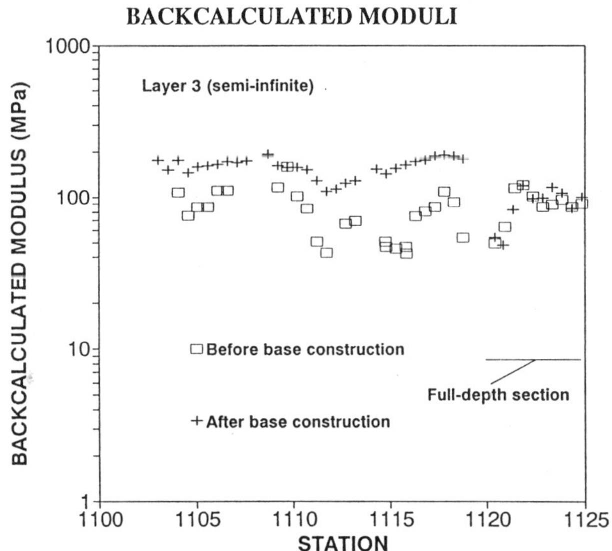

FIG. 11--Comparison of layer 3 moduli from sections 1-4, before and
after base construction.

CONCLUSIONS

On the basis of the analysis conducted thus far it is reasonable to
conclude the following:
- The variability studies seem to suggest that, at least for the
  embankment at Mn/ROAD, a relatively large amount of deflection
  variability can be expected even at short sample spacing.
- Within the variability encountered the backcalculated and laboratory
  moduli appear to compare well.
- The surface condition of the soil has a noticeable impact upon the
  variability of the measured deflections especially the inner sensors
  D_0 and D_1.
- The backcalculated moduli obtained from post-base construction tests
  are larger and less variable than those prior to construction owing
  to decreased stresses in the subgrade and better surface conditions.
- Over the range of moisture contents observed in the undisturbed
  specimens the degree of saturation had little impact on the
  laboratory resilient modulus.

ACKNOWLEDGMENTS

This work was funded by the Minnesota Department of Transportation and
the University of Minnesota Center for Transportation Studies. Their
support is gratefully acknowledged. The authors would like to express
their sincere appreciation to the staff of the Physical Research and
Physical Testing sections of Mn/DOT for conducting the FWD tests and to
the Mn/DOT Foundations Laboratory for the lab tests. RST-Sweden
provided the backcalculation program used in the study.

REFERENCES

Houston, S. L, and Perera, R., September/October 1991, "Impact of
    Natural Site Variability on Nondestructive Test Deflection
    Basins," ASCE Journal of Transportation Engineering, Vol. 117,
    No. 5, pp. 550-565.

Houston, W. N., Mamlouck, M. S., and Perera, R. W. S., March/April 1992,
    "Laboratory versus Nondestructive Testing for Pavement Design,"
    ASCE Journal of Transportation Engineering, Vol. 118, No. 2, 1992,
    pp. 207-222.

Janoo, V. C., and Berg, R. L, 1990, "Predicting the Behavior of Asphalt
    Concrete Pavements in Seasonal Frost Areas Using Nondestructive
    Techniques," Report 90-10, U.S. Army Corps of Engineers Cold
    Regions Research and Engineering Laboratory.

Lee, S. W., Mahoney, J. P., and Jackson, N. C., 1988, "Verification of
    Backcalculation of Pavement Moduli," TRR No. 1196, Transportation
    Research Board, pp. 85-95.

Newcomb, D. E, February 1987, "Comparisons of Field and Laboratory
    Estimated Resilient Moduli of Pavement Materials," Proceedings,
    Association of Asphalt Paving Technologists, Vol. 56, pp. 91-106.

Newcomb, D. E., Benke, R., and Cochran, G. R., 1991, "Minnesota Road Research Project - An Overview," ASCE Cold Regions Engineering, <u>Proceedings of the Sixth International Specialty Conference</u>, pp. 463-471.

Thompson, M. R., and Robnett, Q. L., January 1979, "Resilient Properties of Subgrade Soils," <u>ASCE Transportation Engineering Journal</u>, Vol. 105, No. TE1, pp. 71-89.

Washington State Transportation Center, April 1987, University of Washington, Washington State Department of Transportation, Evercalc User's Guide 1.1.

Nenad Gucunski[1]

DETECTION OF MULTI-COURSE SURFACE PAVEMENT LAYERS BY THE SASW METHOD

REFERENCE: Gucunski, N., **"Detection of Multi-Course Surface Pavement Layers by the SASW Methods,"** Nondestructive Testing of Pavements and Backcalculation of Moduli (Second Volume), ASTM STP 1198, Harold L. Von Quintas, Albert J. Bush, III, and Gilbert Y. Baladi, Eds., American Society of Testing and Materials, Philadelphia, 1994.

ABSTRACT: The ability of the SASW method in detection of layer thicknesses and moduli of multi-course pavements has been examined by numerical simulation of the test. The results are represented by the response of a pavement system to a vertical circular loading in the wave number and spatial domains, from which Rayleigh wave dispersion curves were derived using procedures equivalent to those in the field. The simulation shows that for common thicknesses and material properties of pavement surface layers, and the frequency range of accelerometers, the SASW method should enable detection of individual courses. The paper also discusses the influence of receiver positioning and an applied filtering criteria on the obtained dispersion curve.

KEYWORDS: SASW method, pavements, nondestructive testing, seismic methods, layered systems, Rayleigh waves, dispersion, numerical solution

The Spectral-Analysis-of-Surface-Waves (SASW) method is a seismic method for in situ evaluation of elastic moduli and layer thicknesses of layered systems, like pavements. The method is based on the phenomena of dispersion of Rayleigh waves in layered systems. It has been successfully used for evaluation of road profiles where the surface asphalt-concrete or concrete layer was treated as a single layer (Nazarian and Stokoe 1984 and 1986; Nazarian et al. 1983 and 1988). In practice asphalt-concrete pavements are made of two courses, the lower course of medium texture, and the top of fine-aggregate mix. Similarly, sheet asphalt or asphalt-concrete can lay over a concrete base. In those cases it may be necessary to define the properties and thicknesses of both layers.

The objective of the research was to examine the ability of the SASW method in detection of layer thicknesses and moduli of asphalt-concrete and concrete surface layers consisting of several courses, considering the practical limitations of the instrumentation in use and the widely accepted reduction procedures. This was done through the numerical simulation of the SASW test.

---

[1]Assistant Professor, Department of Civil and Environ. Engineering, Rutgers University, Piscataway, NJ 08855.

THE SASW TEST AND ITS NUMERICAL SIMULATION

The objective of the SASW test is to determine the experimental Rayleigh wave dispersion curve, and from it, utilizing the inversion process, define the elastic moduli profile of the pavement. The typical SASW test setup and the basic equations used in the evaluation of the phase velocity are presented in Fig. 1, where β represents the phase difference between two receivers at frequency f. A detailed description of the method can be found in Gucunski and Woods (1991), and Nazarian and Stokoe (1986).

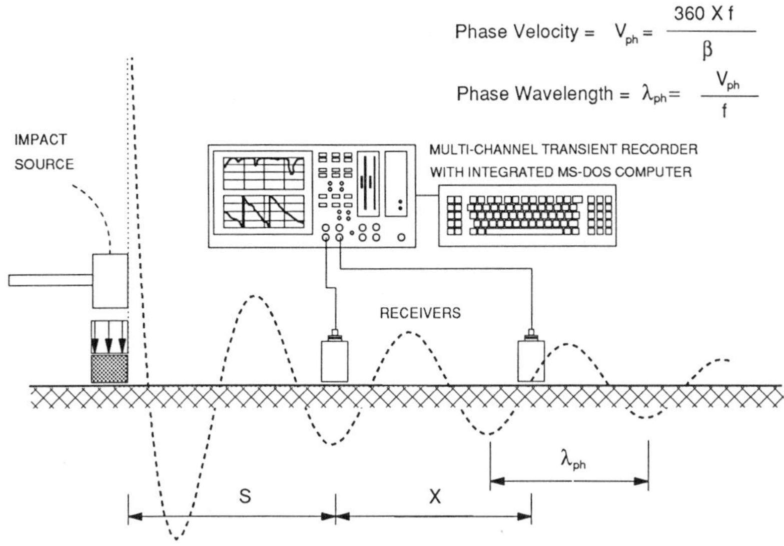

FIG. 1--Schematic of experimental arrangement for SASW test.

The SASW test can be described as an axisymmetric problem in which the impact source is represented by a circular loading at the center of the system, while the response of receivers is represented by displacements, velocities or acceleration of the surface at distance r from the source, as shown in Fig. 2. Vertical surface displacement function $w_0(r)$ for a particular frequency can be written as

$$w_0(r) = -pR_0 \int_0^\infty J_1(kR)\, J_0(kr)\, w_0(k)\, dk \tag{1}$$

where J_0 and J_1 represent Bessel functions of the first kind of order zero and one, respectively, k the wave number, and $w_0(k)$ the vertical surface displacements in the wave number domain. The equation indicates that is a result of a superposition of both body and surface waves, and waves which can be described as quasi static waves. For the undamped system Rayleigh waves represent singular points of function $w_0(k)$, where k is a real number. The detailed description of the numerical simulation of the SASW test can be found in Gucunski and Woods (1992).

CASE STUDY

Three cases were studied, utilizing the numerical simulation of

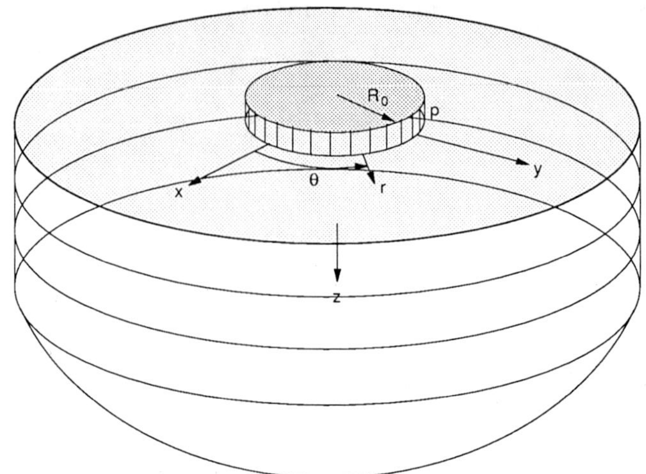

FIG. 2--An axisymmetric model of a layered system with circular loading.

the SASW test on a pavement. The three cases, as shown in Fig. 3,
include: a single course surface layer (case 1), a two-course surface
layer with the stiffer upper course (case 2), and a two-course surface
layer with the stiffer lower course (case 3). All the layers have
Poisson's ratio 0.35 and damping ratio 0.02. The mass density of the
surface layer courses was chosen as 2500 kg/m^3, of the base 2100 kg/m^3,
and of the half-space 1920 kg/m^3.

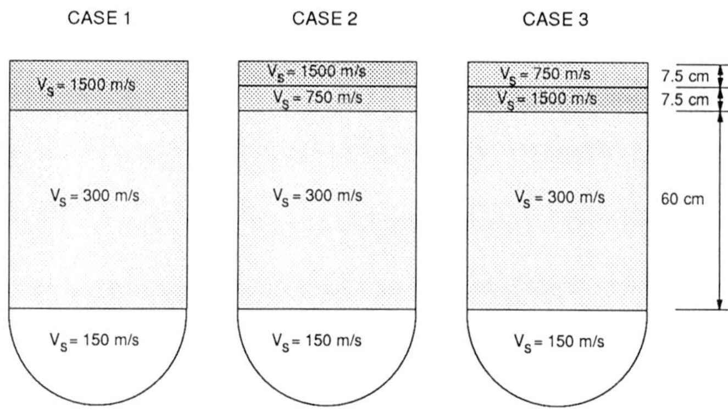

FIG. 3--Analyzed pavement profiles.

    The dispersion curves for all cases were developed using the
procedure equivalent to the one in an actual SASW test. The first step

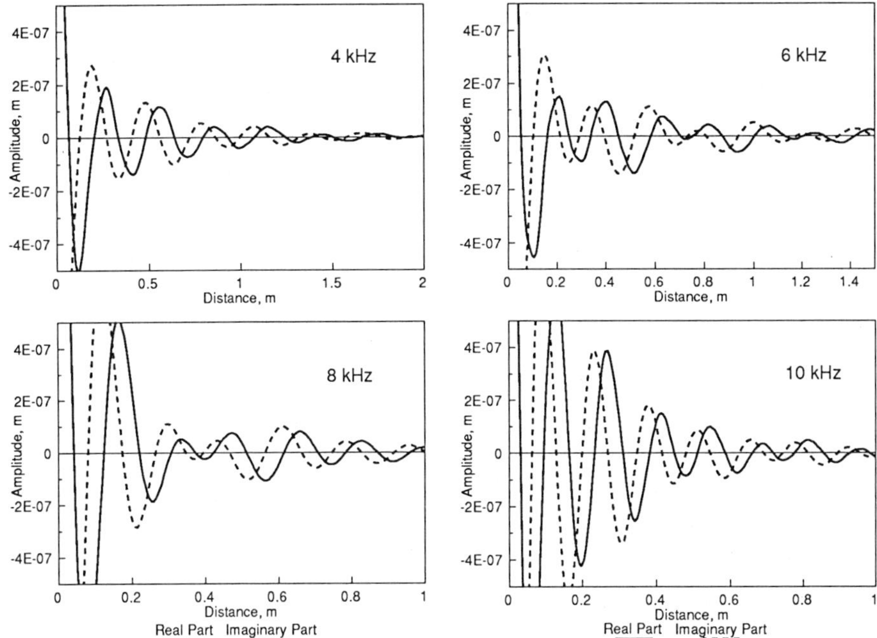

FIG. 4--Surface vertical displacements in the spatial domain for case 1.

of the test represents generation of elastic waves by applying an impact at the surface. Figures 4, 5 and 6 show the generated waves for cases 1, 2 and 3 for p=5 MPa and R_0=1.5 mm, at four different frequencies. Significant variations in the wavelength and the amplitude of the surface displacement with distance can be observed for case 1 at 6 and 8 kHz, for case 2 at 10 and 13 kHz, and for case 3 at 2 and 4 kHz. The figures clearly indicate that the generated wave can be dominated by more than a single Rayleigh mode. The frequencies in Figs. 4, 5, and 6 were selected from observations of wave fields in the wave number domain represented by Figs. 7, 8 and 9. The wave field represents function $w_0(k)$ for a wide range of frequencies generated by the source of the same kind as for those in Figs. 4, 5 and 6. Amplitudes of generated waves in Figs. 7, 8 and 9 are normalized so for each particular frequency the maximum value is 1. The normalized amplitude is presented as a function of frequency f, wavelength λ_{ph} and phase velocity V_{ph}, instead of k and f. These variables are related as

$$k = \frac{2\pi f}{V_{ph}} , \qquad \lambda_{ph} = \frac{V_{ph}}{f} \qquad (2)$$

Symmetric (longitudinal) and antisymmetric (flexural) mode branches (Ewing et al. 1957), typical for dispersion in a single-layer infinite plate, can be identified in Figs. 7, 8 and 9. The two modes can be also seen in Fig. 10 as cross-sections of the wave field for case 1 at 6 kHz, case 2 at 10 kHz, and case 3 at 2 kHz. In all three cases, for the first peak or the antisymmetric branch, both the real and imaginary parts have about equal magnitude. This is an indication of the pure propagation nature of the wave. On the other hand, while the real part of the second peak or the symmetric branch takes the maximum, the imaginary part

approaches zero. This is an indication of a mixed propagating and
oscillating nature of the response in this phase velocity range.

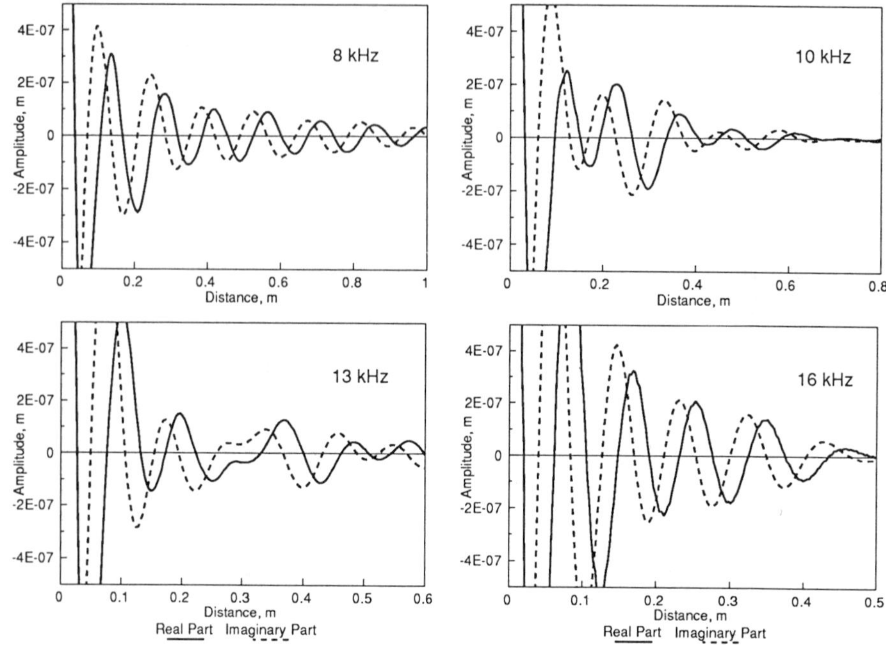

FIG. 5--Surface vertical displacements in the spatial domain for case 2.

The second step of the SASW test is the generation of the
dispersion curve according to equations included in Fig. 1. The
dispersion curve is evaluated for several receiver spacings and two
directions, and filtered according to some filtering criteria. The
earliest filtering criteria was suggested by Heisey et al. (1982)

$$\frac{\lambda_{ph}}{3} < X < 2\lambda_{ph} \tag{3}$$

All the curves are finally statistically combined to derive an average
dispersion curve, as described by Nazarian and Stokoe (1983).

Dispersion curves for six receiver spacings filtered according to
Heisey's criteria, and the average dispersion curve as a function of
frequency, for all three cases are presented in Fig. 11. The two numbers
in the legend represent the distance between the near and far receivers
and the source in meters, respectively. The greatest variation of the
phase velocity with the variation of receiver spacing can be noticed in
frequency ranges that match those where the wave fields from Figs. 7, 8
and 9 are characterized by multiple peaks. Similar study was conducted
by Roesset et al. (1989) indicating even larger variations in phase
velocities than those presented herein. The average curve in this figure
is derived as an average of dispersion curves with respect to the same
frequency.

Figure 12 represents dispersion curves and the average dispersion

curve as a function of wavelength. The averaging in this case, typical
for SASW testing, is done with respect to the same wavelength. In
addition, the average dispersion curve obtained from averaging with
respect to the frequency is plotted. Even though averaging with respect
to the wavelength represents averaging of waves that were detected at
different frequencies, and from that point the procedure may be
considered improper, only minor differences between the two average
dispersion curves can be noticed.

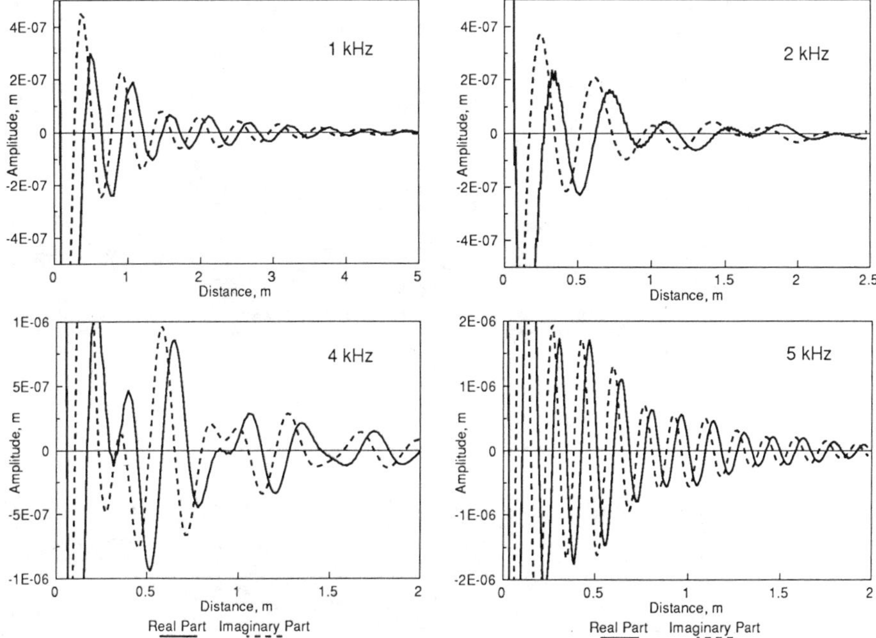

FIG. 6--Surface vertical displacements in the spatial domain for case 3.

The dispersion curves from Figs. 11 and 12 were also filtered
according to criteria proposed by Gucunski and Woods (1992)

$$\lambda_{ph} < X < 4\lambda_{ph} \tag{4}$$

and averaged according to the same procedure. The results are presented
in Fig. 13. Minor variations between dispersion curves for individual
receiver spacings than in the cases where Heisey's criterion was applied
can be noticed. Also, the average dispersion curve obtained from
averaging with respect to the wavelength and with respect to the
frequency indicate only small differences.

Finally, dispersion curves for all three cases, and for both
Heisey's (H) and Gucunski and Woods (G) filtering criteria, are plotted
as a function of frequency and wavelength in Fig. 14, respectively. The
differences between the average dispersion curves are attributed to
different positions of receivers for the two filtering criteria and the
consideration of both body and surface waves in the definition of the
dispersion curve. For the same frequency the receivers are according to
Heisey's criterion positioned much closer to the source than for the

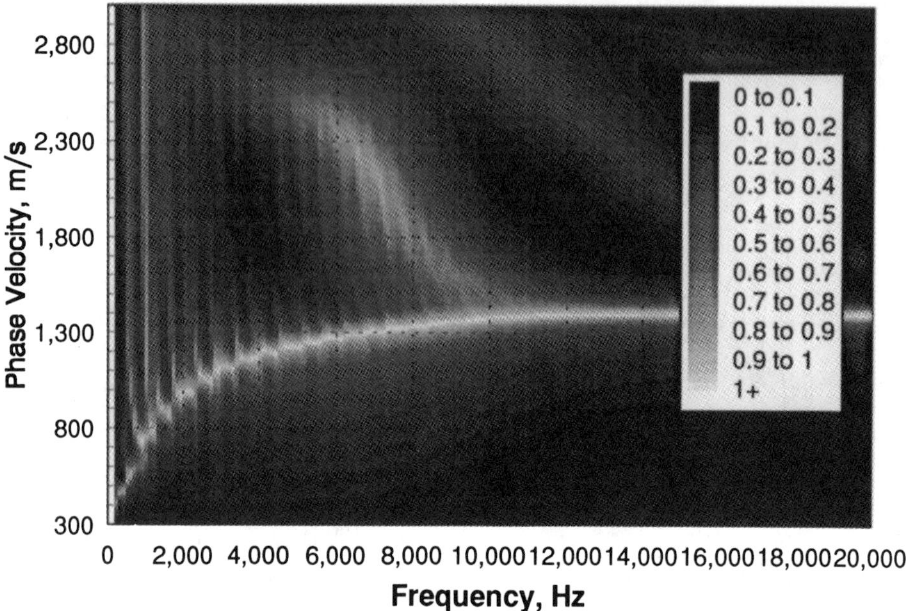

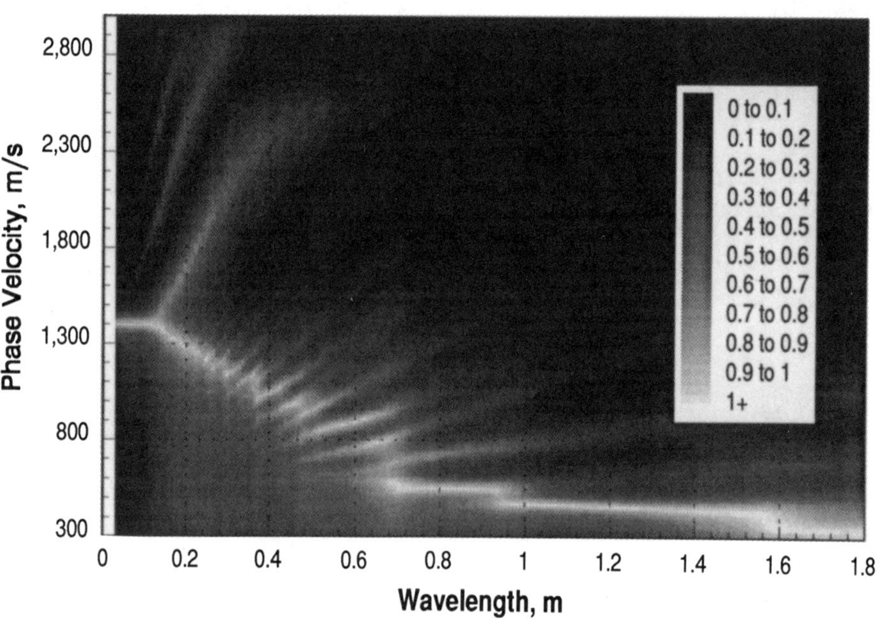

FIG. 7--Wave field generated by a circular source for case 1.

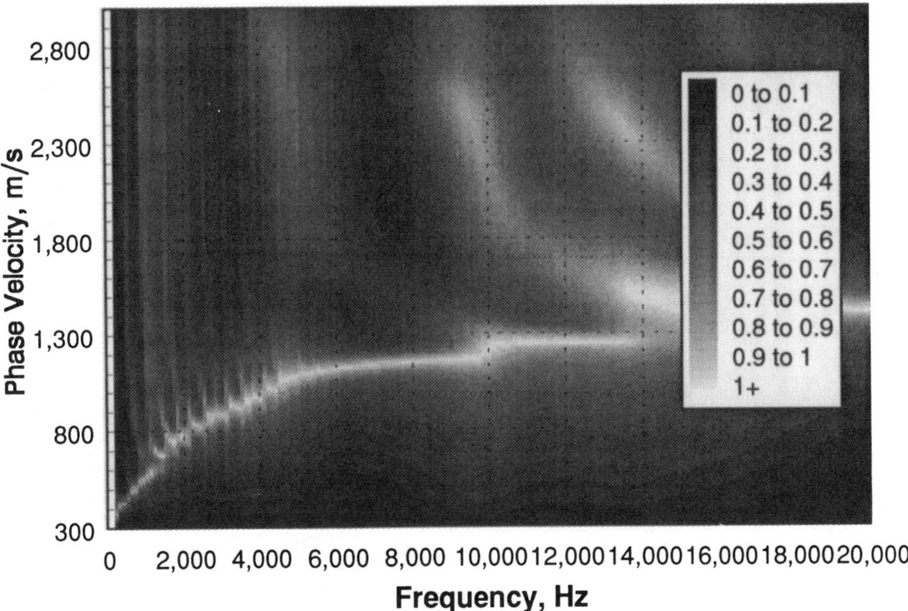

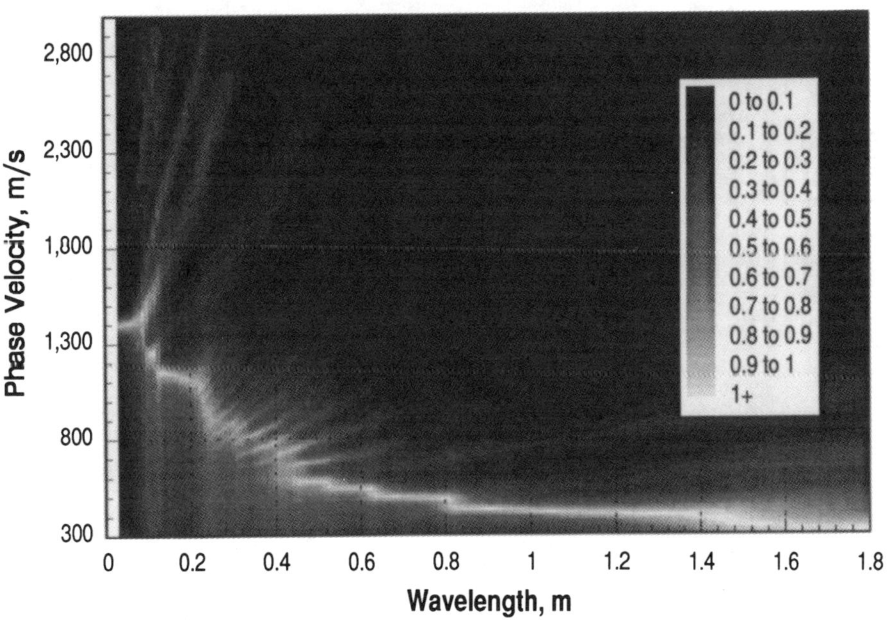

FIG. 8--Wave field generated by a circular source for case 2.

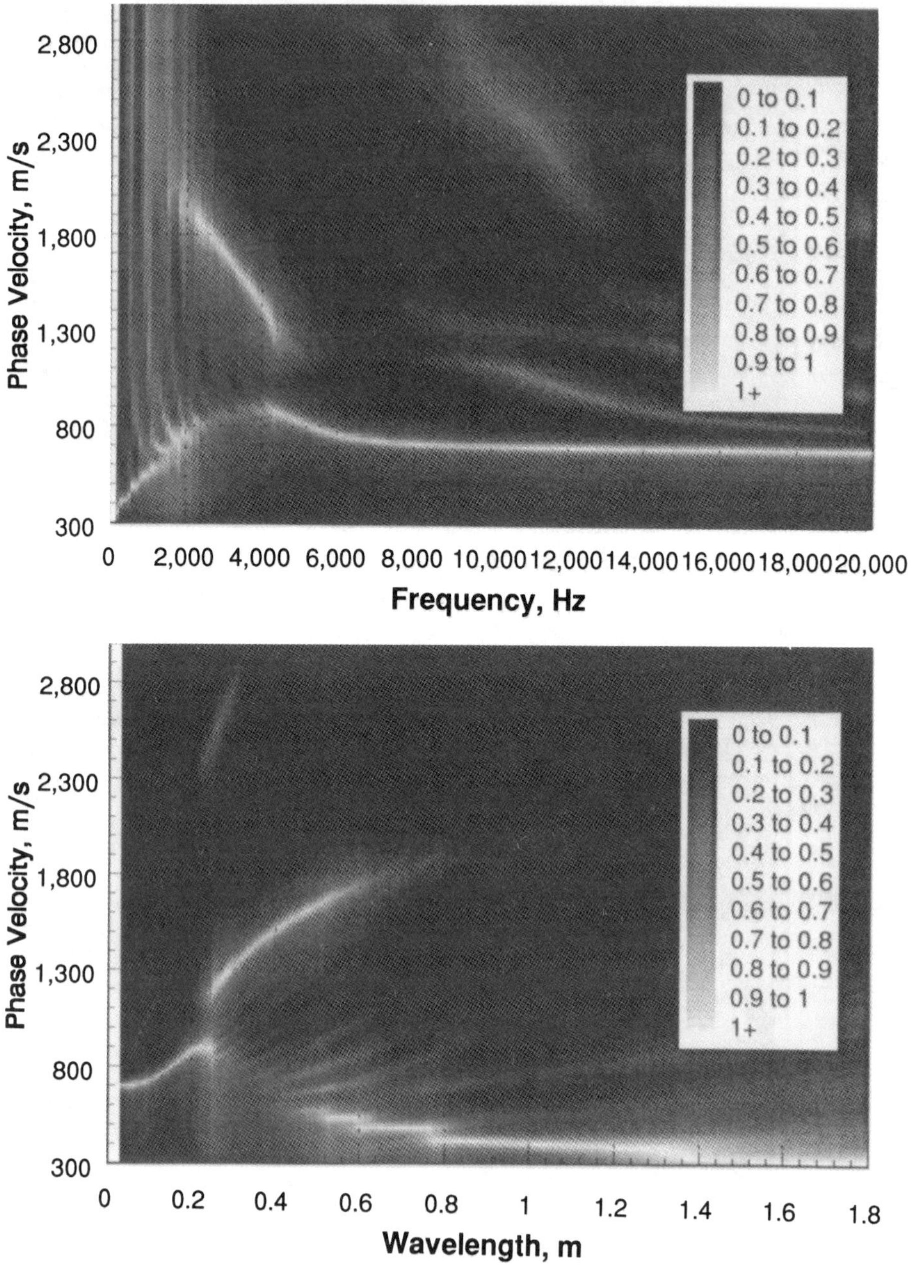

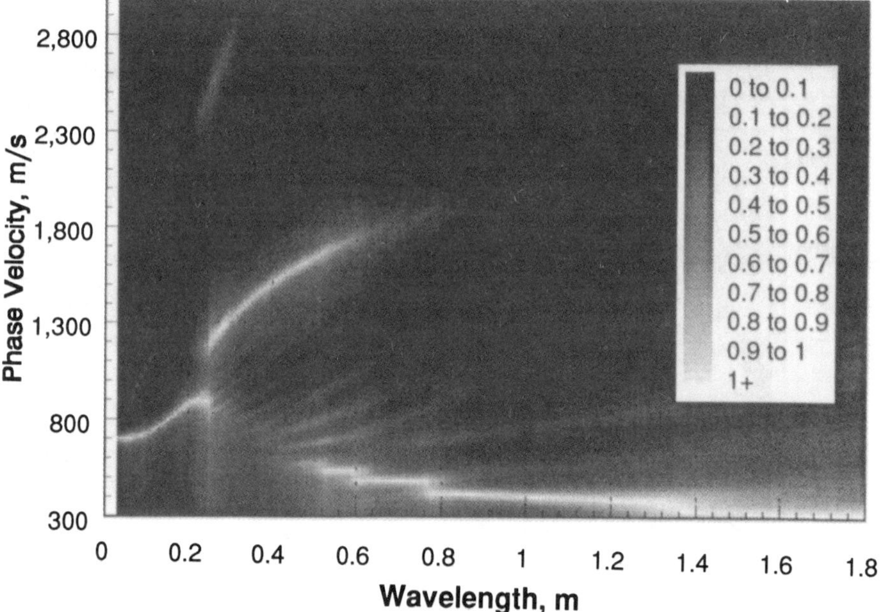

FIG. 9--Wave field generated by a circular source for case 3.

other criterion. Therefore phase velocities can differ due to different attenuation characteristics of body and surface waves. Possible existence of multiple Rayleigh modes, as previously described, can additionally affect the difference.

CONCLUSIONS

The conclusions of the study are:

1. The SASW test is capable of detecting the existence of an asphalt-concrete or concrete surface layer consisting of courses of different moduli by utilizing currently available high frequency miniature accelerometers.
2. The obtained dispersion curve depend on the selection of the filtering criterion, especially in frequency ranges where two or more Rayleigh modes dominate the wave field.
3. Averaging of dispersion curves with respect to the frequency and with respect to the wavelength provide similar results. Averaging with respect to the frequency is preferred for the reason described in the paper.

REFERENCES

[1]   Ewing, W.M., Jardetzky, W.S. and Press, F., 1957, _Elastic Waves in Layered Media_, McGraw-Hill, Inc., New York, NY.

[2]   Gucunski, N., and Woods, R.D., 1991, "Instrumentation for SASW Testing," _Recent Advances in Instrumentation, Data Acquisition and Testing of Soil_, Geotechnical Special Publication No. 29, ASCE, pp. 1-16.

[3]   Gucunski, N., and Woods, R.D., 1992, "Numerical Simulation of the SASW Test," _Soil Dynamics and Earthquake Engineering Journal_, Vol. 11, No. 4, pp. 213-227.

[4]   Heisey, J.S., Stokoe, K.H.II, Hudson, W.R., and Meyer, A.H., 1982, "Determination of In Situ Shear Wave Velocities from Spectral Analysis of Surface Waves," Research Report No. 256-2, Center for Transportation Research, The University of Texas at Austin.

[5]   Nazarian, S., Stokoe, K.H. II and Hudson, W.R., 1983, "Use of Spectral Analysis of Surface Waves Method for Determination of Moduli and Thicknesses of Pavement Systems," _Transportation Research Record_, No. 930, pp. 38-45.

[6]   Nazarian, S. and Stokoe, K.H.II, 1984, "Nondestructive Testing of Pavements Using Surface Waves," _Transportation Research Record_, No. 993, pp. 67-79.

[7]   Nazarian, S. and Stokoe, K.H. II, 1986, "Use of Surface Waves in Pavement Evaluation," _Transportation Research Record_, No. 1070, pp. 132-144.

[8]   Nazarian, S., Stokoe, K.H.II, Briggs, R.C. and Rogers, R., 1988, "Determination of Pavement Layer Thicknesses and Moduli by SASW Method," _Transportation Research Record_, No. 1196, pp. 133-150.

[9]   Roesset, J.M. Chang, D.-W., Stokoe, K.H.II and Aouad, M., 1989, "Modulus and Thickness of the Pavement Surface Layer from SASW Tests," _Transportation Research Record_, No. 1260, pp. 53-63.

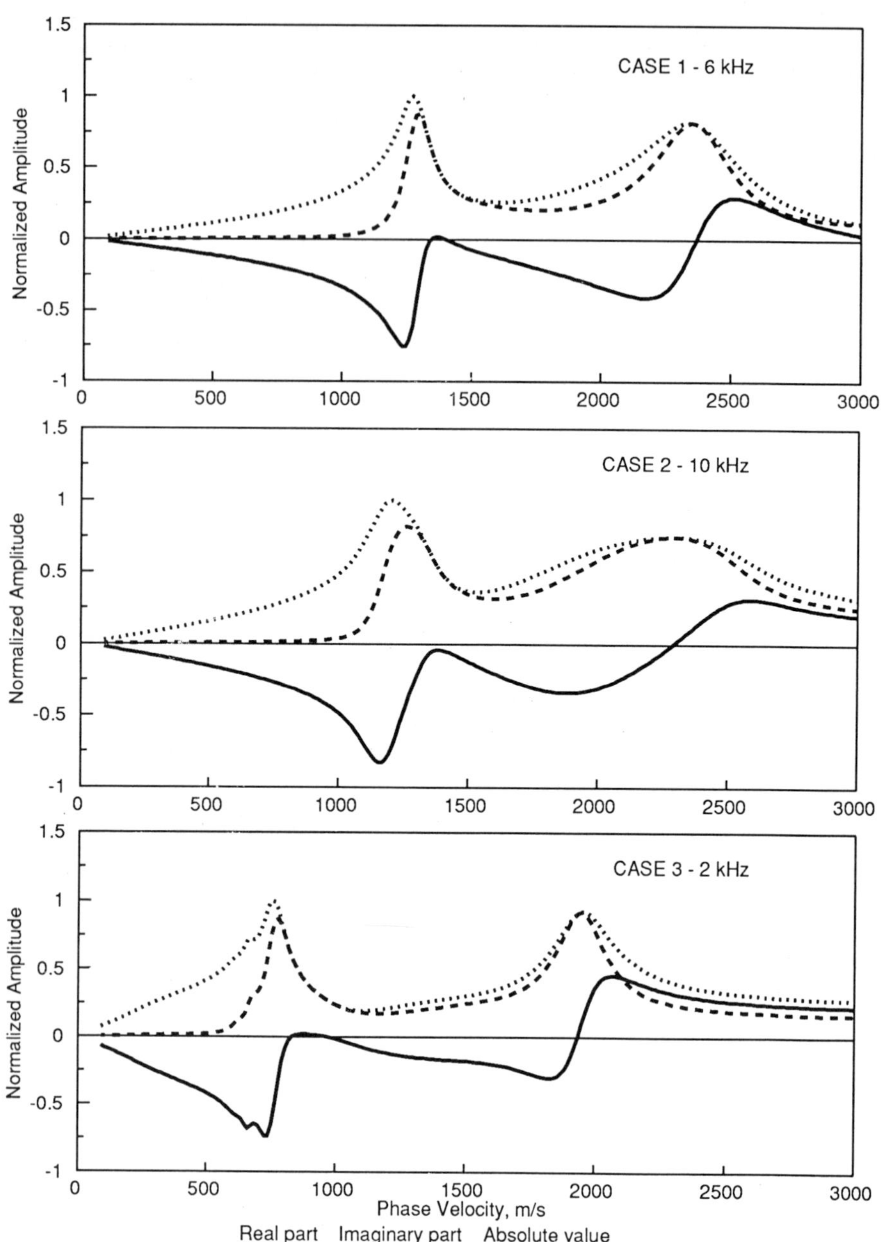

FIG. 10--Normalized surface vertical displacements in the wave number domain for cases 1, 2 and 3.

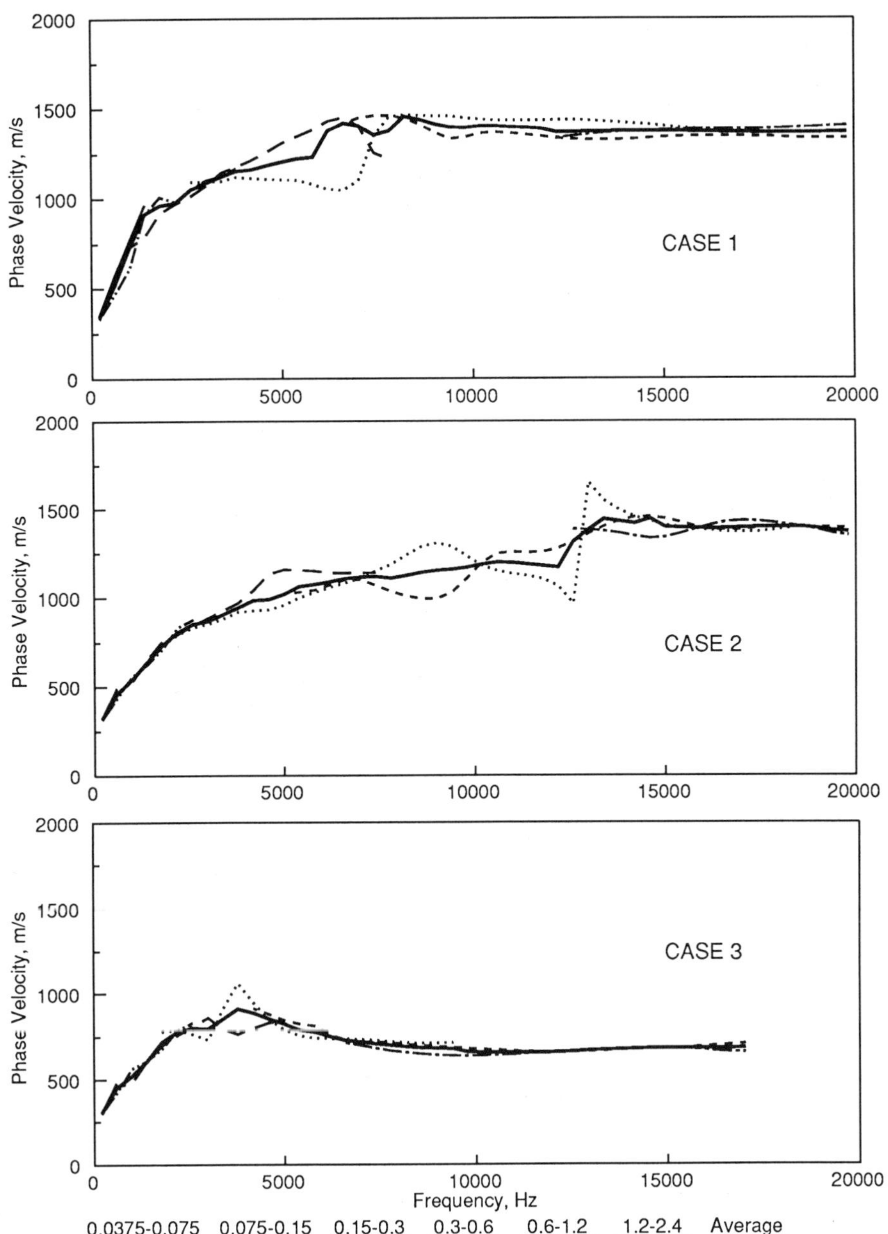

FIG. 11--Dispersion curves for six receiver spacings filtered by Heisey's criteria and the average dispersion curve.

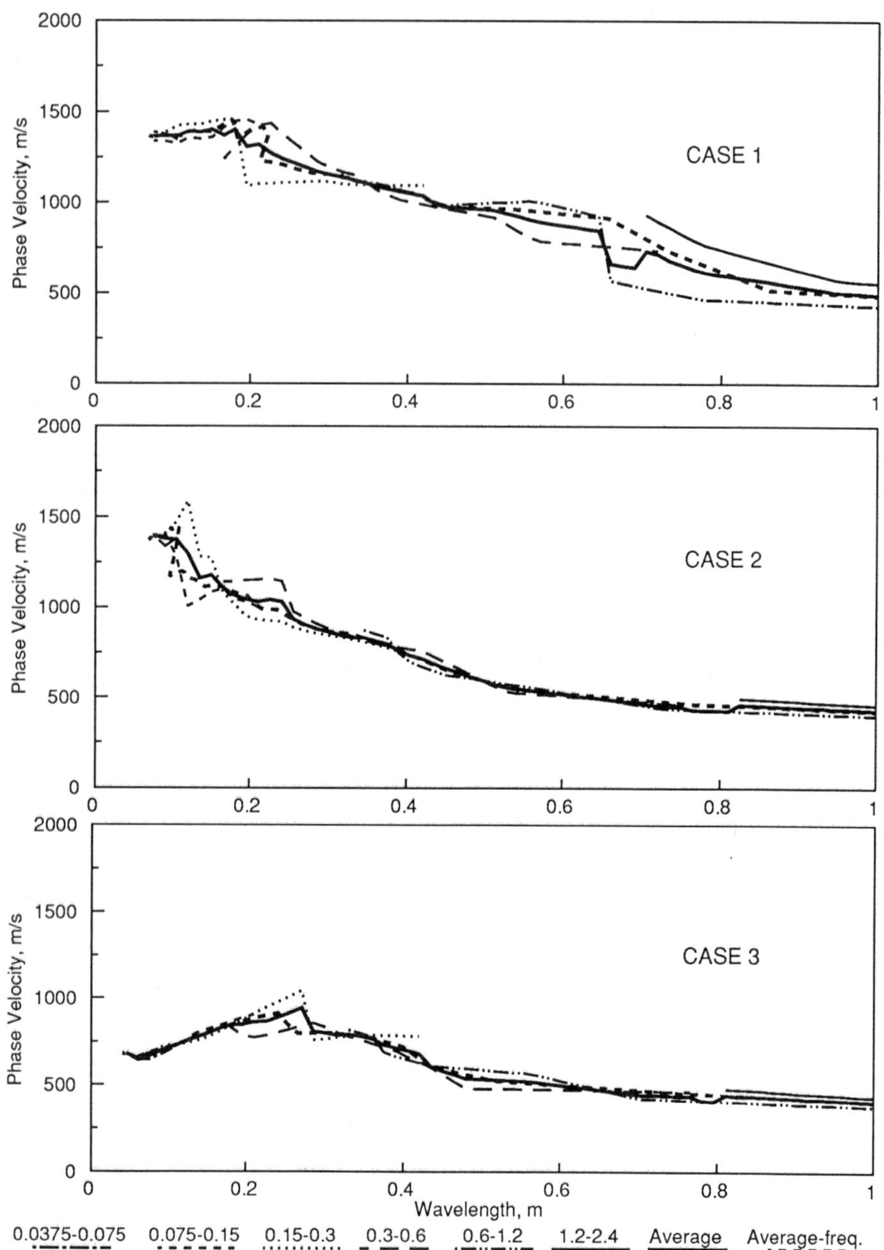

FIG. 12--Dispersion curves for six receiver spacings filtered by
Heisey's criteria and average dispersion curves.

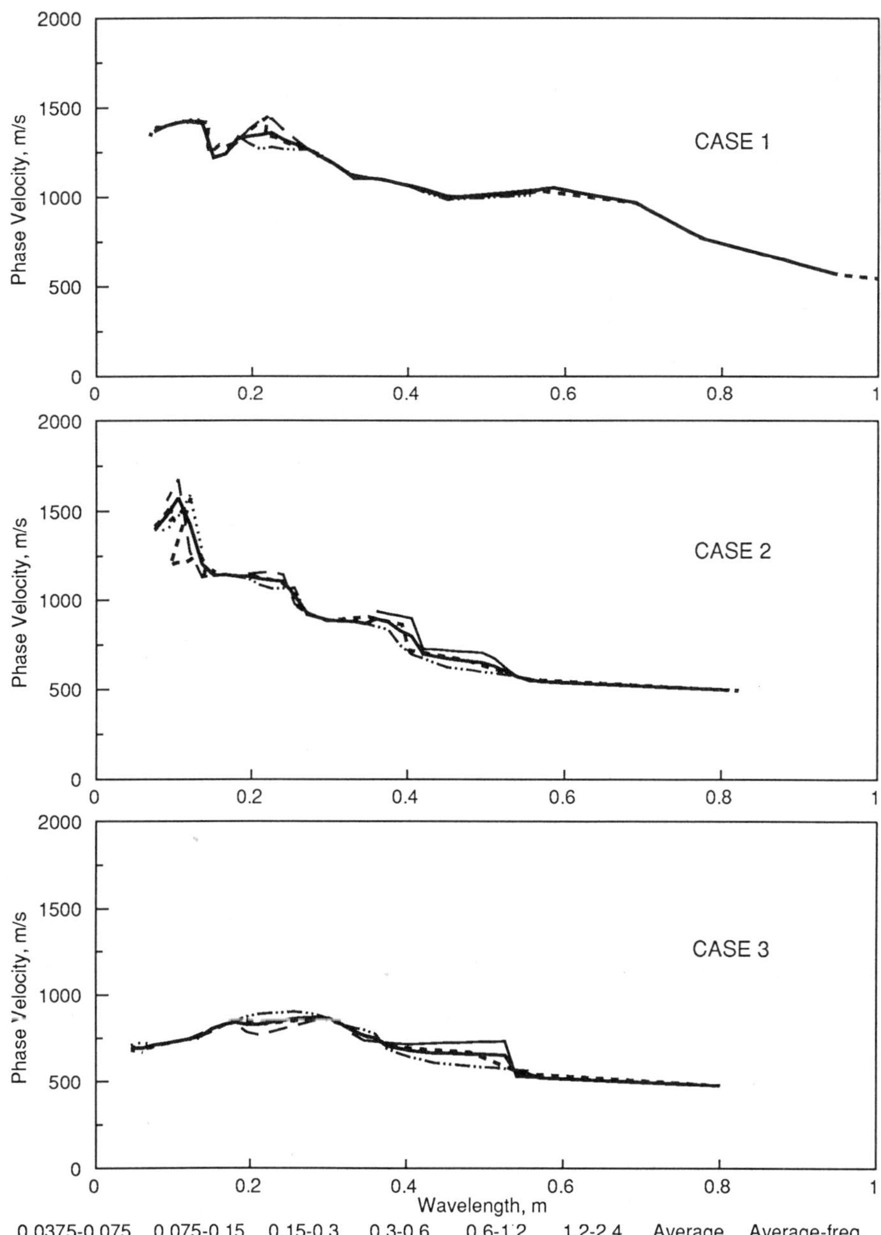

FIG. 13—Dispersion curves for six receiver spacings filtered by
Gucunski and Woods criteria and average dispersion curves.

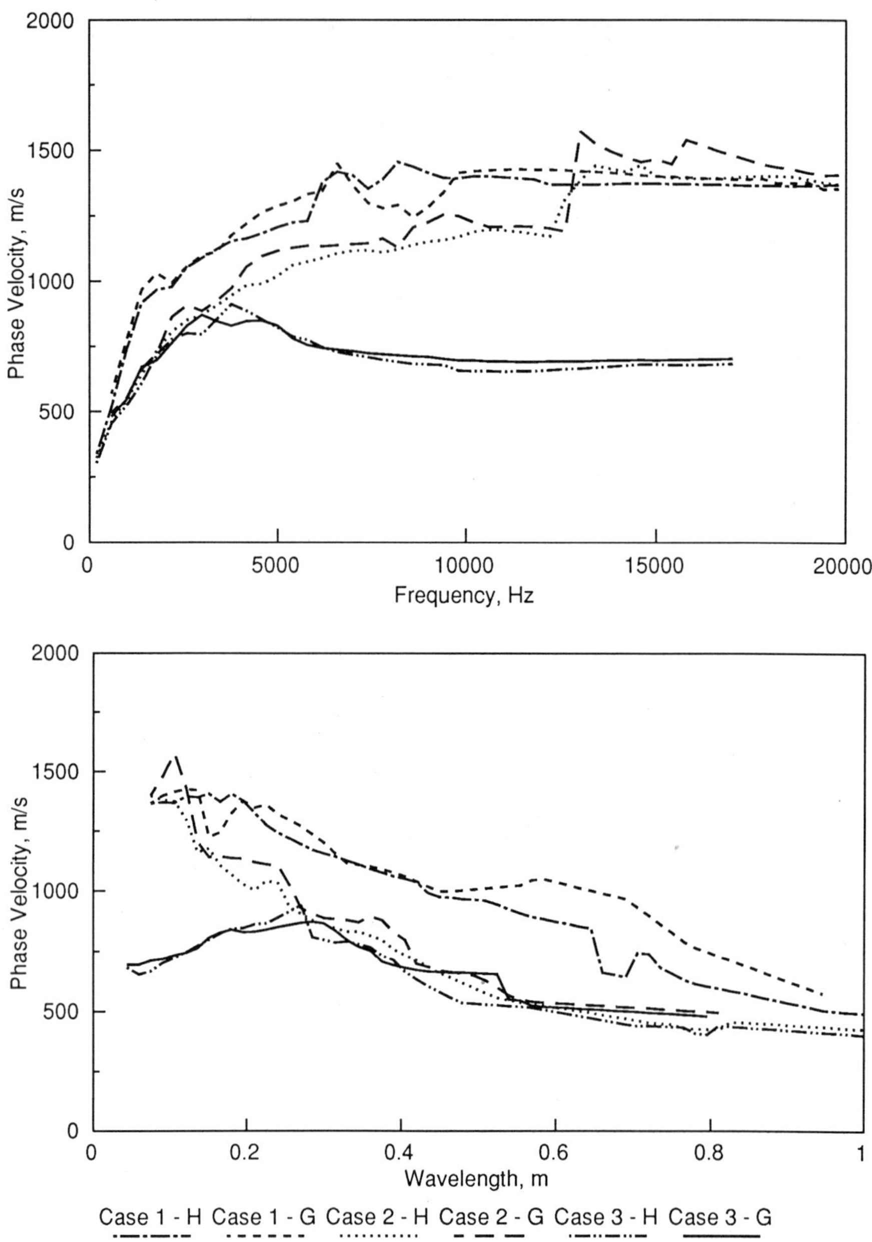

FIG. 14--Average dispersion curves for cases 1, 2, and 3 based on filtering criteria by Heisey (H), and Gucunski and Woods (G).

Damon J. Jackson[1], Michael R. Murphy[2], Andrew Wimsatt[3]

STRATEGIES FOR THE APPLICATION OF THE FALLING WEIGHT DEFLECTOMETER TO EVALUATE LOAD TRANSFER EFFICIENCY AT JOINTS IN JOINTED CONCRETE PAVEMENT

REFERENCE: Jackson, D. J., Murphy, M. R., and Wimsatt, A., "Strategies for the Application of the Falling Weight Deflectometer to Evaluate Load Transfer Efficiency at Joints in Jointed Concrete Pavement," Nondestructive Testing of Pavements and Backcalculation of Moduli (Second Volume), ASTM 1198, Harold L. Von Quintas, Albert J. Bush, III, and Gilbert Y. Baladi, Eds., American Society of Testing and Materials, Philadelphia, 1994.

ABSTRACT: The Falling Weight Deflectometer data for load transfer efficiency at joints in Jointed Concrete Pavement (JCP) must be properly analyzed and interpreted in order to prevent misleading results. Some parameters affecting pavement deflection are the pavement layer moduli, Poisson's ratios, thicknesses, and layer interaction. All of these must be considered to accurately "backcalculate" moduli from deflections. Joints add even more variables which affect the deflection measurement such as loss of support due to pumping at or near the joint. It is this parameter which will be discussed in this paper. First, a methodology for determining load transfer efficiency at a joint with subbase support will be given. Second the effects of loss of subbase support will be discussed and a methodology, with qualifications, given to determine load transfer at a joint that has loss of subbase support.

KEYWORDS: Loss of support, joint efficiency, falling weight deflectometer, geophones

---

[1,2, and 3] Pavement Engineers, (D-8Pav), Texas Department of Transportation, 125 E. 11th St., Austin, Texas 78701-2483.

BACKGROUND

In Texas, the Dynatest 8000 Falling Weight Deflectometer is used to measure deflections at various locations on rigid pavement structures in order to provide data for moduli backcalculation, to determine load transfer coefficients at joints and cracks and to detect loss of support at slab edge, corners, and joints. Several different methodologies have been evaluated for detecting loss of support or computing load transfer coefficients, however these methods have not been found to provide a satisfactory explanation of some sets of deflections readings found in the field. As a result of this evaluation, a technique for computing load transfer coefficients for jointed concrete pavements has been developed which is the subject of this paper.

LOAD TRANSFER EFFICIENCY AT A JOINT

Load transfer efficiency is generally taken to be the ratio of deflections measured at equal distances on each side of a joint. If the two deflections are approximately equal, then generally 100 % load transfer is assumed [1]. It should be noted that non-equal deflections at equal distances from the load should not be automatically interpreted as the result of loss of load transfer without first determining if loss of support (loss of subbase support) exists beneath the slab at or near the joint. In order to measure deflections upstation and downstation of a joint the sensor locations on the FWD must be modified by placing a sensor behind the load plate. Typically, sensor W3 is mounted 305 mm (12 in) behind the load plate on a sensor bar extension and the remaining sensors W2 through W7 are spaced at 305 mm (12 in) increments along the sensor bar in front of the load plate. When the load plate is placed on the upstation side of the joint, the W3 sensor will also be upstation of the joint and sensors W2 through W7 will be downstation of the joint. When the load plate is placed on the downstation side of the joint, then the W3 sensor alone will be upstation of the joint. In the methodology which has been developed, at least one drop on each side of the joint is necessary in order to determine the least load transfer coefficient. Referring to Figs. 1 and 2, two ratios are computed using the W2 and W3 deflections:

$$LT_{us} = USR \times 100 = (W2/W3)_{\text{load upstation}}$$

(% Load Transfer)   (Upstation Ratio)
(upstation)

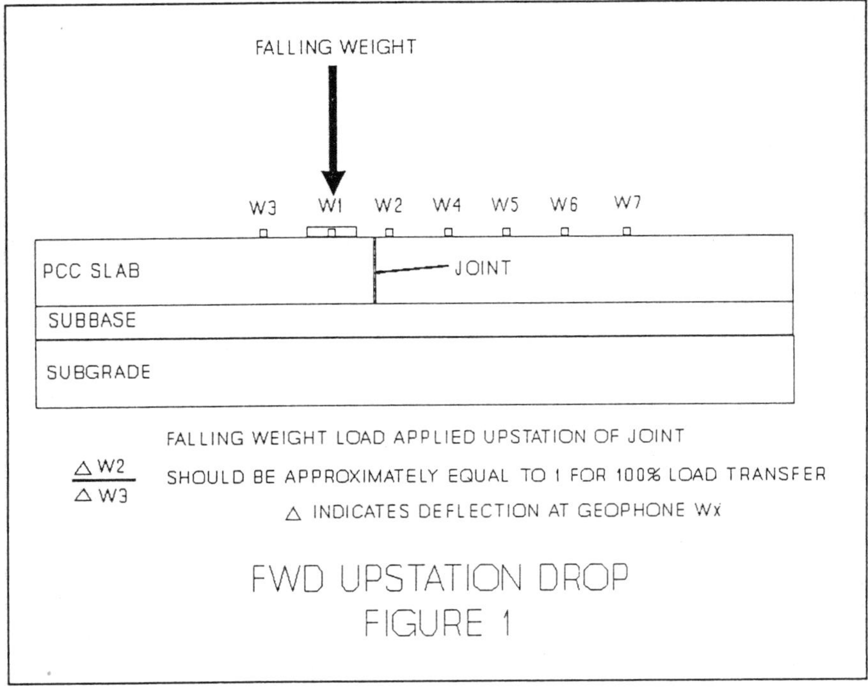

$$\underset{\substack{(\% \text{ Load Transfer}) \\ (\text{downstation})}}{LT_{ds}} \quad = \quad \underset{(\text{Downstation Ratio})}{DSR \times 100} \quad = \quad (W3/W2)_{\text{load downstation}}$$

$$\underset{(\% \text{ Load Transfer joint})}{LT_{joint}} \quad = \text{ the smallest of } LT_{us} \text{ or } LT_{ds}$$

Using these % load transfer equations, the following set of criteria have been established to evaluate the various conditions seen in the field:

1. If USR is approximately equal to DSR then symmetry of load transfer can be assumed.

2. If USR is approximately equal to DSR and both USR and DSR are approximately equal to 1.0 then symmetry and 100% load transfer can be assumed in most all cases.

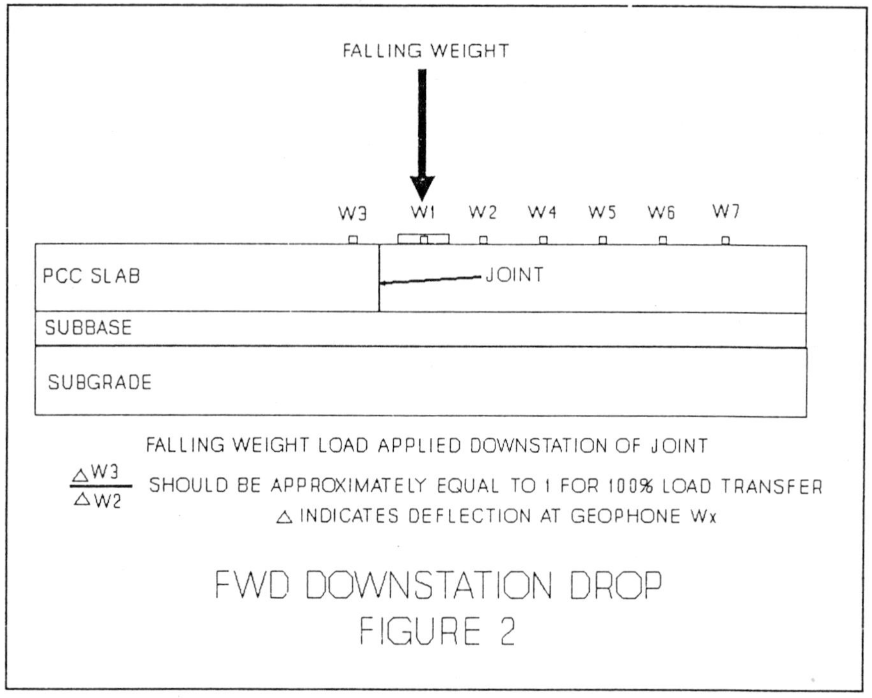

3. If USR does not equal DSR and USR is approximately equal to 1 or the DSR is approximately equal to 1, assume either less than 100% load transfer in one direction and/or possibly that a loss of support exists under the joint.

4. If USR does not equal DSR and neither USR nor DSR is approximately equal to 1, then assume possibly that a loss of support exists under the joint and/or possibly less than 100% load transfer in one direction.

5. If USR is approximately equal to DSR and both USR and DSR are much less than 1, then assume possibly that a loss of support exists under the joint and possibly less than 100% load transfer in both directions.

Asymmetrical load transfer can occur at a joint which has areas of loss of support beneath it and/or a non-vertical crack which has little or no load transfer. [1] Referring to Fig. 3, a load placed on the upstation side of the joint for condition A, would show that load transfer takes place. A

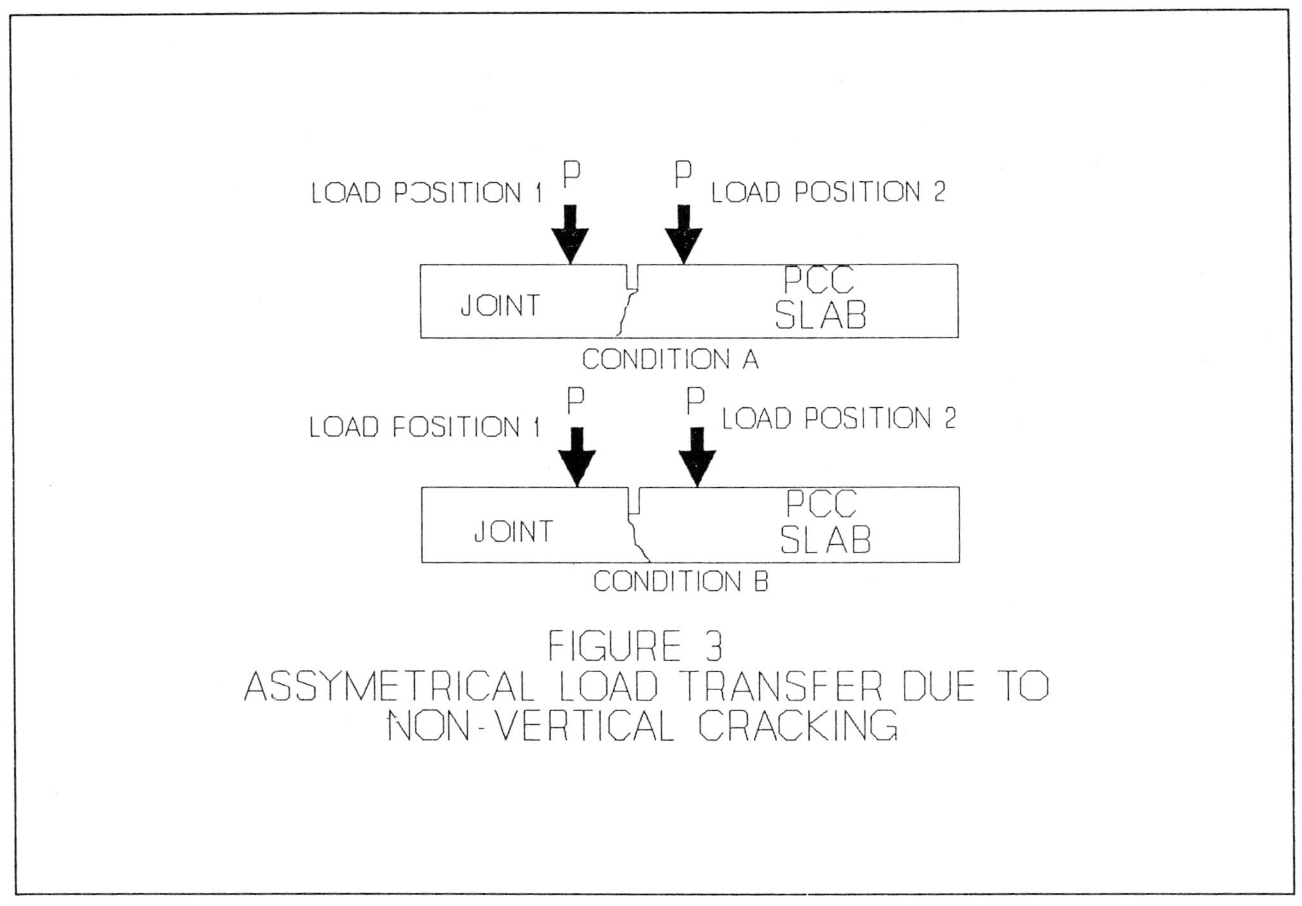

FIGURE 3
ASSYMETRICAL LOAD TRANSFER DUE TO
NON-VERTICAL CRACKING

load placed downstation of the joint would show that less than 100 % load transfer takes place because of the loss of interlock at the joint. The exact opposite is true for condition B.

LOAD TRANSFER AT A JOINT WITH LOSS OF SUPPORT

The detection of loss of support under concrete pavements is important since loss of support creates a condition of nonuniform stress distribution. This nonuniform stress distribution produces areas of stress concentration beneath the slab within the remaining subbase present around the area of support loss. This increase in stress in certain areas beneath the slab can lead to faulting and cracking. Also, since a loss of support results in a greater deflection of the slab under load than would occur if the slab was fully supported, an undetected loss of support can lead to an erroneous estimation of material properties using modulus backcalculation techniques.

Various methodologies have been developed for detecting loss of support under portland cement concrete pavements using NDT equipment. In detecting loss of support, there are two main concerns 1) determining if loss of support is evident (exists) and 2) determining the extent of the loss of support. Since, as stated previously, a loss of support usually results in greater deflections than would occur on a fully supported slab, the magnitude of deflections and shape of the deflection bowl are typically used to signal a potential loss of support. Two techniques which have been developed for loss of support detection using the FWD include Crovetti and Darter's procedures [2] and the methodology developed by Ricci et al [1]. A brief summary of these two procedures is given in the following commentary.

CROVETTI AND DARTER'S PROCEDURE

Actually two methodologies have been developed by Crovetti and Darter to detect loss of support under a PCC pavement. The first methodology involves a straight forward plotting of FWD load vs Deflection for each of at least three load levels. Once these points have been plotted, a best fit line is drawn through the points and compared with a line which would be formed by simply connecting the points. A substantial difference in the shape of these two lines indicates a potential loss of support. By extending the best fit line so that it intersects the horizontal axis, an intercept value can be obtained which is useful in indicating the size of the area where loss of support exists.   An intercept value less than 0.05 mm (0.002 in) signifies that no loss of support is present; intercept values above 0.05 mm (0.002 in) signify the presence of a loss of support with the size of the area where loss of support exists increasing with increasing intercept value.

Crovetti and Darter's second methodology is more detailed and can be used to indicate both the presence or absence of a loss of support and the approximate size (plan dimension) of the area where loss of support is present.   The reader is referred to reference [2] for a more detailed discussion of this methodology.   It is important to note that in this methodology, the sensor reading directly under the plate and the adjacent three sensor deflections are used to compute elastic modulus values at the center of the slab which are later used in the analysis.

RICCI ET AL

In this methodology, the deflection basin is used to detect the presence of a loss of support by computing the slopes of lines connecting various sensors.   Two statistics M and Q are computed, using the following relationships:

$$M = Arc\ tan\ [6/(W_1 - W_2)]$$

$$Q = Arc\ tan\ [(W_2 - W_7)/24]$$

where:

M =   The angle of a line, as measured from the vertical, connecting a plot of the second and first sensor readings.

Q =   The angle of a line, as measured from the horizontal, connecting a plot of the second and seventh sensor readings

W_i = The ith sensor in the group.

The presence or absence of a loss of support is indicated by the magnitudes of the values Q and M as follows:

1.   If Q is greater than 18 the presence of a loss of support is indicated.

2.   When Q is greater than 18, the smaller M is the larger the loss of support.

3.   At a joint, if Q is less than 10 and M is greater than 70, then full load transfer can be assumed.

SUMMARY AND CONCLUSIONS

The technique described herein, discusses a practical, field implementable approach to determining % Load Transfer at a joint. This value is a ratio of the deflections taken at equal distances on opposite sides of the Falling Weight Deflectometer (FWD) load plate.

It should be noted that the percent load transfer value can also be influenced by loss of support underneath the concrete pavement (slab). Therefore, the deflection data should also be analyzed for loss of support.

REFERENCES

1.   Ricci, E. A., Meyer A.H., Hudson, W. R., and Stokoe, K. H. II, "The Falling Weight Deflectometer for Nondestructive Evaluation of Rigid Pavements", Report FHWA/TX-86/44+387-3F Center for Transportation Research, The University of Texas at Austin for the Texas Department of Transportation, November 1985.

2.   Crovetti, J. A. and Darter, M. I., "Void Detection for Jointed Concrete Pavements", _Transportation Research Record 1041_, Transportation Research Board, National Research Council, Washington, D.C., 1985.

3.   Panak, J. J. and Matlock, H., "A Discrete-Element Method of Analysis for Orthogonal Slab and Grid Bridge Floor Systems", TxDOT Research Report 56-25 Center for Transportation Research, The University of Texas at Austin for the Texas Department of Transportation, May 1972.

4.   Tayabji, S. D. and Colley, B. E., "Analysis of Jointed Concrete Pavements", Report FHWA/RD-86/041 Construction Technology Laboratories, Portland Cement Association, February 1986.

Chung-Lung Wu[1], and Mang Tia[2]

FIELD TESTING AND STRUCTURAL EVALUATION OF SELECTED CONCRETE PAVEMENT SECTIONS IN FLORIDA

REFERENCE: Wu, C. Lu., and Tia, M., "Field Testing and Structural Evaluation of Selected Concrete Pavement Sections in Florida," <u>Nondestructive Testing of Pavements and Backcalculation of Moduli (Second Volume), STP 1198</u>, Harold L. Von Quintas, Albert J. Bush, III, and Gilbert Y. Baladi, Eds., American Society for Testing and Materials, Philadelphia, 1994.

ABSTRACT: Twenty-four existing concrete pavement sections covering a wide range of performance levels, ages and types of subbase, surface layers and joints were selected for evaluation in order to determine the causes of the problems in the pavements with poor performance. Each section was evaluated by a condition survey, falling weight deflectometer (FWD) tests, laboratory tests on core samples, and a structural analysis. FWD tests were preformed at both midday and midnight. Loads were applied at four different positions on the slab, namely, slab center, slab corner, edge center and joint center. The measured FWD deflections at various locations were then used along with the results of the laboratory tests on the cored samples and a finite element program, FEACONS IV, to estimate the pavement parameters. With the pavement modeled by the estimated pavement parameters, the maximum stresses caused by critical temperature-loading conditions were computed using the FEACONS IV program. The computed maximum stresses were then compared with the flexural strength of the concrete to evaluate the structural adequacy of each test section.

The critical stress analysis method, which was used in this study, was shown to give good prediction of the structural performance of the test sections. The use of a very stiff subbase is not beneficial to the performance of a concrete pavement, as indicated by the results of the critical stress analyses as well as field surveys. High elastic modulus of the concrete and long slab length were also shown to produce adverse effects to concrete pavement performance.

KEYWORDS: concrete pavement, falling weight deflectometer, condition survey, FEACONS, econocrete base, critical stress analysis, elastic joints.

Concrete pavements in Florida have exhibited widely different performance. Some have performed satisfactorily and have served beyond their design life. Examples of the concrete pavements with good performance records are portions of I-4, I-95, U.S. 29 in Escambia County and Econocrete Test Road (U.S. 41) near Fort Myers, and many others. However, some have poor performance records and have shown severe signs of distress and failure (such as pumping, faulting, and cracking) before the end of their intended service life. Two cases of concrete pavements with extremely poor performance records are the concrete section of I-10 and the section of I-75 with an econocrete base in Manatee and Sarasota counties. The premature failure of these pavements might be attributed partially to the unexpected high traffic loads on the pavement sections. However, in consideration of the fact that some of the well-performing pavements also have high traffic loads on them, the problems might be attributed to the inadequacies of their designs. In order to design concrete pavements to meet today's demand of traffic volume and frequency, it is necessary to evaluate the general performance of existing pavements and to identify the causes of the problems associated with them.

---

[1] Paving Engineer Construction Technology Laboratory, 5420 Old Orchard Road, Skokie, IL 60077-1030.

[2] Professor, Department of Civil Engineering, University of Florida, Gainesville, FL 32611.

Twenty-four concrete pavement sections were selected for evaluation. They cover a wide range of performance level, age, and structural design, and are representative of the various typical pavements in Florida. These 24 sections include eight sections selected from I-10, two sections selected from I-4, three sections selected from I-75 in Hillsborough County, five sections selected from I-75 in Sarasota and Manatee Counties and six sections selected from the Econocrete Test Road on U.S. 41 near Fort Myers. Each test section was evaluated by a condition survey, falling weight deflectometer (FWD) tests, laboratory tests on core samples, and a structural analysis. Based on the results of these evaluations, the performance of these test sections was assessed and the causes of problems (if any) were determined. Some of the test sections have been tested in different years and different seasons. This was done to evaluate the deterioration rate of the pavements with different section designs.

BACKGROUND

Several different concrete pavement section designs have been used in Florida for the past few decades. The variations in the design included the use of different materials for subbase and subgrade, the use of skewed joints, and the use of composite surface layers, etc. This section presents the pavement section designs used in several major highways in Florida. The types of distress and failure associated in these pavements are also briefly described.

I-10 Concrete Pavement (FDOT 1976, 1983)

Two types of section designs were used on I-10 concrete pavements. Figure 1 shows the typical design used in the western portion of I-10 pavements. It had a 20-cm (8-in.) or 23-cm (9-in.) thick jointed concrete pavement with a 30-cm (12-in.) thick conventional stabilized subgrade. On the eastern portion of I-10 concrete pavements, cement- and lime-treated subbase was used because of the clay embankment materials. The typical section design is shown in Figure 2. It had a 23-cm (9-in.) thick jointed concrete pavement with a 15-cm (6-in.) thick cement-treated subbase and a 15-cm (6-in.) thick lime-treated subbase.

For the entire concrete section of I-10, the joints were evenly spaced at 6.1-m (20-foot) intervals. With only a few exceptions, most of the joints were undoweled. No special subdrain or edgedrain was used on any of the pavement sections.

The major types of pavement distress and failure found on I-10 concrete highway are (1) pumping, (2) faulting, (3) transverse cracking, (4) corner cracking, (5) diagonal cracking, and (6) deterioration of shoulder material. Signs of pumping are prominent on most sections of I-10 that are exhibiting failure or distress. It is believed that pumping has led to most of the other distresses on I 10.

I-75 Concrete Pavement in Sarasota and Manatee Counties (FDOT 1986)

The typical design used on this 46.4-Km (29-mile) concrete pavement is shown in Figure 3. The pavement section had 23-cm (9-in.) thick concrete slabs laid unbounded over a 15-cm (6-in.) thick econo-crete base, and had an econocrete shoulder. The concrete slabs had skewed joints and random slab lengths of 5.2, 6.7, 7.0, and 4.9-m (17, 22, 23, and 16 feet). Dowel bars at transverse joints and tie bars at longi-tudinal joints were used. Additionally, a strip of filter fabric was placed under the econocrete shoulder and elastomeric and silicone sealants were used at the joints. No edgedrain or subdrain was used. No joint was provided for the econocrete base, with the exception of the construction joints which did not match with the concrete slab joints.

In April, 1982, within about a year after this pavement had been opened to traffic, the Florida Department of Transportation (FDOT) Bureau of Materials & Research personnel reported a substantial number of cracked slabs and signs of pavement pumping on this pavement. The common types of distress or failure observed on this pavement are (1) transverse cracking near the center of the slabs, mostly in the longer slabs, (2) corner cracking at the acute corners of the slabs, and (3) pavement pumping. Some longi-tudinal cracking was also observed.

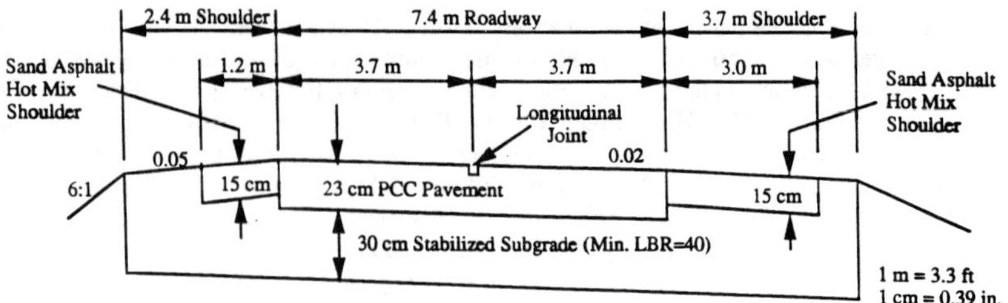

Figure 1 A Typical Pavement Section on Western Portion of I-10 Concrete Pavement

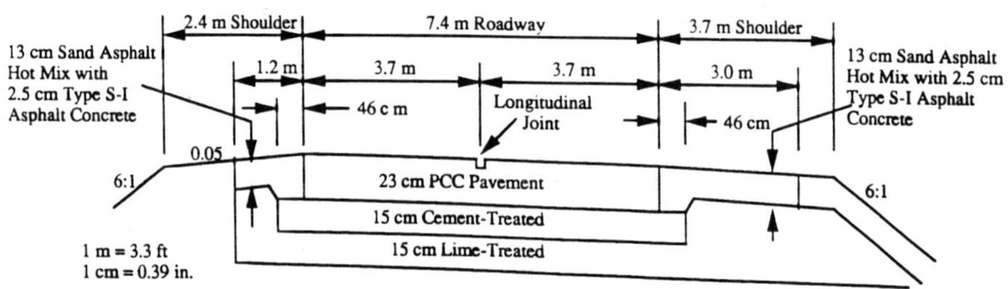

Figure 2 A Typical Pavement Section on Eastern Portion of I-10 Concrete Pavement

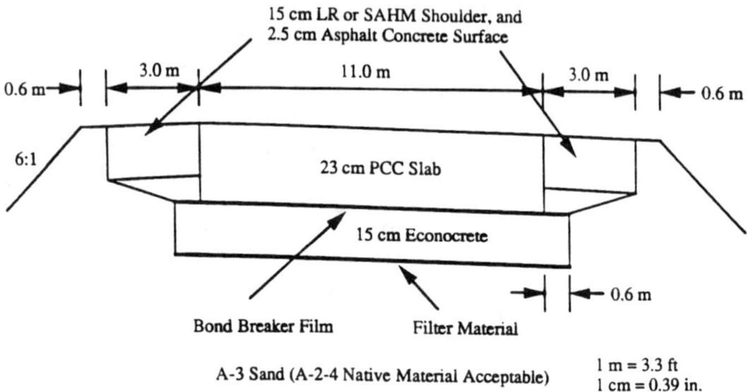

Figure 3 A Typical Pavement Section on I-75 in Sarasota and Manatee Counties

I-95 Concrete Pavement in Duval County (FDOT 1985)

The I-95 concrete pavement in Duval County (19.2 km) was constructed in 1966 through 1967. The typical pavement section has 23-cm (9-in.) of concrete slab and 30-cm (12-in.) or 15-cm (6-in.) of conventional stabilized subgrade. This pavement has shown satisfactory performance. The deterioration on this pavement was due to traffic loads and age. The Present Serviceability Index (PSI) ranges from 2.1 to 4.2.

I-4 Concrete Pavement in Hillsborough County (FDOT 1985)

The I-4 concrete pavement in Hillsborough County (41.6 km) was constructed in 1973. The typical section has a slab thickness ranging from 20-cm (8-in.) to 25-cm (10-in.), and a subbase made of either 15-cm (6-in.) of limerock or 30-cm (12-in.) of conventional stabilized soil (minimum LBR = 75). The concrete slab joints are doweled and spaced at even intervals of 6.1-m (20 feet). This pavement has performed satisfactorily. The pavement is still in good condition except for a few locations where truck traffic is extremely heavy. These heavy traffic locations have some badly cracked slabs. There is no sign of pumping.

I-75 Concrete Pavement in Hillsborough County

The I-75 concrete pavement in Hillsborough County (17.6 km) is the most recently built concrete interstate highway in Florida. The construction of the first project in this pavement was started in 1981. The typical section is similar to that of I-75 in Sarasota County, with the exception that the econocrete shoulder is made of a higher quality aggregate and the concrete slab thickness is 30-cm (12-in.). Similar to the I-75 concrete pavement in Sarasota County, the section in the first project has 15-cm (6-in.) of econocrete base, skewed joints spaced at 5.2, 6.7, 7.0, and 4.9 m (17, 22, 23, and 16 feet) intervals, dowel bars, tie bars, and no subdrain.

Due to the problems on the I-75 concrete pavement in Sarasota County, FDOT modified the design for the rest of the projects on this pavement. For the remaining projects under construction, the design was modified as follows:

(1)   The outside lane of concrete (adjacent to the shoulder) was widened from 3.7-m (12 feet) to 4.3-m (14 feet).
(2)   The econocrete was widened by 0.6-m (2 feet).
(3)   The econocrete shoulder was reduced from 3.3-m (10 feet) to 2.4-m (8 feet) in width. The variable shoulder thickness was changed to a constant thickness of 15-cm (6-in.).
(4)   A drainage blanket with six inches of #57 stone with a filter fabric was to be constructed beneath the shoulder for the full width.

For the projects which had not been contracted out at the time when the decision to change the design was made, a new design was used. This new design is shown in Figure 4. The design features are listed below:

(1)   Concrete slab thickness of 33-cm (13-in.).
(2)   Fifteen cm (6-in.) of stabilized subgrade using #57 stone, instead of econocrete base.
(3)   Forty-six cm (18-in.) of A-3 material with a maximum of 7 percent passing #200 sieve, and 76 cm (30-in.) of A-3 material with a maximum of 10 percent passing #200 sieve.
(4)   Skewed joints spaced at 4.0, 5.8, 5.5, and 3.7-m (13, 19, 18 and 12 feet) intervals.
(5)   Edgedrains.

A portion of this concrete pavement was open to traffic in late 1984. No major distress has been observed on this pavement yet. A few hairline transverse cracks have been observed on the first project, which used the unmodified design, before it was open to traffic. In early 1986, a decision was made to

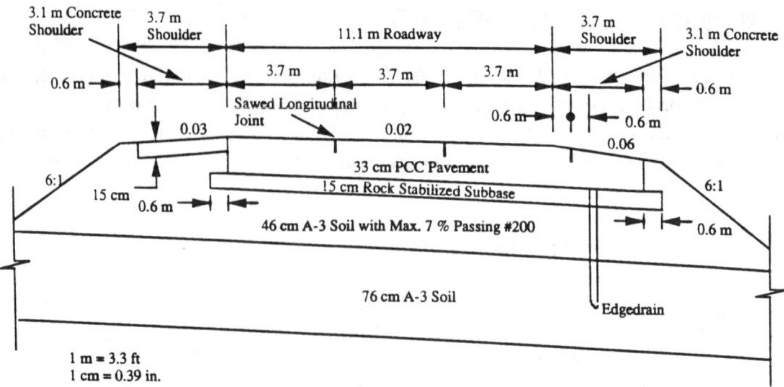

Figure 4 A Typical Pavement Section on I-75 in Hillsborough County

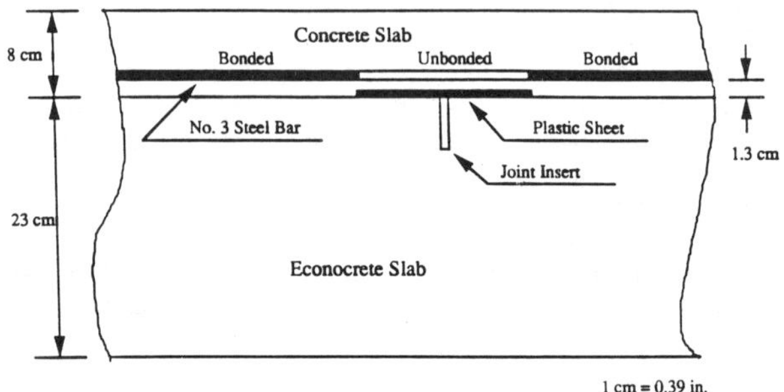

Figure 5 Elastic-Jointed Composite Pavement with Continuous Reinforcement

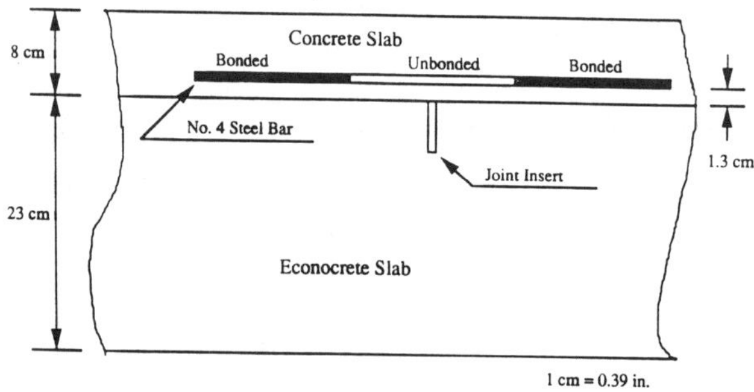

Figure 6 Composite Pavement with Elastic Doweled Joints

saw additional transverse joints across the centers of the 6.7- and 7.0-m (22- and 23-foot) slabs in this project to make short slabs of 3.4- and 3.5-m (11- and 11.5-feet) in length, before this project was open to traffic. The performance of this new concrete pavement with the various new design features still remain to be seen.

<u>Florida Econocrete Test Road</u> (FDOT 1980, Larsen & Mayfield 1984)

An experimental road consisting of thirty-three different pavement sections was constructed on U.S. 41 north of Ft. Myers by FDOT in 1977 to study the performance of various composite pavement systems with econocrete base. Of the thirty-three test sections, thirty had concrete surfaces, while the other three had asphalt surfaces.

These various concrete test sections can be grouped into three major types according to their temperature-load response characteristics. These three major types of pavement sections are (1) composite concrete pavements with an unbonded interface, (2) monolithic composite concrete pavements (with a bonded interface) and (3) elastic jointed concrete pavements. The response and performance of these three types of concrete pavements were evaluated in this study. This section presents the basic designs of these three pavement types.

Five test sections of the Econocrete Test Road could be grouped into the first pavement type. Three test sections had a 8-cm (3-in.) continuously reinforced concrete surface laid unbonded over a 23-cm (9-in.) econocrete base. Two test sections had a 8-cm (3-in.) fiber-reinforced concrete surface laid unbonded over a 20-cm (8-in.) econocrete base. All five sections had a 15-cm (6-in.) type "A" modified stabilized subgrade.

All five sections showed extensive cracking within six months after they were open to traffic. The concrete surfaces were removed and the pavements were all overlaid with asphalt.

Fourteen test sections of the Econocrete Test Road could be classified as monolithic composite concrete pavements. They were characterized by a thin concrete surface of 5- or 8-cm (2- or 3-in.) thick bonded to an econocrete layer of 20- or 23-cm (8- or 9-in.) thick. The concrete layer was placed before the econocrete had achieved its initial set. Additional bonding between the layers was further enhanced by scarifying the surface of the econocrete layer with a steel tine comb.

The basic feature of elastic joints in continuous reinforced concrete pavements is characterized by an alternate bonded and unbonded interface between reinforcement and concrete (Larsen 1978). Upon casting, no preformed joints are provided on the concrete layer. Instead, crack initiators are installed at midpoint of the unbonded area. The function of crack initiators is to ensure that cracks will occur at specified positions.

After the setting of concrete and dissipation of heat of hydration, the concrete contracts. Some of the early volume changes may be treated as plastic deformation and will not induce strains in the concrete or the reinforcement. However, after the bond strength between the concrete and the steel has been built up, the continuous volume decrease of the concrete within the bonded area will be resisted by the bonded steel. This will cause compressive strain in the bonded steel and tensile strain in the concrete. On the other hand, no resistance of concrete volume decrease is provided by the steel within the unbonded area, and therefore, the maximum concrete tensile strain occurs within this area. The differential of tensile strains in concrete and the installation of crack initiators will eventually cause the concrete to be cracked at mid-length of the unbonded area. These cracks are treated as joints. After the joints are formed, the functions of the unbonded steel are to tie the cracked slabs together and to transfer the load between adjacent slabs. It should be noted that the strain in the unbonded steel will always be opposite to the deformation of the concrete. Thus, it is to function as a latent spring.

The principle of elastic joints was used in ten of the test sections of the Econocrete Test Road. These sections had a composite surface layer made up of either a 8-cm (3-in.) concrete layer laid over a 23-cm (9-in.) econocrete layer, or a 5-cm (2-in.) fiber-reinforced concrete layer over a 23-cm (9-in.) econocrete layer, and were characterized by an alternate bonded and unbonded interface between the concrete and the econocrete. All sections had a 15-cm (6-in.) Type "A" modified stabilized subgrade.

Three sections had continuous reinforcement in the 8-cm (3-in.) concrete layer with an alternate bonded and unbonded interface between the steel and the concrete. A plastic sheath was used over the length of the unbonded steel. The concrete and econocrete had an alternate bonded and unbonded interface. A 1.3 mm thick plastic sheet was placed between the concrete and the econocrete to produce an unbonded interface. A plastic joint insert, which was to act as a crack initiator, was inserted into the econocrete layer at the mid-point of the unbonded area. Figure 5 shows the basic features of these sections.

Four sections utilized elastic dowels, which represent a refinement or modification of the elastic joint principle. In these sections, steel reinforcement in the 8-cm (3-in.) concrete layer was not continuous and was placed only across the anticipated cracks and long enough to hold the cracks tightly together. Joint inserts (crack initiators) were installed in the econocrete at various specified intervals, and the elastic dowels were installed over these inserts. No unbonding between the concrete and the econocrete layers was attempted. Figure 6 depicts the basic features of these sections.

Three sections partially incorporated the elastic joint principle. These sections had a 5-cm (2-in.) fiber-reinforced concrete layer laid over a 23-cm (9-in.) econocrete layer. The anticipated cracks were to be held together by the steel fiber reinforcement instead of the steel rebars. The fiber-reinforced concrete and the econocrete had an alternate bonded and unbonded interface. A wax-base curing compound was used as a bond-breaker to produce the unbonded interface. A plastic joint insert was installed into the econocrete layer at the midpoint of the unbonded area. Figure 7 shows the basic features of these sections. These three test sections showed severe longitudinal cracks followed by severe spalling within 6 months after they were open to traffic. They were, subsequently, replaced with an asphalt overlay.

FIELD AND LABORATORY TESTING PROGRAM

Twenty-four sections of concrete pavement, designated as Sections 1 through 24, in Florida were selected for field testing and evaluation. Preliminary identification of these test sections was first done through the information obtained from various FDOT documents. The pertinent pavement information used to make the initial selection included (1) the type and thickness of surface layer, (2) the type of subbase, (3) the type and spacing of joints, (4) the age of the pavement, and (5) the present serviceability index (PSI). The initial selection was followed by a visual inspection of the prospective test sections to ascertain the actual pavement conditions and to determine the exact locations of the slabs to be tested. The final selection was made such that they would cover a wide range of performance levels, ages, subbase types, slab thickness and joint types.

These 24 sections include eight sections selected from I-10, two sections selected from I-4, three sections selected from I-75 in Hillsborough County, five sections selected from I-75 in Sarasota and Manatee Counties and six sections selected from the Econocrete Test Road near Fort Myers. Table 1 displays the basic information of these test sections.

A detailed condition survey and Falling Weight Deflectometer (FWD) test were conducted on each pavement test section. Cores were also taken from the test sections.

<u>Falling Weight Deflectometer Test</u>

A Falling Weight Deflectometer (FWD) was used to evaluate the structural performance of the test sections. Basically, a FWD test involves delivering an impact load to a pavement surface by dropping a mass from a specified height onto a circular loading plate connected by a set of springs, and measuring

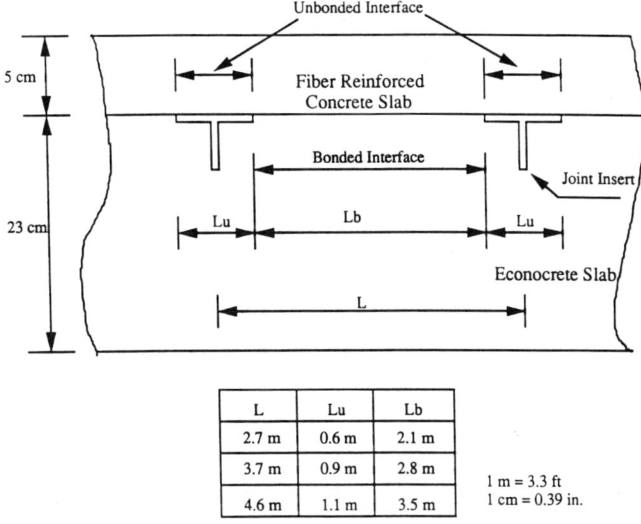

L	Lu	Lb
2.7 m	0.6 m	2.1 m
3.7 m	0.9 m	2.8 m
4.6 m	1.1 m	3.5 m

1 m = 3.3 ft
1 cm = 0.39 in.

Figure 7 Elastic-Jointed Composite Pavement Using Fiber Reinforcement

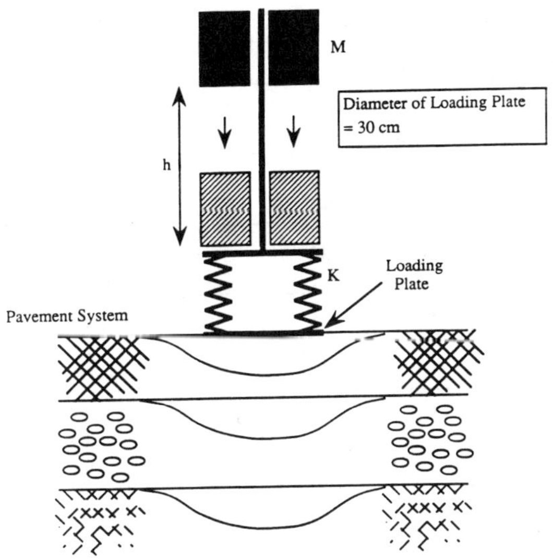

Figure 8 Schematic Presentation of a Falling Weight Deflectometer

Table 1 Basic Information on the Test Sections

Section	Surface Layer	Subbase	Joint	Year of Construction	PSI
1	23 cm Concrete Slab	15 cm Stabilized (LBR = 40) & 15 cm Stabilized (LBR = 20)	Dowelled, 6.1 m spacing	1973	3.35
2	23 cm Concrete Slab	15 cm Cement-Treated	Plain, 6.1 m spacing	1974	0.45
3	23 cm Concrete Slab	15 cm Cement-Treated	Plain, 6.1 m spacing	1974	0.45
4	23 cm Concrete Slab	30 cm Stabilized (LBR = 40)	Plain, 6.1 m spacing	1974	2.9
5	20 cm Concrete Slab	15 cm Stabilized (LBR = 40) & 12 cm Stabilized (LBR = 20)	Plain, 6.1 m spacing	1970	3.75
6	20 cm Concrete Slab	15 cm Stabilized (LBR = 40) & 15 cm Stabilized (LBR = 20)	Plain, 6.1 m spacing	1970	1.8
7	23 cm Concrete Slab	15 cm Cement-Treated	Plain, 6.1 m spacing	1976	2.4
8	23 cm Concrete Slab	15 cm Cement-Treated	Plain, 6.1 m spacing	1976	2.85
9	23 cm Concrete Slab	15 cm Limerock (LBR = 40)	Doweled, 6.1 m spacing	1973	3.15
10	23 cm Concrete Slab	30 cm Stabilized (LBR = 75)	Doweled, 6.1 m spacing	1973	2.85
11	33 cm Concrete Slab	15 cm Stabilized & 122 cm A-3	Doweled & Skewed, 4.0-, 5.8-, 5.5-, & 3.7-m intervals	1984	Good Condition
12	30 cm Concrete Slab	15 cm Econocrete	Doweled & Skewed, 5.2-, 6.7-, 7.0-, & 4.9-m intervals	1984	Good Condition
13	30 cm Concrete Slab	15 cm Econocrete	Doweled & Skewed, 5.2-, 6.7-, 7.0-, & 4.9-m intervals	1984	Good Condition
14	23 cm Concrete Slab	15 cm Econocrete	Doweled & Skewed, 5.2-, 6.7-, 7.0-, & 4.9-m intervals	1981	Good Condition

Table 1--continued

Sec-tion	Surface Layer	Subbase	Joint	Year of Construction	PSI
15	23 cm Concrete Slab	15 cm Econocrete	Doweled & Skewed, 5.2-, 6.7-, 7.0-, & 4.9-m intervals	1981	Badly Cracked
16	23 cm Concrete Slab	15 cm Econocrete	Doweled & Skewed, 5.2-, 6.7-, 7.0-, & 4.9-m intervals	1981	Good Condition
17	23 cm Concrete Slab	15 cm Econocrete	Doweled & Skewed, 5.2-, 6.7-, 7.0-, & 4.9-m intervals	1981	Badly Cracked
18	23 cm PCC Slab	15 cm Econocrete Subbase	Doweled & Skewed, 5.2-, 6.7-, 7.0-, & 4.9-m spacing	1981	Good Condition
19	86 m PCC Slab & 23 cm Econo-crete Slab	15 cm Type "A" Modified Stabilized	Plain & Skewed, 4.6-m spacing	1977	4.2
20	8 cm PCC Slab & 23 cm Econo-crete Slab	15 cm Cement-Treated	Plain, 4.6-m spacing	1977	4.2
21	8 cm PCC Slab & 23 cm Econo-crete Slab	15 cm Type "A" Modified Stabilized	Doweled 6.1-m spacing	1977	4.3
22	8 cm PCC Slab & 23 cm Econo-crete Slab	15 cm Type "A" Modified Stabilized	Elastic Doweled (1) 6.1-m spacing for Fall Test (2) 3.7-m spacing for Winter Test	1977	4.25
23	8 cm CRC Slab & 23 cm Econo-crete Slab	15 cm Type "A" Modified Stabilized	Elastic Joint 3.7 m spacing	1977	4.1
24	5 cm FRC Slab & 23 cm Econo-crete Slab	15 cm Type "A" Modified Stabilized	Elastic Joint 4.6-m spacing	1977	3.95

NOTE: N.A. Not Available
     1 m = 3.28 feet
     1 cm = 0.39 inch

the maximum deflections caused by the load at specified positions. Figure 8 shows a schematic presentation of the FWD system.

Each test section consisted of either two or three sets of slabs. Except Section 18, each set of slabs consisted of either two or three slabs which were one slab apart from one another. In the running of FWD test on a slab, the FWD vehicle was usually positioned on an adjacent slab and tended to heat up the adjacent slab substantially. Thus, the separation of two test slabs by at least one slab between them was necessary to avoid the heating of the test slabs. When two sets of slabs were selected in a section, they were within half a mile from one another.

FWD tests were run at midday between 11 a.m. and 5 p.m., and at midnight between 11 p.m. and 5 a.m. These are the two times in a day when a pavement slab is at its two extreme curling conditions. At midday, the temperature differential in the slab (which is equal to the temperature at the top of the slab minus that at the bottom of the slab) tends to be positive, and the slab tends to curl down at the edges and joints. This is an ideal time to run the FWD test for evaluation of joints and edges, since the slab is most likely to be at full contact with the subgrade at the edges and joints at this time. At midnight, the temperature differential tends to be negative and the slab tends to curl down at the slab center. This is an ideal time to run the FWD test at the center of the slab for evaluation of the condition of the concrete slab and the subgrade.

For Sections 1 through 17, the FWD loads were applied to a slab at four positions, namely (1) slab corner, (2) edge center, (3) joint center, and (4) slab center. Three geophone configurations were used. Figure 9 and Figure 10 show these loading configurations.

However, because of time constraint and the fact that deflections obtained from various positions are fully useful only if the slab is in full contact with the subgrade at the loading position, the loading configuration was modified for some sections. For the six sections on Econocrete Test Road (Sections 19 through 24), the FWD loads were only applied to the slab center in the night time tests and to the edge center and joint center in the day time tests.

Section 18 consisted of fourteen consecutive slabs. The FWD loads were applied to each slab at five positions, namely (1) the corner in an adjacent slab, (2) slab corner, (3) edge center, (4) joint center, and (5) slab center. Geophones were placed only in longitudinal direction for each load position. Tie bars were cut during the test. The FWD tests were performed both before and after the cut of tie bars.

Four different load levels were used for each load test. The four load levels were pressure of 300, 550, 750, and 100 kPa (44, 80, 109, and 145 psi) on a circular loading plate of 30-cm (12-in.) in diameter, which corresponded to total loads of 21.2, 38.9, 53.0, and 70.7 kN (4.9, 9.0, 12.3, and 16.4 kips), respectively. The use of several load levels was necessary in order to determine the linearity of the FWD load-deflection characteristics. A non-linear load-deflection relationship usually indicates that the slab is not in full contact with the subgrade.

During the period of FWD tests, temperature at various depths of the concrete slab and the air temperature were measured by thermal couple. On I-10, the temperatures at top and bottom of the slabs were measured. On the Econocrete Test Road and I-75, the temperature was measured at five different depths in the slab. They were at 1.3, 6.4, 11.4, 16.5, and 21.6-cm (0.5, 2.5, 4.5, 6.5, and 8.5 in.) from the top surface.

<u>Field Condition Survey</u>

A thorough condition survey was performed on each of the test sections. This included (1) an assessment of the drainage condition, (2) checking for the presence and misalignment of dowel and tie bars, (3) measuring joint and edge faults, and (4) identification of the types and severity levels of distress.

Drainage condition was assessed by (1) visual observation of the roadside ditches (to see if there was any accumulation of water), (2) visual inspection of the nearby catch basins (to see if they were clear

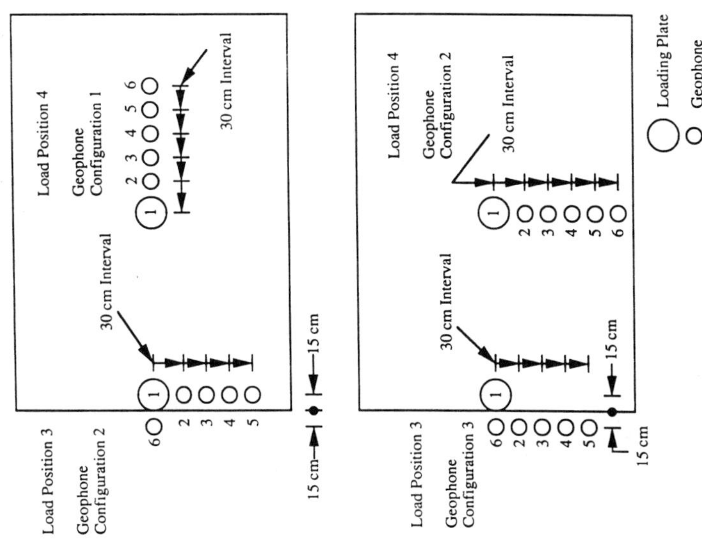

Figure 10 Geophone Configurations for Load Positions 3 and 4

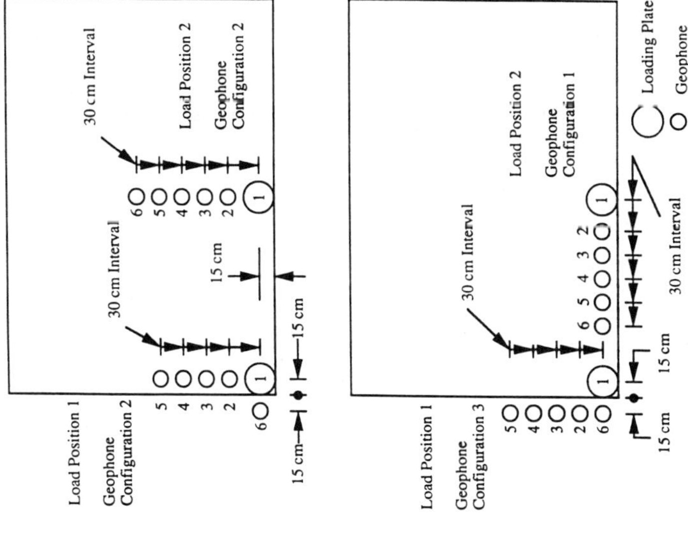

Figure 9 Geophone Configurations for Load Positions 1 and 2

of obstructions), (3) visual observation of the surrounding soil (to see if there was any accumulation of water), (4) observation of the flow of water during a heavy rain, and (5) checking for the presence of sub-drains (such as edgedrains).

Dowel and tie bars were checked with the aid of a metal detector. Any horizontal misalignment of the dowel or the tie bars could be detected.

Joint and edge faults were measured by means of a fault meter which had a sensitivity of 0.05 inch.

Identification of types and severity levels of distress was performed in accordance with the guidelines in FHWA's Highway Pavement Distress Identification Manual (Smith et al 1979). The types of distress which were to be identified include (1) blow-up, (2) corner break, (3) depression, (4) D cracking, (5) joint faulting, (6) joint sealant damage, (7) lane-shoulder heave or dropoff, (8) land/shoulder separation, (9) longitudinal cracking, (10) shoulder deterioration, (11) patch deterioration, (12) pumping (13) scaling or map cracking, (14) spalling, (15) swell, (16) transverse cracking, (17) diagonal cracking, and (18) popouts.

<u>Laboratory Testing of Core Samples</u>

For each of the 24 sections, three or four corings were made from the center of the slab. One sample was also taken from the shoulder of each section for Sections 1 through 17. The depth of each coring was at least 8-cm (3-in.) into the embankment material. The diameter of the cores was either 10-cm or 15-cm (4-in. or 6-in.). The cored cylindrical concrete and econocrete samples were evaluated for their compressive strength, splitting tensile strength, and modulus of elasticity in the laboratory. The subbase samples were not in intact forms and no test was run on them except for identification of the material. The subgrade samples were evaluated in the sieve analysis, liquid limit and plastic limit tests, and classified according to the AASHTO Soil Classification System.

Compressive strength tests on the concrete samples were performed in accordance with the ASTM Standard Test Method C 39 for Compressive Strength of Cylindrical Concrete Specimens. Splitting tensile strength tests were performed in accordance with the ASTM Standard Test Method C 496 for Splitting Tensile Strength of Cylindrical Concrete Specimens. For determination of the elastic modulus of concrete, three 1.3-cm (half-inch) strain gages (Gage type EA-06-500 BH-120 by Micro-Measurements) were adhered to the side of a compressive strength test specimen, in the vertical direction and at 120° from one another. The strain gages were connected to and calibrated with a Vishay/Ellis strain indicator system. Strain readings from the three strain gages were recorded at every 22.2 kN (5000 pound) load interval, as the concrete specimen was loaded in the standard compressive strength test. The means of the three strain readings were used in plotting the stress-strain diagrams which were used to determine the modulus of elasticity. The slope of the line from the origin to the stress-strain curve at one-half of the compressive strength on the stress-strain diagram was taken to be the modulus of elasticity. This modulus value is essentially the same as the secant modulus at one-half of the compressive strength f_c.

The elastic modulus of econocrete was determined in a similar way, except that a compressometer with a gage length of 5-cm (2-in.) (Type SC-1000-2 AB by Tinius Olsen) was used to measure the deformation of the loaded econocrete cylinder. The deformation-force curve was automatically plotted on a graph paper by the Tinius Olsen model MM Flat-Bed XY Recorder. The deformation-force plot was then converted to a stress-strain plot and the elastic modulus of the econocrete was then calculated using this curve.

Sieve analyses of the subgrade soil samples were performed in accordance with the ASTM Standard Test Method C 136 for Sieve Analysis of Fine and Coarse Aggregates. Liquid limit and plastic limit tests were performed in accordance with ASTM Standard Test Method D 4318 for Liquid Limit, Plastic Limit, and Plasticity Index of Soils. The results of the sieve analysis, liquid limit and plastic limit tests were used to classify the subgrade soil samples in accordance with the AASHTO Soil Classification System.

FEACONS IV COMPUTER PROGRAM

The finite-element computer program, FEACONS IV, was used to model the pavement systems in this study. In this program, a jointed concrete pavement is modeled as a three-slab system (13, 14) as shown in Figure 11. Since the analysis of the concrete pavement response generally involves the determination of deflections and stresses on a slab, whose response characteristics are influenced mainly by its two adjacent slabs, it is usually adequate to model a concrete pavement as a three-slab system.

A concrete slab is modeled as an assemblage of rectangular plate bending elements with three degrees of freedom at each node. If a stiff base/subbase is used in a permanent system, it is modeled as a thin plate by using the same rectangular plate bending elements as used for the concrete slabs.

The subgrade is modeled as a liquid or Winkler foundation which is modeled by a series of vertical springs at the nodes. Subgrade voids are modeled as initial gaps between the slab and the springs at the specified nodes. A spring stiffness of zero is used where a gap exists. Load transfers across the joints between two adjoining slabs are modeled by shear (or linear) and torsional springs connecting the slabs at the nodes of the elements along the joint. Frictional effects at the edges are modeled by shear springs at the nodes along the edges.

PROCEDURE FOR THE STRUCTURAL ANALYSIS

Basically, the procedure for the structural analysis involves (1) estimating the pavement parameters of each test slab, (2) computing the maximum stresses in each slab caused by various combinations of thermal and loading conditions and (3) assessing the structural adequacy of each slab from the values of the pavement parameters and the ratio of maximum induced stress to the flexural strength of the concrete.

The important parameters in a pavement system to be determined include the (1) elastic modulus of concrete (E_c), (2) elastic modulus of econocrete (E_e), if used, (3) subgrade stiffness (K_s), (4) edge stiffness (K_e), (5) linear joint stiffness (K_l) and (6) torsional joint stiffness (K_t). The elastic moduli of concrete and econocrete were determined by laboratory testing of core samples. The subgrade stiffness (K_s) was determined by matching the measured deflection basin with the theoretical deflection basin caused by a 40-kN (9-kip) FWD load applied at the center of the slab.

Similar methods were used to estimate the other pavement parameters. However, several different FWD load positions were used. To determine the edge stiffness (K_e), the 40-kN (9-kip) FWD load was applied at the edge center. The FWD load was applied at the joint center to determine the values of linear and torsional joint stiffness (K_l & K_t) (Tia et al, 1989).

Once the estimated pavement parameters were determined, they were then used to model these pavement sections using the FEACONS IV computer program, and maximum stresses in the slab caused by five combinations of thermal and loading conditions were computed. These five combinations of thermal and loading conditions were (1) a 40-kN (9-kip) load at the slab center and no temperature differential in the slab, (2) a 40-kN (9-kip) load at the joint center, and no temperature differential, (3) a 40-kN (9-kip) load at the edge center and no temperature differential, (4) a 98-kN (22-kip) axle load at the edge center and a temperature differential of $+11.1°C$ ($+20°F$) in the slab, and (5) a 98-kN (22-kip) axle load and a temperature differential of $-5.6°C$ ($-10°F$) in the slab. A coefficient of thermal expansion of 3.3 x $10^{-6}/°C$ (6 x $10_{-6}/°F$) was assumed for the concrete in the FEACONS analyses.

The ratio of the maximum stress to flexural strength of concrete or econocrete was computed for each of the loading conditions and used to assess the structural adequacy of the pavement sections.

Four of the eight test sections on I-10 (Sections 1 through 4) were tested twice in two different years. The evaluation results could be compared with each other. Thus, the rates of deterioration of these pavement sections could be evaluated. The test sections on Econocrete Test Road were also tested twice.

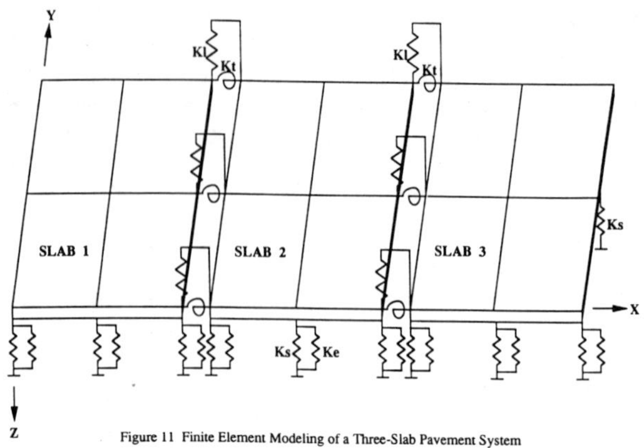

Figure 11 Finite Element Modeling of a Three-Slab Pavement System

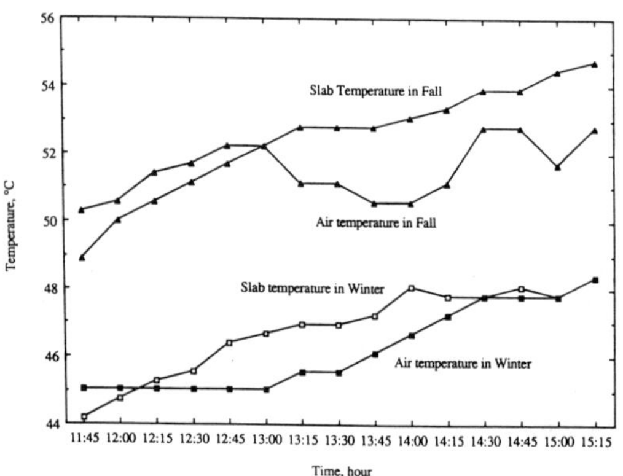

Figure 12 Air and Average Slab Temperature During the Day on the Econocrete Test Road

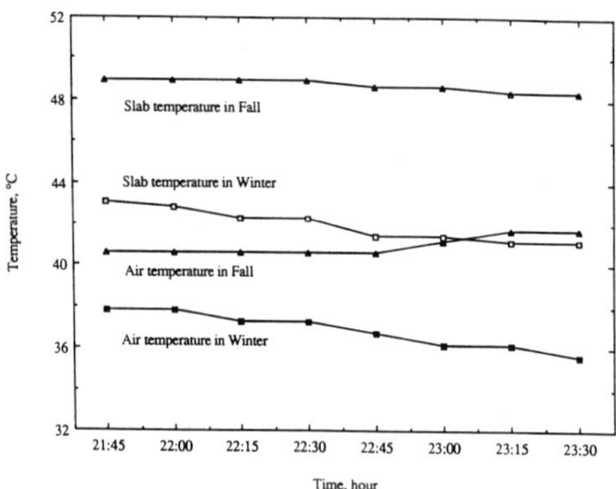

Figure 13 Air and Average Slab Temperature During the Night on the
Econocrete Test Road

The first set of tests was run in Fall, 1986 while the second one was run in Winter, 1987. A comparison between the results of evaluation obtained from these two sets of tests was made to determine the seasonal effects on the pavement structural performance. An evaluation was made on the performance of these test sections, which had various different designs.

EVALUATION OF TEST SECTIONS

This section presents the results of the evaluation of the 24 pavement sections. The presentation of the evaluation results is made in five groups, namely, the test sections on (1) I-10 (Test Sections 1 through 8), (2) I-4 (Test Sections 9 & 10), (3) I-75 in Hillsborough County (Test Sections 11 through 13), (4) I-75 in Sarasota and Manatee Counties (Test Sections 14 through 18), and (5) Econocrete Test Road (Test Sections 19 through 24). Each test section was evaluated by a field condition survey and a structural analysis.

<u>Evaluation of the Test Sections on I-10</u>

Results of Condition Survey

Eight test sections from I-10 were selected for evaluation in this study (see Table 1 for the basic information on these sections). Of the eight sections, four sections were badly deteriorated with a present serviceability index (PSI) of less than 2.5. They were sections 2, 3, 6, and 7. Sections 1 and 8 were in marginal condition with severe signs of distress. Sections 4 and 5 were in good condition with good riding quality. The results of condition survey on these eight test sections are summarized and displayed in Table 2.

Sections 1 through 4 have been tested twice during this study. Both of the two sets of results are displayed. It can be noted that Section 1, which had a 23-cm (9-in.) concrete slab and a conventional stabilized subbase, was in marginal condition with severe signs of pumping when it was surveyed in the first year. However, the signs of pumping were reduced considerably and the pavement condition were improved in the second year. These improvements might be due to the fact that sidedrains have been installed along the edges and the joints have been resealed. It was also noted that this section had a high water table and accumulations of water could be noted in the natural side ditches and the surrounding ground. Section 1 had dowel joints. However, some of the dowel bars were found to be implanted in the slab near the joint but not across the joint. This fact enabled the comparison of the performance of the doweled slabs and that of the undoweled slabs in the same section. Results of the condition survey and FWD tests did not show the doweled slabs to be better than the undoweled slabs for this test section.

Sections 2, 3, and 7 were badly deteriorated sections. They had these common characteristics: (1) a poor drainage condition, (2) no subdrain, (3) severe signs of pumping, and (4) a cement-treated subbase. The signs of pumping in Section 2 were not as severe as those in the other two sections, due to the fact that the pavement had recently been undersealed and the shoulder had been patched or overlaid with an asphalt concrete, but it was evident that it had had severe pumping problems and was still having pumping problems. It was noted, in Test Section 3, that while this pavement section was badly deteriorated in general, there were a few good slabs at locations of higher evaluation where the drainage condition was better. When comparing the condition survey results obtained from the second year with those obtained in the first year, it is noted that the signs of pumping on Section 2 and 3 were reduced and the drainage conditions were also improved. This might be attributed to the fact that the shoulders had been patched along the pavement edges on these two sections. It should be pointed out that Section 2 tested in the second year was located at about 1500 feet away from that tested in the first year. This was because there was a long span of badly cracked slabs prior to Section 2 when it was tested in the second year. However, they were still in the same area and had the same section design. Within that span, only eight slabs out of 130 consecutive slabs were found uncracked.

Table 2 Results of Condition Survey of Test Sections on I-10

TEST SECTION	DISTRESS	DRAINAGE CONDITION	REMARKS
1[a]	1. Severe signs of pumping. 2. Medium/severe shoulder deterioration. 3. Medium shoulder dropoff (with the shoulder lower than the edge by as much as 0.5 in.). 4. Light/medium faulting of transverse joint (with a fault of up to 0.15 in.). 5. Infrequent transverse cracks at the center of the slab, and corner cracks.	1. High water table. 2. Accumulations of water in the natural side ditches and surrounding ground. 3. No subdrain.	1. Fair riding quality. 2. Some of the dowel bars were misplaced. They were found to be implanted in the slab near the joint but not across the joint.
1[b]	1. Light signs of pumping. 2. Medium/severe faulting of transverse joint (with a fault up to 0.35 in.) 3. 10 to 20 popouts per square yard.	1. Good drainage condition. 2. High water table. 3. Sidedrain present.	1. Sidedrains have been installed along the edges recently. 2. Joints have been resealed. 3. Signs of pumping are now reduced. 4. No cracks were observed for 25 consecutive slabs.
2[a]	1. Severe and frequent transverse, diagonal corner and longitudinal cracks. 2. Medium signs of pumping. 3. Medium shoulder dropoff (up to 0.9 in.). 4. Medium faulting of transverse (up to 0.15 in.). 5. Medium deterioration of shoulder patch.	1. No subdrain. 2. Adequate drainage for surface water, there was no accumulation of surface water noted.	1. Poor riding quality. 2. The slabs had been undersealed with cement grout. Each slab was noted to have six grout holes in it. 3. The shoulder had been patched or overlaid with asphalt concrete.
2[b]	1. Light signs of pumping. 2. Medium land/shoulder joint separation. 3. Medium faulting of transverse joints (up to 0.2 in.). 4. Light/medium patch deterioration. 5. 20 popouts per square yard.	1. Good drainage condition. 2. No subdrain.	1. Start of a good section after miles of badly cracked slabs. 2. Shoulder has been patched along edges. 3. No cracks observed for 25 consecutive slabs.

Table 2--continued

TEST SECTION	DISTRESS	DRAINAGE CONDITION	REMARKS
3[a]	1. Frequent corner and transverse cracks. 2. Severe signs of pumping. 3. Medium shoulder dropoff (up to 0.7 in.). 4. Medium faulting of transverse joint up to 0.15 in.). 5. Medium deterioration of shoulder patch. 6. Medium scaling.	1. No subdrain. 2. Accumulations of water (in the natural side ditches noted at locations of lower elevation).	1. Poor riding quality. 2. The slabs were in good conditions at locations of higher elevation.
3[b]	1. Light signs of pumping. 2. Medium faulting of transverse joints (up to 0.2 in.). 3. Light shoulder drop off. 4. One transverse crack was observed in 20 slabs. 5. Some popouts of concrete.	1. No subdrain 2. Accumulation of water in the natural side ditches.	1. Some cracks were found in the shoulder. 2. Shoulder has been patched along the edges. 3. Hairline cracks were observed all over the slabs.
4[a]	1. Hairline cracks all over the slabs. 2. Medium shoulder dropoff (up to 0.8 in.). 3. Some popouts in the concrete. 4. Light/medium signs of pumping.	1. Good drainage condition.	1. Good riding quality. 2. Shoulder joints were sealed.
4[b]	1. Light signs of pumping. 2. Light corner break (one was observed in 15 slabs). 3. Light longitudinal cracks (two were observed in 15 slabs). 4. Light/medium faulting of transverse joints (up to 0.15 in.). 5. Some popouts of concrete	1. Good drainage condition.	1. Joints were sealed along the edges. 2. Hairline cracks were all over the slabs.

Table 2--continued

TEST SECTION	DISTRESS	DRAINAGE CONDITION	REMARKS
5	1. Some hairline cracks in the concrete. 2. Some reflective cracks in the asphaltic shoulder.	1. Good drainage condition. 2. No sign of pumping.	1. Very good riding quality. 2. 8-inch slabs.
6	1. Longitudinal cracks on some slabs. 2. Some diagonal or transverse cracks. 3. Map cracking and crazing.	1. Good drainage condition. 2. No sign of pumping.	1. Poor riding quality. 2. 8-inch slabs.
7	1. Medium/severe signs of pumping. 2. Some transverse, corner and longitudinal cracks. 3. Medium faulting of transverse joint. 4. Medium shoulder dropoff. 5. Map cracking and crazing.	1. No subdrain. 2. Some accumulations of water in the side ditches.	1. Poor riding quality.
8	1. Medium/severe signs of pumping. 2. Some longitudinal, diagonal, transverse and corner cracks. 3. Joint spalling. 4. Medium faulting of transverse joint. 5. Medium should dropoff.	1. No subdrain. 2. Some accumulations of water in the side ditches.	1. Poor riding quality.

NOTE: [a] Tests conducted in October, 1985.
   [b] Tests conducted in February, 1987.
   1 inch = 2.54 cm

Sections 4 and 5 were the two good sections on I-10. They had these common characteristics: (1) a good drainage condition, and (2) hairline cracks on the slab. The hairline cracks were visible only upon close examination and did not affect the riding quality of the pavement. These hairline cracks might have resulted from the old age of the pavement and might have helped to relieve the thermal-load-induced stresses in the slabs. The good drainage condition might explain the good performance of this section of highway.

Section 6 was the only badly deteriorated test section which did not have a cement-treated subbase or a pumping problem. Section 6 and Section 5 were at about the same location and had the same design, with a slab thickness of 8 inches and a conventional stabilized subbase. However, Section 6 was on the eastbound lane while Section 5 was on the westbound lane. Only the total two-way ADT at that location could be obtained from R.C.I. computer database and the one-way ADT was approximated by dividing the total ADT by two. It was noted by actual observation that the eastbound traffic was much heavier than the westbound traffic. This would explain the more extensive deterioration of Section 6 as compared with Section 5.

Section 8, which was also experiencing severe pumping problems, was in marginal condition. This section had the same design as that of Section 7 (with a 15-cm (6-in.) cement-treated subbase) and a similarly poor drainage condition. However, Section 8 had a slightly lower traffic volume (with an ADT of 4,500) as compared with that of Section 7 (with an ADT of 5,500). The slightly less deteriorated condition of Section 8 as compared with Section 7 could be attributed to the slightly lower traffic volume on Section 8.

The predominance of pavement pumping observed on the pavement sections with cement-treated subbases and without any subdrain may be attributed to the fact that the cement-treated subbase is not a free-draining material. Water cannot drain freely through the cement-treated subbase. This condition was observed during a rainy day. Water was observed to accumulate in the gaps along the edges of the slabs, rather than draining through the subbase beneath. The accumulation of water between the slabs and the subbase layers will accelerate the pumping of the subbase materials.

The drainage of water from the roadway surface to the roadside ditch seemed to be adequate. There were no high accumulations of water in the ditches during a heavy rain. The problem seemed to be with the drainage of water through the subbase. Water which happened to flow through the cracks, edges and joints of the slab to the subbase would be trapped between the slab and the subbase for some time before it could drain out of the pavement system.

Results of Structural Evaluation

Since complete FWD and laboratory data were obtained from only five test sections (Section 1 through 5), a structural evaluation was performed on each of the five test sections. A representative slab from each section was used in this analysis. Table 3 shows the estimated pavement parameters of these five sections. It can be noted that Sections 2 and 3, which had cement treated subbases and Section 4 which had a thick (30-cm) stabilized subbase, had much higher subgrade stiffness (K_s). In the first year, Section 1, which had doweled joints, had higher joint shear stiffness (K_l). However, a slip of distance 0.25-cm (0.1-in.) was needed to model the looseness in the dowel bars, and thus the joint stiffnesses were in actuality much lower. Also, the linear joint stiffness reduced considerably when the tests were performed in the following year and did not show higher joint stiffness as compared with the other sections.

By comparing the results obtained from the two sets of tests conducted in two different years, it can be noted that, for Section 1, the subgrade stiffness (K_s) had increased from 60 MN/m³ (0.221 kci) to 72 MN/m³ (0.267 kci). This may be attributed to the fact that edgedrains had been installed in this section. The edgedrains might have kept the subbase drier and subsequently increased its stiffness. For the other three sections, the subgrade moduli had all decreased. However, the installation of edgedrains did not seem to improve the joint stiffness. All four sections had shown a decrease in their joint stiffness.

Table 3 Estimated Pavement Parameters of Test Sections on I-10

TEST SECTION	K_s (kN/m³)	E_c (MPa)	K_l (MPa)	K_t (kN)	K_e (MPa)	SLIP (cm)	SUBGRADE VOIDS
1[a]	60	28104	48265	222400	55	0.25	---
2[a]	189	26380	310	222400	172	---	---
3[a]	292	22788	---	---	55	---	---
4[a]	303	26042	2413	4448	97	---	---
5[a]	120	27428	1379	13344	53	---	0.025 cm deep and 30 cm wide along the joint
1[b]	73	28104	207	445	28	---	---
2[b]	124	26380	90	8896	128	---	---
3[b]	255	22788	138	89	110	---	---
4[b]	165	26042	1379	890	55	---	---

Note: [a] Tests conducted in October, 1985

[b] Tests conducted in February, 1987

K_s = Subgrade modulus, E_c = Concrete modulus, K_l = Linear joint stiffness
K_t = Torsional joint stiffness, K_e = Edge stiffness
1 cm = 0.394 in., 1 MN/m³ = 3.684 pci, 1 MPa = 0.145 ksi, 1 kN = 0.225 kip

Table 4 displays the maximum computed stresses in the slab caused by the five loading conditions for these five test sections. Along with these maximum computed stresses, the estimated flexural strength of concrete and the ratio of the maximum computed stress to the estimated flexural strength are also displayed. It can be noted that the computed maximum stresses for the different test sections are, in general, fairly close to one another. The most critical loading condition is the combination of a positive temperature differential in the slab and a load at the center of the slab. The combination of a 98-kN (22-kip) axle load and a temperature differential of +11.1°C (+20°F), which represents a severe loading condition in Florida, produces a maximum stress ranging from 3.1 MPa (456 psi) to 3.9 MPa (561 psi) for the five test sections tested in the first year, and from 3.3 MPa (484 psi) to 4.0 MPa (581 psi) for the four sections tested in the following year. From the evaluation results, Section 1, which had doweled joints, had the highest computed stress for both of the two tests. The maximum stresses for this section are 3.9 MPa (561 psi) and 4.0 MPa (581 psi) (or 91% and 94% of the estimated flexural strength of concrete, respectively). This maximum stress occurs at the transverse centerline. If the applied axle load was slightly higher than 98 kN (22 kips), the flexural strength of the concrete could be exceeded and a transverse crack near the center of the slab could be produced. This explains the occurrence of the center transverse cracks on Test Section 1. It can also be noted that the other maximum computed stresses for Section 1 were higher than those for the other four sections. Considering the variation of the field and laboratory data and the small amount of difference in the computed stress, one cannot conclude that Section 1 was structurally worse than the other four sections. However, it can be concluded that Section 1, which had dowel joints, did not have a better structural performance than the other sections, which did not have doweled joints.

FWD tests were run on Sections 1 and 2 in the fall and in the winter seasons during the first year of the study. Analysis of the FWD data indicated that the subgrade moduli of these two sections were slightly higher in the winter than in the fall. This might be due to the drier subgrade condition in the winter. The maximum computer thermal-load-induced stresses in these two sections were slightly lower for the winter condition.

It can be noted from the results of the structural evaluation that these test sections were not substantially different from one another in structural performance. The long-term performance of these pavement sections, which was affected greatly by moisture damage, could not be determined merely by a structural analysis. A structural analysis, however, can be used to determine the structural adequacy of a pavement system subjected to a combination of severe applied loads and temperature changes.

It can also be observed from Table 4 that the maximum thermal-load-induced stresses had increased for the four sections (Sections 1 through 4) over the period of one year. It had increased from 3.9 MPa (561 psi) to 4.0 MPa (581 psi) for Section 1, from 3.3 MPa (484 psi) to 3.6 MPa (523 psi) for Section 2, from 3.1 MPa (456 psi) to 3.3 MPa (484 psi) for Section 3, and from 3.4 MPa (492 psi) to 3.6 MPa (526 psi) Section 4. The ratio of the maximum computed stress to the estimated flexural strength of concrete for Section 1 had increased from 91% to 94%. The increase of the stress-strength ratio was 6%, 5%, and 6% for Sections 2, 3, and 4, respectively.

Evaluation of the Test Sections on I-4

Results of Condition Survey - Two test sections from I-4 were selected for evaluation in this study. They were designated as Test Sections 9 and 10 (see Table 1 for the basic information on these sections). These are two well-performing sections. Despite the lack of subdrains, the drainage condition seemed to be good and there was no sign of pavement pumping. Section 9 had a limerock subbase, while Section 10 had a conventional stabilized subbase. These free-draining subbase materials in combination with a sandy soil embankment provided a good natural subdrainage for these pavement sections. The states of deterioration, as noted by the crazing of concrete and the few cracks on the slabs, were due to the heavy traffic loads and the age of the pavement. Section 10, which had much higher traffic (with an ADT of 36,400), was more deteriorated than Section 9 (with an ADT of 22,000).

Table 4 Maximum Computed Stresses for Test Sections on I-10

TEST SECTION	ESTIMATED FLEXURAL STRENGTH OF CONCRETE (kPa)	MAXIMUM COMPUTED STRESS (kPa) DUE TO:				
		40-kN LOAD AT SLAB CENTER	40-kN LOAD AT JOINT CENTER	40-kN LOAD AT EDGE CENTER	$\Delta T = +11.1°C$ 98-kN AXLE LOAD AT EDGE CENTER	$\Delta T = -5.6°C$ 98-kN AXLE LOAD AT SLAB CORNER
1[a]	4240	924 (22%)[c]	1434 (34%)	1441 (34%)	3868 (91%)	-1310 (31%)
2[a]	4675	779 (17%)	1103 (24%)	1117 (24%)	3323 (71%)	-1345 (29%)
3[a]	4123	717 (17%)	---	1186 (29%)	3144 (76%)	---
4[a]	4730	724 (15%)	862 (18%)	1151 (24%)	3392 (71%)	-1186 (25%)
5[a]	5357	841 (16%)	1089 (20%)	1372 (26%)	3641 (68%)	-1296 (24%)
1[b]	4240	993 (23%)*	1434 (34%)	1579 (37%)	4006 (94%)	-1489 (35%)
2[b]	4675	910 (19%)	1441 (31%)	1262 (27%)	3606 (77%)	-1386 (30%)
3[b]	4123	772 (19%)	1214 (29%)	1179 (29%)	3337 (81%)	-1338 (32%)
4[b]	4730	848 (18%)	1048 (22%)	1365 (29%)	3627 (77%)	-1317 (28%)

Note: [a] Tests conducted in October, 1985.
 [b] Tests conducted in February, 1987.
 [c] Percent of the estimated flexural strength of concrete.
 1 kPa = 0.145 psi, 1 kN = 0.225 kip, 1°C = 1.8°F

Results of Structural Evaluation - A structural evaluation was performed on the two test sections on I-4 using the same developed procedures (as described in an earlier section). The estimated pavement parameters and the maximum computed stresses due to the five loading conditions are displayed in Table 5. The maximum computed stress caused by the combination of a 98-kN (22-kip) axle load and a temperature differential of +11.1°C (+20°F) was 75% of the estimated flexural strength of concrete for Section 9.

Evaluation of the Test Sections on I-75 in Hillsborough County

Results of Condition Survey - The three test sections on I-75 in Hillsborough County (designated as Sections 11, 12 & 13) had the three most recently adopted designs in Florida. Section 11 had a slab thickness of 33-cm (13-in.), 15-cm (6-in.) of stabilized subbase, edgedrains and skewed joints spaced at 4.0, 5.8, 5.5, and 3.7 m (13, 19, 18, and 12 feet) intervals. Section 12 had a slab thickness of 30-cm (12-in.), 15-cm (6-in.) of econocrete base, skewed joints spaced at 5.2, 6.7, 7.0, and 4.9 m (17, 22, 23, and 16 feet) intervals and no subdrain. Section 13 had a slab thickness of 30-cm (12-in.), 15-cm (6-in.) of econocrete, a drainage blanket beneath the shoulder and skewed joints spaced at 5.2, 6.7, 7.0, and 4.9 m (17, 22, 23, and 16 feet). All of these sections had econocrete shoulders, doweled bars at transverse joints and tie bars at longitudinal joints. At the time of field condition survey and FWD test in early 1986, Section 12 had not been open to traffic, while Sections 11 and 13 had been open to traffic for only about one year.

Sections 11 and 13 were in excellent condition and no sign of distress was noted. A few hairline transverse cracks were noted on Section 12. However, these hairline cracks were hardly noticeable and were infrequent. For all of these three sections, no severe distress or failure was foreseen for the near future. However, the long-term performance of those sections could only be seen with time.

Results of Structural Evaluation - A structural evaluation was performed on these three test sections. The FWD data and the laboratory test results were used to estimate the pavement parameters. The skewed joints in these sections were analyzed as regular transverse joints with the proper adjustment in the positions of the wheel loads. When a single axle load was to be applied at the slab corner or at the joint center, one wheel load was placed on one side of the joint while the other load was placed on the other side, to model the actual loading condition. It was realized that the analysis results might be slightly in error, however, they should give rough estimates of the actual stresses in the slabs.

Table 6 displays the estimated pavement parameters and the maximum computed stresses due to the five loading conditions. It can be noted that Sections 12 and 13, which had an econocrete subbase, had much higher moduli than that of Section 11, which had a stabilized subgrade. There was no slip in the doweled joint, and no subgrade void. This indicated that the joints and the subgrades were in good condition. The maximum computed stresses due to a 40-kN (9-kip) load with no temperature effect in these three sections are much lower than those in the test sections on I-10 and I-4. This was due to the thicker slabs used. The maximum computed stresses due to the combination of a 98-kN (22-kip) axle load and a temperature differential of +11.1°C (+20°F) in these three sections are also relatively lower than those in the I-10 and I-4 test sections. The maximum computed stresses to strength ratio for this loading condition were 58% and 63% for Sections 11 and 12, respectively.

Evaluation of the Test Sections on I-75 in Sarasota and Manatee Counties

Results of Condition Survey - Five test sections from the concrete pavement on I-75 in Sarasota and Manatee Counties were selected for evaluation in this study. They were designated as Sections 14, 15, 16, 17, and 18. These five test sections were located in five different construction projects. Sections 14, 16, and 18 were located in three well-performing projects, while Sections 15 and 17 were located in two poorly-performing projects. The concrete pavement of I-75 in Sarasota and Manatee Counties consists of twelve projects. Three of these twelve projects have shown excellent performance while the other nine projects have shown exceptionally poor performance. Since all of these twelve projects used essentially

Table 5 Estimated Pavement Parameters and Maximum Computed Stresses for Test Sections on I-4

TEST SECTION	K_s (MN/m³)	E_c (MPa)	K_l (MPa)	K_t (kN)	K_e (MPa)	SLIP (cm)	SUBGRADE VOIDS
9	235	24139	68950	4448	66	0.025	0.025 cm deep and 30 cm wide along the joint
10	307	24257	---	---	---	---	---

TEST SECTION	ESTIMATED FLEXURAL STRENGTH OF CONCRETE (psi)	MAXIMUM COMPUTED STRESS (kPa) DUE TO:				
		40-kN LOAD AT SLAB CENTER	40-kN LOAD AT JOINT CENTER	40-kN LOAD AT EDGE CENTER	$\Delta T = +11.1°C$ 98-kN AXLE LOAD AT EDGE CENTER	$\Delta T = -5.6°C$ 98-kN AXLE LOAD AT SLAB CORNER
9	4344	745 (17%)[a]	834 (19%)	1276 (29%)	3275 (75%)	-1089 (25%)
10	4378	717 (16%)	---	---	---	---

Note: [a] Percent of the estimated flexural strength of concrete.
K_s = Subgrade modulus, E_c = Concrete modulus, K_l = Linear joint stiffness
K_t = Torsional joint stiffness, K_e = Edge stiffness
1 cm = 0.394 in., 1 MN/m³ = 3.684 pci, 1 MPa = 0.145 ksi
1 kN = 0.225 kip, 1 kPa = 0.145 psi, 1°C = 1.8°F

Table 6 Estimated Pavement Parameters and Maximum Computed Stresses for Test Sections on I-75 in Hillsborough County

TEST SECTION	K_s (MN/m^3)	E_c (MPa)	K_l (MPa)	K_t (kN)	K_e (MPa)	SLIP (cm)	SUBGRADE VOIDS
11	159	27366	345	57824	193	0	0
12	304	25663	---	---		---	---
13	226	23608	---	---	---	---	---

TEST SECTION	ESTIMATED FLEXURAL STRENGTH OF CONCRETE (psi)	MAXIMUM COMPUTED STRESS (kPa) DUE TO:				
		40-kN LOAD AT SLAB CENTER	40-kN LOAD AT JOINT CENTER	40-kN LOAD AT EDGE CENTER	$\Delta T = +11.1$°C 98-kN AXLE LOAD AT EDGE CENTER	$\Delta T = -5.6$°C 98-kN AXLE LOAD AT SLAB CORNER
11	4447	441 (10%)[a]	683 (15%)	676 (15%)	2558 (58%)	-1055 (24%)
12	4468	455 (10%)	---	710 (16%)	2806 (63%)	---
13	4096	469 (13%)	---	---	---	---

Note: [a] Percent of the estimated flexural strength of concrete. The longest slab length for each section was used in the analysis.
K_s = Subgrade modulus, E_c = Concrete modulus, K_l = Linear joint stiffness
K_t = Torsional joint stiffness, K_e = Edge stiffness
1 cm = 0.394 in., 1 MN/m^3 = 3.684 pci, 1 MPa = 0.145 ksi
1 kN = 0.225 kip, 1 kPa = 0.145 psi, 1°C = 1.8°F

the same design, the understanding of the difference between these two groups of projects would help the determination of the causes of problems on this pavement.

Results of the condition survey indicated that Sections 14, 16, and 18 were in excellent condition with no sign of pavement pumping. Sections 15 and 17 were badly cracked and were similar in the types of distress noted. Almost all of the 6.7- and 7.0-m (22- and 23-foot) slabs had transverse cracks near the center of the slab. A few slabs had corner cracks at the acute corner of the slab. It was also noted that very few of the 4.9- and 5.2-m (16- and 17-foot) slabs were cracked. There were signs of pumping of the econocrete base which were identified by streaks of white stain extending from the cracks or edges of the slabs.

There was a speculation that the slabs in the well-performing projects might have been accidentally bonded to the econocrete base. Corings obtained from the centers of the slabs, however, showed that the slabs were not bonded to the econocrete for all these five sections.

Results of Structural Evaluation - Complete FWD and laboratory data were obtained from only three test sections, namely Sections 14, 15, and 18, and thus the structural evaluation was performed only on these three sections. The pavement parameters were estimated from the laboratory test and FWD test data. Here again, the skewed joints were analyzed as regular transverse joints with the proper adjustment in the positions of the applied load as described earlier in this section. The longer slab lengths of 6.7 m (22 feet) and 7.0 m (23 feet) were used in the analyses.

In the testing of Section 18, tie bars between the shoulder and the outside lane were cut in order to study the effects of tie bars on the structural performance of pavement. FWD tests were run on the slabs both before and after the cutting of the tie bars. Fourteen slabs were tested. However, only results from two slabs were fully analyzed. The one with a slab length of 6.7 m (22 feet) is presented here.

Table 7 displays the estimated pavement parameters and the maximum computed stresses due to the five loading conditions. It should be pointed out that the elastic modulus of concrete for Section 18 was obtained from results of core sample tests performed by the FDOT Bureau of Materials and Research (15). The flexural strength of the concrete was estimated as 15 % of the mean compressive strength of the core samples from this project.

It can be noted that the concrete modulus for Section 15 was much higher than that for Section 14 and Section 18, while the flexural strength of concrete for these three sections was close to one another. The maximum computed stress due to the combination of a 98-kN (22-kip) axle load and a temperature differential of $+11.1°C$ ($+20°F$) in Section 15 was 3.9 MPa (563 psi) and was about 20 % higher than that in Sections 14 and 18. The high computed stresses in the longer slabs could be the cause for the occurrence of the transverse cracks on the longer slabs in Section 15. Even though Sections 14 and 18 had the same design as that of Section 15, the lower elastic modulus of the concrete helped to lower the thermal-load-induced stresses in the slabs in Section 14. It can also be noted that the edge stiffness decreased slightly after the tie bars were cut. However, the maximum computed stresses were not affected significantly by the cutting of the tie bars.

Evaluation of the Test Sections on the Econocrete Test Road

Results of Condition Survey - Six test sections were selected from the Econocrete Test Road for evaluation. They were designated as Test Sections 19 through 24. The detailed information on the design used on each section has been displayed in Table 1. The results of the condition survey are summarized in Table 8. All of the six sections were in very good condition. Although no subdrain was used in these six sections, they all appeared to have good drainage and no sign of pumping was observed. Section 20 had a 15-cm (6-in.) cement-treated subbase, which was not a free-draining material. However, in this section, it was found that joint seals were in very good condition and no crack existed. This might have prevented water from coming into the pavement system and, thus, minimized the possibility of occurrence of

Table 7 Estimated Pavement Parameters and Maximum Computed Stresses for Test Sections on I-75 in Sarasota and Manatee Counties

TEST SECTION	K_s (MN/m³)	E_c (MPa)	K_l (MPa)	K_t (kN)	K_e (MPa)	SLIP (cm)	SUBGRADE VOIDS
14	255	24588	10343	17792	18	0	0.025 cm deep and 30 cm wide along the joint
15	229	31772	10343	17792	7	---	---
18	312	24201	1379	2224	22 [18][b]	---	0.005 cm deep, 30 cm wide along the joint

TEST SECTIONS	ESTIMATED FLEXURAL STRENGTH OF CONCRETE (kPa)	MAXIMUM COMPUTED STRESS (kPa) DUE TO:				
		40-kN LOAD AT SLAB CENTER	40-kN LOAD AT JOINT CENTER	40-kN LOAD AT EDGE CENTER	$\Delta T = +11.1°C$ 98-kN AXLE LOAD AT EDGE CENTER	$\Delta T = -5.6°C$ 98-kN AXLE LOAD AT SLAB CORNER
14	4523	738 (16%)[a]	841 (19%)	1124 (25%)	3241 (72%)	-1110 (25%)
15	4813	772 (16%)	---	1317 (27%)	3882 (81%)	---
18	4985	752 (15%)	979 (20%) [972 (20%)]	1117 (23%) [1158 (23%)]	3213 (64%) [3206 (64%)][b]	-1076 (22%) [-1076 (22%)]

Note: [a] Percents flexural strength are in parentheses.
[b] After cutting of tie bars.
K_s = Subgrade modulus, E_c = Concrete modulus, K_l = Linear joint stiffness
K_t = Torsional joint stiffness, K_e = Edge stiffness
1 cm = 0.394 in., 1 MN/m³ = 3.684 pci, 1 MPa = 0.145 ksi
1 kN = 0.225 kip, 1 kPa = 0.145 psi, 1°C = 1.8°F

Table 8 Results of Condition Survey on Test Sections on the Econocrete Test Road

TEST SECTION	DISTRESS	DRAINAGE CONDITION	REMARKS
19	1. Light scaling in the concrete.	1. Good drainage condition. 2. No subdrain. 3. No signs of pumping.	1. Very good rideability. 2. Surrounding grounds are of lower elevation.
20	1. Light separation between land and shoulder.	1. Good drainage condition. 2. No subdrain. 3. No signs of pumping.	1. Very good rideability. 2. Surrounding grounds are of lower elevation. 3. Accumulation of water in the natural side ditches.
21	1. Light separation between lane and shoulder. 2. Light scaling in concrete.	1. Good drainage condition. 2. No subdrain. 3. No signs of pumping.	1. Very good rideability. 2. Surrounding grounds are of lower elevation. 3. Accumulation of water in the natural side ditches.
22	1. One crack observed in the section. 2. Light spalling along the crack.	1. Good drainage condition. 2. No subdrain. 3. No signs of pumping.	1. Very good rideability. 2. Elastic doweled joints. 3. Some joints were missing.
23	1. Light spalling of slab at the edge at one location.	1. Good drainage condition. 2. No subdrain. 3. No signs of pumping.	1. Very good rideability. 2. Elastic joints. 3. Joints were formed in right position.
24	1. Light deterioration of joints. 2. Five cracks found in the section. 3. Some longitudinal cracks near the longitudinal joints.	1. Good drainage condition. 2. No subdrain. 3. No signs of pumping.	1. Very good rideability. 2. All the joints not formed at designed position. 3. Pavement higher than surrounding area.

pumping. All the other sections had a 15-cm (6-in.) stabilized subbase. This free draining material provided a good natural subdrainage for these pavement sections.

Section 22 had elastic doweled joints and Section 23 had elastic joints. From the results of thorough crack surveys (16) conducted by FDOT Bureau of Materials and Research, it was found that the joints were formed at the designated positions.

Test Section 24 had a 5-cm (2-in.) fiber-reinforced concrete layer and had partially incorporated the principle of elastic joints in the design. The joints were found to be formed at random rather than at the designed locations. Longitudinal cracks were formed near the longitudinal joints. This section had very good rideability. However, it could be noted that the steel fibers at the top of the concrete slabs had rusted badly and the concrete had deteriorated to some extent with popouts and scaling.

Results of Structural Evaluation - A structural evaluation was performed on the six test sections on the Econocrete Test Road. As mentioned earlier, these six sections were tested twice (the first time in Fall, 1986 and the second time in Winter, 1987). Thus, the evaluation was based on both sets of test data. A representative slab from each section was used for the analysis.   The estimated pavement parameters for the Fall and Winter tests are displayed in Tables 9 and 10. Tables 11 and 12 list the maximum computed stresses in the slab caused by the five loading conditions, the estimated flexural strength of concrete and econocrete, and the ratios of the maximum computed stress to the estimated flexural strength for these six pavement sections for the Fall test and the Winter test, respectively. Since no core sample was taken from the test site for the Winter test, average values of concrete and econocrete moduli obtained from the cores taken during the Fall test were assumed for the analysis.

From Tables 9 and 10, it can be noted that Section 20 which had a 15-cm (6-in.) cement-treated subbase, had a much higher subgrade modulus (K_s) than those with a 15-cm (6-in.) stabilized subbase. Among the sections with a stabilized subbase, Sections 22 and 23 which had elastic joints, had relatively much higher linear joint stiffness (K_l). It was also observed that the use of conventional dowel joints improved the joint stiffness slightly. (As evidenced by comparing the values of K_l of Section 19 with those of Section 21). However, the improvement in joint stiffness was much greater with the use of elastic joints. Section 20, which had a cement-treated subbase, had the highest linear joint stiffness (K_l) among the six sections, even though it had plain joints.

From Tables 11 and 12, it can be noted that the most critical stress occurred when the pavement is subjected to a combination of positive temperature differential in the slab and a load at the edge center. Therefore, this load was used to evaluate the structural adequacy of these pavement sections. Since the surface layer was composed of two different materials (with concrete at the top and econocrete at the bottom), the maximum computed stresses for both the top and bottom of the surface layer were displayed in Tables 11 and 12. For the Fall test, the maximum stress for the six sections ranged from 2.6 MPa (374 psi) to 3.6 MPa (526 psi) at the top, and from 1.5 MPa (218 psi) to 1.9 MPa (270 psi) at the bottom of the composite slabs. Of these six sections, Section 24, which had a 5-cm (2-in.) fiber-reinforced concrete laid on top of a 23-cm (9-in.) econocrete layer, had the highest stress to strength ratio (85% for Fall test and 91% for Winter test) for the econocrete layer. The high thermal-load-induced stresses might be the cause for the cracks on Section 24. For the other sections, the ratios of maximum stress to flexural strength range from 51% to 80%.

To evaluate the effects of seasonal temperature changes on the pavement performance, the tests were conducted in the Fall and in the Winter, and the air and slab temperatures during the test were recorded. Figure 12 presents the air temperature and average slab temperature during the day for the Winter and the Fall tests. The temperatures during the night are shown in Figure 13. It can be noted that the difference between the Fall and the Winter slab temperatures remained about the same (ranging from 5 to 6.1°C (9 to 11°F) during the day and from 5.8 to 7.2°C (10.5 to 13°F) at night). Even though the mean slab temperatures in the Fall and in Winter were different, the slabs experienced the same amount of temperature variation in both cases.

Table 9 Estimated Pavement Parameters of Test Sections on the Econocrete Test Road - Fall Test.

TEST SECTION	K_s (MN/m³)	E_c (MPa)	E_e (MPa)	K_l (MPa)	K_t (kN)	K_e (MPa)	SLIP (cm)	SUBGRADE VOIDS
19	190	25608	12480	124	2224	66	0	0
20	367	22953	10894	2758	222	41		0.005 cm deep and 30 cm wide along the joint
21	198	23746	9991	241	2224	48	0	0.009 cm deep and 30 cm wide along the joint
22	149	25629	11984	1724	26688	50	0	0
23	163	32076	10922	1724	1334	103	0	0
24	163	36226	11349	---	---	114	0	0

1 cm = 0.394 in., 1 MN/m³ = 3.684 pci, 1 MPa = 0.145 ksi, 1 kN = 0.225 kip.

K_s = Subgrade modulus, E_c = Concrete modulus, K_l = Linear joint stiffness

K_t = Torsional joint stiffness, K_e = Edge stiffness

Table 10 Estimated Pavement Parameters of Test Sections on the Econocrete Test Road - Winter Test.

TEST SECTION	K_s (MN/m³)	E_c (MPa)	E_e (MPa)	K_l (MPa)	K_t (kN)	K_e (MPa)	SLIP (cm)	SUBGRADE VOIDS
19	198	22767	13128	974	4454	55	0	0.006 cm deep and 30 cm wide along the joint
20	367	20502	10894	5516	445	34	0	0.008 cm deep and 30 cm wide along the joint
21	185	23746	9991	234	4448	45	0	0.005 cm deep and 30 cm wide along the joint
22	176	26132	11756	1724	53388	59	0	0.013 cm deep and 30 cm wide along the joint
23	182	29848	96252	2275	2224	93	0	0.006 cm deep and 30 cm wide along the joint
24	174	34179	11666	---	---	33	0	0

1 cm = 0.394 in., 1 MN/m³ = 3.684 pci, 1 MPa = 0.145 ksi, 1 kN = 0.225 kip.

K_s = Subgrade modulus, E_c = Concrete modulus, K_l = Linear joint stiffness

K_t = Torsional joint stiffness, K_e = Edge stiffness

Table 11 Maximum Computed Stresses for Test Sections on Econocrete Test Road - Fall Test

TEST SECTION	ESTIMATED FLEXURAL STRENGTH OF CONCRETE (kPa)	MAXIMUM COMPUTED STRESS (kPa) DUE TO:				
		40-kN LOAD AT SLAB CENTER	40-kN LOAD AT JOINT CENTER	40-kN LOAD AT EDGE CENTER	$\Delta T = +11.1°C$ 98-kN AXLE LOAD AT EDGE CENTER	$\Delta T = -5.6°C$ 98-kN AXLE LOAD AT SLAB CORNER
19 CONC.	5206	-648 (13%)[a]	-1027 (20%)	-1027 (20%)	-2786 (54%)	+1014 (19%)
ECONO.	2655	+434 (16%)	+ 690 (26%)	+ 690 (26%)	+2000 (70%)	- 676 (25%)
20 CONC.	4296	-565 (13%)	- 703 (16%)	- 958 (22%)	-2579 (60%)	+ 869 (20%)
ECONO.	2220	+372 (17%)	+ 462 (21%)	+ 634 (29%)	+1696 (76%)	- 572 (26%)
21 CONC.	3613	-641 (18%)	-1000 (28%)	-1062 (29%)	-2655 (73%)	+ 986 (27%)
ECONO.	2889	+400 (14%)	+ 621 (21%)	+ 655 (23%)	+1648 (57%)	- 614 (21%)
22 CONC.	5178	-669 (13%)	- 827 (16%)	-1083 (21%)	-2779 (54%)	+ 896 (17%)
ECONO.	2765	+441 (16%)	+ 545 (20%)	+ 710 (26%)	+1820 (66%)	- 586 (21%)
23 CONC.	4530	-765 (17%)	- 945 (21%)	-1110 (25%)	-2717 (60%)	+ 993 (22%)
ECONO.	2179	+427 (20%)	+ 531 (24%)	+ 621 (28%)	+1503 (69%)	- 558 (26%)
24 CONC.	5178	-945 (18%)	---	-1365 (26%)	-3627 (70%)	---
ECONO.	2151	+455 (21%)	---	+ 690 (32%)	+1827 (85%)	---

Note: [a] Percent of the estimated flexural strength of concrete.

1 kN = 0.225 kip, 1 kPa = 0.145 psi, 1°C = 1.8°F

Table 12 Maximum Computed Stresses for Test Sections on Econocrete Test Road - Winter Test

TEST SECTION	ESTIMATED FLEXURAL STRENGTH OF CONCRETE (kPa)	MAXIMUM COMPUTED STRESS (kPa) DUE TO:				
		40-kN LOAD AT SLAB CENTER	40-kN LOAD AT JOINT CENTER	40-kN LOAD AT EDGE CENTER	$\Delta T = +11.1°C$ 98-kN AXLE LOAD AT EDGE CENTER	$\Delta T = -5.6°C$ 98-kN AXLE LOAD AT SLAB CORNER
19 CONC.	5206	-607 (12%)[a]	-1020 (20%)	- 979 (19%)	-2641 (51%)	+ 979 (19%)
ECONO.	2655	+441 (17%)	+ 745 (28%)	+ 717 (27%)	+1924 (72%)	- 710 (27%)
20 CONC.	4296	-545 (13%)	- 655 (15%)	- 931 (21%)	-2455 (57%)	+ 814 (19%)
ECONO.	2220	+379 (17%)	+ 455 (20%)	+ 648 (29%)	+1710 (77%)	- 565 (25%)
21 CONC.	3613	-648 (18%)	- 986 (27%)	-1076 (30%)	-2648 (73%)	+ 993 (27%)
ECONO.	2889	+400 (14%)	+ 614 (21%)	+ 669 (23%)	+1641 (57%)	- 614 (21%)
22 CONC.	5178	-676 (13%)	- 896 (17%)	-1076 (21%)	-2848 (55%)	+ 945 (18%)
ECONO.	2765	+434 (16%)	+ 572 (21%)	+ 690 (25%)	+1662 (60%)	- 607 (22%)
23 CONC.	4530	-738 (16%)	- 924 (20%)	-1103 (24%)	-2944 (65%)	+ 972 (21%)
ECONO.	2179	+407 (19%)	+ 503 (23%)	+ 607 (28%)	+1744 (80%)	- 531 (24%)
24 CONC.	5178	-883 (17%)	---	-1489 (29%)	-3737 (72%)	---
ECONO.	2151	+462 (21%)	---	+ 779 (36%)	+1965 (91%)	---

Note: [a] Percent of the estimated flexural strength of concrete.

1 kN = 0.225 kip, 1 kPa = 0.145 psi, 1°C = 1.8°F

By comparing the Fall and the Winter test results, it can be noted that, for five of the six sections (except Section 21), the subgrade modulus was higher in the Winter than in the Fall. Other pavement parameters, such as K_1, K_t, and K_e, were noted to change from Fall to Winter, but no definite trends could be observed. Thus, with the consideration of the variation of field and laboratory data, it can be stated that there was no significant difference between the performance of these pavement sections in the Fall and that in the Winter.

CONCLUSIONS

The main conclusions from this study can be summarized as follows:

1. Pavements with a cement-treated or econocrete subbase and without any subdrain (such as side-drain or edgedrain) generally show severe signs of pumping and poor performance, as seen from the conditions of I-10 and I-75 in Sarasota and Manatee Counties.
2. Pavements with a limerock or stabilized subbase on a sandy soil subgrade show good drainage characteristics even without the incorporation of any subdrain, as seen from the conditions of I-4.
3. The use of retrofit edgedrains and sidedrains appears to be effective in improving the drainage condition of a pavement. It can increase the subgrade stiffness by keeping the subgrade drier.
4. The use of a thick subbase with proper drainage characteristics (such as a 12-in. stabilized sub-base) along with edge joint seals in a pavement appears to give good performance.
5. The combination of long slab lengths, high elastic modulus of concrete, and stiff subbase caused the thermal-load-induced stresses in the lower slabs to be excessive and the occurrence of transverse cracks near the center of the slabs.
6. Pavement sections with elastic joints have substantially higher joint stiffness than those with doweled or plain joints.
7. The use of tie bars increases the pavement edge stiffness. However, it has little effect on the maximum thermal-load-induced stresses in the concrete slabs.
8. Pavements with bonded composite slabs, made up of a concrete layer laid bonded over an econocrete layer, have shown good performance from performance records and results of analysis. The use of a bonded interface between the concrete and the econocrete layers is definitely beneficial to the performance of the pavements.
9. The structural analysis procedure is found to be useful in evaluating the structural adequacy of pavement sections. However, other information such as drainage condition is also needed to assess the long-term performance of a pavement section.

ACKNOWLEDGEMENTS

The Florida Department of Transportation (FDOT) is gratefully acknowledged for providing the financial support, field testing equipment, and personnel that made this investigation possible. The technical advices provided by Mr. Torbjorn J. Larsen and Dr. Jamshid M. Armaghani are appreciated.

REFERENCES

FDOT, December 1976, "Rigid Pavement Investigation - I-10 Leon and Jefferson Counties," Technical Report No. 76-8A, Structural Materials and Research Section, Bureau of Materials and Research, Florida Department of Transportation.

FDOT, November 1980, "Florida Econocrete Test Road Post Construction and Material Report," Research Report FL/DOT/BMR-80/221, Bureau of Materials and Research, Florida Department of Transportation.

FDOT, May 1983, "An Evaluation of the Development and Current Condition of Interstate 10," Technical Report, Florida Department of Transportation.

FDOT, April 1985, "Roadway Characteristics Inventory Planning Manual," Bureau of Transportation Statistics, Florida Department of Transportation.

FDOT, March 1986, "Report on Concrete Pavement in Sarasota & Manatee Counties," Florida Department of Transportation.

Larsen, T. J., 1978, "Elastic Jointed Pavement Design," Technical Report ST 78-1, Bureau of Materials and Research, Florida Department of Transportation.

Larsen, T. J. and S. F. Mayfield, August 1984, "Florida Econocrete Test Road Performance Evaluation," Research Report FL/DOT/BMR-84-288, Florida Department of Transportation.

Smith, R. E., M. I. Darter and S. M. Herrin, March 1979, "Highway Pavement Distress Identification Manual for Highway Condition and Quality of Highway Construction Survey," U.S. Department of Transportation/Federal Highway Administration.

Tia, M., J. M. Armaghani, C. L. Wu, S. Lei, and K. L. Toye, 1987, "FEACONS III Computer Program for an Analysis of Jointed Concrete Pavements," TRB, Transportation Research Record, No. 113, pp. 12-22.

Tia, M., C. L. Wu, B. E. Ruth, D. Bloomquist, and B. Choubane, August 1989, "Field Evaluation of Rigid Pavements for the Development of a Rigid Pavement Design System - Phase IV," Final Report, Project No. 4910450424912, Department of Civil Engineering, University of Florida.

Dennis R. Hiltunen[1] and Reynaldo Roque[1]

BACKCALCULATION OF SYSTEM PARAMETERS FOR JOINTED RIGID PAVEMENTS

REFERENCE: Hiltunen, D. R., and Roque, R., "Backcalculation of System Parameters for Jointed Rigid Pavements," Nondestructive Testing of Pavements and Backcalculation of Moduli (Second Volume) ASTM STP 1198, Harold L. Von Quintas, Albert J. Bush, III, and Gilbert Y. Baladi, Eds., Amercian Soceity for Testing and Materials, Philadelphia, 1994.

ABSTRACT: A procedure has been developed to backcalculate parameters required for the mechanistic evaluation of jointed rigid pavements. The two paramters are the coefficient of dowel/concrete interaction (G), and the coefficient of subgrade reaction (k) in the vicinity of the joint. The procedure involves two primary steps. First, actual FWD tests are conducted at the joints of the pavement system, and the deflection transfer efficiency and the surface deflection under the FWD load are determined. Next, a finite element model of the pavement system is used to determine the relationship between deflection transfer efficiency and surface deflection. This relationship is developed for a range of G and k values, holding all other input parameters fixed. With the two FWD measurements and the theoretical relationship between the two, the G and k parameters can be determined for the pavement system under evaluation.

KEYWORDS: backcalculation, coefficient of dowel/concrete interaction, coefficient of subgrade reaction, deflection transfer efficiency, falling weight deflectometer, finite element method, joint

---

INTRODUCTION

The use of mechanistic-based models to predict the structural response of existing pavement systems has continued to increase. These prediction models normally require fundamental properites of the in situ pavement materials as input. Nondestructive deflection testing in conjunction with modulus backcalculation have played key roles in estimating these in situ properties. For jointed rigid pavement systems, the majority of the distresses observed in the field are near the joints. Thus, the response of the pavement system near the joints is of significant interest.

Current nondestructive deflection measurements for backcalculation purposes are most commonly obtained at the center of the slabs of jointed rigid pavements, and away from the joints. While deflection tests are often conducted across the joints, little use is made of these measurements beyond a simple calculation of deflection transfer efficiency.

---

[1]Assistant Professor, Department of Civil Engineering, The Pennsylvania State University, University Park, Pennsylvania 16802.

A backcalculation procedure has been developed to provide more detailed information of pavement condition and behavior near the joints in rigid pavement systems. The procedure is based upon surface deflection measurements across the joints and analytical pavement response analysis using the ILLI-SLAB finite element program. The purpose of this paper is to descibe in detail the procedure developed. The use of the procedure on data collected from seven rigid pavement sites in Pennsylvania will also be presented.

MODELING JOINTED RIGID PAVEMENT SYSTEMS

The primary purpose of pavement response models is to predict the structural responses of the pavement (e.g., stresses, strains, and deflections) to applied loads. Elastic layer theory is one of the most widely used pavement response models. Because this model considers each layer to extend to infinity horizontally, it is only suited for predicting the responses of jointed rigid pavement systems far away from the joints, yet it is the response near the joints that often is of most interest. For this reason, finite element response models have been developed for jointed rigid pavement systems. These models can account for finite slab lengths and various conditions at the slab interfaces, e.g., dowel bars and aggregate interlock. While it may be desirable to model the true three-dimensional behavior, two-dimensional models are generally accepted as adequate. Of the numerous two-dimensional models currently available, the ILLI-SLAB program (Ioannides [1984]) is probably the most versatile and widely used for jointed rigid pavement analysis.

As is generally true, an increase in model sophistication usually is accompanied by an increase in the number of required inputs. Finite element models require more input parameters than do elastic layer models. A summary of the input requirements for the ILLI-SLAB program is shown in table 1. Many of the required inputs can be readily determined for most problems, e.g, slab geometry, dowel size and spacing, etc. The material parameters are generally more difficult to ascertain.

An extensive sensitivity analysis conducted by Kilareski, Ozbeki, and Anderson (1984) determined that, over the expected ranges of all parameters, the material parameters most controlling the structural response of jointed rigid pavement systems near the joints are the coefficient of dowel/concrete interaction (G) and the coefficient of subgrade reaction (k). The modulus of elasticity of the concrete (E) was found to have a lesser influence on pavement response. The remaining material properties can generally be assumed using typical values from the literature.

Current nondestructive testing (NDT) techniques attempt to estimate the values of k and E. These estimates are generally based upon deflection tests conducted at the center of the slabs and away from the joints. They may not represent accurate system parameters for modeling the behavior near the joints, i.e., the "effective" values for these parameters may be different near the joints than at the center of the slabs. Current NDT techniques do not provide estimates for G, and this parameter is typically assumed even though its value has a strong influence on pavement response near the joints. Given these deficiencies, a new method was developed to provide better estimates of these critical system parameters.

BACKCALCULATION PROCEDURE

The backcalculation procedure is based upon surface deflection measurements across the joints using a falling weight deflectometer (FWD), and analytical pavement response analysis using the ILLI-SLAB finite element program. The step-by-step procedure is as follows:

1.   Perform FWD tests at the joints as shown in figure 1. For each test, record the FWD load, and the deflections at the center of the load plate and on each side of the joint.

Table 1. Input Requirements for the ILLI-SLAB Program

<u>Slab Geometry</u>

- Length
- Width
- Thickness
- Joint width (if joints are present)
- Number of lanes (and if tied together)

<u>Subgrade Properties</u>

- k = coefficient of subgrade reaction, or
- E = Young's modulus
- μ = Poisson's ratio (if E is used)

<u>Concrete Slab Properties</u>

- E = modulus of elasticity
- μ = Poisson's ratio
- Unit weight

<u>Dowel Properties (transverse and longitudinal joints)</u>

- Diameter
- E = modulus of elasticity
- Spacing between dowels
- μ = Poisson's ratio
- G = coefficient of dowel-concrete interaction

<u>Load Properties</u>

- Average contact pressure
- Size of loaded area

<u>Temperature-Related Properties (if curling analyzed)</u>

- Temperature differential between top and bottom of slab
- α = coefficient of thermal expansion/contraction for concrete

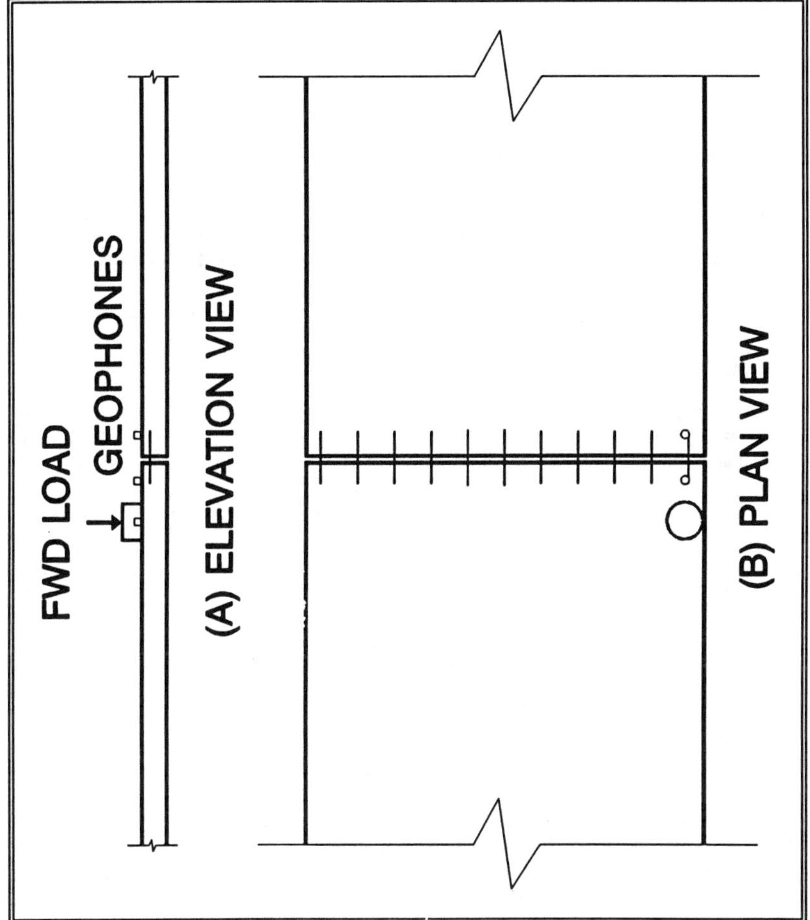

Figure 1. Schematic of FWD Joint Test

Consistent placement of the FWD measurement system should be followed, and the placement geometry should be documented.

2.   Normalize the measured deflections to a standard FWD load level (e.g., 9000 lb [40 kN]), and calculate the deflection transfer efficiency across the joint (or simply joint efficiency) as the ratio of the leave slab deflection to the approach slab deflection expressed as a percentage (see figure 2).

3.   Develop an ILLI-SLAB finite element model for the pavement system subjected to the standard FWD load. The finite element model will require values for each of the inputs listed in table 1. An appropriate finite element mesh must be generated, an example of which is shown in figure 3. Ioannides (1984) provides detailed guidelines for the generation of a mesh for the ILLI-SLAB program. The mesh should be developed such that nodal points are located at each of the FWD geophone locations.

4.   Run the finite element model for a matrix of values of G and k, holding all other inputs constant. The modulus of elasticity of the concrete (E) can be assumed from typical values as desribed later in this paper. Alternatively, E can be obtained from FWD tests at the center of the slab, although in the authors' experience these values can be very erratic, or from laboratory tests on cores extracted from the pavement.

5.   For each of the finite element model runs, determine the surface deflection at the nodal points corresponding with the FWD geophone locations, and calculate the deflection transfer efficiency.

6.   Develop a plot of deflection transfer efficiency versus surface deflection at the center of the FWD load from the data generated above. This will establish two families of parallel curves for constant values of G and k, respectively, as shown in figure 4.

7.   Locate the point on the plot developed in step 6 that corresponds to the measured deflection transfer efficiency and surface deflection from the FWD test.

8.   Interpolate as necessary between the two pairs of parallel curves that enclose the point located in step 7 to determine the G and k values of the pavement system under test. These G and k values, used in conjunction with the other input parameters, will provide a finite element model that matches as close as possible the actual response of the pavement system under the FWD load.

FIELD TEST RESULTS

The backcalculation routine described above was conducted on data collected from seven jointed rigid pavement sites in Pennsylvania in conjunction with a project for the Pennsylvania Department of Transportation (PennDOT) (Hiltunen, et al. [1991]). FWD tests were conducted at each site on approximately 25 slabs/joints, and the data were reduced according to step 2 above. The results of the data reduction are shown in table 2. Shown in this table for each site are the maximum, minimum, and average values from the approximately 25 tests for the deflection transfer (joint) efficiency and the surface deflection under the center of the FWD load.

ILLI-SLAB finite element models were developed for each site using slab geometry and dowel bar information collected from PennDOT records. The finite element mesh, including the location of FWD load, is shown in figure 3 for site 7. Based upon a study of portland cement concrete properties as part of the overall project (Hiltunen, et al. [1991]), the modulus of elasticity of the concrete was assumed constant at 4 million psi (27.6 GPa).

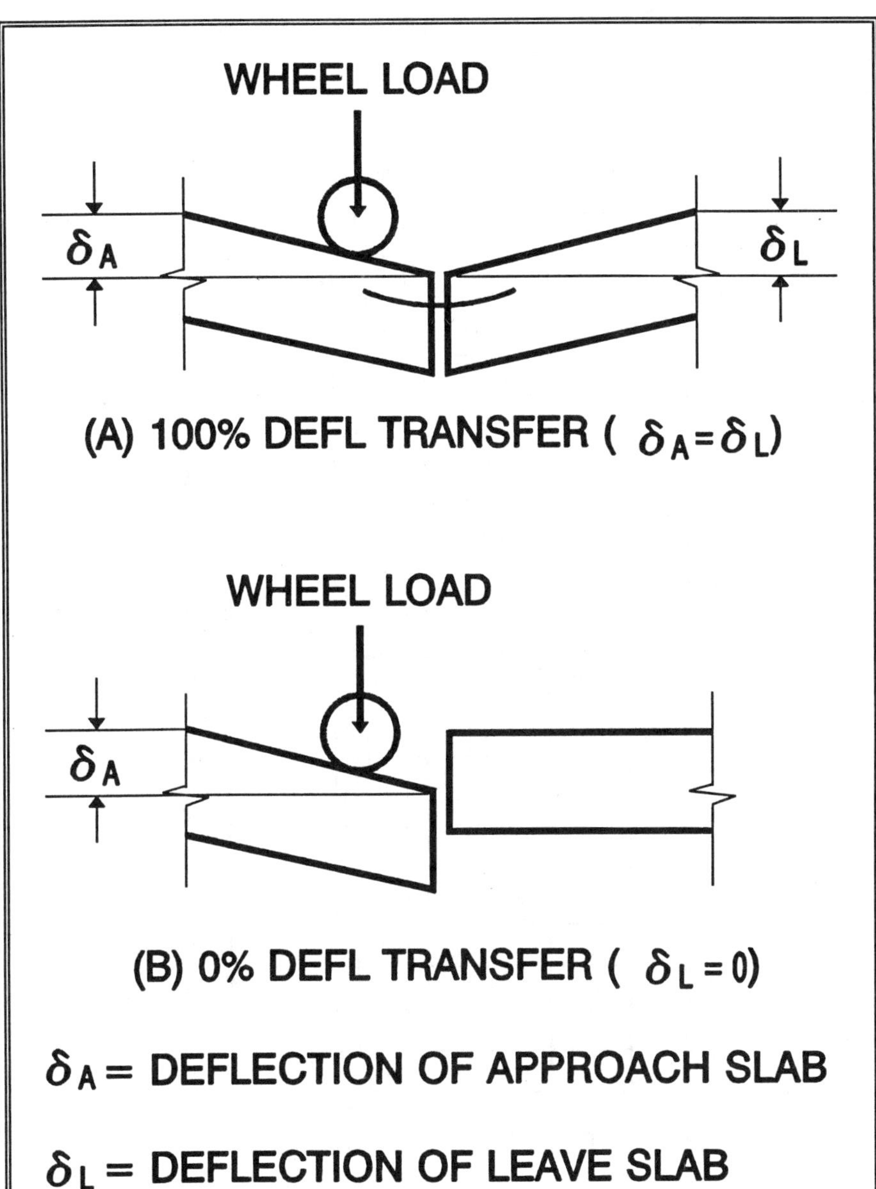

Figure 2. Schematic of Joint Deflection Transfer Efficiency

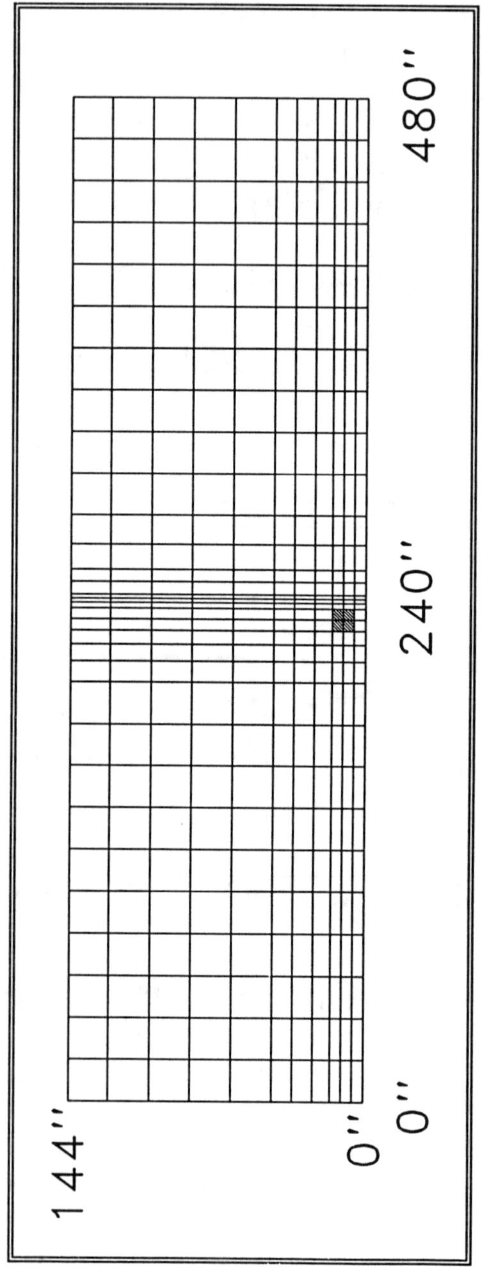

Figure 3. Finite Element Mesh for Site 7 (1" = 1 in. = 25.4 mm)

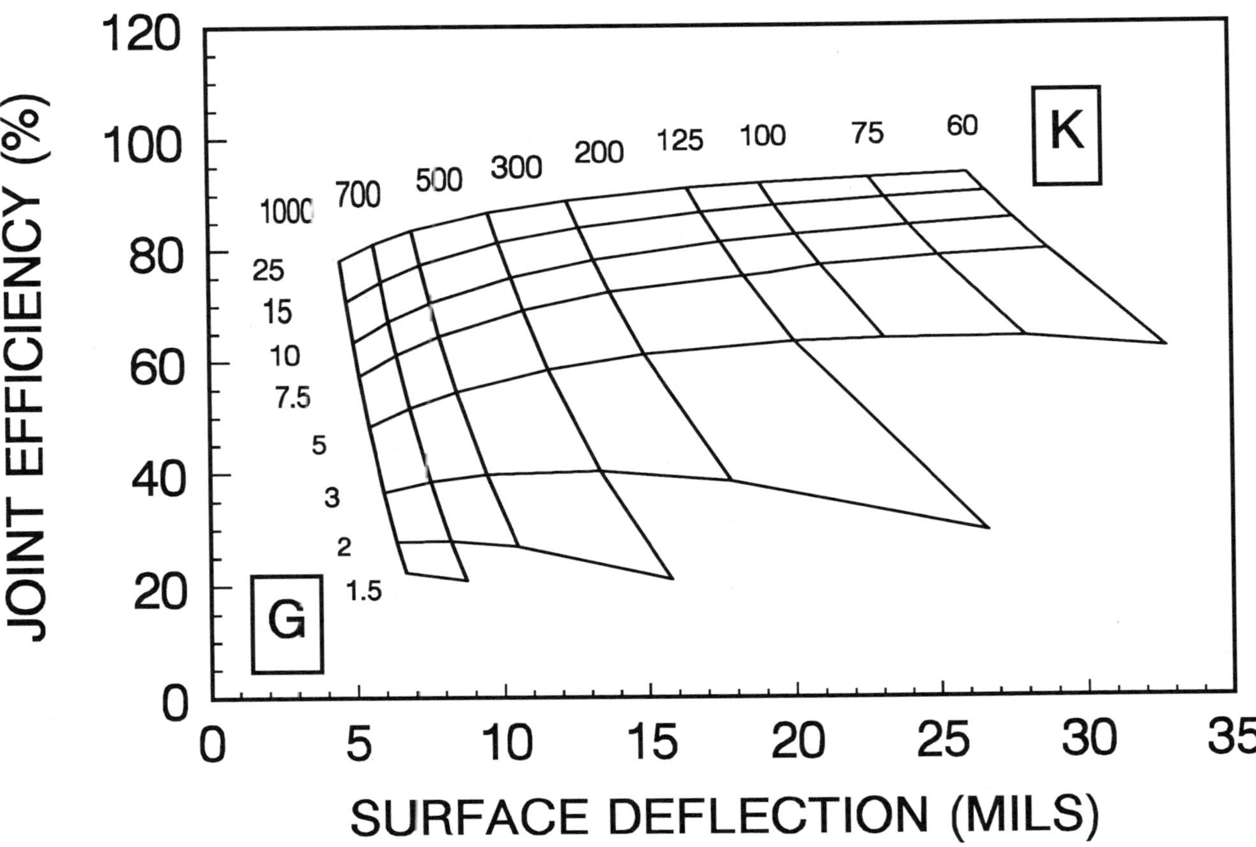

Figure 4. Joint Efficiency vs. Surface Deflection for Site 7 (Please See Table 3 for Unit Definitions for G and k) (1000 mils = 25.4 mm)

Table 2. FWD Joint Deflection Test Data Summary

Site No.	Case	Surface Deflection (mils)	Joint Efficiency (%)
5	Maximum	8.51	116.0
	Minimum	3.79	78.0
	Average	5.44	88.3
6	Maximum	4.47	92.0
	Minimum	2.35	77.0
	Average	3.33	84.2
7	Maximum	24.79	90.0
	Minimum	11.31	79.0
	Average	17.97	84.3
9	Maximum	7.33	110.0
	Minimum	2.99	37.0
	Average	5.02	91.5
10	Maximum	4.57	139.0
	Minimum	2.49	46.0
	Average	3.28	67.6
11	Maximum	9.73	86.0
	Minimum	4.19	56.0
	Average	6.09	67.6
12	Maximum	11.49	65.0
	Minimum	5.64	25.0
	Average	9.08	48.9

Note: 1000 mils = 25.4 mm

For each test site a series of finite element model runs was conducted over a range of G and k values. From these runs, plots of joint efficiency versus surface deflection were generated for each site, an example of which is shown in figure 4 for site 7. For each site, two pairs of backcalculated G and k values were then determined: one using the average joint efficiency and average surface deflection, and the other using a "worst-case" condition, which was defined to be the minimum joint efficiency and the maximum surface deflection. These backcalclated parameters are shown in table 3. It is observed that very reasonable values for both G and k are obtained in most cases. The values are very consistent with what is typically expected for these parameters in real pavement systems. It is also observed that these parameters are not constant for all jointed rigid pavements. In fact, they are strong indicators of the in situ condition of the joint, as Kilareski, Ozbeki, and Anderson (1984) have demonstrated.

INFLUENCE OF CONCRETE MODULUS

The influence of the modulus of elasticity of the concrete (E) was briefly discussed above. In terms of pavement structural response near the joint, E has a relatively small influence as compared with G and k. However, it may seem inappropriate to some individuals to simply assume E constant at a typical value. Therefore, a limited investigation was conducted to study the influence of E on both the backcalculation and pavement response modeling processes.

For site 7, finite element model runs were conducted to develop the relationship between joint efficiency and surface deflection for seven values of E between 2 and 6 million psi (13.8 and 41.3 GPa), including the 4 million psi (27.6 GPa) case presented in figure 4. Using the FWD measured joint efficiency and surface deflection from table 2, G and k values were backcalculated for each value of E. The backcalculations were done for both the average and worst-case conditions as before. The results are shown in table 4 and are plotted in figures 5 and 6. It is observed that indeed the values of the backcalculated G and k parameters do depend on the value of E, though the dependence is not strong above 4 million psi (27.6 GPa).

Next, a finite element model of the pavement system subjected to an 18-kip (80-kN) single axle load located near the joint was developed, from which the maximum tensile stress in the concrete (σ) was determined for each of the 14 G, k, and E triplets. These results are found in table 4 and are plotted in figure 7. It is observed that the stresses are approximately equal for E between 4 and 6 million psi (27.6 and 41.3 GPa), while they decrease as E decreases below 4 million psi (27.6 GPa). It appears that the stresses are relatively constant in the range of E of high quality paving concrete.

In summary, values of backcalculated G and k parameters do depend upon the modulus of elasticity of the concrete (E). However, the response of the pavement system in terms of stress appears to be relatively independent of E, as long as corresponding G and k values are used with each E. This really confirms the findings of the sensitivity study reported above. Therefore, it is possible to develop a reasonable finite element model to predict jointed rigid pavement response by fixing the concrete modulus at a typical value, and backcalculating the corresponding G and k parameters from FWD measurements.

USES, LIMITATIONS, AND FUTURE DEVELOPMENT

As discussed throughout this paper, the G and k parameters backcalculated following the procedure described herein can be used as inputs for finite element structural response predictions of jointed rigid pavement systems. Such a prediction model would be a key component of a mechanistic-based overlay design procedure, for example. In addition, the G and k parameters could be used in a

Table 3. Summary of G and k Backcalculated Parameters

Site	G (pi x 10^6)		k (pci)	
	Average	Worst	Average	Worst
5	4.50	1.35	250	135
6	5.00	2.25	700	450
7	1.20	0.75	115	77
9	12.00	0.60	400	400
10	2.50	0.85	1600	1200
11	1.20	0.60	800	475
12	0.41	0.19	480	460

Notes: 1 pi = 1 lb/in. = 1.75 N/cm

1 pci = 1 lb/in.^{3} = 0.272 N/cm^3

Table 4. Influence of Concrete Modulus (Site 7)

E (psi x 10^6)	G (pi x 10^6)		k (pci)		σ (psi)	
	Average	Worst	Average	Worst	Average	Worst
2.0	2.00	1.20	170	112	152.9	204.7
3.0	1.44	0.92	144	91	164.5	218.6
3.5	1.36	0.84	125	83	167.2	223.6
4.0	1.20	0.75	115	77	173.0	233.1
4.5	1.17	0.72	112	72	172.1	232.9
5.0	1.15	0.70	105	68	171.9	231.5
6.0	1.00	0.63	96	62	177.7	237.7

Notes: 1 psi = 6.89 kPa

1 pi = 1 lb/in. = 1.75 N/cm

1 pci = 1 lb/in.^{3} = 0.272 N/cm^3

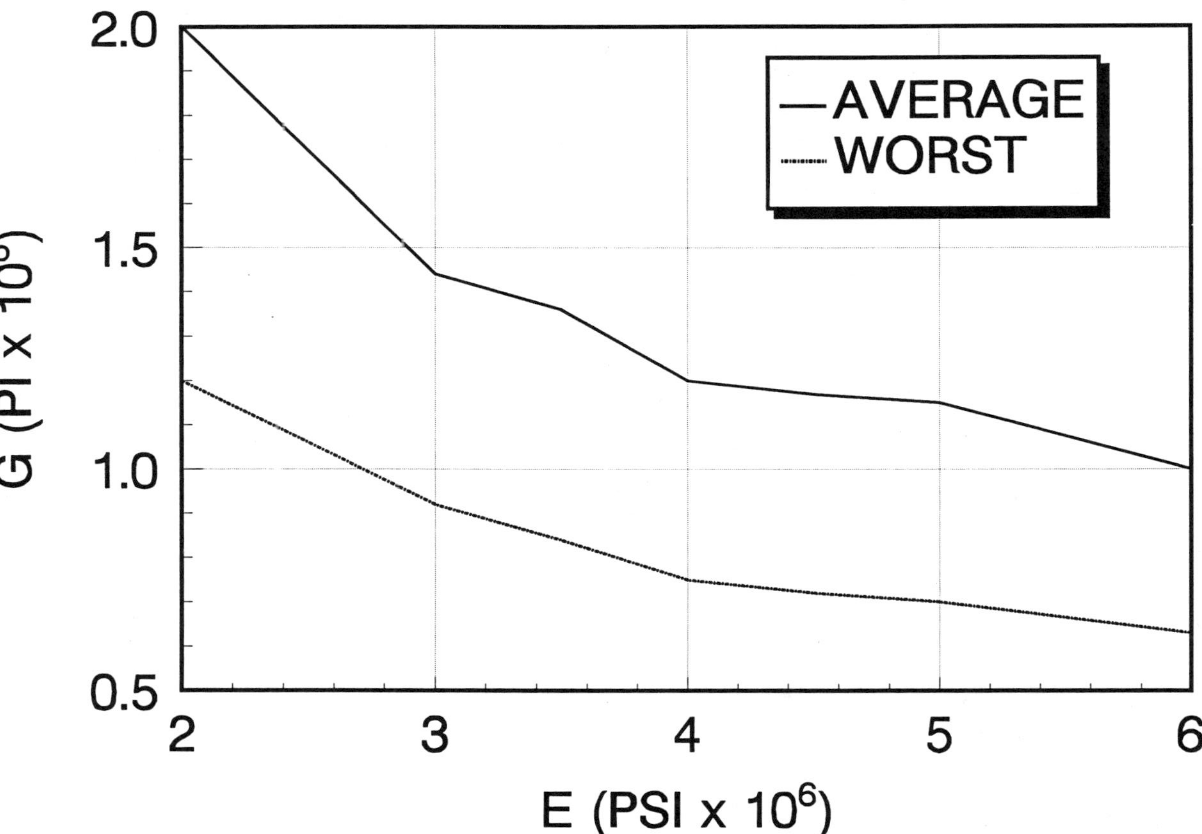

Figure 5. Coefficient of Dowel/Concrete Interaction versus Concrete Modulus for Site 7 (1 pi = 1.75 N/cm, 1 psi = 6.89 kPa)

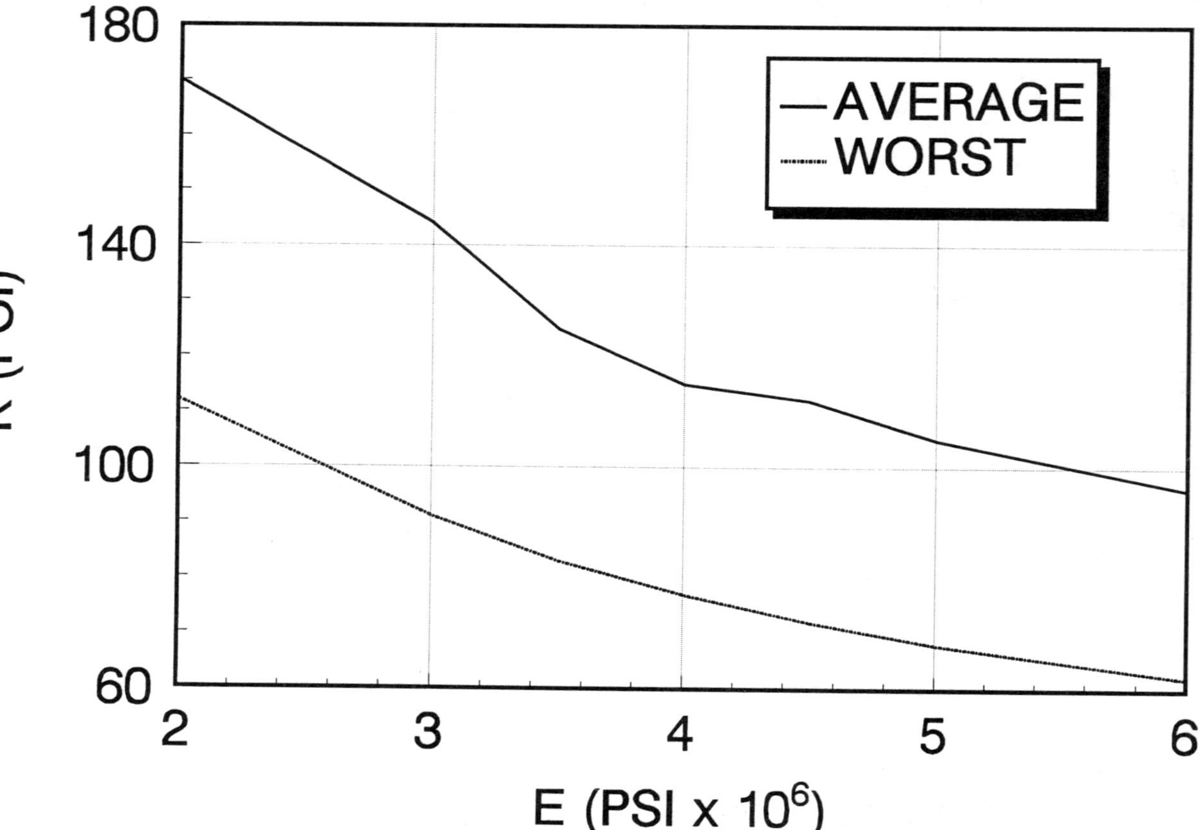

Figure 6. Coefficient of Subgrade Reaction versus Concrete Modulus for Site 7 (1 pci = 0.272 N/cm^3, 1 psi = 6.89 kPa)

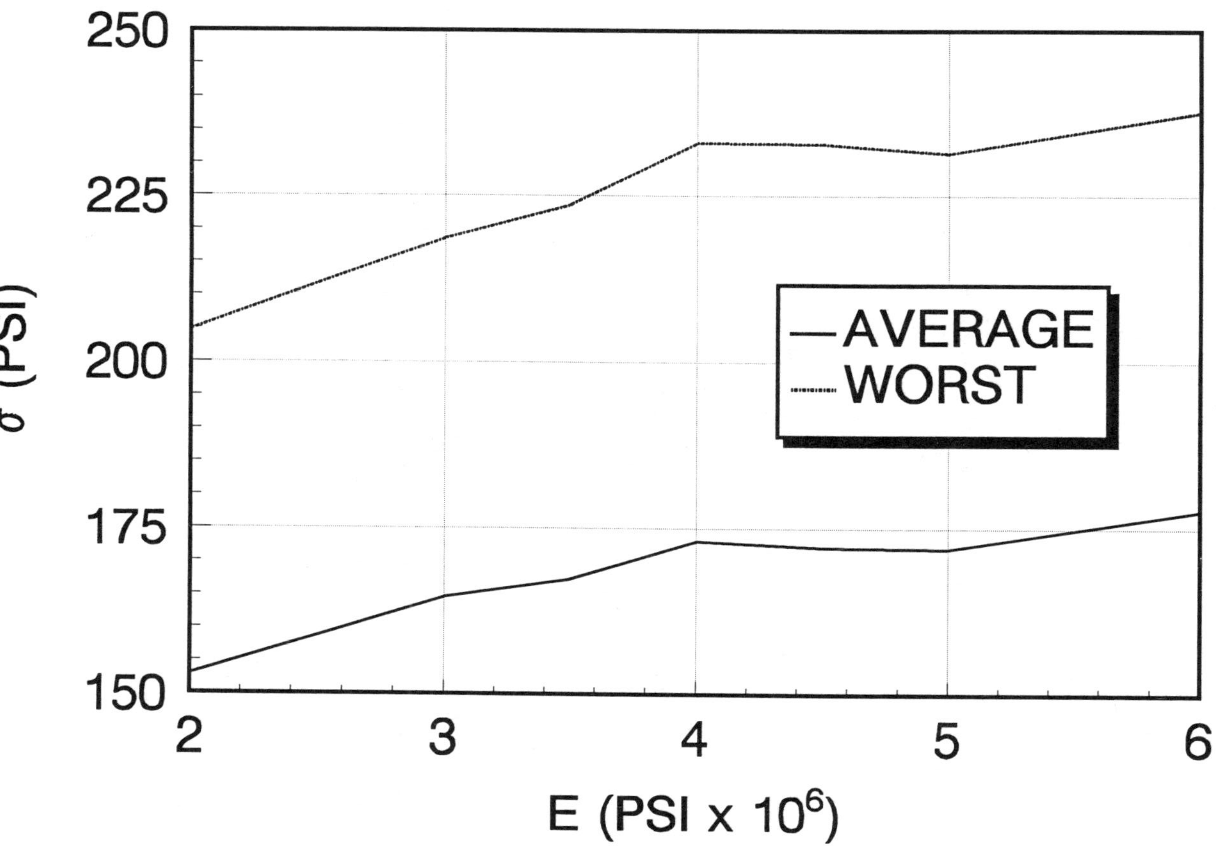

Figure 7. Maximum Tensile Stress versus Concrete Modulus for Site 7 (1 psi = 6.89 kPa)

pavement management system to assess in situ joint condition. Kilareski, Ozbeki, and Anderson (1984) have described such a procedure.

The backcalculation procedure presented herein obviously would be difficult to implement for production testing and evaluation. It would be necessary to develop a catalog of joint efficiency versus surface deflection relationships for different pavement geometries. However, it is felt that an efficient technique could be developed for production use by adding a computerized iteration routine to the procedure. It should also be noted that the purpose of this paper was to present the details of the backcalculation procedure, and not to suggest that a production level technique is available. It would not be appropriate to develop a production level technique until the professional community reviews and evaluates the details of the procedure.

The obvious future development of the backcalculation procedure is the automation described above. In addition, it would also be prudent to independently verify the accuracy and reasonableness of the backcalculated G and k parameters. From the limited studies conducted to date, they appear to be reasonable, but more rigorous verification is desireable.

CONCLUSIONS

A procedure has been developed to backcalculate parameters required for the mechanistic evaluation of jointed rigid pavements. The two paramters are the coefficient of dowel/concrete interaction (G), and the coefficient of subgrade reaction (k) in the vicinity of the joint. Previous pavement response models have either ignored the influence of joints on pavement response, or have assumed values for these parameters when they indeed vary with the in situ condition of the joints. The procedure is based upon FWD measurements and finite element modeling of the pavement system. Use of the procedure on seven pavement systems in Pennsylvania has demonstrated that the procedure is viable.

ACKNOWLEDGEMENTS

The work reported herein was supported in part by the Pennsylvania Department of Transportation under Project 89-06. This support is gratefully acknowledged. In addition, the authors wish to thank D. L. Coudriet, J. R. James, and X. Wu for their assistance in reducing the FWD data, in performing the finite element model runs, and in preparing tables and figures for the manuscript.

REFERENCES

Hiltunen, D. R., Stoffels, S. M., Kilareski, W. P., Coudriet, D. L., and James, J. R. (1991), "Development of Pavement Design Guide Procedures for Pennsylvania," Final Report to Pennsylvania Department of Transportation for Research Project 89-06, November.

Ioannides, A. M. (1984), "Analysis of Slabs-On-Grade for a Variety of Loading and Support Conditions," Ph.D. Dissertation, University of Illinois at Urbana-Champaign.

Kilareski, W. P., Ozbeki, M. A., and Anderson, D. A. (1984), "Fourth Cycle of Pavement Research at the Pennsylvania Transportation Research Facility; Vol. 4, Rigid Pavement Joint Evaluation and Full Depth Patch Designs," Report No. FHWA/PA-84-026, Pennsylvania Department of Transportation, Harrisburg, PA, December, 248 pp.

James A. Crovetti,[1] Mercedes R. Tirado-Crovetti[2]

EVALUATION OF SUPPORT CONDITIONS UNDER JOINTED CONCRETE PAVEMENT SLABS

REFERENCE: Crovetti, J. A., and Tirado-Crovetti, M. R., "Evaluation of Support Conditions Under Jointed Concrete Pavement Slabs," Nondestructive Testing of Pavements and Backcalculation of Moduli (Second Volume), ASTM STP 1198, Harold L. Von Quintas, Albert J. Bush, III, and Gilbert Y. Baladi, Eds., Amercian Soceity for Testing and Materials, Philadelphia, 1994.

ABSTRACT: The evaluation of support conditions under jointed concrete pavement slabs is an important input into pavement analyses. Closed-form solutions are available to determine support conditions under the central portion of the slab; however, these solutions assume interior loading (infinite slab) and slab-on-grade conditions. This paper provides corrections necessary to extend these solutions to finite slab sizes. Also provided are closed form solutions for determining foundation support conditions under slab edges and corners using deflection measurements obtained at these locations.

KEYWORDS: falling weight deflectometer, backcalculation, foundation support value.

The performance of jointed concrete pavements (JCP) is intimately tied to the uniformity of support provided to the slabs by the foundation materials. Loss of support, or nonuniformity of support, may result in significant increases in slab stresses, ultimately leading to premature failure (cracking) of the pavement slabs. Loss of support may result from environmental action (i.e., curling and warping) or from physical changes in foundation materials (i.e., pumping, saturation, densification, etc.), either of which may be limited or avoided through proper materials selection and/or pavement design. Designers may take actions to limit the magnitude or duration of support loss (i.e., joint spacing, joint reinforcement, etc.); however, verification of support uniformity throughout the pavement's design life typically requires structural testing and analysis of inservice pavements.

---

[1]Assistant Professor, Department of Civil and Environmental Engineering, Marquette University, Milwaukee, WI, 53233.

[2]Project Manager, Engineering and Research International, 1401 Regency Drive East, Savoy, IL 61874.

Pavement analyses are greatly enhanced by the availability of closed-form solutions for the computation of deflections and stresses in pavement slabs subjected to loadings. Westergaard (1926, 1939, 1948) and Losberg (1960) provided equations for interior, edge and corner loading conditions which represent the forward calculation of pavement response due to applied external loads, assuming a fully supported slab. These equations provide the necessary framework for the development of rigorous backcalculation methods which can be used to analyze in situ pavement systems. Ioannides (1990) and Hall (1991) present such methods which allow for the backcalculation of foundation support values and elastic moduli of the PCC layer based on surface deflections measured with conventional nondestructive deflection testing (NDT) equipment at interior load positions.

In general, the use of the classical equations or existing backcalculation methods includes the following assumptions:

1.	The slab is acting as a plate supported uniformly by a dense-liquid or elastic solid foundation.
2.	There are no man-made layers between the slab and the foundation.
3.	In the case of interior loads, the test slab is of sufficient dimension such that any free edges, cracks or joints are far enough away as not to influence the deflected shape of the slab.
4.	In the case of edge loadings, the test slab is of sufficient dimension such that cracks or joints are far enough away as not to influence the deflected shape of the slab.
5.	In the case of corner loadings, the slab is of sufficient dimensions such that cracks or joints are far enough away as not to influence the deflected shape of the slab.

This paper presents new techniques for backcalculating in situ subgrade support values acting at three critical pavement locations, interior, edge, and corner, based on surface deflection obtained with typical NDT devices. The uniformity of support is determined by comparison of edge and corner support values to interior conditions. The scope is limited to the dense-liquid foundation model, except in those instances where comparison with elastic solid foundation modeling is appropriate. The techniques presented are based on previous research by Ioannides and Hall, with the inclusion of necessary adjustments to account for slab size effects. A more detailed analysis of this topic is presented by Crovetti (1993).

INTERIOR FOUNDATION SUPPORT

As stated, the backcalculation of pavement parameters requires an understanding of the forward analysis of pavement response. As such, classical works will be reviewed briefly to provide this background.

Westergaard (1948) provides an equation for the calculation of maximum surface deflection under a circular load for the dense liquid foundation model. This equation may be written in non-dimensional form as follows:

$$\Delta_i = 0.125 * \left[1 + \left(\frac{1}{2\pi}\right) \left(\ln\left(\frac{\gamma a}{2\ell}\right) - \frac{5}{4} \right) \left(\frac{a}{\ell}\right)^2 \right] \qquad (1)$$

where: Δ_i = Nondimensional deflection term = $\delta_i k\ell^2/P = \delta_i D/P\ell^2$
             δ_i = Maximum surface deflection
             k = Modulus of subgrade reaction
             ℓ = Radius of relative stiffness
             P = Applied load
             D = Slab bending stiffness modulus = $E_c H_c^3$ / $[12*(1-\mu_c^2)]$
             E_c = Young's modulus of PCC slab
             H_c = Slab thickness
             μ_c = Poisson's ratio of slab
             a = Radius of applied load
             γ = 1.7810725± ($\ln \gamma$ = Euler's Constant = 0.57721566490)

The radius of relative stiffness, ℓ, is related to slab and foundation parameters as follows:

$$\ell = \sqrt[4]{\frac{E_c H_c^3}{12(1-\mu_c^2)k}} = \sqrt[4]{\frac{D}{k}} \qquad (2)$$

Regression analysis of Eqn.1, for a/ℓ = 0.06 to 0.40 (encompasses all expected field values), yields a somewhat more manageable form of the equation as follows:

$$\Delta_i = .12538 - 0.0082 \left(\frac{a}{\ell}\right) - 0.02746 \left(\frac{a}{\ell}\right)^2 \qquad (3)$$

$$R^2 = 0.99987$$

Westergaard (1948) and Losberg (1960) provide equations for the determination of surface deflections at any radial distance from the loaded area. These equations may be solved for various radial distances to provide inputs which can be used to calculate descriptive parameters of the deflected surface shape. One such parameter is the deflection basin AREA (Hoffman and Thompson 1981) which is calculated based on surface deflections at 0, 30.48, 60.96, and 91.44 cm (0, 12, 24, and 36 in.) from the load center (load radius = 150 mm). As originally defined, the basin AREA represents the trapezoidal cross-sectional area of the deflected surface normalized by the maximum deflection.

Figure 1 illustrates an extension of this basin AREA concept to any sensor configuration, calculated with the general equation:

$$AREA = \left(\frac{1}{\delta_1}\right) \sum_{i=1}^{n} (R_{i+1} - R_i) \frac{(\delta_i + \delta_{i+1})}{2} \qquad (4)$$

where: R_i = Radial distance between sensor i and the load center
             δ_i = Surface deflection measured at sensor i

Regardless of sensor grouping used, the theoretical maximum value of AREA is equal to the radial distance between the first and nth sensor

used and the appropriate units for AREA are length. Due to the
proliferation of design/analyses methodologies which utilize AREA terms
calculated based on differing sensor groupings, care must be exercised
to ensure that the AREA calculation is consistent with the chosen
methodology.

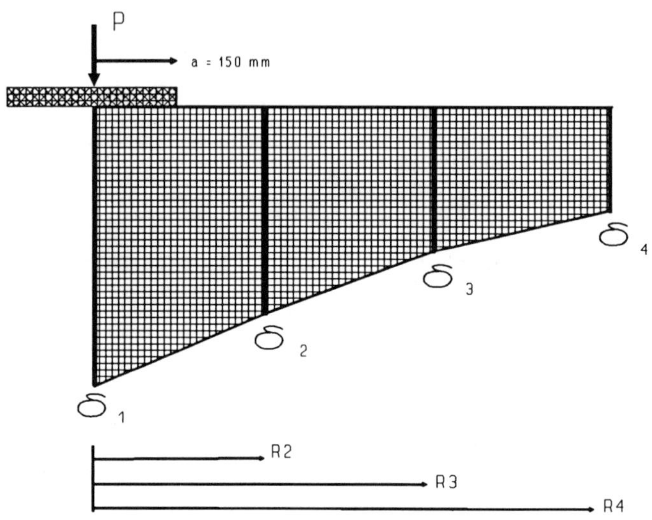

Figure 1: Deflection Basin AREA Concept.

     Ioannides, et.al.(1989) demonstrated that unique relationships
exist between AREA and the radius of relative stiffness, each dependent
on the load radius and sensor grouping. **Figure 2** illustrates one such
relationship, developed for a load radius of 150 mm and a four sensor
grouping with spacings of 0, 30.48, 60.96, and 91.44 cm from the load
center.
     Using **Figure 2**, AREAs appropriately calculated from surface
deflections may be used to backcalculate the radius of relative
stiffness of the pavement system. This relationship is also imbedded
within a backcalculation program provided by Ioannides (program ILLI-
BACK).
     Hall (1991) provides regression equations for the backcalculation
of radius of relative stiffness based on AREA calculated from
theoretical deflections obtained with the Losberg equations. For the
dense liquid foundation model, the equation may be written in SI units
as:

$$\ell = \left[\frac{\ln\left(\dfrac{91.44 - AREA}{4603.189}\right)}{-2.069416} \right]^{4.387009} \tag{5}$$

where: ℓ = Radius of relative stiffness, cm
        AREA = Deflection basin AREA calculated with 4 sensors spaced
             at 0, 30.48, 60.96 and 91.44 cm (Load Radius = 15 cm)

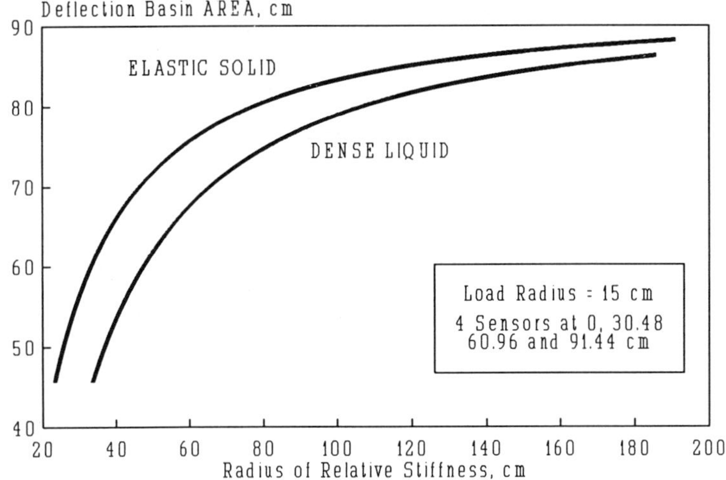

Figure 2: AREA vs Radius of Relative Stiffness.

After backcalculation of ℓ, the modulus of subgrade reaction, k, and slab bending stiffness modulus, D, may be independently backcalculated from maximum surface deflections using **Eqn. 1 or 3**. When backcalculating k, the non-dimensional deflection term becomes $\Delta_i = \delta_i k \ell^2 / P$. When backcalculating D, the non-dimensional deflection term becomes $\Delta_i = \delta_i D / P \ell^2$.

In summary, existing methods for backcalculating k-value and D, based on surface deflections obtained under interior loadings, may be summarized as follows:

1. Measure surface deflections at 0, 30.48, 60.96 and 91.44 cm from the load center, using an applied radius of 15 cm and applied load, P.
2. Calculate deflection basin AREA **(Eqn.4)**.
3. Backcalculate radius of relative stiffness, ℓ **(Fig.2 or Eqn. 5)**.
4. Backcalculate k-value and/or D **(Eqn. 1 or 3)** using maximum measured deflection, δ_1, applied load P, and backcalculated ℓ as inputs.

The summarized procedure is based on ultimately backcalculating k and D (Step 4) using only maximum surface deflection. Theoretically, k can be determined from any sensor position (Ioannides 1989). However, for the purposes of this paper, which is aimed at determining the support provided to the slab at various locations (interior, edge and corner), only maximum deflection is used. The reasons for this choice are as follows:

1.   Ease of application.
2.   Maximum deflection is least affected, relatively, by
      reductions in slab size.
3.   Sensors currently in use for measuring surface deflections
      include an error which is made up of a proportional
      component, typically less than ±2% of the indicated reading,
      and a fixed component, typically ± 2 microns (± 0.08 mils).
      Assuming each sensor behaves similarly, the fixed component
      of error will diminish in relative magnitude as deflection
      values increase.
4.   Errors in backcalculated k-values, due to slab size effects
      discussed subsequently, will remain basically constant
      throughout the range of sensors used to determine AREA,
      therefore additional calculations will not identify this
      problem.

As stated previously, equations presented to this point are based
on the assumption of interior loading conditions (infinite slab size).
Ioannides (1984) indicates that for the dense liquid foundation model,
the minimum slab dimension, length or width, must be at least eight
times the radius of relative stiffness, ℓ (i.e., $L/\ell \geq 8$). While ℓ may
assume a wide range of values, a value of 90 cm ± 30 cm may be assumed
to encompass the vast majority of in-service highway pavements. For
airfield pavements, average ℓ values of 120 cm (± 30 cm) may be more
appropriate.

In the case of highway pavements, the average ℓ value of 90 cm
would require a slab to have a minimum dimension of 7.2 m (8*ℓ) in order
to satisfy interior loading conditions. As most highway pavement slabs
are constructed to a width of 3.6 to 4.3 m ($4 \leq L/\ell \leq 4.8$), it is
apparent that this important condition will be violated for the majority
of in-service highway pavements. In the case of airfield pavements, the
average ℓ would require a slab to have a minimum dimension of 9.6 m in
order to satisfy interior loading conditions. As most airfield pavement
slabs are constructed to a width of 3.8 to 7.6 m ($3.2 \leq L/\ell \leq 6.3$), it
is again apparent that this important condition will be violated for the
majority of in-service airfield pavements.

While it is recognized that "true" interior loading conditions may
seldom be achieved in the field, the following important questions arise
as to the effect of slab dimensions on the integrity of backcalculated
interior pavement parameters:

1.   How is deflection basin AREA affected by reduction in slab
      size?
2.   How is maximum deflection affected by reduction in slab
      size?
3.   Will load transfer across joints or cracks compensate for
      reductions in slab dimensions so that interior conditions
      are more closely attained?

To answer these questions, a sensitivity analysis of the
backcalculation equations was coupled with finite element modeling of 2-
layer pavement systems using the ILLI-SLAB computer program. During
computer modeling, input ℓ values between 60 and 150 cm, in increments
of 30 cm, were used. Both single square slabs (varying length) and

multiple rectangular slabs (constant width, varying length) were
investigated. The single square slab analysis will be detailed in this
paper with rectangular slab results included as appropriate. In all
trials, a 30 cm square load (equivalent a = 172 mm) was used and surface
deflections were calculated at 0, 30.48, 60.96 and 91.44 cm from the
load center. Slab dimensions were varied from 2.1 to 12.2 m, which
resulted in L/ℓ (W/ℓ) values ranging from 1.4 to 20. For L/ℓ values
greater than approximately 8, little effect of slab size was noted.
However, for L/ℓ value less than 8, both maximum surface deflection and
deflection basin AREA values increased. These results are in agreement
with those reported by Ioannides (1984).

Slab size effects are quantified herein by the use of error
ratios, calculated as the ratio of (back)calculated values (maximum
deflection, AREA, ℓ) to theoretical values (infinite slab size). As
illustrated in **Figures 3 and 4**, maximum deflection error ratios are more
pronounced than AREA error ratios. Additionally, the variation in basin
AREA is dependent on ℓ values (**Fig.4**) while variations on maximum
deflection values are essentially insensitive to ℓ values (**Fig. 3**).

Figure 5 illustrates the error ratios on ℓ values backcalculated
from AREA. Note: The AREA value used is uncorrected for slab size
effects. As shown in **Fig. 5**, the backcalculated ℓ error ratio is
relatively insensitive to input ℓ values and closely parallels the
deflection error ratios shown in **Fig. 3**.

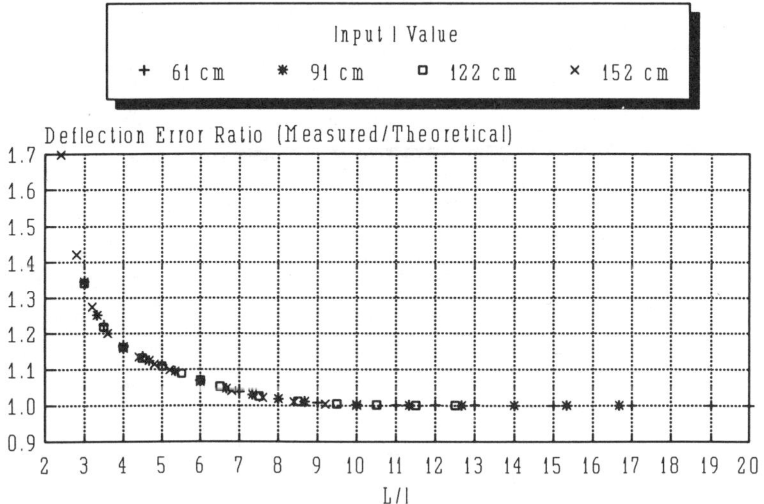

**Figure 3: Maximum Deflection Error Ratios (Single Square Slab -
Interior Loading).**

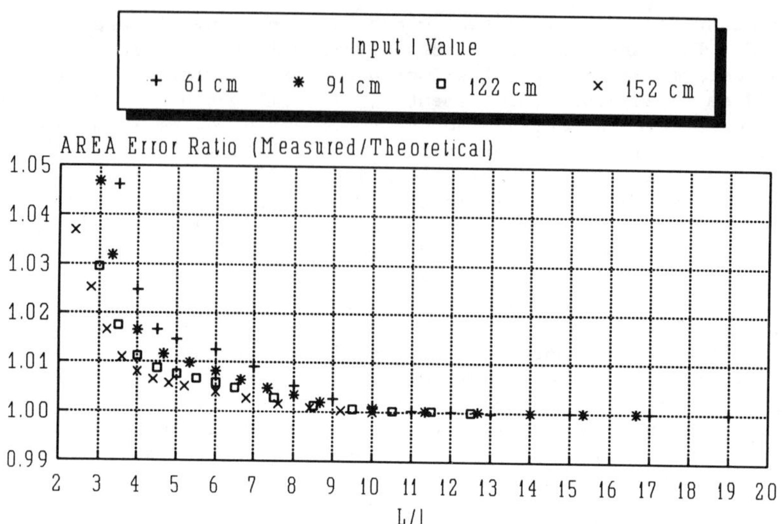

Figure 4: Deflection Basin AREA Error Ratios (Single Square Slab - Interior Loading).

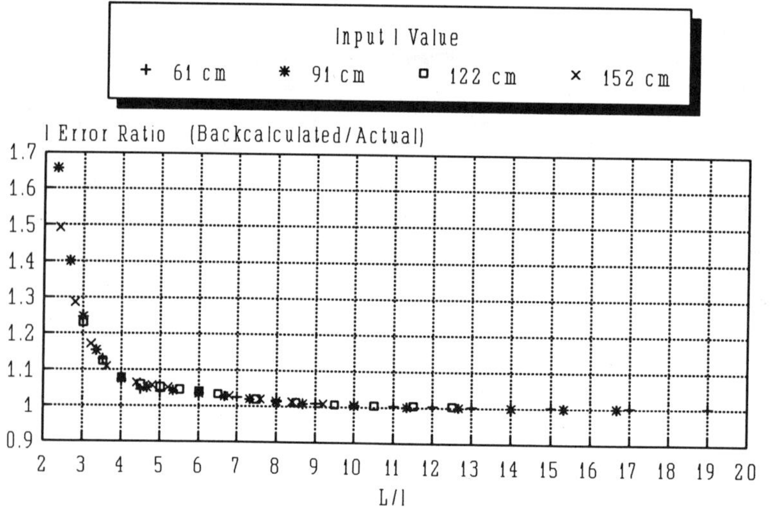

Figure 5: Backcalculated l Error Ratios (Single Square Slab - Interior Loading).

The trends exhibited in **Figures 3 and 5** were combined to quantify their ultimate effects on k- and D values backcalculated using **Equations 1 or 3**. **Figure 6** illustrates the error ratios due to slab size effects. As shown, significant underestimation of k-values and overestimation of D values may occur for small L/ℓ values.

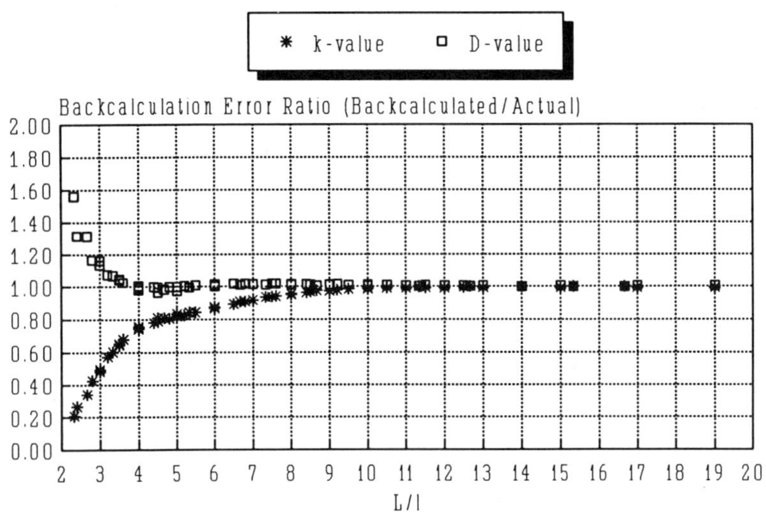

Figure 6: **Backcalculated Error Ratios on k-value and D (Single Square Slab - Interior Loading).**

To compensate for these potential errors, the following new method is proposed for use in backcalculating interior k- and D values from surface deflection measurements on single square slabs:

1. Measure surface deflections, calculate basin AREA (**Eqn. 4**), and backcalculate an estimated ℓ value, ℓ_{est} (**Eqn. 5**).
2. Calculate the L/ℓ_{est} ratio using in situ slab dimensions and backcalculated (estimated) ℓ value.
3. Determine appropriate correction factors for ℓ_{est} and maximum deflection based on L/ℓ_{est}.
4. Calculate adjusted ℓ, ℓ_{adj}, and adjusted maximum deflection, δ_{adj} by multiplying ℓ_{est} and measured maximum deflection, δ_1, respectively, by the appropriate correction factor.
5. Backcalculate interior k-value and D using **Eqn.** 1 or 3 with ℓ_{adj} and δ_{adj} as input values replacing ℓ_{est} and δ_1.

Correction factors for ℓ_{est} and maximum deflection, δ_1, were developed as a function of in situ slab dimensions, L, and ℓ_{est} values initially backcalculated from measured AREA. This was necessitated due to the fact that the true L/ℓ ratio cannot be determined a priori. The correction factors, illustrated in **Figure 7**, are of the following general form:

$$CF = 1 - K_1 \exp\left[K_2\left(\frac{L}{\ell_{est}}\right)^{K_3}\right] \tag{6}$$

where: CF = Correction factor
            L = Minimum slab dimension
            ℓ_{est} = ℓ value initially backcalculated from measured AREA
            K_1, K_2, K_3 = regression constants

Regression constants for the correction factors are as follows:

	Estimated ℓ value	Maximum Deflection
K_1	5.29875	1.06817
K_2	-2.17612	-0.66914
K_3	0.49895	0.84408
R-squared	0.9930	0.9958

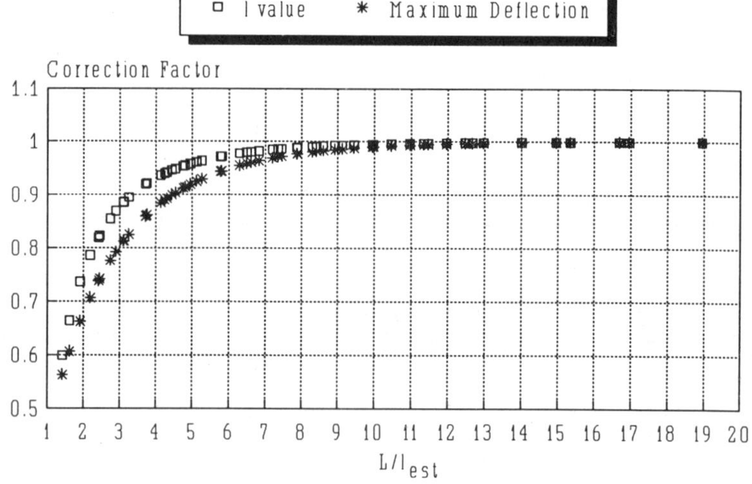

Figure 7: Correction Factors for Maximum deflection and Backcalculated Radius of Relative Stiffness (single Square Slab - Interior Loading).

Figure 8 illustrates the improvement in k- and D value backcalculation which can be attained throughout the range of L/ℓ values investigated for single square slabs under interior loading. (Compare to **Fig. 6**)

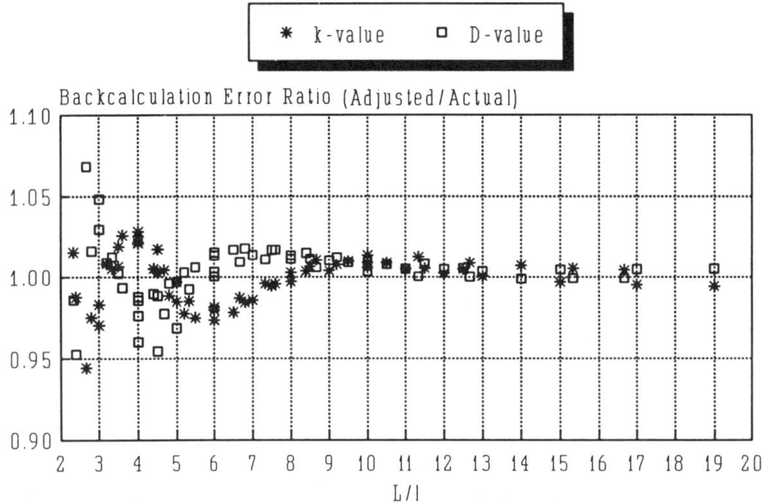

Figure 8: Backcalculation Error Ratios on k-value and D With Corrections Applied (single Square Slab - Interior Loading).

EDGE AND CORNER FOUNDATION SUPPORT

The foundation support available along slab edges and corners can also be determined using deflection basin AREA measurements, obtained under corner and edge loading conditions, following methods similar to those detailed for the interior loading condition. However, the applicability would be restricted to analysis of free edge and/or corner conditions, as load transfer across slab edges and corners (tied PCC shoulder, transverse contraction joints) alters basin AREA measurements in a manner which cannot be compensated for with conventional sensor placements.

Finite element modeling of edge and corner loading indicates that total edge/corner deflection, δ_t, defined as the sum of maximum _measured_ deflection, δ_1 (at the center of the load plate on the loaded slab), and the deflection across the joint, δ_u (at a point on the unloaded slab which is symmetrically opposite from the location of maximum deflection measurement) remains a relative constant, regardless of available deflection load transfer. Utilizing this consistency in total edge/corner deflection, one can approximate the _free_ edge/corner deflection by this _total_ deflection.

The benefit of this free edge/corner deflection approximation is highlighted by the availability of closed form solutions for the calculation of free edge/corner deflections as a function of load radius, a, and slab bending stiffness, D. Westergaard (1948) provides an equation for the calculation of free edge deflection a point coincident with the center of a circularly loaded area placed tangentially to the free edge (i.e., where maximum deflection measurements are obtained with NDT devices) as follows:

$$\Delta_e = \sqrt{\frac{2+1.2\mu_c}{12(1-\mu_c^2)}} \left[1-(0.76+0.4\mu_c)\left(\frac{a}{\ell}\right) \right]^2 \tag{7}$$

where: Δ_e = Nondimensional free edge deflection at load center
$\Delta_e = \delta_e k\ell^2/P = \delta_e D/P\ell^2$

Assuming $\mu_c=0.15$, this equation can be simplified as:

$$\Delta_e = 0.4311 - 0.707 \left(\frac{a}{\ell}\right) + 0.2899 \left(\frac{a}{\ell}\right)^2 \tag{8}$$

Eqn. 8 may be written in an alternate form, expanding the nondimensional deflection term and substituting $\ell^4 = D/k$, as follows:

$$1 = \left(\frac{P}{\delta_e \sqrt{D\,k}}\right)\left[.4311-.707a\left(\sqrt[4]{\frac{k}{D}}\right) +.2899a^2\left(\sqrt{\frac{k}{D}}\right) \right] \tag{9}$$

The use of **Eqn. 9** allows for the direct backcalculation of edge k-value, for any given free edge deflection (δ_e), applied load (P), and slab stiffness modulus (D). If the assumption is made that slab thickness and elastic modulus is constant (constant D) throughout each individual slab (slab elastic modulus is previously backcalculated based on interior loadings) **Eqn. 9** contains only one unknown term, the edge k-value, thus allowing for this direct, albeit cumbersome backcalculation. For edge deflections measured across joints, total edge deflection (δ_t) would be used to approximate the free edge deflection (δ_e).

For the case of corner loadings, Westergaard (1926) provides an approximate equation which is applicable for the calculation of extreme corner deflection resulting from a tangential circular load. During typical NDT at corner locations, the closest sensor to this extreme corner location is the one which is placed at the center of the load plate. As such, regression analysis of finite element results was used to develop an equation for the calculation of nondimensional free corner deflection, Δ_c, at a point coincident with the center of the circularly loaded area. This equation, similar in form to the edge loading equation (**Eqn.8**), is as follows:

$$\Delta_c = 1.1815 - 2.0745 \left(\frac{a}{\ell}\right) + 1.263 \left(\frac{a}{\ell}\right)^2 \tag{10}$$

For the purposes of backcalculation, **Eqn.10** may be written in a form similar to **Eqn.9**, as follows:

$$1 = \left(\frac{P}{\delta_c \sqrt{D\,k}}\right)\left[1.1815 - 2.0745a\left(\sqrt[4]{\frac{k}{D}}\right) + 1.263a^2\left(\sqrt{\frac{k}{D}}\right)\right] \qquad (11)$$

As with edge loadings, when deflections are measured across joints total corner deflection (δ_t) would be used to approximate the free corner deflection (δ_c).

Small slab size effects may also result in an increase in measured deflections at both the edge and corner locations. **Figure 9**, developed using results obtained during finite element modeling, illustrates the effects of slab size on edge and corner deflections.

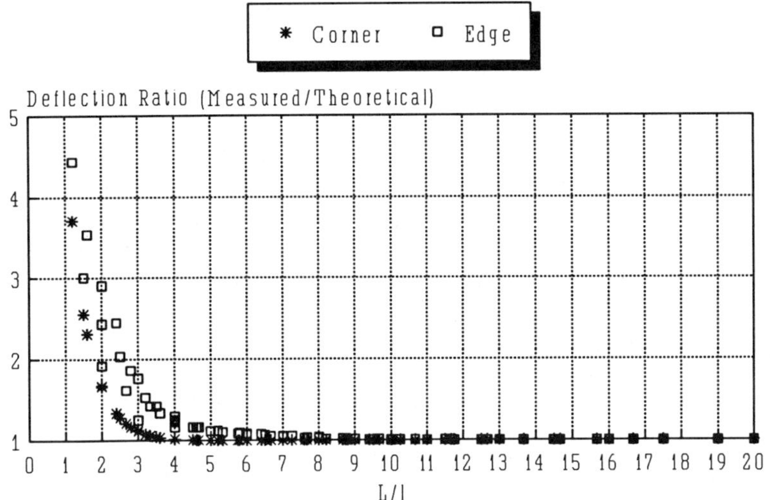

Figure 9: Deflection Ratios for Edge and Corner Loads (Single Square Slabs).

As shown in **Fig. 9**, a minimum L/ℓ ratio of approximately 8 is required for edge deflections to match theoretical (infinite slab) deflections while a minimum L/ℓ of approximately 4 is required for corner deflections. Equations for determining correction factors for edge/corner deflection measurements were developed following procedures similar to those outlined for the interior deflections. Thus, **Eqn.6** may be used to determine appropriate correction factors with the substitution of regression constants as follows:

	Edge	Corner
k_1	1.4115	0.9651
k_2	0.3569	0.1834
k_3	1.3282	2.2734
R-squared	0.9977	0.9999

For the purposes of correction factor calculation it is suggested to replace L/ℓ_{est} with L/ℓ_{adj} in **Eqn. 6**, using the ℓ_{adj} value obtained during interior analysis along with known slab dimensions.

UNIFORMITY OF SUPPORT

The uniformity of support provided to the slab may be altered due to a variety of factors, including:

1.   Erosion of base/subgrade materials,
2.   Densification of base layers,
3.   Curling/warping of the PCC slab.

The ultimate effect of each of the above listed factors may be to create partially unsupported slabs. Subsurface erosion and/or densification, which may be considered as irreversible phenomena, typically occurs near slab edges and corners, where load related deflections and surface water infiltrations are highest. Slab curling and warping are cyclic phenomena resulting from temperature and moisture variations throughout the slab depth, respectively. At times when the top of the slab is cooler/drier than the bottom, the slab will tend to curl/warp upwards along the edges and corners, with the amount of curl/warp increasing with distance from the geometric center of the slab. During times when the top of the slab is warmer/wetter than the bottom, the slab edges and corners will tend to curl/warp downwards. Under certain conditions of large gradients and/or stiff foundations, upward curling/warping will produce unsupported areas around the periphery of the slab while downwards curling/warping may actually result in loss of contact between the slab and foundation in the central slab region.

The use of surface deflections to identify locations of poor support, commonly termed "voids", has been investigated (Uddin, et al. 1987, Crovetti and Darter 1984) with mixed conclusions as to the appropriateness of the results. Based on the findings of this research, it is the authors' belief that surface deflections obtained at interior, edge and corner locations *can* be used as comparative measures, provided the objective is to determine uniformity of support rather than to quantify exact void dimensions.

The equations which have been presented to this point are all based on the assumption of full support. As such, a linear trend of load versus deflection is expected, with flatter slopes representing weaker foundations, for any given slab bending stiffness modulus. When deflection tests are conducted over unsupported or poorly supported edges and corners, due to warping/curling and/or subgrade voids, the linearity of deflection response is altered and maximum measured deflection is increased for any given load level. **Figure 10** illustrates this point for a corner load condition.

A factorial analysis of finite element results for loaded, partially supported slabs reveals the following general trends:

1.   When support loss is due to upwards warping/curling, deflection response will be composed of two essentially linear portions, the first representing unsupported

cantilever behavior and the second parallelling fully supported behavior. The breakpoint between these occurs at a deflection value approximately equal to the magnitude of curling.

2.   When support loss is due to subsurface "voids", deflection response will behave in a manner similar to the warped/curled slabs, however linear portions may be more distorted as void dimension and/or foundation k-value increases.

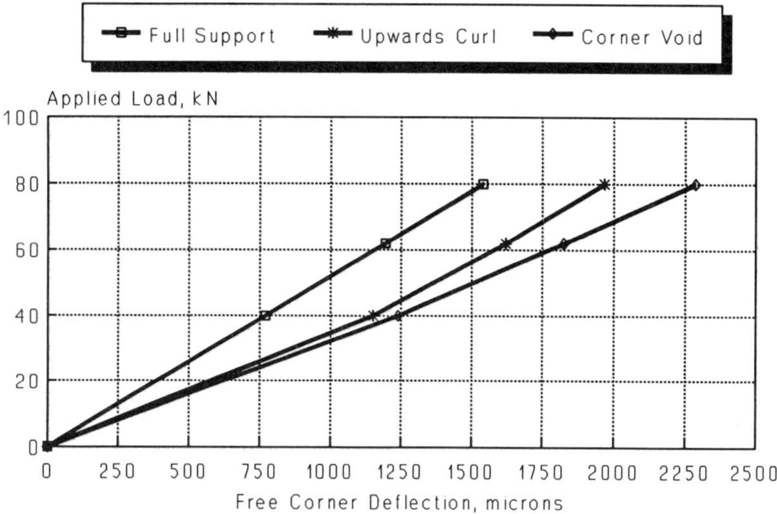

Figure 10: Load vs Deflection for Various Corner Support Conditions.

Considering the above trends, incremental deflection measurements may be analyzed in conjunction with individual measurements to better define (non)uniformity of foundation support. During the incremental analysis, the increase in measured deflection, Δ_δ, resulting from an increase in applied load, Δ_P, is of importance. These incremental values, Δ_δ and Δ_P, are used as inputs during backcalculation, replacing δ and P, respectively, in **Eqns 9 and 11**. At present, it is recommended to use at least three test load levels, approximately equal to 1.0, 1.5 and 2.0 times the design load levels, resulting in two load increments (1.0 to 1.5 and 1.5 to 2.0 times the design load).
Finite element analysis of slab deflections resulting from incremental loads placed over areas with foundation voids or upwards curling revealed the following trends:

1.   Individual analysis of load/deflection combinations resulted in backcalculated k-values consistently lower than interior

values with the magnitude of this discrepancy increasing as void size or curl magnitude increases. Additionally, for all cases the magnitude of the discrepancy decreases as load level increases.

2. Incremental analysis of deflections resulted in backcalculated k-values which converged to interior values if the load is applied over areas with small voids or slight curling. For loads placed over areas with large voids or significant curling, convergence was not achieved.

Recognizing these trends, the complete analysis of edge/corner deflections can be summarized as follows:

1. If the edge/corner k-values backcalculated during individual analysis are lower than interior values but backcalculated k-values obtained during incremental analysis using the first and/or subsequent load increments closely approximates the interior k-value, it can be assumed that poor support exists during testing and is most likely the result of warping/curling and/or small void areas.

2. If the edge/corner k-values backcalculated during both individual and incremental analyses are significantly less than interior values, poor support exists during testing and is most likely due to relatively large subsurface "voids" and/or significant curling.

While these incremental analysis conclusions are not without exception, they typically will allow for differentiation between support conditions which are usually unavoidable (i.e., warping/curling) and those which indicate subsurface deficiencies (i.e., densification, pumping, etc.) which must be corrected and/or avoided in future pavement designs.

For illustration purposes, **Table 1** provides the results of individual and incremental analyses applied to the corner deflection trends exhibited in **Fig. 10**.

SUMMARY AND CONCLUSIONS

This paper has presented techniques for use in backcalculating slab bending stiffness moduli from interior loading and moduli of subgrade reaction at interior, edge and corner locations for single square slabs (SSS) of any dimension. These techniques, which allow for the assessment of uniformity of support beneath individual concrete pavement slabs, represent an extension to available closed-form solutions which assume "infinite" slab behavior. As shown, backcalculation errors approaching 80% (**Fig.6**) may be introduced when using conventional techniques for analyzing small slabs.

Due to content limit constraints, results have been detailed for only SSS systems, supported by a dense-liquid foundation. The appropriate choice of foundation support model (i.e., dense-liquid or elastic solid) represents perhaps the most controversial of decisions. Based on research conducted by the authors, the following general statements are offered:

1.   If the slab is <u>uniformly</u> supported by an elastic solid
      foundation but deflections are analyzed assuming a dense-
      liquid foundation, backcalculated edge/corner k-values would
      typically be higher than interior values.

2.   If the slab is supported <u>nonuniformly</u> by an elastic solid
      foundation but deflections are analyzed assuming a dense-
      liquid foundation, backcalculated edge/corner k-values may
      still exceed interior values.

3.   If deflections are analyzed assuming a dense-liquid
      foundation and backcalculated edge/corner k-values are less
      than interior values, the assumption of poorly or
      unsupported edges/corners would be valid regardless of the
      "correct" foundation model.

A more detailed analysis of variable field conditions, including
slab size and shape, joint load transfer efficiency, temperature
gradients, and choice of foundation support model is provided by
Crovetti (1993).

TABLE 1--Individual and incremental backcalculation analysis

	Load Level	P kN	Dc microns	k^1 kPa/cm
Full Support Individual Analysis	1	40	770	543
	2	62	1194	543
	3	80	1540	543
Full Support Incremental Analysis	1-2	22	424	543
	2-3	18	346	543
Upward Curling Individual Analysis	1	40	1152	270
	2	62	1620	320
	3	80	1966	355
Upward Curling Incremental Analysis	1-2	22	468	455
	2-3	18	346	543
Corner Void Individual Analysis	1	40	1240	235
	2	62	1825	260
	3	80	2285	270
Corner Void Incremental Analysis	1-2	22	585	310
	2-3	18	460	330

[1] Corner k values backcalculated using Eqn. 11 with $D=3.8 \times 10^7$ MPa-cm^3
   and a=15cm. Interior k value = 543 kPa/cm.

REFERENCES

Crovetti, J.A. and Darter, M.I., "Void Detection Procedures," Appendix
    to Final Report, NCHRP Project 1-21, June 1984.

Crovetti, J.A., "Evaluation of Jointed Concrete Pavements Incorporating
    Open Graded Permeable Bases," Ph.D. Thesis, University of Illinois
    at Urbana-Champaign, Dept. of Civil Engineering, 1993.

Hall, K.T., "Performance, Evaluation, and Rehabilitation of Asphalt-
    Overlaid Concrete Pavements," Ph.D. Thesis, University of Illinois
    at Urbana-Champaign, Dept. of Civil Engineering, 1991.

Hoffman, M.S. and Thompson, M.R., "Mechanistic Interpretation of
    Nondestructive Pavement Testing Deflections," University of
    Illinois Report No. UILU-ENG-81-2010, University of Illinois at
    Urbana-Champaign, June, 1981.

Ioannides, A.M., "Analysis of Slab-On-Grade for a Variety of Loading and
    Support Conditions," Ph.D. Thesis, University of Illinois at
    Urbana-Champaign, Dept. of Civil Engineering, 1984.

Ioannides, A.M., Barenberg, E.J., and Lary, J.A., "Interpretation of
    Falling Weight Deflectometer Results Using Principles of
    Dimensional Analysis," _Proceedings_, 4th International Conference
    on Concrete Pavement Design and Rehabilitation, Purdue University,
    1989, pp. 231 to 247.

Ioannides, A.A., "Dimensional Analysis in NDT Rigid Pavement
    Evaluation," _Transportation Engineering Journal_, American Society
    of Civil Engineers, Vol. 116, No. TE1, 1990.

Losberg, A., _Structurally Reinforced Concrete Pavements_,"
    Doktorsavhandlingar Vid Chalmers Tekniska Hogskola, Goteborg,
    Sweden, 1960.

Uddin, W., Hudson, W.R., Elkins, G.E., and Reilley, K.T., "Evaluation of
    Equipment for Measuring Voids Under Pavements," FHWA Report No.
    FHWA-TS-87-229, September 1987.

Westergaard, H.M., "Computation of Stresses in Concrete Roads," HRB
    Proceedings, 5th Annual Meeting, Vol. 5, Part I, 1926, pp. 90-112.

Westergaard, H.M., "Stresses in Concrete Runways of Airports," HRB
    Proceedings, 19th Annual Meeting, 1939, pp. 197-202.

Westergaard, H.M., "New Formulas for Stresses in Concrete Pavements of
    Airfields," Transactions, American Society of Civil Engineers,
    Vol. 113, 1948, pp. 425-439.

Soheil Nazarian[1], Srinivasa Reddy[2] and Mark Baker[3]

DETERMINATION OF VOIDS UNDER RIGID PAVEMENTS USING IMPULSE RESPONSE METHOD

REFERENCE: Nazarian, S., Reddy, S., and Baker, M., "Determination of Voids Under Rigid Pavements Using Impulse Response Method," Nondestructive Testing of Pavements and Backcalculation of Moduli (Second Volume), ASTM STP 1198, Harold L. Von Quintas, Albert J. Bush, III, and Gilbert Y. Baladi, Eds., Amercian Soceity for Testing and Materials, Philadelphia, 1994.

ABSTRACT: Presence of voids and loss of support are the major factors contributing to the premature distress of rigid pavements. To detect voids, nondestructive testing techniques are normally utilized. One of the more promising methods in this area is the Impulse Response (IR) method. A careful survey of the literature indicates that the method is employed in a rather subjective manner and is not based upon a solid theoretical background. In this paper, a method for reducing the impulse response data is presented. The outcomes of field and laboratory testing are also reported. From these results, the proposed methodology seems quite simple and has the potential for becoming an effective tool for detecting voids in rigid pavements.

KEY WORDS: nondestructive testing, impulse response, pavements, void detection, rigid, concrete

---

Deterioration of pavements is the main problem concerning present day highway engineers. One important element which contributes to the initiation of rigid pavement deterioration is voids or loss of support under the pavements. This is detrimental because of the subsequent reduction in the fatigue life of the pavement. To detect voids, nondestructive testing techniques are normally utilized.

Detection of voids based on the deflection methods is explained in Uddin et al. (1983). Several dynamic deflection devices including the Dynaflect, the Road Rater, and the Falling Weight Deflectometer have been used to detect voids. It has been found that the deflections are significantly influenced by temperature differentials in the slab especially at the pavement edge and corner. Based upon extensive field results, Ricci et al. (1985), enumerated problems that may exist in the evaluation of rigid pavements using the deflection-based equipment.

The IR method (also known as Transient Dynamic Response method and Mechanical Impedance method) has also been used to evaluate the subgrade support of concrete slabs by several consulting groups. According to

---

[1] Associate Professor, [2] Graduate Research Assistant, [3] Research Faculty, Center for Geotechnical and Highway Materials Research, University of Texas at El Paso, El Paso, Texas 79968-0516.

these groups, the method can identify comparatively good, questionable, and void/poor support conditions of slabs. However, the state-of-practice is not robust and sometimes yields unsatisfactory results. Some improvements in reducing and interpreting the impulse response test results are presented. The main objective of this paper is to demonstrate the feasibility and sensitivity of this improved IR method to determine voids below pavements.

A detailed description of field procedure is discussed herein. Fundamental assumptions made to interpret the results of this testing technique are discussed. The step-by-step analysis procedure is also demonstrated. Results of field and laboratory testing programs are reviewed. Based upon limited field testing, the proposed methodology seems quite promising as an effective tool for detecting voids under rigid pavements.

THEORETICAL BACKGROUND

<u>Fundamental Principles</u>

To conduct the IR test in situ, the surface of the pavement is impacted. The response of the pavement as well as the impacted energy to the pavement are measured with a geophone and a load cell, respectively. The force and displacement signals are captured, digitized and processed to develop a flexibility spectrum (i.e. the ratio of the displacement and load as a function of frequency). The flexibility spectrum is utilized to diagnose the existence of voids.

To analytically model the impulse response testing procedure, several parameters should be considered. Fundamentally, the problem can be categorized as a flexible slab with finite dimensions placed on top of a layered elastic half-space. To construct such a model, a sophisticated finite element algorithm should be utilized. The dynamic nature of the loading must also be considered. The pattern of wave propagation, in terms of trapped and transmitted energy, contributes to the response of the system. Such an elasto-dynamic three-dimensional model is of great value in understanding the impulse response testing. However, the computation time necessary for solving this problem in practical cases is prohibitive.

At the other end of the spectrum, the most simplified solutions have been achieved by assuming that the slab is rigid, round, and uniformly loaded. Gazetas (1983) has summarized many of these approaches. The slab placed on a supporting soil is approximated by a single-degree-of-freedom mass-spring-dashpot system. The stiffness of the spring and the damping of the dashpot are directly related to the properties of the material under the slab as well as the dimensions of the slab. This approach can be implemented quite rapidly. However, the solution may be oversimplified and the results may not be representative of actual field cases.

A hybrid approach has been utilized to model the flexibility of the slab. The dynamic nature of the problem has been considered utilizing the circular rigid slab mass-spring-dashpot system approach. However, this solution has been modified in two aspects. First, the shape factors suggested by Dobry and Gazetas (1986) are incorporated to account for the rectangular shape of the slab. Second, a newly-developed factor considering the variation in the stiffness of a slab as a function of the location of impact has also been used. These parameters are described in the following sections.

<u>Dynamic Response of a Slab</u>

Many factors affect the dynamic response of a slab. The major parameters to consider are: 1) modulus of subgrade, 2) Poisson's ratio of subgrade, 3) planar dimensions (length and width) of the slab, and 4) mass density of the soil.

The dimensions of the slab are typically known. The values of
Poisson's ratio and mass density are typically assumed and not measured.
The most significant parameter is the modulus of subgrade--one of the
parameters of interest in this study.

Other site-specific parameters affect the dynamic response of a
slab as well. These parameters include: 1) depth to bedrock, 2) depth
of embedment of the slab, 3) shape of foundation, 4) the ratio of
modulus of base and subgrade, 5) damping properties of soil, 6)
anisotropy of the soil, and 7) the rigidity of the slab.

The depth to bedrock and anisotropic properties of subgrade may
affect the measured response of the slab. However, for simplicity, they
are not considered in this study. Typically paving slabs are not
embedded; therefore, this parameter was also ignored. The shape of the
slab affects its response and was incorporated here. Finally, the
rigidity of the slab, which is of utmost importance in detecting voids
and foundation softening was also considered.

Dynamic analysis of a slab can be carried out using the dynamic
flexibility, $A(\omega)$, as a function of frequency, ω (Gazetas 1983). For a
harmonic excitation, $F = F_{max}e^{i\omega t}$, the vertical displacement can be
defined as $D = D_{max}e^{i\omega t+\phi}$. The dynamic flexibility is defined as the
ratio between the displacement, D, and the input force, F, of the slab.

Many researchers (see Gazetas 1983 for historical development)
have established an analogy between the dynamic response of a SDOF
oscillator and a slab-soil system. The key characteristics of SDOF as
pertinent to this study are shown in Fig. 1. Three parameters are
required to describe a SDOF system -- natural frequency, damping ratio
and gain factor. The last two can be replaced by the static amplitude
and the peak amplitude. These three parameters are collectively called
the modal parameters of the system. To successfully conduct tests,
these three parameters should be accurately determined.

The flexibility spectrum is defined as the ratio of the Fourier
transform of the system response and system input, with all initial
conditions being zero. The flexibility spectrum is, therefore:

$$A_v(\omega) = \frac{D(\omega)}{F(\omega)} = \frac{A_s}{1 - \left(\dfrac{\omega}{\omega_n}\right)^2 + i2\beta\dfrac{\omega}{\omega_n}} \tag{1}$$

where A_s = static flexibility, ω_n = undamped natural frequency, and β
= damping ratio.

According to Lysmer et al. (1965) the static flexibility, A_s, for
a rigid circular slab can be written as:

$$A_s = \frac{1 - \nu}{4GR} \tag{2}$$

where G = shear modulus of the subgrade, $R = (A/\pi)^{0.5}$ = equivalent
radius, A = area of the slab, and ν = Poisson's ratio of soil (assumed).

<u>Considerations for Shape of Slab</u>

Most slabs are not circular in shape. The usual practice for
transforming any irregular shape of the slab to a circular slab is by
taking an equivalent circle of the same area (translational modes) or by
taking an equal moment of inertia (rotational modes) (Whitman and
Richart 1967). However, Dobry and Gazetas (1986) suggest that this
practice is not necessarily correct, and that the aspect ratio L/B can
have an important influence on dynamic stiffness and damping. Based
upon analytical and experimental studies, they suggested a shape

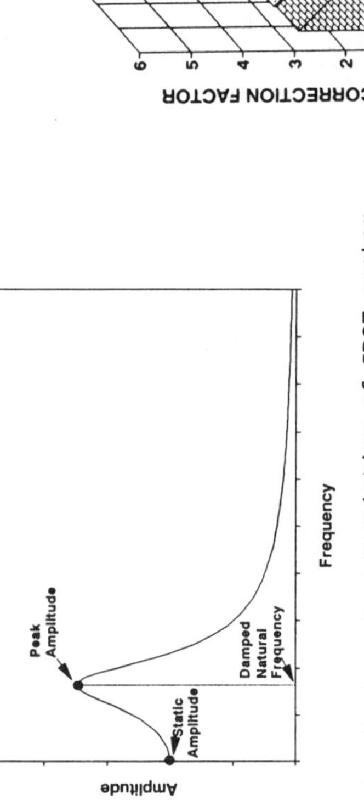

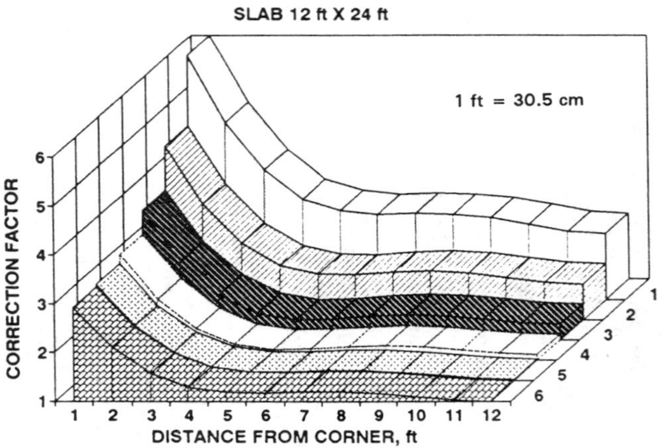

FIG. 3--Variation in static flexibility with distance.

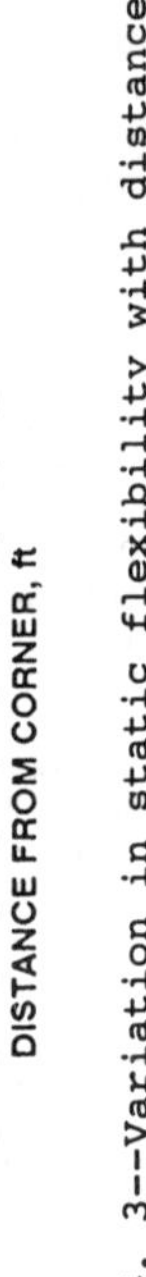

FIG. 1--The key characteristics of SDOF system.

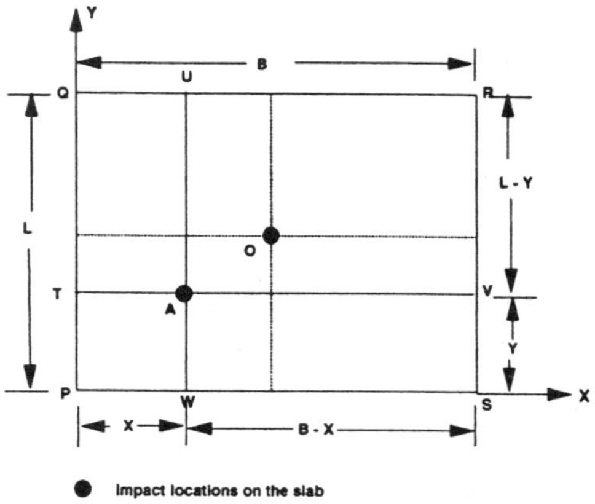

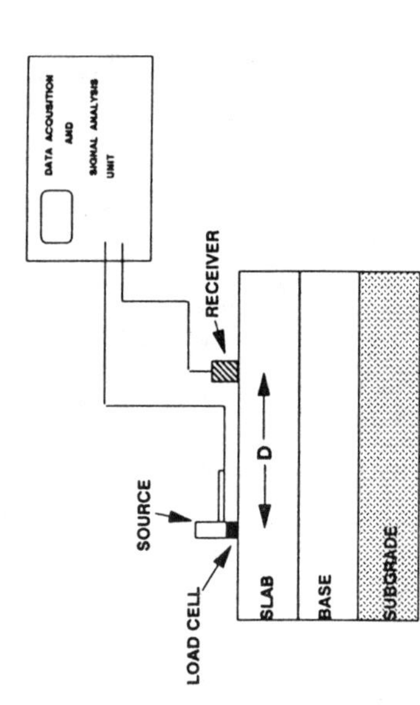

FIG. 4--Schematic of Impulse Response test.

FIG. 2--Planar view of a typical slab.

correction factor, S_v. This correction factor is simply applied to the solution obtained from the equivalent circular slab. The dimensionless static stiffness parameter (shape factor) for vertical vibration is obtained from:

$$S_v = 0.73 + 1.54 \left(\frac{A}{L^2}\right)^{0.75} \tag{3}$$

where A is the area of the slab, and L is the length of the slab. For a rectangular shaped slab, $A/L^2 = B/L$, where B is the actual width of the slab.

The modulus of subgrade, G, is then calculated from (Dobry and Gazetas 1986):

$$G = \left(\frac{1 - \nu}{2\,LA_s S_v}\right) \tag{4}$$

Considerations for Point of Impact

The calculations using the equations in the above sections are done assuming that the slab is rigid. In practice, however, most slabs are flexible. Therefore, it is necessary to consider the effects of this flexibility on the stiffness of paving slabs.

An intact flexible slab, as shown in Fig. 2, is impacted at two points. Point C corresponds to the center of the slab and Point A to an arbitrary selected point with coordinates x and y relative to a corner of this slab. It is intuitive that the closer the point of impact is to a corner of the slab, the higher the flexibility will be. The maximum flexibility is measured at the corner of the slab and the minimum flexibility (maximum stiffness) is obtained at the center of the slab. This variation in flexibility with the point of impact introduces a large ambiguity in the interpretation of the impulse response test results.

The variation in the flexibility of a typical slab with the location of impact is shown in Fig. 3. The flexibility ratio of at a given point is obtained by dividing the flexibility at that point by the flexibility of the center. The flexibility of the corner is as much as six times greater than that of the center. It is obvious that if this parameter is not included in the interpretation of the results only significantly large voids may be detected. A correction factor has been developed as part of this study and is described here.

To overview the method, the goal is to transform the measured flexibility at a given point to the "equivalent flexibility" at the center of the same slab. In this manner, if the equivalent flexibility of a measured point is close to that of the center, it can be assumed that the slab is intact at the measured point. Conversely, if a large difference exists between the actual flexibility of the center and the equivalent flexibility of the point, the slab will probably contain a void. In the rest of this section, the discussion will be focused towards the determination of equivalent flexibility.

According to Timoshenko and Goodier (1951), if the relative flexibility of the corner or center of a flexible slab (as compared to a rigid one) can be determined from:

$$I_f = I_1 + \left(\frac{1 - 2\nu}{1 - \nu}\right) I_2 \qquad (5)$$

The influence factors I_1 and I_2 can be computed using equations given by Steinbrenner (1954). These two influence factors are functions of the length and the width of the slab as well as the depth to a rigid layer in the subgrade underlying the slab. Given the fact that the subgrade is assumed to be an elastic half-space, the selection of depth parameter is rather ambiguous. This value was selected based upon an extensive finite element study specifically performed for calibrating the hybrid model. The detailed derivation of this parameter can be found in Reddy (1992).

The correction factor needed to determine the equivalent flexibility can be determined from:

$$I_s = \frac{I_{fxy}}{I_{fc}} \qquad (6)$$

where I_{fxy} = Flexibility factors for co-ordinate (x,y) from Equation 5 = $I_{f1} + I_{f2} + I_{f3} + I_{f4}$, I_{f1} = Influence factor for rectangle PTAW, I_{f2} = Influence factor for rectangle TQUA, I_{f3} = Influence factor for rectangle URVA, I_{f4} = Influence factor for rectangle AVSW, and I_{fc} = Flexibility factor for center point from Equation 5.

The factor I_s is basically the ratio of the relative flexibility of Point A (with coordinates x,y) to the relative flexibility of the center of the slab. In more practical terms, by dividing static flexibility by I_s the equivalent flexibility is obtained. It should be noted that Equation 5 is only applicable to the corner or the center of the slab. By utilizing the superposition principles, the formula was applied to the four rectangles having Point A in common.

PRACTICAL ASPECTS

To implement the impulse response testing properly, several steps should be taken. These steps are:
1) conducting field tests, to determine the response of the system,
2) extracting the modal parameters from the response,
3) determining engineering parameters from the modal parameters, and
4) predicting the existence of possible defects from the engineering parameters.

<u>Field Testing</u>

The schematic of the impulse response test on pavements is illustrated in Fig. 4. Typical test equipment includes an impulse hammer, a geophone (vertical velocity transducer), and a dynamic signal analyzer. The stress wave energy is imparted to the pavement with the impulse hammer. The impulse hammer is a handheld or machine mounted weight, which is instrumented with a load cell to measure the force of an impact. Due to the impact, a spherical wavefront is generated. When this wavefront arrives at the interface between the surface layer and another material (such as the subgrade), a portion of the energy is transmitted into the second material and the remainder is reflected back into the surface layer. In the case of voids almost all of the energy is reflected. After the impact, the load time-history of the hammer blow and the deformation time-history are simultaneously monitored and measured using the dynamic signal analyzer.

Typical force and velocity time-history records captured simultaneously on a concrete slab are shown in Fig. 5. The force time history shown in Fig. 5a can be clearly divided into three distinct regions. First, an initial zone, to the left of point A with approximately zero amplitude, corresponds to the so called pre-triggering event. The pre-triggering portion of a record illustrates the relative level of background noise before the actual hammer hit. The second section, located between Points A and B, corresponds to the actual imparted load due to the hammer hit. The duration of impulse is typically about 2 msec. Finally, the section to the right of Point B illustrates the background noise again.

The velocity time history (Fig. 5b) shows a pre-trigger delay similar to that of the force time history. However, after the so-called initial response due to the impact, the steady-state response of the system is evident. The amplitude of this steady-state response decays with time. The decay in amplitude is generally a function of the degree of contact between the slab and the soil beneath.

In the next step, the two time records shown in Fig. 5 are Fourier-transformed to obtain the frequency-domain representation of these records. The Fourier transform allows us to decompose a complicated time record into several hundred steady-state sinusoidal waveforms.

The velocity spectrum is divided by the force spectrum to obtain the mobility (response) spectrum. The mobility spectrum, which is the particle velocity normalized to the impact force as a function of frequency, is a complex valued quantity. To determine the flexibility spectrum, the mobility spectrum is integrated. A flexibility spectrum, corresponding to the time records shown in Fig. 5, is plotted in Fig. 6. The flexibility spectrum demonstrates the characteristics of a single-degree-of-freedom (SDOF) system. To determine the modal parameters, a curve is fitted to the flexibility spectrum. On a linear scale, the magnitude of the flexibility contains one prominent peak at 62.5 Hz. Basically, the flexibility spectrum is quite similar to the response of a lightly damped system.

DETERMINATION OF MODAL PARAMETERS

To obtain the modal parameters, an appropriate curve fitting process is required. The main goal is to determine the static flexibility, peak flexibility and the frequency at which the peak occurs (damped natural frequency).

A curve-fitting algorithm, which uses the coherence function as a weighing function is used (Richardson and Formenti 1982). For an SDOF, the curve can be theoretically described as:

$$A_v(\omega) = A_s \frac{\sum_{i=1}^{m} (S - Z_i)}{\sum_{i=1}^{n} (S - P_i)} \tag{7}$$

where $A_v(\omega)$ = flexibility at a frequency, S = Laplace operator = $j(2\pi f)$, Z_i = ith zero, P_i = ith pole, n, m = number of poles and zeros (2 for a SDOF), and A_s = gain factor.

Once the poles and zeros are known, damping ratio and natural frequency can be calculated. For an underdamped SDOF system, the two poles are complex conjugates of one another. The real and imaginary

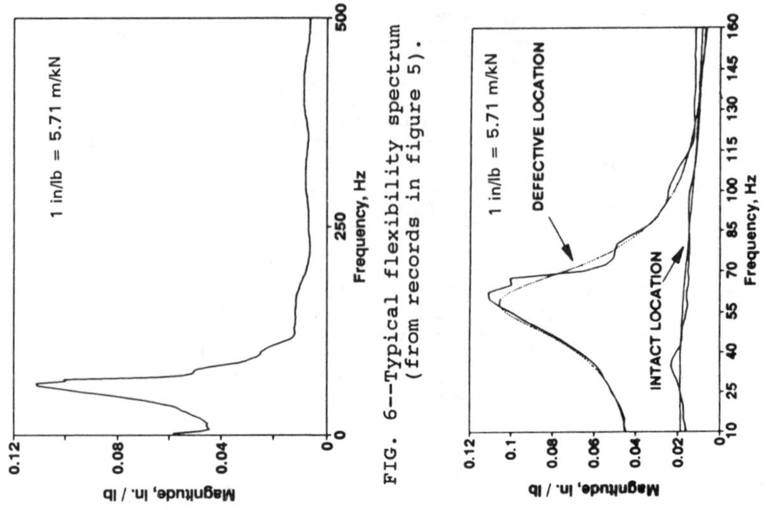

FIG. 6--Typical flexibility spectrum (from records in figure 5).

FIG. 7--Typical flexibility spectra from intact and defective pavements.

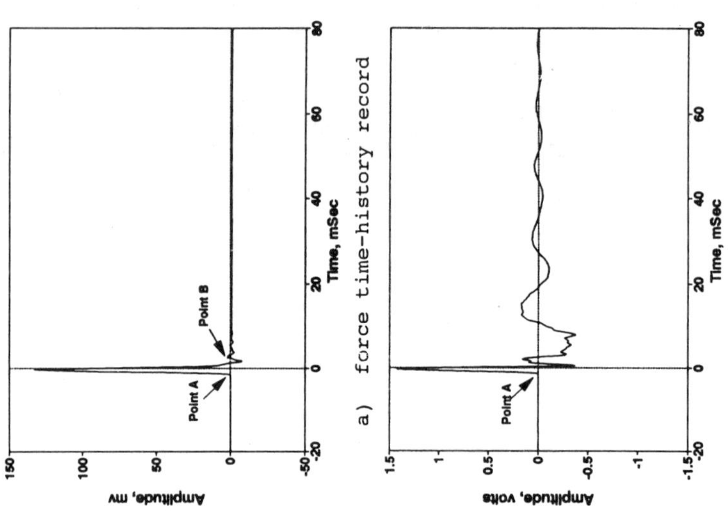

a) force time-history record

b) velocity time-history record

FIG. 5--Typical time domain records from an impulse response test.

components of these poles are related to the damping ratio and natural frequency of the systems by:

$$\text{Real } (P_i) = \omega_n D \tag{8}$$

$$\text{Imag } (P_i) = \omega_n (1 - D^2)^{0.5} \tag{9}$$

These two relationships can be manipulated to:

$$\beta = \sqrt{\frac{R^2}{1 + R^2}} * 100 \tag{10}$$

and

$$\omega_n = ABS\left[\frac{Real\, P_i}{D}\right] \tag{11}$$

where R = Real (P_i)/Imag (P_i). Typical records demonstrating the curve fitted data at an intact point and a point containing void are shown in Fig. 7. The two curves exhibit significantly different trends.

ENGINEERING INFORMATION EXTRACTED

Two engineering parameters are obtained from the modal parameters -- modulus of subgrade and damping ratio of the system. These two parameters are then utilized to characterize the existence of several distress precursors such as voids below the pavement. As indicated before, the modulus of subgrade, G, is calculated from:

$$G = \frac{(1-v)}{[2\, LA_{eqv} S_v]} \tag{12}$$

where A_{eqv} = (A_s/I_s) is the equivalent flexibility.

INTERPRETATION

Two numerical values are utilized to interpret the state of a given slab, damping ratio and modulus of subgrade. Table 1 shows a broad relationship between the quality of the slab and the two parameters.

The damping ratio, which typically varies between zero and 100 percent, is an indicator of the degree of resistance of the slab to movement. A slab that is in contact with the subgrade demonstrates a highly-damped behavior and typically has a damping ratio of greater than 70 percent. A slab containing an edge void would demonstrate a damping ratio on the order of 10 to 40 percent (depending on the extent of the void). A loss of support located in the middle of the slab will have a damping of 30 to 60 percent.

The subgrade modulus is related to two parameters: rigidity (combination of slab modulus and thickness) and the support from the subgrade. A high subgrade modulus obtained from Equation 13 is an indication of either a relatively rigid slab or a good subgrade support.

Conversely, a low subgrade modulus can be due to either a void under the slab or a weak slab.

To properly interpret the results, the combination of these two parameters should be utilized. As indicated in Table 1, a high modulus value with a low damping ratio can be interpreted as a defect-free, intact slab. A high modulus value with a high damping ratio typifies a good slab with not much support from the subgrade. If the modulus is low but the damping ratio is also low, a void should be expected. Finally, a low modulus, with a high damping ratio corresponds to a poor slab over a good support. Conversely, a small modulus value indicates a poor support or low quality concrete.

TABLE 1--<u>Relation between damping and normalized modulus to the existence of Distress</u>.

Normalized Modulus	Damping Ratio	
	Low	High
Low	Void	Fair slab and poor support
High	Good slab and good support	Strong slab and weak support

CASE STUDIES

To gain more confidence in the methodology presented here, a series of field and laboratory investigations was carried out. The results of these are discussed next.

EXPERIMENTAL SECTION

A small slab was constructed in the proposed location at a new parking lot near the UTEP campus. The resident soil was 1.8 to 3.0 m (6 to 10 ft) of a clayey fill placed on top of a weathered andesite. Before the construction of the model, the modulus profile of the resident material was determined with the SASW tests. Based upon these test results, the first 1.2 m (4 ft) of the material has a modulus of 70 MPa (10 ksi). Below a depth of 1.8 m (6 ft), the material has a modulus of about 520 MPa (75 ksi) corresponding to the andesite.

A plan view of the slab is shown in Fig. 8. A 1.22 m X 2.44 m X 14 cm (4 ft X 8 ft X 5.5 in.) slab was poured. Two 0.3 cm (1/8 in.) thick plates were placed at two edges of the slab. Both plates were thoroughly greased before placement. The first plate (Defect A in Fig. 7) was 30 cm (1 ft) square; the second plate (Defect B) was 61 cm (2 ft) square. A third defect (Defect C) was placed in the middle of the slab. This defect was made of stiff styrofoam. The plates were hammered out 72 hours after the placement of the concrete. Unfortunately, the hammering of the plate resulted in the complete separation of the slab and the subgrade in the lateral directions. This eliminated the possibility of comparing the results over the Defects A and B with the control (intact) section. However, the overall repeatability of the testing technique could be determined.

The slab was tested for three consecutive weeks. The first series of tests was carried out after three days of curing. At this time, the modulus was measured as 10 GPa (1460 ksi) (using seismic methods). The second and third series were carried out after 10 days (modulus of 31 GPa (4450 ksi)) and 17 days (modulus of 38 GPa (5470 ksi)), respectively.

The repeatability of the impulse response tests was investigated by comparing moduli of subgrade measured at three different times. The

center defect could not be detected because the modulus of the stiff
styrofoam was close to the modulus of the subgrade. Shown in Fig. 9 is
the variation in modulus along a longitudinal line 15 cm (0.5 ft) from
the edge of the slab (Line a). The modulus values are in good agreement
from all three sets. The values from the first set are consistently
lower because the slab was tested when the concrete was three days old.
The results from the other two sets are similar. As indicated before,
the rigidity of the slab plays a role in the calculated subgrade
modulus. This is well reflected in this study as the modulus of
concrete increases, the subgrade modulus calculated increases.

Based upon an extensive finite element analysis, Reddy (1992),
proposed the following relationship for relating variation in modulus
with the variation in static flexibility:

$$\frac{A_{s1}}{A_{s2}} = \frac{3.079 - 0.254 \, lnE_1}{3.079 - 0.254 \, lnE_2} \qquad (13)$$

where A_s and E are the static flexibility and modulus of the slab
respectively. If Equation 13 is applied to the results from this study,
a 30 percent increase in the flexibility should be obtained when the
results from the tests after 3 days are compared with those of 10 days.
This value is close to that obtained from the experimental result (see
Fig. 9). Equation 13 predicts a 5 percent difference in the results
obtained after 10 days and 17 days. This is well within accuracy and
precision of the method. As such the two results cannot be
distinguished.

The damping ratios obtained from the three sets of data are
compared in Fig. 10. The damping ratios (except for one point) are
quite similar. Once again demonstrating the consistency and
repeatability of the method.

FIELD TESTING

A series of diagnostic tests was carried out on a section of
Highway US 59 in Polk County, Texas. The highway consisted of two 3.7 m
(12 ft) wide lanes. A 3.0 m (10 ft) wide shoulder along the outside
lane and a 1.2 m (4 ft) wide shoulder along the median. The pavement at
this section consisted of about 18 cm (7 in.) of AC, over an 18 cm (7
in.) layer of PCC, over subgrade soil. The section had experienced
extensive distress. It had been extensively repaired before the overlay
was placed. The reason for testing this pavement section was to
determine if the deterioration and loss of support under the slab could
be measured.

Three adjacent slabs were tested. The three sections were
selected based upon the results obtained from the FWD deflection basins
measured early in the morning on the day that tests were carried out.
The slabs are called Slab 13, Slab 14 and Slab 15 for sake of
simplicity. The condition of these three slabs at the time of testing
is shown in Fig. 11. Slab 14 contained a reflective crack initiated
from the south edge and disappeared roughly 1.5 m (5 ft) from the south
edge. The pattern and location of the crack were in good agreement with
those of a repaired crack in the PCC layer before the placement of the
overlay. No other distress was evident from the surface.

Six lines, marked as 1 through 6 in Figure 11, were tested. For
each line, tests were started 60 cm (2 ft) from one edge and were
terminated 60 cm (2 ft) from the other edge to avoid the tapered edges.
At this time, reduction and interpretation of data over a tapered edge
is not possible. Tests were carried out at 60 cm (2 ft) intervals.

Line 1 corresponds to the middle of Slab 15. No distress was
apparent at this point. The variation in subgrade modulus with spacing

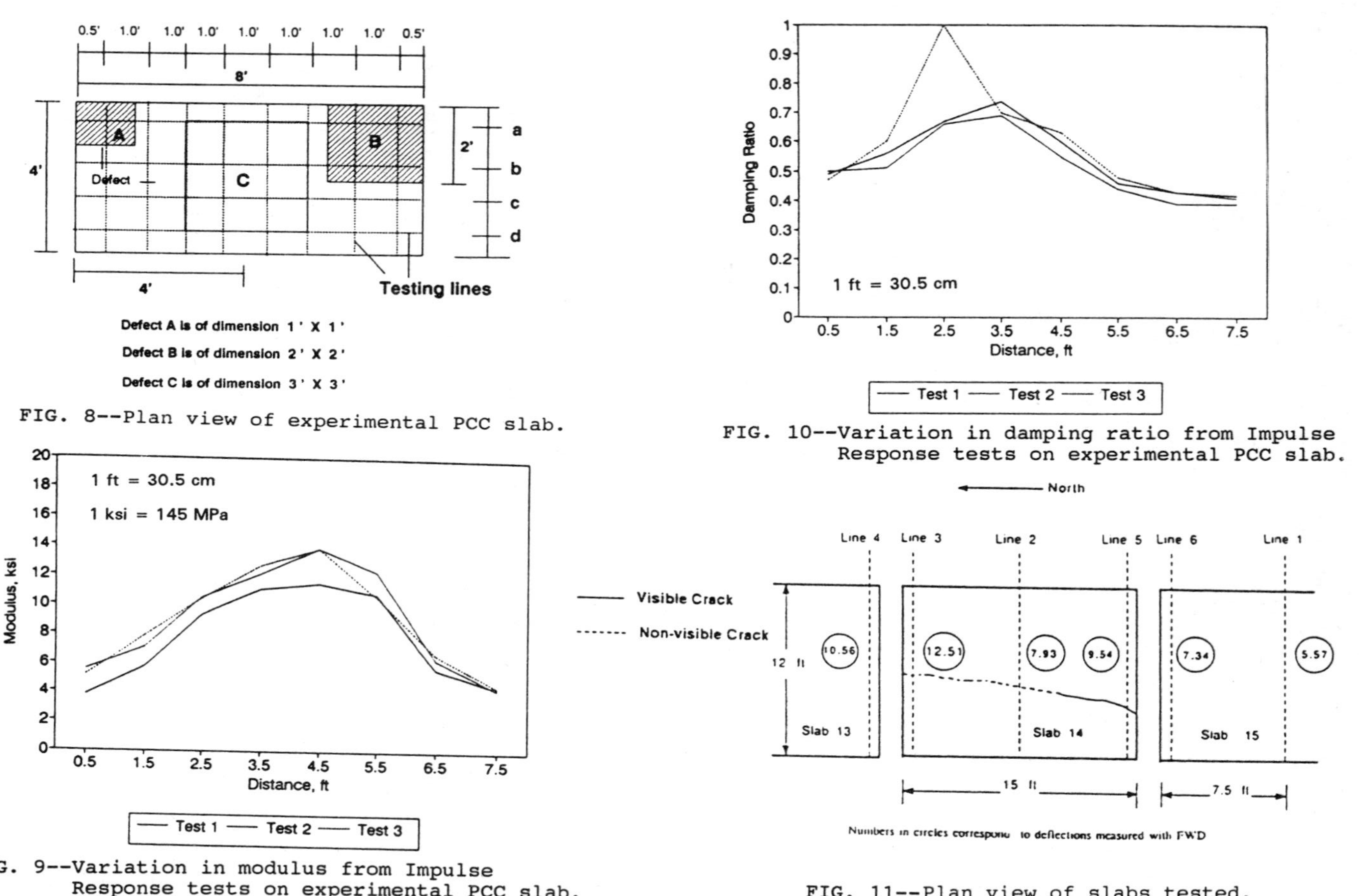

FIG. 8--Plan view of experimental PCC slab.

FIG. 10--Variation in damping ratio from Impulse Response tests on experimental PCC slab.

FIG. 9--Variation in modulus from Impulse Response tests on experimental PCC slab.

FIG. 11--Plan view of slabs tested.

is shown in Fig. 12. "Edge" corresponds to the edge away from the
median. It can be seen that the modulus is almost constant and small
variation is evident in the data.

Modulus variation along line 2 is also shown in Fig. 12. Decrease
in moduli of more than 50 percent is evident about 1.8 m (6 ft) from the
edge. This region is in good agreement with the location of the repaired
crack under the overlay. Once again, the crack was not exposed at this
point.

Shown in Fig. 13 is the variation in modulus along line 3 (i.e.
northern edge of Slab 14). The result from this line is relatively
similar to the midslab. The existence of the weak pavement along the
center of the slab was predicted correctly; even though not exposed at
the surface.

Line 4 corresponds to the south edge of Slab 13. This line is
about 30 cm (1 ft) away from Line 3. However, the two lines are located
on the edges of two adjacent slabs. Moduli obtained along this line are
also shown in Fig. 13. This side of the joint is far stiffer and in
better contact with the subgrade. Reduction in modulus around the
midpoint of the slab is apparent. In our opinion, this corresponds to
the weakening of the subgrade at this point. Even though not yet
manifested we believe that cracks should appear at this section shortly.

Line 5 corresponds to the southern edge of Slab 14. As indicated
before, reflective cracks were evident at this point. Variation in
modulus with distance is shown in Fig. 14. The extremely low modulus at
the spacing of 1.2 m (4 ft) corresponds to the location where the crack
was exposed. The extremely high modulus along the sides of this crack
indicates that the crack had propagated through the concrete. The
cracked concrete was seated intimately over the subgrade.

Finally, Line 6 corresponds to the northern edge of Slab 15 (i.e.
the opposite side of the joint from Line 5). Based upon measured moduli
(see Fig. 14), this slab is in good contact with subgrade and small
variation in modulus is apparent.

SUMMARY AND CONCLUSIONS

A new methodology for quantifying and interpreting the impulse
response data is introduced. Detailed explanation of the Impulse
testing is included as well. This new methodology is used in testing
several experimental models and actual field sites.

Several preliminary conclusions can be drawn, based on
experimental and field testing. Impulse response testing is quite
successful in detecting voids. When the effects of the shape of the
slab and the location of impact on flexibility spectrum are taken into
consideration, the test results improve significantly.

ACKNOWLEDGEMENTS

This work is funded by the Strategic Highway Research Program
(SHRP) under Project H-104B. Mr. Brian Cox is the Project Manager. The
opinions expressed here are those of the authors and do not necessarily
reflect the views of SHRP.

REFERENCES

Dobry, R., and Gazetas, G., 1986, "Dynamic Response of Arbitrary Shaped
Foundations," _Journal of Geotechnical Engineering, ASCE_, Vol. 112, No.
2, New York, NY, pp. 109-135.

Gazetas, G., 1983, "Analysis of Machine Foundation Vibrations: State of
the Art," _Soil Dynamics and Earthquake Engineering_, Vol. 2, No. 1, pp.
1-42.

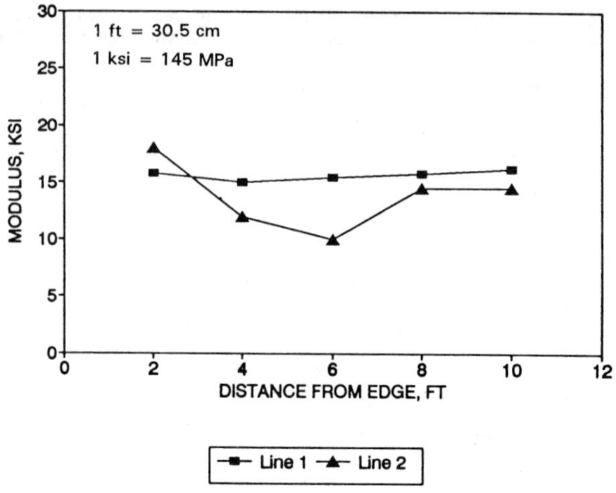

FIG. 12--Variation in modulus along Lines 1 and 2.

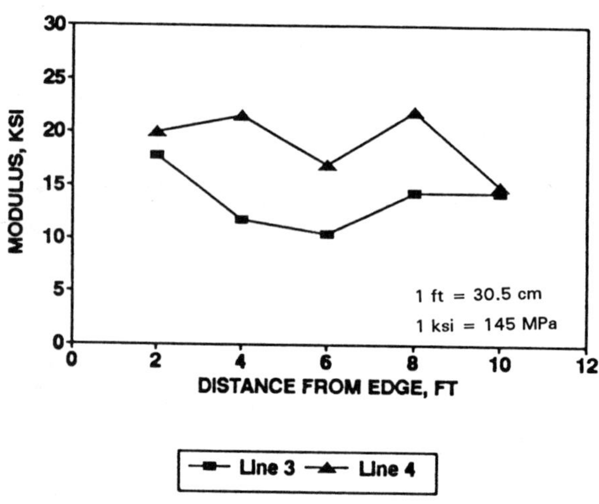

FIG. 13--Variation in modulus along Lines 3 and 4.

Lysmer, J., 1965, "Vertical Motion of Rigid Footings," Ph.D. Dissertation, University of Michigan, Ann Arbor, Michigan.

Reddy, S., May 1992, "Improved Impulse Response Testing Theoretical and Practical Validations," Master's Thesis, The University of Texas at El Paso, 335 pp.

Ricci, A. E., Meyer, A. H., Hudson, W. R., and Stokoe, K. H., II, November 1985, "The Falling Weight Deflectometer for Nondestructive Evaluation of Rigid Pavements," Research Report No. 387-3F, Center for Transportation Research, The University of Texas at Austin.

Richardson, M. H., and Formenti, D. L, 1982, "Parameter Estimation from Frequency Response Measurements Using Rational Fraction Polynomials," _Proceedings, First International Modal Analysis Conference_, Orlando, Florida, pp. 167-181.

Steinbrenner, W., October 1954, _Tafeln zur setzungsberechnung_, Die Strasse, Vol. 1, pp. 121-124.

Timoshenko, S., and Goodier, J. N., 1951, _Theory of Elasticity_, 2nd edition, McGraw Hill Book Co., New York, 506 pp.

Uddin, W., Torres, V., Hudson, W. R., Meyer, A.H., and McCullough, B.F., October 1983, Dynaflect Testing for Rigid Pavement Evaluation," Report No. CTR256-6F, Center for Transportation Research, The University of Texas at Austin.

Whitman, R. V., and Richart, F. E., Jr. 1967, "Comparison of Footing Vibration Tests with Theory," _Journal of Soil Mechanics and Foundations_, ASCE, Vol. 93, No. SM6, pp. 143-168.

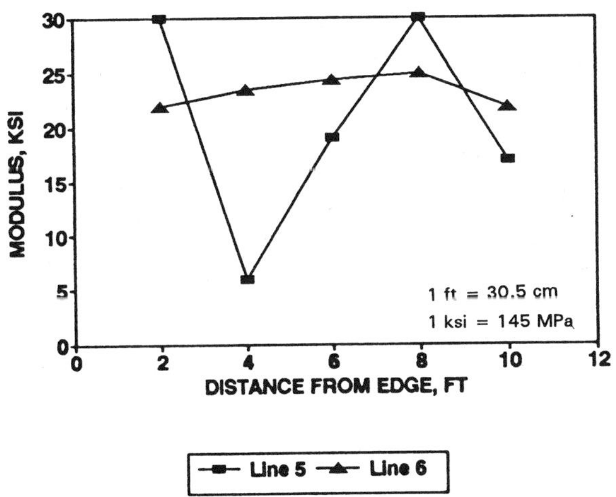

FIG. 14--Variation in modulus along Lines 5 and 6.

Waheed Uddin[1] and W. Ronald Hudson[2]

EVALUATION OF NDT EQUIPMENT FOR MEASURING VOIDS UNDER CONCRETE PAVEMENTS

REFERENCE: Uddin, W., and Hudson, R., **"Evaluation of NDT Equipment for Measuring Voids Under Concrete Pavements,"** Nondestructive Testing of Pavements and Backcalculation of Moduli (Second Volume), ASTM STP 1198, Harold L. Von Quintas, Albert J. Bush, III, and Gilbert Y. Baladi, Eds., Amercian Soceity for Testing and Materials, Philadelphia, 1994.

ABSTRACT: The partial loss of subgrade support associated with voids under portland cement concrete (PCC) pavements leads to increased deflections and increased load stresses. This can cause a significant reduction in the fatigue life of the pavement. This paper presents the findings of a comprehensive research study to establish the performance capabilities and limitations of the following methods and equipment for detection and measurement of voids beneath PCC pavements: (a) Proof Roller, (b) Deflection Equipment, (c) Ground Penetrating Radar Equipment, and (d) Transient Dynamic Response (TDR) Equipment. The records of grout quantity, based on field tests conducted by the highway agencies of several states during undersealing projects, were the prime source to verify the presence of voids.

All methods required extensive manual data analysis and output interpretation. Deflection methods are the least satisfactory because pavement deflections are significantly influenced by daily and seasonal variation of temperature. The TDR method is very labor intensive and relies on subjective interpretation. Radar methods hold good promise if the data interpretation and processing is enhanced.

KEY WORDS: pavement, PCC, voids, deflection, radar, dynamic, grout

BACKGROUND

The partial loss of subgrade support under portland cement concrete (PCC) pavements associated with voids leads to increased load stresses. This can cause a significant reduction in the fatigue life of a pavement. Void detection is important to rigid pavement condition evaluation, yet it is one of the most uncertain aspects of field

---

[1]Assistant Professor, Department of Civil Engineering, The University of Mississippi, University, MS 38677

[2]Professor of Transportation Engineering, The University of Texas at Austin, Austin, Texas 78712

testing. Voids beneath the pavement can vary in depth from as small as one thousandth of an inch, causing partial loss of support, to a much larger depth of several inches. The effect of any depth of void is detrimental on the performance of the PCC pavement since field deflections often measure 0.025 inches (0.63 mm) or less. Generally, the idea of measuring voids under a pavement leads one to the assumption that large cavities will be measured. However, void size (depth) may vary from the delamination state adjacent to joints to a considerably larger dimension. Estimation of the dimension of a void area beneath PCC pavements is important to calculate the quantity for undersealing work.

Unlike deflection testing for structural evaluation, which is a well defined area of pavement measurement, voids beneath PCC pavements are poorly defined and measurement results are difficult to evaluate properly. Void detection at the present time is in its technical infancy as was x-ray evaluation of the human body some 30 years ago. Nevertheless, the detect-ion of voids beneath concrete pavements is an important consideration in planning maintenance treatments for restoration of support to the slab and extending the life of a pavement. Void detection is also necessary to develop improved models for the prediction of loss of support and effec-tive k (modulus of subgrade reaction) over time which will lead to improved and more reliable design of PCC pavements.

Many of the devices used for measuring voids under pavements are significantly different in operating principles. Knowledge of the capabilities and limitations of available pavement condition monitoring devices can assist in the selection of devices to achieve consistent measurements. During 1985-1987 the Federal Highway Administration (FHWA) conducted a comprehensive pavement evaluation equipment study, "Pavement Condition Monitoring Methods and Equipment," to establish a better understanding of the procedures and devices used to monitor and evaluate pavements [1, 2, 3]. This study consisted of three equipment categories; (1) deflection measuring equipment, (2) distress survey equipment and methods, and (3) void detection devices. This paper describes the evaluation of equipment for measuring voids beneath portland cement concrete (PCC) pavements.

METHODS AND EQUIPMENT USED TO DETECT VOIDS

Visual Surveys

The visual surveys consist primarily of locating deposits of pumped material along the shoulder with the assumption the adjacent slab area has a void underneath it. The drainage characteristics of shoulder and pavement, both longitudinally and transversely, has an impact on the location of the deposits. Other visual clues of void presence include: ejection of water and fine material as trucks cross over the pavement, faulting of transverse joints (it can also be caused by differential settlement of slabs), and corner cracking. Visual surveys may be useful in augmenting the other techniques, but Birkhoff and McCullough [4] found that visual surveys alone provided a probability of only 50 percent for successfully detecting a void.

<u>Proof-Rolling and Static Deflection Equipment</u>

Proof-rolling with a heavy load is a traditional method to determine the presence of voids and the need for undersealing concrete pavements. This is accomplished by measuring the movement of a slab under a specified load on a slab corner. Axle loads of 18-kip (80 kN) or more are used by agencies for this purpose. A modified version of the Benkelman Beam has been used to measure total movement and the differential lift between the slabs in pre-grout surveys as well as to monitor slab lift during under-sealing projects [5]. Arkansas has used a 25-ton roller for the load [3].

Testing every joint between midnight and 10 a.m. minimizes problems with joint lock-up due to thermal expansion, and conversely measures at the point of lowest load transfer. A reading is made at the corner of a joint or crack, and then the load is rolled to a point near the joint or crack and another reading is made. Georgia specifications [6] state that undersealing is needed if, with an 18-kip axle, deflections in the range of 0.015 inch (0.28 mm) to 0.03 inch (0.76 mm) are produced. Post-undersealing deflections that exceed the pre-testing trigger level are undersealed again. CALTRANS found that 18-kip (80 kN) deflection measure-ments do not correlate with the need for, or extent of, undersealing [7].

<u>Dynamic Deflection Equipment</u>

Several types of nondestructive testing (NDT) dynamic deflection equipment have been used to detect voids, including Dynaflects, Road Raters, and Falling Weight Deflectometers. Descriptions and operating principles of these devices are presented elsewhere [1]. In a Texas study, Torres and McCullough [8] found a combination of visual survey and normalized outputs of Sensors 1 and 5 of the Dynaflect to be an effective means of void detection. Uddin et. al. [9] recommend that; (1) the Dynaflect be placed about one foot (0.305 m) from the pavement edge when testing for voids, (2) a temperature correction be applied to Sensor 1 deflections, (3) deflection profiles measured at the centerline of the outside lane and at one foot from the outside edge be plotted, and (4) the areas susceptible to voids be marked on a relative basis.

Dynaflect deflections are also used in Ohio and Indiana for detecting voids under concrete pavements. Indiana measures deflections with a Dynaflect at mid-slab over a one mile length to get a base deflection measure. These values are averaged and compared to values obtained by testing on jointed pavements in the outer wheel path at approximately 100-foot (30 m) intervals, depending upon joint and crack location. Continuously reinforced concrete (CRC) pavements are measured at exactly 100-foot (30 m) intervals. Areas with deflections greater than the base deflection level are undersealed [10]. Heavier deflection devices may be more desirable for void detection because of the significant temperature effects on PCC slab curling. The following void detection procedures utilizing deflection data were developed in NCHRP Project 1-21 [11];

1. A rapid field testing procedure for on-the-spot indication of the existence of a void. It requires deflection measurements at slab corners under a sequence of three different load levels.

2. A procedure that uses deflection test and a void detection plot to estimate the horizontal size of the void at each joint tested.

Radar Technology

Radar equipment transmits electromagnetic energy which is reflected from the target. The use of radar technology for measuring voids under pavement appears promising. The short-pulse ground penetrating radar (GPR) has been used to identify the subsurface structure of soil, and to locate underground utilities, pipelines and tunnels. With refinements in antenna and other hardware developments, the technology has been applied to pavement. At the present time, the output of a radar unit requires the interpretation of a trained specialist. Problems in the interpretation of outputs have been reported when voids are full of water. Improvements could be made in the analysis software to obtain repeatable and accurate results.

Operation of Radar System

A radar system primarily consists of the following components: (a) Transmitter, (b) Isolator, (c) Antenna, (d) Receiver, (e) Signal processor, and (f) Display. The following section describes various components of a radar system [12].

a) The function of the transmitter is to generate a known waveform. The name "short pulse radar" is derived from the fact that the transmitted waveform of such a system is actually a very narrow pulse, which typically might last on the order of one-billionth of a second.

b) During the transmit cycle, the isolator provides a direct path from the transmitter to the antenna so that the great majority of the radiated energy is directed into the ground.

c) The electromagnetic wave transmitted by the antenna travels in the radiated direction until it strikes a discontinuity in the electro-magnetic properties of the media. Such discontinuity is almost always associated with a material change in the media. The first such discontinuity is associated with the air-ground interface. At this interface, a portion of the incident wave is scattered back away from the ground and a portion propagates into the ground. Next, the wave traveling within the ground strikes the next discontinuity (pavement-base interface or pavement void area). The portion of the wave reflected by the discontinuity (void under pavement) forms the basis for target detection and assessment.

d) Shortly after the pulse transmission, the isolator connects the radar receiver to this same antenna port for signal reception. In this receiving mode, the antenna collects the electromagnetic energy in the return reflection, or echo, and delivers it to the receiver. The receiver amplifies signals for subsequent processing and reduces the signal contamination produced by electrical noise and reflections ("clutter") from objects that are not of interest to the radar operator. Of the two primary signal contaminants, noise and clutter, clutter is more severe in typical ground-penetration applications and is the principle limitation of the radar system's ability to detect faint target echoes.

e) Following signal reception, the output of the receiver is then passed to the signal processor which extracts the desired information from the received signal.

f) The final radar system component is the display, which presents the information contained in the radar return signal in an appropriate format for interpretation by the operator. Figure 1 illustrates the interpretation of radar data based on computer simulation [12].

    Both Donohue & Associates, Inc., and Gulf Applied Radar have developed dedicated vehicles with radar systems for detection of voids under pavements. The non-contact antenna units are pushed along the pavement surface at low speeds during the radar survey.

GULF Applied Radar System   The radar system developed by the Gulf Applied Research Corporation of Houston, Texas is called RODAR Pavement Evaluation System [12]. The RODAR is a totally self-contained system. It has been used in several States for void detection, pre- & post-grout surveys, and delamination. The system is mounted on a conventional van and carries two non-contact antennas behind the vehicle, anywhere from 6 to 14 inches (0.15 to 0.36 m) off the ground. A crew of two qualified personnel operates the system at 10 mph (16 kmph) in routine operations. The system measures the depth and thickness of layers to approximately 15 inches (0.38 m), depending on the pavement materials. This system is capable of estimating void size from a horizontal stand-point, but the volume can only be roughly approximated [12]. The radar truck is also equipped with (1) a fifth wheel, which is the basic locating device, (2) a painting device that paints a reference mark approx-imately every 1000 feet (305 m) and (3) a standard video camera. The data is monitored by the operator and reduced in the field. A color graphics-based system had been developed to aid in data interpretation.

DONOHUE'S Remote Sensing Van   The remote sensing van, developed by Donohue & Associates of Waukesha, Wisconsin, is equipped with both the ground-penetrating radar and the infrared remote-sensing equipment. The major use of infrared thermography has been the detection of delamina- tions. The van also carries a video camera which is used to record surface conditions [13]. The greater resolution of the higher frequency transducers gives the ability to discriminate between closely spaced objects and interfaces. In operation, the unit travels at 2 to 4 mph (3 to 6 kmph). The electromagnetic pulse is repeated at a predetermined rate and the resultant stream of radar data is fed to the chart recorder where a continuous hard-copy profile of the data is produced as the transducer is moved along the surface. The data is also recorded on magnetic tape and monitored on an oscilloscope [13].

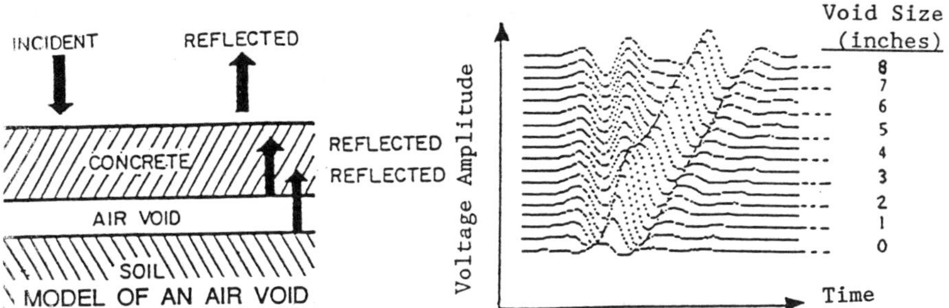

FIG. 1--Plot of calculated void reflection waveforms received by a
short-pulse radar as a function of void thickness [12].

Transient Dynamic Response (TDR) Method

    The Transient Dynamic Response (TDR) method is being used in the
United States by Test consult CEBTP Ltd. The TDR method is based on the
analysis of vibrations by the measurement of mechanical impedance [14].
This method consists of striking a load cell placed on the surface of
the pavement with a small instrumental hammer and recording the velocity
signals with a geophone held against the pavement. The geophone and
load cell are connected in the field to a spectrum analyzer and a
computer which computes a frequency response curve. The support
condition for a particular test point can be determined by the
continuous increases of the curve in the 0 to 800 Hz frequency range. A
void is identified by the down slope in this region. The test is
performed on several points in the slab marked as a grid pattern. The
final results are reported as maps, as illustrated in Figure 2. Mapping
of the horizontal extent of the void is done by determining the
interface between full support and loss of support on the surface of the
slab, which is manually mapped by a technician in the field. It is a
labor intensive method requiring the closing of a lane with the
technician walking around and testing different points on the slab. The
data interpretation is rather subjective and requires the expertise of
specially trained and experienced personnel.

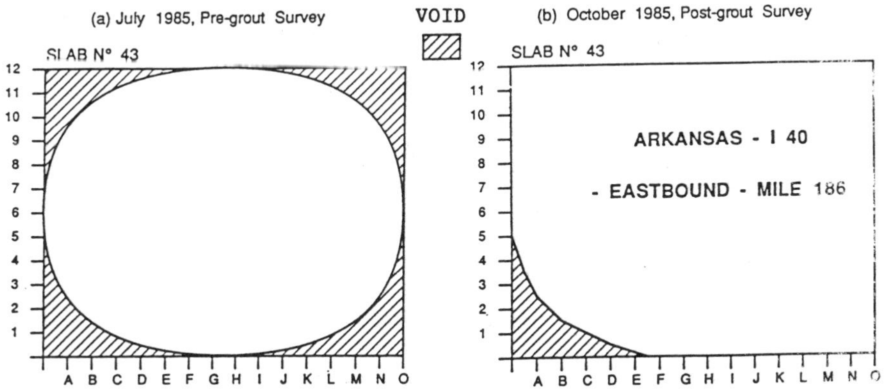

FIG. 2--Maps of void areas by the TDR method [20].

FEASIBILITY OF EQUIPMENT EVALUATION BASED ON FIELD STUDIES

Destructive Methods

The successful evaluation of void detection devices depends, in part, on our ability to verify the presence of voids on the test sites by independent means. The study of these various devices in a side-by-side test is somewhat more difficult than deflection testing. This is due to the difficulty in determining the actual location of voids beneath a pave-ment by independent means other than these devices. Some researchers have excavated trenches along the side of concrete pavements, cored through the pavement or removed portions of the slab to observe the voids.   These methods have had mixed success in observing the voids due to disruption of the base material and the difficulty in trying to determine where the voids were once the slab is removed. The methods are expensive, time consuming and disruptive to traffic.

Test Slab with Known Void Location and Sizes

Concrete test slabs built specifically for research with known void configurations, for example test slabs built at The University of Texas Balcones Research Center [15] and the Florida Department of Transportation research facility [16], can be used to determine if a device can disting-uish void shape and locations. The problem of selection of other sites is tied to the ability to determine actual void locations independently. The only other available independent means of verification of voids are the destructive methods mentioned earlier.

Undersealing Projects

One approach to selection of sites, in which independent verification of void presence can be made, is to try to find sites which are scheduled for some form of concrete pavement restoration (CPR), parti-cularly undersealing. It may be possible to join in a cooperative effort with a contractor and State DOT on a job scheduled for maintenance before undersealing to confirm the sites selected to be undersealed, and after undersealing has been performed to confirm that voids have been filled. In a cooperative effort the information was shared on a study being performed during 1986-1987 for the Federal Highway Administration under contract DTFH61-85-C-00120, "Materials and Methods for Undersealing Concrete Pavements" [17]. The shared information included field data for void detection and undersealing from various projects performed in several States. In addition, field data from studies performed by the Florida Department of Transportation was also acquired.

Office Evaluation of Selected Equipment

The following criteria were established to evaluate and compare the selected equipment; (a) principle of operation, (b) vehicle requirements, (c) operating characteristics, (d) field data collection and processing, and (e) office data processing. The researchers of this study [3] and the FHWA undersealing study [17] contacted several States

requirements, (c) operating characteristics, (d) field data collection and processing, and (e) office data processing. The researchers of this study [3] and the FHWA undersealing study [17] contacted several States where concrete pavement stabilization and void detection work has been performed in recent years. A summary of these surveys is presented in Table 1.

Table 1--<u>A summary of user surveys on the use of void detection equipment.</u>

States	Proof Rolling Visual	Benkel-man Beam	Dynaflect	FWD	Gulf Radar	Donohue Radar	CEBTP TDR
Arkansas	25 tons *					X	X
Florida		X		X	X*	X	
Georgia	X*						
Indiana			X*				
Illinois				X**			
North Carolina		X		X	X		
Oklahoma	X						X*
Tennessee	X				X		

* Preferred equipment ** And Road Rater
California: Deflection methods are not considered appropriate; radar was not satisfactory.

SYNTHESIS OF FIELD STUDIES

Undersealing of concrete pavements to fill up voids existing between the concrete slabs and the underlying base and subbase layer is generally an important rehabilitation technique for any PCC pavement restoration project. The estimate of undersealing quantity has been traditionally based on the use of historical averages for this type of work. However, in most cases, this crude estimate is far off the actual quantities. Deflection measurements by heavy proof rolling equipment or dynamic deflection equipment have been used by several States [3] to detect voids. These States include Georgia, Illinois, Indiana, Tennessee, Florida, California, Arkansas and North Carolina. Some States have made concentrated efforts to evaluate deflection tests and other available methods like ground penetrating radar and impulse test methods. Data provided by the Florida Department of Transportation [16, 18] and data collected in the FHWA undersealing study [17] are synthesized in this section to evaluate the effectiveness of various equipment used to detect and measure voids under PCC pavements. The Oklahoma data is taken from a published paper [19].

<u>Arkansas Field Study</u>

In the second half of 1985, several miles of Interstate 40 (I-40) in the eastbound and westbound directions, located in Prairie County, Arkansas, were surveyed by the Arkansas Highway and Transportation Department for undersealing operations [20]. The pavement in this study consisted of a 10-inch (254 mm) jointed reinforced concrete pavement (JRCP). The pavement was 24 feet (7.3 m) wide with two lanes in each direction. Doweled expansion joints were spaced at 45-foot (13.7 m) intervals and undoweled stress relief sawed joints were spaced at 15-foot (4.6 m) intervals. Several test sections were identified on this site. Data collected on the outside lane [20] from this section is discussed here for the purpose of comparing the various methods.

There were 89 transverse joints in the first test section. Pressure grouting record for the undersealing of this test section showed that 63% of all transverse joints accepted grout. The specification allowed for lifting of the slab not to exceed 0.125 inch (3.18 mm) total accumulative movement, measured at the outside joint corner.

Data Interpretation and Conclusions   The interpretation of the data obtained from field tests using proof rolling with a 51,500 lb (229 kN) roller, the Donohue Radar, CEBTP-TDR and grout record indicated that; (a) the deflection test data and TDR data appeared to be influenced by temperature variations; both daily and seasonal, (b) proof rolling data appeared to be inadequate for measuring voids, (c) the TDR method (pre-grout survey) showed 100% of the slabs as having voids and that 77% of the grouted slabs improved after grouting, (d) the Donohue-Ground Penetrating Radar showed 91% of the grouted slabs having voids during the pre-grout survey while no voids were noted for the grouted slabs during the post-grout survey. Figure 2 shows maps based on the data before and after grouting work for Slab #43. None of these methods were capable of determining the average thickness of the void area.

Florida Field Studies

The Florida Department of Transportation (DOT) has performed several concrete pavement rehabilitation projects on Interstate Highway 10 (I-10) during 1981-1986. The Florida DOT performed three major studies to evaluate ground penetrating radar and FWD devices in order to compare their performance in the field and identify a reliable method to locate voids.

o 1983 - A test slab measuring 12 feet (3.7 m) by 100 feet (30.5 m) with known void sizes beneath the slab was constructed in Gainsville and tested by GPR and FWD devices.

o 1984 - A section of I-10 near DeFuniak Springs was selected to evaluate the effectiveness of GPR (both Donohue & Associates and Gulf Applied Radar) and FWD devices for locating voids beneath the in-service PCC pavement.

o 1985/1986 - Gulf GPR Equipment and deflection surveys were performed on two sections of I-10 in Walton County to detect voids beneath PCC pavements and to compare the results of these methods.

The Florida DOT has generously provided all deflection data and radar survey reports from their two studies of in-service pavements [16 and 18]. Overall, the GPR radar equipment gave better estimates of voids beneath PCC pavements in these Florida studies than the deflection methods. Deflection tests were found to be inadequate and were affected by temperature variations within the slab.

<u>Data Interpretation and Conclusions</u>. To compare the methods, joints in the three test sections were selected where most of the data from every method was available. There was a concentration of loaded-axle deflection test "failures" in the early part of each morning's test [18]. Similar results were shown for the FWD tests. Deflections at slab corners were much higher at night than during the day. For this study, several of the FWD tests were analyzed using the NCHRP procedure [11]. Voids were indicated by radar and verified by cores at several spots where test rolling did not indicate failure. Results of the post-grout radar survey indicated the same voids seen in the pre-grout survey (which did not fail test rolling). The test rolling data was taken in January in the early morning when the slab could show upward curl at curves and edges due to negative temperature differentials. A recommendation was made to conduct a radar survey before undersealing [18] and it was concluded that if a radar survey had been performed prior to the undersealing contract, it could have been used to determine; (a) the number of slabs with indications of significant voids, (b) locations of these voids and the need for undersealing, and (c) a better estimate of grout quantities.

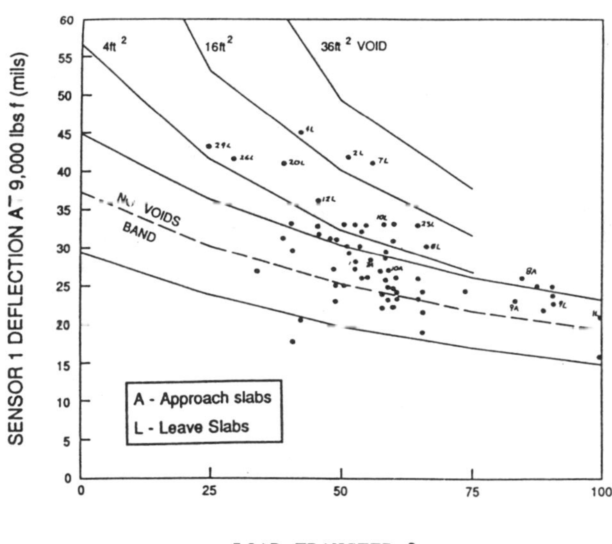

FIG. 3--Detailed analysis of FWD deflection data based on the NCHRP 1-21 method, North Carolina Study (21).

North Carolina Field Study

The North Carolina Department of Transportation (NCDOT) conducted
a field study of void detection equipment in 1986 as a part of their
pavement stabilization research project on I-40, West of Statesville,
North Carolina. The jointed pavement had 34 slabs in the test section.
Each slab was 30 feet (9.1 m) long. NCDOT conducted Benkelman Beam and
FWD tests and Gulf Applied Radar surveys in this study before and after
undersealing. Figure 3 illustrates an example of the FWD deflection
analysis for void detection. In addition, the FHWA undersealing study
researchers [17] collected Dynaflect deflection data along with the
other tests. Grouting flow and quantities were recorded and five slabs
were removed by NCDOT to examine the area beneath the slab for voids and
grout lenses. The NCDOT field data [21] was synthesized by the
researchers of this study.

Data Interpretation and Discussion. From grout data records and slab
removals, it was estimated that 21 out of 34 slabs (62%) accepted a
significant quantity of grout. Nine slabs did not accept any grout and
the rest had very little grout. As shown in Table 2, the radar
predictions appear to be more reasonable than the deflection methods
used in this study. Radar plots also indicated locations and
approximate size of voids in every slab.

Table 2--Comparison of equipment performance based on pre- and post-
grout surveys.

	Equipment	Pre-Grout Survey (slabs with void)	Post Grout Survey (Results)
(1)	Benkelman Beam	91%	65% (Slabs with grout or no voids)
(2)	FWD (NCHRP 1-21 procedure)	94%	88% (Slabs with grout or no voids)
(3)	Dynaflect (The University of Texas procedure)	64%	74% (slabs with grout or no voids)
(4)	Gulf Radar	88%	74% (26 slabs with grout)
(5)	Grout/Slab Removal Record	----	74% (25 slabs with grout)

Oklahoma Field Study

The Oklahoma Department of Transportation (OKDOT) selected a 13
mile (20.9 km) section on US 69 near Muskogee for stabilization in 1985
[19]. Three sites were identified in this section. Ninety consecutive
slabs from each of these test sites were tested for void detection by an
18-kip (80 kN) single-axle load deflection method and the TDR method.
The section consisted of a 9-inch (229 mm) plain PCC pavement with sawed
transverse joints at 15 feet (4.6m).

Within each site, 9 slabs were selected for repeat testing using

the TDR method and 15 slabs were selected for grouting and retest. One
slab from each site was removed and several cores were taken to verify
the results. Deflection measurement were made before and after
undersealing, using the single axle 18-kip (80 kN) truck and two dial
gauges. All deflection readings with the truck on site ranged from 5 to
10 mils (0.13 to 0.25 mm), indicating no voids. The TDR method was used
to test and map all slabs on the first site. The following day nine
slabs were retested and then tested again on the third day. The area of
void did not change significantly during the repeat testing. Fifteen
slabs which were stabilized with grout were also retested. The slab
with voids (predicted by the TDR method) accepted grout, and cores as
well as slab removal confirmed that the voids existed as mapped.

<u>Data Interpretation</u>.   The OKDOT report [<u>19</u>] concluded that: (a) the
axle-load deflection method can be meaningless because of low
deflections on some sites and (b) the TDR method proved successful for
void prediction and verification of grouting although it is labor-
intensive, costly and subject to environmental conditions. However, the
OKDOT report also mentioned that the areas of voids (mapped by the TDR
method) decreased with grouting but the voids reappeared when traffic
resumed. This may imply that void prediction by the TDR method may be
greatly influenced by environmental factors like temperature variation
within the slab.

Tennessee Field Study

     Deflection and radar surveys were used to identify voids beneath
PCC pavements on a site where undersealing for concrete pavement
rehabilitation was carried out in 1985 by the Tennessee Department of
Transportation(TDOT). The site was located on I-64 near Knoxville,
Tennessee. The highway consisted of 10 miles (16 km) of plain jointed
concrete pavement with transverse skewed joints at 18 to 25 feet (5.5 to
7.6 m) spacing. Several sections on eastbound and westbound lanes were
selected and tested. Before undersealing, proof rolling with a 21 ton
(187 kN) single-axle vehicle was conducted. This data was not
available. Gulf Applied Radar performed the radar survey. Grout could
not be pumped satisfactorily [<u>22</u>].

<u>Data Interpretation and Discussion</u>   Copies of Gulf's outputs showed
plans of lanes with the void marked. It was reported that voids found
by radar could be a result of honeycombing in the PCC slab which was
discovered when several slabs were removed [<u>22</u>].

SUMMARY AND CONCLUSIONS

     The primary thrust of this study was on evaluation of
nondestructive test methods for void detection, including deflection
measuring equipment (proof-roller, Benkelman Beam, Dynaflect, Falling
Weight Deflectometer, and Road Rater), Ground Penetrating Radar (GPR)
Equipment (Donohue and Gulf), and the Transient Dynamic Response (TDR)
method. Data from field studies conducted in Arkansas, Florida, North
Carolina, Oklahoma, and Tennessee in conjunction with pavement
undersealing work were gathered and reviewed. The effectiveness of each

method in predicting voids was evaluated using the synthesis of field studies, user surveys and a set of performance evaluation criteria. The evaluations presented in this report were influenced by several practical limitations, including budget and time constraints. There is no proven practical method to independently validate or prove void size and location. Major findings of each method are summarized below.

<u>Deflection Measuring Devices</u>   Use of the Beckelman Beam or dial gauges to measure pavement deflections under a heavy axle load or proof-roller has been a conventional method of identifying voids. Comparison of deflection profiles and load-deflection plots using data collected with dynamic deflection devices are improvements to the traditional approach. Deflections of PCC pavements, however, are significantly influenced by the following factors: (1) roadbed soil condition (subgrade support), (2) strength of base or subbase, (3) temperature differential within the slab, and (4) daily and seasonal temperature variation. The prediction of voids may be unreliable because of these factors. In addition, these methods are not able to determine the size and extent of voids.

<u>Radar Devices</u>   Short-pulse Ground Penetrating Radar (GPR) holds good promise for reliable and cost-effective measurements of voids beneath PCC pavements. It is a rapid test method that does not require extensive traffic control and lane closures. Preliminary data interpretation can be made in the field. Results are apparently not affected by environmental factors or subgrade support conditions. Both Donohue and Gulf Applied versions of the mobile test vehicles featuring radar equipment and non-contact antennas appear to produce reasonable results. The data interpretation is complex, requiring specially trained technicians. The use of a color video monitor by Gulf Applied Radar has enhanced the data analysis procedure. "Honeycombing" in the concrete can influence the results. The following improvements are needed in data interpretation techniques; (1) a user friendly analysis software that can provide reliable answers to location and size of voids, areas of partial support and layer thickness, (2) repeatability and accuracy, supported by independent measurements and (3) electronic data transfer from output files to the primary condition database. Some improvements in data analysis are reported in a recent study [23].

<u>TDR Method</u>   The TDR method is based on analytical techniques for wave propagation problems and produces void maps which are easy to understand and use in the field for predicting void locations, sizes, and boundaries. However, field experience with this equipment is limited in the United States. The method is laborious and time consuming in its present state and requires expert judgement. In addition, the results are apparently influenced by slab curl and environmental factors.

ACKNOWLEDGEMENT

The cooperation of the Departments of Transportation of the States of Florida, Arkansas, North Carolina, Tennessee and Oklahoma is greatly appreciated. Cooperation and assistance of ARE Inc staff is acknowledged. The citation or illustration of trade names of commercially available equipment does not constitute any official

endorsement or approval. The contents of this paper reflect the views
of the authors who are responsible for the facts, findings and the
accuracy of the data presented herein. The contents do not necessarily
reflect the policies of the agencies cited in the paper.

REFERENCES

[1]   Hudson, W.R., G.E. Elkins, W. Uddin, and K.T. Reilly, "Evaluation
       of Pavement Deflection Equipment," Report No. FHWA-TS-87-208, by
       ARE Inc for the Federal Highway Administration, March 1987.

[2]   Hudson, W.R., G.E. Elkins, W. Uddin, and K.T. Reilly, "Improved
       Methods and Equipment to Conduct Pavement Distress Surveys,"
       Report No. FHWA-TS-87-213 prepared by ARE Inc for Federal Highway
       Administration, April 1987.

[3]   Uddin W., W.R.Hudson, G.E. Elkins, and K.T. Reilly, "Evaluation of
       Equipment for Measuring Voids Under Pavements," Report No. FHWA-
       TS-87-229 prepared by Are Inc for the Federal Highway
       Administration, September 1987.

[4]   Birkoff, J.W. and B.F. McCullough, "Detection of Voids Underneath
       Continuously Reinforced Concrete Pavements," Research Report 177-
       18, Center for Transportation Research, The University of Texas at
       Austin, August 1979.

[5]   Thornton, J.B. and W. Gulden, "Pavement Restoration Measures to
       Precede Joint Resealing,"TRR 752, Transportation Research Board,
       Washington, D.C.,1980.

[6]   "Special Provision, Modification of the Standard Specifications,"
       Section 450 Pressure Grouting Portland Cement Concrete Pavement,
       Department of Transportation, State of Georgia, 1983.

[7]   Wells, G.K., "Grout Subsealing of PCCP," Business, Transportation,
       and Housing Authority, State of California, September 1985.

[8]   Torres, Francisco and B. Frank McCullough, "Void Detection and
       Grouting Process," Research Report 249-3, Center for
       Transportation Research, The University of Texas at Austin,
       February 1982.

[9]   Uddin, W., V. Torres-Verdin, W.R. Hudson, A.H. Meyer, and B.F.
       McCullough, "Dynaflect Testing for Rigid Pavement Evaluation,"
       Report No. CTR 256-6F, Center for Transportation Research, The
       University of Texas at Austin, October 1983.

[10]   Yoder, S.R., "Rehabilitation by Undersealing," Proceedings of the
       Continuously Reinforced Concrete Pavements Workshop, sponsored by
       the Louisiana Department of Transportation & Development, Office
       of Highways in cooperation with FHWA, New Orleans, Louisiana,1987.

[11]   Crovetti, J.A. and M.I. Darter, "Appendix C - Void Detection
       Procedures,"Appendix to Final Report, NCHRP Project 1-21, 1985.

Procedures,"Appendix to Final Report, NCHRP Project 1-21, 1985.

[12] Gulf Applied Radar "Equipment Bulletins - RODAR," Gulf Applied Research, Marietta, Georgia, 1987.

[13 "A nondestructive Method for Determining the Thickness of Sound Concrete on Older Pavement," prepared for Donohue & Associates, Inc. for the Iowa Department of Transportation, December 1983.

[14] Test Consult CEBTP, "Equipment Brochure -- The Transient Dynamic Response Method," Chestire, England.

[15] Ricci, E.A., A.H. Meyer, W.R. Hudson, and K.H. Stokoe II, "The Falling Weight Deflectometer for Nondestructive Evaluation of Rigid Pavements," Report 387-3F, Center for Transportation Research, The University of Texas at Austin, November 1985.

[16] Florida Department of Transportation, "Reports related to the Voids Detection Equipment Used on Undersealing Project at I-10," 1984.

[17] McCullough, B.F., A.H. Meyer, R.P. White, P.R. Flannagan, and G.E. Elkins, "Materials and Methods for Undersealing Concrete Pavements," Draft Interim Report FH69, prepared by ARE Inc for the Federal Highway Administration, February 1987.

[18] "Concrete Slab Pavement Testing Report and Evaluation, performed on I-10, Walton County, State Project No. 60002-3416," prepared by Metric Engineering, Inc., Panama City, Florida for the Florida Department of Transportation, 1986.

[19] Pederson, C.M. and L.J. Senowski, "Slab Stabilization of Portland Cement Concrete Pavements," Transportation Research Record 1083, Transportation Research Board, Washington, D.C. 1986.

[20] Arkansas Highway and Transportation Department, "Data and Reports on Undersealing Project," provided to ARE Inc researchers for the undersealing study sponsored by FHWA, under Contract DTFH61-85-C-00120, 1986.

[21] North Carolina Department of Transportation, "Data of Void Detection Tests on the Undersealing Project," provided to ARE Inc researchers for the undersealing study sponsored by FHWA under Contract DTFH61-85-C-00120, 1986.

[22] Tennessee Department of Transportation, "Data of Void Detection Tests on the Undersealing Project," provided to ARE Inc for the undersealing study sponsored by FHWA under Contract DTFH61-85-C-00120, 1986.

[23] Maser, K.R. and Tom Scullion, " Automated Pavement Subsurface Profiling Using Radar: Case Studies of Four Experimental Field Sites," TRR 1344, Transportation Research Board, Washington, D.C., 1992.

Proposed Standard Guide

Richard W. May[1] and Harold L. Von Quintus[2]

THE QUEST FOR A STANDARD GUIDE TO NDT BACKCALCULATION

REFERENCE: May, R. W., and Von Quintus, H. L., **"The Quest for a Standard Guide to NDT Backcalculation,"** Nondestructive Testing of Pavements and Backcalculation of Moduli (Second Volume), ASTM STP 1198, Harold L. Von Quintas, Albert J. Bush, III, and Gilbert Y. Baladi, Eds., Amercian Soceity for Testing and Materials, Philadelphia, 1994.

ABSTRACT: This paper provides a brief history of the efforts of ASTM Subcommittees D18.10 and D04.39 in working toward a standard guideline for the backcalculation of Nondestructive Testing Deflection data. For nearly six years, this document has been under development and seven drafts have been balloted at the subcommittee level. Currently, this process has been temporarily tabled until after the symposium; because, it is felt by many that this technology is still evolving. The main emphasis of this work is to record the latest version of the draft standard which represents the combined input of the people working on these subcommittees. It is hoped that this paper will elicit comments from the many engineers working in this area so that their constructive criticism can improve upon this guideline and allow it to move forward in the ASTM balloting process.

KEYWORDS: backcalculation, nondestructive testing, flexible pavements, moduli, deflections, layered elastic, pavement evaluation, standard guide

---

BACKGROUND

The goal of this paper is to document the notable effort by several individuals who have been trying to assemble a document that would standardize "a procedure" that is continuing to evolve. In addition, we hope to stimulate discussion on the many controversial issues that are involved and possibly, reach some kind of overall compromise as to the recommended state-of-the-art at this time. On January 5, 1986, Von Quintus distributed the first draft of a "Standard Practice for Calculating In-Situ Resilient Modulus of Pavement Materials". Numerous negatives were received and

---

[1] Senior Engineer, Asphalt Institute, P.O. Box 14052, Lexington, KY 40512-4052.

[2] President, Brent Rauhut Engineering, Inc., 8240 Mopac Expressway, Austin, TX 78759.

considerable discussion followed, relative to whether Subcommittee D18.10 on Bearing Tests of Soils In Place should continue to work on this standard.

The struggle continued and Draft number 3 was issued January 6, 1988 to Subcommittees D18.10 and D04.39 on Nondestructive Testing of Pavement Structures as a "Standard Guide for Calculating In-Situ Equivalent Layer Elastic Modulus of Pavement Materials". Eight negative votes were received out of the 32 ballots returned (less than 50 percent of the total number of ballots were returned) and many of these were extremely difficult to satisfy. There were many people who felt it was simply "too early" to try to prepare such a guide, that we needed to "give the state-of-the-art time to advance". It was suggested that we postpone the work on this guide until after the First International Symposium on Nondestructive Testing of Pavements and Backcalculation of Moduli (Bush and Baladi, 1989) held by ASTM in Baltimore, Maryland on June 29-30, 1988. After a review of the 49 papers and the comments made at the symposium, it was hoped that a new task force could decide whether such a guide was feasible at this time.

After the symposium, May, the new task force chairman, issued the fourth draft of the "Standard Guide for Calculating In Situ Equivalent Elastic Moduli of Pavement Materials Using Elastic Layer Theory" on March 24, 1989. After some minor revisions based on Subcommittee D18.10 comments, the fifth draft was distributed to both D04.39 and D18.10 and thoroughly discussed at the June 1990 meeting in San Francisco; nine negatives were received out of the 35 ballots that were returned. After a total reworking of the document, the sixth draft was prepared September 18, 1990. Out of 47 ballots returned, only four negatives votes were received; however, the comments again required that the document be entirely redrafted. Finally, a seventh draft was prepared on April 24, 1992; a total of eight negative votes and four affirmatives with comments were received out of 51 ballots.

This brief synopsis of the history of this draft standard serves as a record of the futile efforts that have led us to this juncture.

The difficulty with a complex draft standard at this stage is that resolving the many issues of comprehensive negatives often generates other negative votes in the next round. However, editorial comments are decreasing and points of contention are becoming more focused. Also, it appears that as the state-of-the-art advances, points of controversy change and personal opinions change as we learn from our experiences with doing backcalculation. To satisfy all of the points with all of the voters may require that all participate in a full-day workshop so that mutual compromises could be reached prior to reballoting.

This paper presents the latest unballoted draft of the standard guide with all of the comments incorporated in the document from the balloting of the seventh draft. The authors are presenting the current thinking of ASTM Subcommittees D18.10 and D04.39 and soliciting constructive criticism.

SCOPE

This guide presents the concepts for calculating the in situ equivalent layer elastic moduli which can be used for pavement evaluation, rehabilitation and overlay design. The resulting equivalent elastic moduli calculated from the deflection data are

method-dependent and represent the stiffnesses of the layers under a
specific Nondestructive Deflection Testing (NDT) device at that
particular test load and frequency, temperature, and other
environmental and site-specific conditions. Adjustments for design
load, reference temperature, and other design-related factors are not
covered in this guide. The intent of this guide is not to recommend
one specific method, but to outline a general approach for estimating
the in situ elastic moduli of pavement layers.

This guide is applicable to flexible pavements and in some
cases, rigid pavements (i.e. interior slab loading), but is
restricted to the use of layered elastic theory (Yoder and Witczak,
1975) as the analysis method. It should be noted that the various
available layered elastic computer modeling techniques (Van
Cauwelaert, et al., 1989) use different assumptions and algorithms
and that results may vary significantly. Other analysis procedures,
such as finite element modeling, may be used, but modifications to
the procedure are required.

Note 1:   If other analysis methods are desired, Van
            Cauwelaert, et al. (1989) can provide some guidance.

TERMINOLOGY

In addition to ASTM Standard Definition of Terms and Symbols
Relating to Soil and Rock Mechanics (D 653), the following
definitions are specific to this standard guide:

<u>Deflection Sensor</u>

The term that shall be used in this guide to refer to the
electronic device(s) capable of measuring the vertical movement of
the pavement and mounted in such a manner as to minimize angular
rotation with respect to its measuring plane at the expected
movement. Sensors may be of several types, such as seismometers,
velocity transducers, or accelerometers.

<u>Deflection Basin</u>

The idealized shape of the deformed pavement surface due to a
cyclic or impact load as depicted from the peak measurements of five
or more deflection sensors.

<u>Resilient Modulus of Elasticity (M_r)</u>

A laboratory test measurement of the behavior of a field-
obtained material sample (either an intact core or a recompacted
specimen) used to approximate the in situ response. Specifically as
shown below, M_r is the applied cyclic deviator stress (σ_1, vertical
stress minus σ_3, horizontal stress) divided by the recoverable axial
strain that occurs when a confined or unconfined and axially loaded
cylindrical material specimen is loaded and unloaded. The Resilient
Modulus is a function of load duration, load frequency, and number of
cycles.

$$M_r = \sigma_d/e_r$$

where:

σ_d = The applied deviator ($\sigma_1-\sigma_3$) stress, and

e_r = The recoverable (resilient) axial strain.

<u>Equivalent Elastic Modulus</u>

The effective in situ modulus of a material, which characterizes
the relationship of stress to strain, specific to the conditions that
existed at the time of NDT testing, that is determined by
backcalculation procedures for an assigned layer of known or assumed
thickness. The collection of all of these layer moduli will produce,
within reasonable limits, the same surface deflections as measured at
various distances from the center of the load when entered into an
layered elastic pavement simulation model analogous to that used in
backcalculation.

<u>Backcalculation</u>

Analytical technique used to determine the equivalent elastic
moduli of pavement layers corresponding to the measured load and
deflections. The analysis may be performed by any of the following
methods: iteration, database searching, closed-form solutions
(currently available only for two layer pavement systems), and
simultaneous equations (using non-linear regression equations
developed from layered elastic analysis output data). The primary
emphasis of this standard will be concerned with the first method;
however, many of the ideas pertaining to the use of the iterative
concept also apply to the other approaches. An iterative analysis
procedure involves assuming moduli values for a layered pavement
structure, computing the surface deflection at several radial
distances from the load, comparing the computed and measured
deflections, and repeating the process, changing the layer moduli
each time, until the difference between the calculated and measured
deflections are within selected tolerance(s) or the maximum number of
iterations has been reached. Alternatively, the analysis procedure
may involve searching through a data base of precalculated deflection
basins computed from a factorial of known layer moduli and
thicknesses until a basin is found which "closely matches" the
measured deflection basin. When analyzing pavement behavior, surface
deflections and other responses are typically calculated (in the
"forward" direction) from layered pavement analysis programs which
use layer moduli as input. In "backcalculation", <u>layer moduli</u> are
selected and adjusted to ultimately compute surface deflections that
best match known surface deflections.

<u>Pavement Materials</u>

The physical constituents that are contained in all of the
various layers of the pavement system; these layers consist of
various thicknesses of placed or stabilized in-place materials for
supporting traffic as well as the native subgrade or embankment
material being protected.

SUMMARY OF GUIDE

A necessary requirement of most overlay or rehabilitation design
procedures is some measure of the in situ or "effective" structural
value of the existing pavement. For years, center-of-load (or
maximum) deflection measurements have been used to determine the
overall structural effectiveness of the existing pavement to carry
load repetitions. The analysis of individual surface deflection
values and the deflection shape or "basin" represents a technique
that can be used to determine separate estimates of the effective
layer properties that collectively describe the overall structural
capacity of the pavement system.
A pavement deflection basin can be induced by a static or
dynamic surface load. Some pavement materials are viscoelastic,

which means these exhibit elastic behavior at high rates of loading
while viscous flow becomes more significant at very slow rates of
loading. For this reason, layered elastic theory is appropriate for
dynamic loading; however, it is difficult to verify whether these
magnitudes of deflection equate to those measured under static
loading. When dynamic loadings are applied, the resulting
displacements registered at each of the deflection sensors are also
dynamic; however, these peak amplitude values do not all occur at the
same time. In a static analysis, such as layered elastic theory,
these peak dynamic deflections are analyzed as if these are
equivalent in magnitude to the deflections that would occur if a load
of "equal" magnitude had been applied statically.

Layered elastic theory is one of the more common analysis
methods being used in the design of flexible pavements and, to a
lesser degree, rigid pavements. This guide is primarily concerned
with the use of layered elastic theory to calculate the layer moduli
in flexible pavements. Various computer programs that use some type
of deflection-matching iterative procedure or database searching
technique have been developed to estimate the pavement material
moduli. As a guide, this standard discusses the various elements of
procedures for calculating and reporting in situ layer moduli of the
pavement cross-section that could then be used in rehabilitation and
overlay design calculations.

Presently, there are two distinct categories of analysis methods
which may be applied to flexible pavements: quasi-static and dynamic.
The quasi-static elastic approaches, discussed in this guide, include
the Boussinesq-Odemark transformed section methods, the numerical
integration layered subroutines, and the finite element methods. As
a general principle, the selection of a method for analyzing NDT data
to determine layer moduli should be compatible with the analysis
procedure that will eventually be used for designing the flexible
pavement rehabilitation. That is, if a particular layered elastic
computer program is to be used in analyzing the pavements for
rehabilitation design purposes, then the same layered elastic
computer program (or its equivalent) should be used as the basis for
determining the material properties from nondestructive testing of
pavements. Similarly, if a finite element procedure is to be used as
a basis for design, it also should be used for analyzing NDT pavement
data. In summary, it is important to consistently use the same
analysis method in both backcalculation and design applications.

The fundamental approach employed in most iterative
backcalculation analysis methods estimating the in situ layer moduli
(see Fig. 1) is that the solution initiates at the outer deflection
sensor location(s) to determine the moduli of the lowest subgrade
layer above the apparent stiff layer. The stiff layer is normally
assigned a fixed modulus. The calculation sequence progresses toward
the center of the basin using the "known" lower layer moduli and the
deflections at smaller radial offsets to calculate the moduli of the
upper layers. This sequence is repeated in an iterative cycle until
a solution is obtained that nearly matches the calculated and
measured deflections. When using the data base searching method, the
sequence may not be the same. In either approach, layer thicknesses
and Poisson's ratios must either be known or assumed. Although the
principles of this approach are applicable to all pavement types
(flexible and rigid), some analysis methods are more appropriate for
specific pavement types and specific NDT devices (Lytton, et al.,
1990). Also, some pavement analysis models are restricted to
pavement structures where the strength of layers decreases with depth
(for example, cement-aggregate mixtures could not be modeled below a
granular base material).

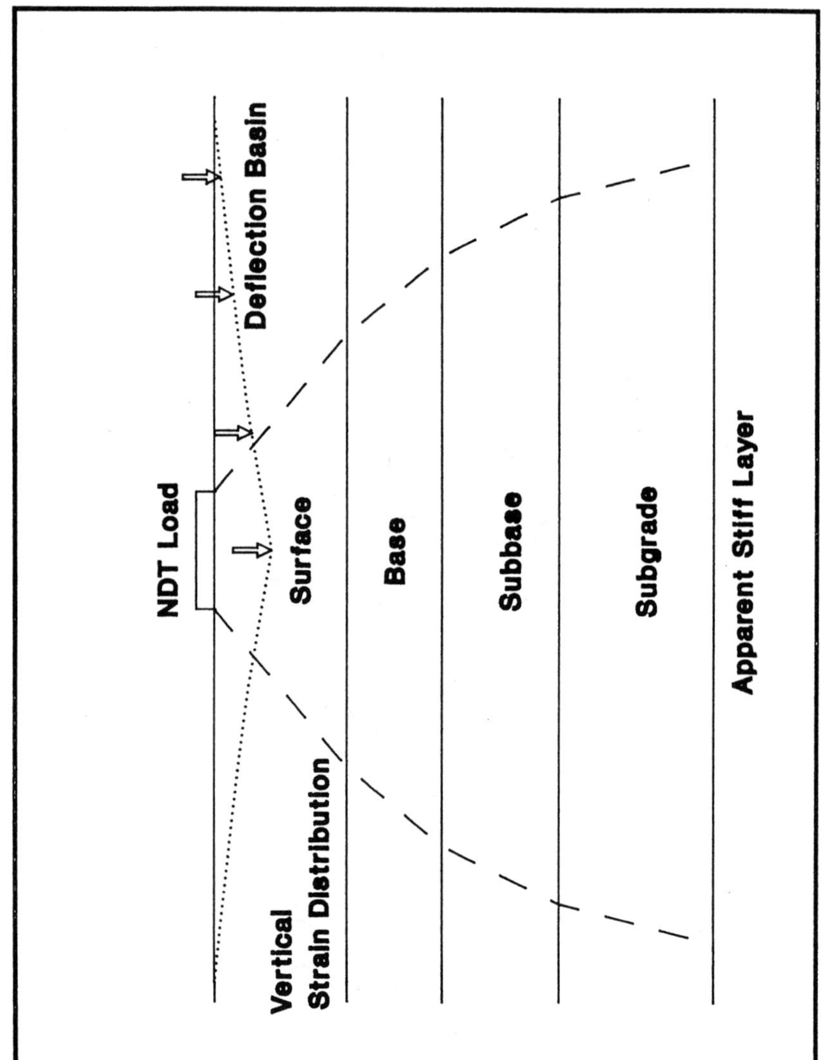

Fig. 1--Depiction of NDT Load Stress Distributed through Pavement

SIGNIFICANCE AND USE

This guide is intended to present the elements of an approach for estimating layer moduli from deflection measurements that may then be used for pavement evaluation or overlay design. To characterize the materials in the layers of a pavement structure, one fundamental input parameter measured in the laboratory and used by some overlay design procedures is the resilient modulus. Deflection analysis provides a technique that may be used to estimate the in situ equivalent layer elastic moduli of a pavement structure as opposed to measuring the resilient moduli in the laboratory of a small and sometimes disturbed sample. For many overlay design procedures that are based on layered elastic theory, the resilient modulus is approximated by this equivalent layer elastic modulus, because the equivalent modulus is determined as an average value for the total layer at the in-situ stress conditions of an actual pavement.

It should be emphasized that layer moduli calculated with this procedure are for a specific loading condition and for the environmental conditions at the time of testing. For these moduli to be used in pavement evaluations and overlay design, adjustments to a reference temperature, season, and design load may be required. These adjustments are not a part of this guide.

The underlying assumption used in the solution is that a unique set of layer moduli exists for the particular loading condition (magnitude and area) and temperature condition, such that the theoretical or calculated deflection basin (using quasi-static layered elastic theory and the assumed static load characteristics of the NDT device) closely approximates the measured deflection basin. In reality, depending on the tolerance allowed in the procedure and the relative number of layers compared to the number of deflection sensors, several combinations of moduli may cause the two basins to "match" (or be within tolerance) reasonably well. A certain degree of engineering judgement is necessary to evaluate these alternative solutions and select the most applicable combination and/or eliminate unreasonable solutions.

There have been several studies that compared the results of various types of equipment and analysis methods; unfortunately, considerable variability has been noted. At this time, no precision estimate has been obtained from a statistically-designed series of tests with different "known" materials and layer thicknesses. The backcalculated results do vary significantly with the various assumptions used in analysis to emulate the actual condition as well as with the techniques used to produce and measure the deflections. Since the guide deals with a computerized analytical method, the repeatability is excellent if the input data and parameters remain the same. The bias of the procedure can not be established at this time. The identity of the "true" in situ modulus, based on resilient modulus testing or some other field or laboratory test, needs to be standardized before the bias of the method can be established.

ANALYTICAL APPROACH

There are several mathematical techniques based on layered elastic theory that may be used to analyze deflection measurements for determining effective layer moduli in a pavement structure.

Note 2:   The user is cautioned against using layer moduli that have been determined from one analysis model in a different model for designing the rehabilitation, because of inherent differences between models. As a general rule, the same model used in overlay or pavement rehabilitation design should also be used in the backcalculation of layer moduli, as discussed

above, unless correlations are developed and
verified.

Regression equations or simplified algorithms developed from
quasi-static layered elastic model computer-generated output may be
used, provided the resulting equivalent layer elastic moduli are used
to recalculate, in the layered elastic model, the deflections at each
point used within the measured deflection basin. The percent error
(between calculated and measured deflection basins) should then meet
the same requirements as discussed later.

PROCEDURE

The following discussion provides general guidelines intended to
assist in the estimation of the structural layer moduli of existing
pavements.

Deflection Testing

The ASTM Standard Guide for Nondestructive Testing of Pavements
Using Cyclic Loading Dynamic Deflection Equipment (D 4602) and ASTM
Standard Test Method for Deflections with a Falling-Weight-Type
Impulse Load Device (D 4694) provide procedures that can be used for
nondestructive deflection testing of pavements using dynamic cyclic
and impact loading deflection equipment, respectively. These test
procedures generally refer to the calibration and operation of
various types of NDT equipment. It should be emphasized that proper
calibration of the sensors is essential for measuring accurate
pavement responses, especially those far away from the load. The
location and spacing of measurements are recommended in ASTM Standard
Guide for General Pavement Deflection Measurements (D 4695).

Delineating Pavement Sections

Plots of deflection parameters as a function of longitudinal
distance or station can be very helpful in defining pavement
subsections with similar characteristics. Longitudinal profile
graphs of both maximum surface deflection and the deflection
measurement furthest from the load should be prepared for the
pavement being evaluated. If the applied load inducing these
deflections varied by more than five percent, the individual
deflections (especially the maximum) should be normalized to a
reference load magnitude to lessen the scatter in the data:

$$\text{normalized deflection} = \text{actual deflection} \times \frac{\text{reference load}}{\text{actual load}}$$

Other deflection basin parameters, such as AREA, may also be plotted
to provide an indication of the variation in overall load
distribution capacity of the pavement. However, the above
normalization process is not necessary or appropriate for the AREA
calculation. A general formula for AREA is defined as follows for
more than one deflection sensor (other definitions exist for specific
numbers of sensors, such as shown in Fig. 2 (Hoffman and Thompson,
1982)); results from different equations may not be comparable):

$$AREA = (Dist_2/2) + \left[\sum_{i=2}^{n-1} \delta_i \times (Dist_i + Dist_{i+1})/(2\delta max)\right] + [Dist_n \times \delta_n /(2\delta max)]$$

where, n is the number of sensors used to measure basin,

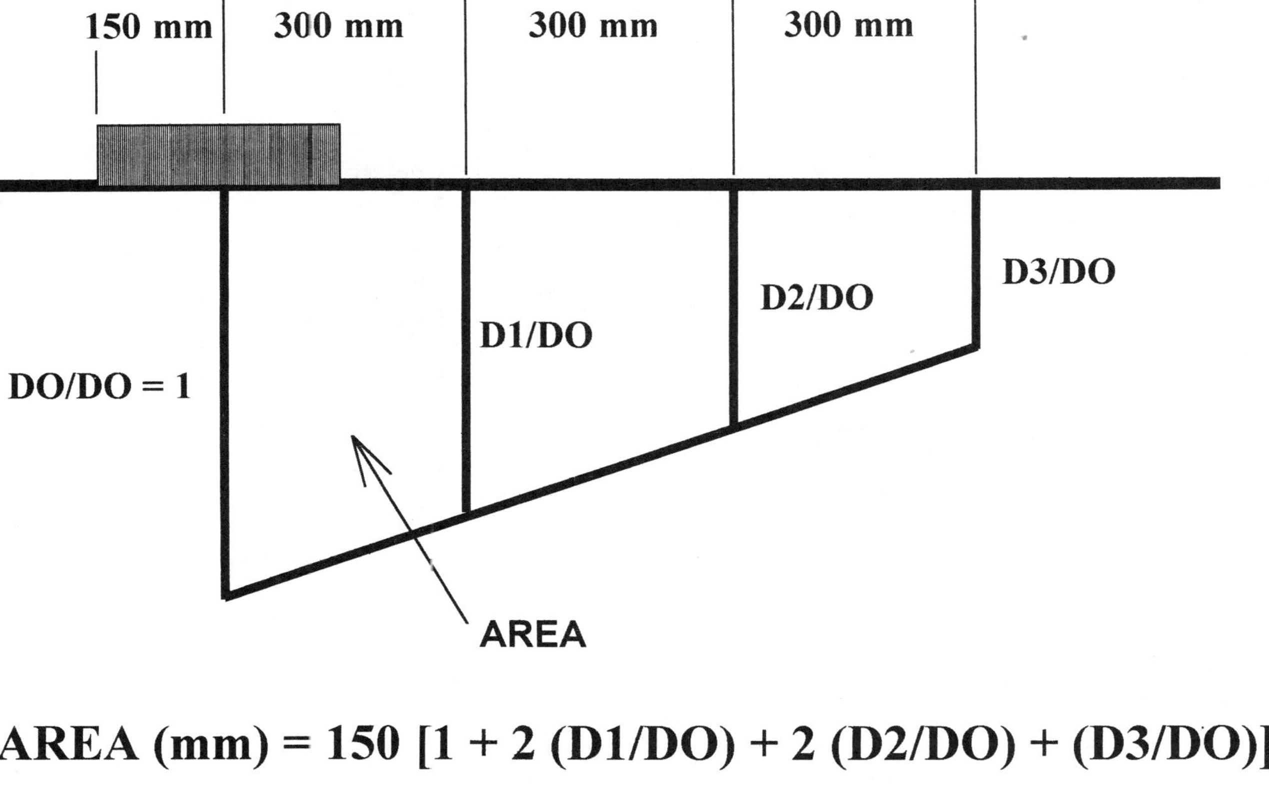

$$\text{AREA (mm)} = 150 \left[1 + 2 \, (D1/DO) + 2 \, (D2/DO) + (D3/DO)\right]$$

Fig. 2--Concept of "AREA" for Structural Capacity

δ_i is the deflection measured with sensor i,
$Dist_2$ is the distance between sensor 2 and 1,
$Dist_n$ is the distance between sensor n and n-1,
$Dist_i$ is the distance between sensor i and i-1,
$Dist_{i+1}$ is the distance between sensor i and i+1, and
δmax is the maximum deflection at the center of the load, measured with sensor 1.

By evaluating these and other longitudinal profiles, pavement segments with significantly different pavement response characteristics can be visually and/or statistically designated as individual subsections.

Note 3:   For some overlay design procedures, results from deflection testing are initially used to designate design sections and aid in evaluating differences in material properties. Deflection data are plotted in the form of a profile by location throughout the length of the pavement section and then separated into subsections with similar deflection basin characteristics. In other procedures, layer moduli are initially calculated for each measured basin and then these moduli or the expected pavement performance based on these moduli are used to delineate uniform subsections.

Note 4:   Subsections with similar deflections, deflection basin characteristics, moduli, and/or expected pavement performance can be statistically checked by using the Student-t test to determine if two sets of data are significantly different.

Under variable topographical or geological conditions, backcalculation of layer moduli for each measurement location may be preferred or even necessary. In more uniform situations, for simplification purposes, an actual "representative" deflection basin could be selected for analysis. However, some site-specific information can be missed and/or additional error introduced. Basins with large differences (greater than two standard deviations within the design section) that may occur could be overlooked by analyzing only a "representative" basin. Locations with notably different deflection magnitudes should be evaluated individually.

Note 5:   If the pavement exhibits only occasional cracks, such as asphalt thermal cracking or concrete joints or cracks, the deflection basins selected for analysis should represent uncracked surfaces (or measurements should be taken with the load and all sensors at least 1.5 m (5 ft) from any cracks), because layered elastic theory does not consider these discontinuities. If the pavement surface has extensive cracking, then these areas should be evaluated as well and the type and severity of cracks should be noted on the report with the backcalculated layer elastic moduli. These kinds of notations may be helpful in explaining the analysis findings for that location. The calculated equivalent moduli will usually reflect the surface condition.

Approximate material classifications and layer thicknesses can be obtained from historical as-built construction records. A pavement coring program will provide more accurate thicknesses, preferably to the nearest 5 mm (0.2 in.) for bound layers or 25 mm (1.0 in.) for unbound layers, and the material type of each layer in the pavement structure, and also check for the existence of a shallow rigid layer (e.g. bedrock).

Note 6:   As a general rule, all material types and layer
            thicknesses recovered from as-built construction
            plans should be verified using field cores and/or
            borings if at all possible. The number of cores
            required per analysis section or project is not a
            part of this standard. Engineering judgement may be
            needed or statistical methods may be utilized (Yoder
            and Witczak, 1975) to determine the number of cores
            required to estimate layer thicknesses to a desired
            level of precision and degree of confidence.
            Thickness variations are dependent on construction
            practice and maintenance activities. However, it
            should be noted that any deviation between the
            assumed and actual in-place layer thicknesses may
            affect the backcalculated layer moduli
            significantly.

For each individual measured or the "representative" measured
deflection basin to be evaluated, the required data are entered into
the preferred analytical technique. The NDT device loading
characteristics, Poisson's ratios and thicknesses of all the assumed
individual layers, deflection values and locations, and possibly
initial estimates of the layer moduli (seed moduli) are included in
the input data set. The Poisson's ratio of the subgrade should be
selected carefully. Small variations in this value may cause
significant differences in the moduli of the upper pavement layers.
Typical ranges of Poisson's ratio values, that may be used if other
values are not available, are the following:

asphalt concrete :	0.30 to 0.40
portland cement concrete :	0.10 to 0.20
unbound granular bases :	0.20 to 0.40[*]
cohesive soil :	0.25 to 0.45[*]
cement-stabilized soil :	0.10 to 0.30
lime-stabilized soil :	0.10 to 0.30

[*] Depending on strain level and degree of saturation.

Note 7:   In programs where seed moduli are required, their
            selection can affect the number of necessary
            iterations, the time required before an acceptable
            solution is achieved and possibly, the final moduli
            that are determined. If an extremely poor selection
            of moduli is made, the analysis may possibly fail to
            find a solution within the specified tolerance
            between calculated and measured deflections. In
            this case, an alternate set of seed moduli may
            provide an acceptable solution before reaching the
            maximum allowable number of iterations. Ordinarily,
            if the tolerance is sufficiently narrow, the final
            moduli that are calculated are not significantly
            affected by the values chosen for the initial set of
            seed moduli.

In addition, many programs require a range of acceptable moduli
values for each of the layers to improve the speed of operation and
to limit the moduli to their approximate practical values.

Thin Layers in Pavements

For upper surface layers that are thin, that is, less than 1/4
of the diameter of the loaded area (for example, 75 mm (3 in.) or
less for a 300 mm (12 in.) loading plate) or layers that are thinner
than the layer directly above, the elastic moduli often cannot be
accurately determined by most backcalculation methods. These thin
layers, if possible, should be combined in assigned thickness with

the same type of material above or below the thin layer, or the moduli of the thin layers can be estimated and assigned as "known" values. For thin asphalt concrete layers (with very few cracks) on unbound granular base courses, the elastic moduli may be measured in the laboratory using ASTM Standard Test Method for Indirect Tension Test for Resilient Modulus of Bituminous Mixtures (D 4123), or mathematically estimated using available regression equations (Asphalt Institute, 1982) or nomographs (Shell International, 1978). The temperature at which the modulus is measured or estimated should correspond to that which existed in the field at the time the deflections were measured. For flexible pavements with single or double bituminous surface treatments, the surface layer is usually combined with the base material in the backcalculation procedure.

Number of Layers

     Based on recommended practice, the number of unknown layers (including subgrade but excluding any fixed apparent stiff layer) to be backcalculated should be no more than five and preferably less. In order to solve for a number of "unknowns" (for example, four layer moduli), as a minimum, that same number and more, if available, of "knowns" (for example, five deflections) should be provided, to better define the basin and reduce the number of possible combinations of moduli that would provide a deflection basin match. Although more deflection points can be derived artificially by interpolating between actual measured points, this is not recommended because additional error can be introduced by not interpreting the correct changes in slope between points. Therefore, if four deflection sensors were used, then a maximum of four unknown layers (three pavement layers and the subgrade) could be used in the structural evaluation. For a pavement where more than three to five layers were constructed, the thicknesses of layers of similar (same type of binder) materials may be combined into one effective structural layer for backcalculation purposes.
     These analysis techniques, in general, iteratively progress toward the center of the deflection basin from the outer edge of the basin in determining these layer moduli. For example, it is possible to estimate (AASHTO, 1986) the minimum distance from the center of the applied load at which the deflection measured at the pavement surface is due primarily to the strain or deflection of the subgrade (see Fig. 1), relatively independent of the overlying layers. Therefore, a measured deflection beyond this distance can be used to solve for the effective subgrade modulus at that stress level directly. For stress-dependent materials, it is advisable that the first sensor beyond this distance be used to solve for the subgrade modulus. Depending on the materials in the pavement structure, it may be necessary to employ non-linear response parameters in the process. Each succeeding deflection point can be attributed to strains that occur in response to the load in successively more layers and it therefore provides some additional "known" information about the upper pavement layers. The effective moduli of these upper layers are then estimated using the closer (to the load) deflections and the previously estimated lower layer moduli.

Estimation of an Apparent Stiff Layer

     Many backcalculation procedures include an apparent stiff (M_r = 700 to 7000 MPa (100,000 to 1,000,000 psi)) layer at some depth into the subgrade. It is intended to simulate either bedrock or the depth where it appears that vertical deflection is negligible. Research has shown that the results of the analysis can be significantly inaccurate by not including such a layer or by not locating this stiff layer near the actual depth, particularly if the actual depth is less than 4.6 m (15 ft). The magnitude of this error is also

affected by the modeling of the subgrade; for example, a nonlinear
stress-dependent (softening) material would also lead to "stiffer"
subgrade layers with depth, or decreasing stress, if included in the
total number of layers.

<u>Tolerances of Deflection Matching</u>

     The accuracy of the final backcalculated moduli is affected by
the tolerance allowed within the procedure for determining a match
between the calculated and measured deflections. Two different
approaches are commonly employed for evaluating this "match". These
are an arithmetic absolute sum (AASE) of percent error and a root
mean square (RMSE) percent error. In both procedures, the engineer
should bear in mind that the significance of random sensor error can
be much greater at the outer sensor locations where the actual
measured deflections are very small; therefore, different tolerance
weighting factors for each sensor may be a consideration.
     An arithmetic absolute sum of percent error, AASE, may be used
to evaluate the match between the calculated and measured deflection
basins and is defined as:

$$AASE = 100 \sum_{i=1}^{n} |(\delta meas_i - \delta calc_i)/ \delta meas_i|$$

     where   n = number of sensors used to measure basin,
                     $\delta meas_i$ = measured deflection at point i, and
                     $\delta calc_i$ = calculated deflection at point i.

The magnitude of tolerance varies with the number of deflection
sensors used to define the basin. It is suggested that the sum of
percent error should not be greater than the following values for the
pavement section to be adequately characterized:

          9-18 percent if nine deflection sensors are used,
          7-14 percent if seven deflection sensors are used, and
          5-10 percent if five deflection sensors are used.

No less than five deflection sensors should be used to describe the
basin.
     A root mean square percent error, RMSE, may also be used to
evaluate the match between the calculated and measured deflection
basins. This measure of error is less dependent on the number of
sensors used to characterize the deflection basin. However, the same
minimal number of deflection sensors (five) as above should be
followed. RMSE is defined as follows:

$$RMSE = 100 \left\{1/n \sum_{i=1}^{n} [(\delta calc_i - \delta meas_i)/ \delta meas_i]^2 \right\}^{0.5}$$

where the parameters are the same as previously defined. A maximum
tolerance limit of 1 to 2 percent on the root mean square error is
recommended.
          Note 8:   If the above requirements for the percent error
                      cannot be met, then conditions may exist which
                      violate the assumptions of layered elastic theory or
                      the actual layer compositions or thicknesses may be
                      significantly different than those used in the
                      model. Additional field material sampling or coring
                      at these locations may provide the means to resolve
                      this problem. If this condition cannot be
                      reconciled, then more complex models which can
                      simulate dynamic loading, material inhomogeneities,

or physical discontinuities in the pavement should be used.

Note 9:   There are several factors that affect the accuracy and applicability of backcalculated layer moduli. Any analysis method that uses an iterative or searching procedure to match measured to calculated deflection basins will result in some error. The magnitude of this error depends on different factors, some of which include: (1) combining different layers into one structural layer, (2) number of deflection points and limitation on number of layers used in the analysis, (3) "noise" or inaccuracies contained in the sensor measurement itself; small deflections that are close in magnitude to the established random error for the sensors, (4) discontinuities such as cracks in the pavement, particularly if located between the load and the sensor, (5) inaccurate assumption of the existence and depth of an apparent stiff layer; depths less than 5 feet may require a dynamic analysis, (6) differences between assumed and actual layer thicknesses; due to inaccurate or unavailable measurements or point-to-point variability, (7) saturated clays directly beneath base materials, (8) extremely weak soils beneath the base and overlying much stiffer soils, (9) non-uniform load pressure distributions at the load-pavement contact area, (10) non-linear, inhomogeneous, or anisotropic materials in the pavement structure (especially the subgrade) and (11) for successive layers, a stiffness ratio (M_r upper layer/ M_r lower layer) less then 0.3.

CONTENTS OF REPORT

The report documenting the backcalculated layer moduli results for each pavement section should include the following:

1)   Identification/location of pavement tested, location of test points analyzed, date and time of deflection testing, file name of original data file, and the backcalculation program (including version number) used.

2)   Details of the NDT device (load range, load footprint, and spacing of all deflection sensors).

3)   The thicknesses, Poisson's ratios (assumed or measured) and material types of each layer in the pavement structure throughout the test section as well as the source of this information. Any differences in construction history or pavement cross-section within the section should be noted if the information is known or available. In addition, any layers that were combined into one structural layer for analysis should be so indicated.

4)   Visual characteristics of the test section. These could include notations on the location of changes in pavement features such as surface appearance or type, transitions from cut to fill, presence of culverts, different soil types, and different shoulder widths. In addition, the locations, types, severity, and extent of pavement distresses such as rutting, washboarding, block cracking, and fatigue cracking should be noted to aid the engineer in understanding any anomalies in the data. The location of the applied

loading relative to any nearby surface distress should also be noted.

5)   The ambient air temperature and pavement surface temperature for each basin measurement. In addition, the average asphalt pavement layer temperature can be obtained by drilling a small hole to the middepth of the asphalt concrete, filling with liquid (oil or water), and measuring the liquid temperature with a thermometer set in the fluid after the reading has stabilized. If this is not possible, some procedures also exist for estimating the pavement temperature as a function of depth using the air temperatures of the previous five days and the current pavement surface temperature (Southgate and Deen, 1969).

6)   The measured load magnitude and measured and calculated deflections for each basin used to backcalculate layer moduli. When a "representative" deflection basin is used, report the range of the actual values measured for each sensor.

7)   The equivalent layer elastic moduli of each structural layer for each backcalculated basin along with the mean and standard deviation for the design section of each layer. In some cases, the results are too few or are not normally-distributed, and other statistical tools may be more appropriate, such as median values, outlier analyses, and frequency distribution plots.

8)   For each layer moduli calculation, the arithmetic absolute sum of percent error or the root mean square percent error between the measured and calculated deflection basins.

SUMMARY

This discussion provides the current status of the draft ASTM Standard Guide for "Calculating In Situ Equivalent Elastic Moduli of Pavement Materials Using Layered Elastic Theory". It represents the eighth draft, unpublished and uncirculated in its present form. We understand that if it were balloted, some more negative votes would surely result. There are a number of contested points which have made this standard appear to be an impossible or at least improbable goal to attain. Some of these items may not be a matter of simply compromising our perspective; but, some could be a question of dealing with incorrect, misleading, or unknown advice.

One of the most debatable issues is how to model a non-linear stress-dependent subgrade material. The characterization from backcalculation, which includes the in-situ overburden condition, provides a better overall "average" than that obtained from a small, usually remolded laboratory specimen. However, since the outer sensor is used for calculating this modulus, the corresponding stress levels can be much lower (and moduli higher) than those directly under the load. This dilemma has led some researchers to resort to dividing the backcalculated result by a factor of three to get the "correct" answer. Even if past performance prediction models were developed using laboratory characterization of the subgrade modulus which tends to give lower values, there must be a better, more scientifically-sound approach to this problem.

A number of other issues have also evolved which will hopefully be addressed at this symposium. These questions are the following:

1. How do we deal with surface cracks in placing the NDT device and in characterizing the pavement material properties (include or avoid)?

2. How many individual layers can we "reasonably" expect to backcalculate in an analysis? What is the minimum number of deflection readings required for this analysis?

3. At what depth does the apparent stiff layer or bedrock become

a critical input parameter?
4. Is the modular ratio (M_r upper layer/ M_r lower layer) of overlying layers a problem?
5. How critical is Poisson's ratio?
6. How critical are seed moduli?
7. What is the best way of handling the many sources of longitudinal variability? Should every deflection basin be evaluated?
8. Should the tolerance of fit vary by deflection sensor?

REFERENCES

American Association of State Highway and Transportation Officials (1986). <u>AASHTO Guide for Design of Pavement Structures 1986</u>.

Asphalt Institute (1982). "Research and Development of The Asphalt Institute's Thickness Design Manual (MS-1) Ninth Edition", Research Report No. 82-2.

Bush, Albert J. III and Baladi, Gilbert Y. (1989). Editors, <u>Nondestructive Testing of Pavements and Backcalculation of Moduli, ASTM STP 1026</u>.

Hoffman, Mario S. and Thompson, Marshall R. (1982). "Backcalculating Nonlinear Resilient Moduli from Deflection Data", Transportation Research Record 852, pp.42-51.

Lytton, R. L., Germann F. P., Chou Y. J., and Stoffels, S. M. (1990). "Determining Asphalt Concrete Pavement Structural Properties by Nondestructive Testing," NCHRP Report No. 327, National Cooperative Highway Research Program.

Shell International (1978). Appendix 2 of the <u>Shell Pavement Design Manual</u>, Shell International Petroleum Company Ltd.

Southgate, H. F. and Deen, R. C. (1969). "Temperature Distribution within Asphalt Pavements and Its Relationship to Pavement Deflection", HRB Record 291, pp 116-131.

Van Cauwelaert, F. J., D. R. Alexander, W. R. Barker, and T. D. White (1989). "A Competent Multilayer Solution and Backcalculation Procedure for Personal Computers, <u>Nondestructive Testing of Pavements and Backcalculation of Moduli, ASTM STP 1026</u>.

Yoder, E. J. and Witczak, M. W. (1975). <u>Principles of Pavement Design</u>, published by John Wiley and Sons, Inc.

Author Index

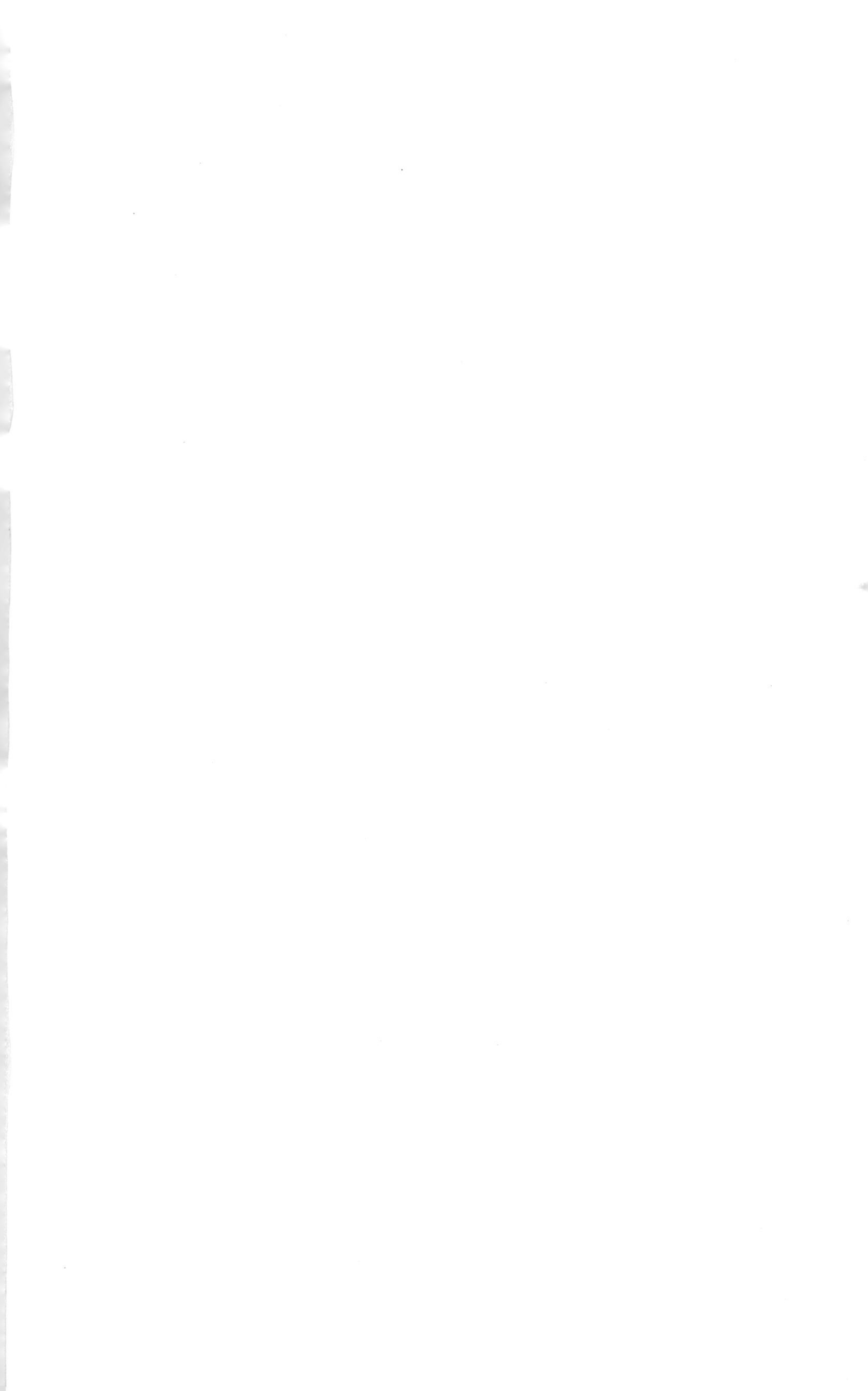